SATURN L-SERIES 2000-04 REPAIR MANUAL

CHILTON'S

Covers all U.S. and Canadian models of Saturn L-series

by Mike Stubblefield

PUBLISHED BY HAYNES NORTH AMERICA, Inc.

Manufactured in USA
© 2004 Haynes North America, Inc.
ISBN-13: 978-1-56392-555-9
ISBN 1-56392-555-9

LIBRARY OF CONGRESS CONTROL NO. 20044787

Haynes Publishing Group
Sparkford Nr Yeovil
Somerset BA22 7JJ England

Haynes North America, Inc
861 Lawrence Drive
Newbury Park
California 91320 USA

ABCDE
FGHIJ
KLM

7T4

Chilton is a registered trademark of W.G. Nichols, Inc., and has been licensed to Haynes North America, Inc.

Contents

INTRODUCTORY PAGES

About this manual – 0-5
Introduction to the Saturn L-series – 0-5
Vehicle identification numbers – 0-6
Buying parts – 0-7
Maintenance techniques, tools and working facilities – 0-7
Jacking and towing – 0-15
Booster battery (jump) starting – 0-16
Conversion factors – 0-17
Fraction/decimal/millimeter equivalents – 0-18
Automotive chemicals and lubricants – 0-19
Safety first! – 0-20
Troubleshooting – 0-21

1
TUNE-UP AND ROUTINE MAINTENANCE – 1-1

2
FOUR-CYLINDER ENGINE – 2A-1
V6 ENGINE – 2B-1
GENERAL ENGINE OVERHAUL PROCEDURES – 2C-1

3
COOLING, HEATING AND AIR-CONDITIONING SYSTEMS – 3-1

4
FUEL AND EXHAUST SYSTEMS – 4-1

5
ENGINE ELECTRICAL SYSTEMS – 5-1

6
EMISSIONS AND ENGINE CONTROL SYSTEMS – 6-1

MANUAL TRANSAXLE – 7A-1 AUTOMATIC TRANSMISSION – 7B-1	**7**
CLUTCH AND DRIVEAXLES – 8-1	**8**
BRAKES – 9-1	**9**
SUSPENSION AND STEERING SYSTEMS – 10-1	**10**
BODY – 11-1	**11**
CHASSIS ELECTRICAL SYSTEM – 12-1 WIRING DIAGRAMS – 12-25	**12**
GLOSSARY – GL-1	**GLOSSARY**
MASTER INDEX – IND-1	**MASTER INDEX**

Mechanic and photographer with a 2002 Saturn Wagon

ACKNOWLEDGMENTS

Technical writers who contributed to this project include Joe Hamilton, Robert Maddox and John Wegmann. Wiring diagrams originated exclusively for Haynes North America, Inc. by Solution Builders.

All rights reserved. No part of this book may be reproduced or transmitted in any form or by any means, electronic or mechanical, including photocopying, recording or by any information storage or retrieval system, without permission in writing from the copyright holder.

While every attempt is made to ensure that the information in this manual is correct, no liability can be accepted by the authors or publishers for loss, damage or injury caused by any errors in, or omissions from, the information given.

About this manual

ITS PURPOSE

The purpose of this manual is to help you get the best value from your vehicle. It can do so in several ways. It can help you decide what work must be done, even if you choose to have it done by a dealer service department or a repair shop; it provides information and procedures for routine maintenance and servicing; and it offers diagnostic and repair procedures to follow when trouble occurs.

We hope you use the manual to tackle the work yourself. For many simpler jobs, doing it yourself may be quicker than arranging an appointment to get the vehicle into a shop and making the trips to leave it and pick it up. More importantly, a lot of money can be saved by avoiding the expense the shop must pass on to you to cover its labor and overhead costs. An added benefit is the sense of satisfaction and accomplishment that you feel after doing the job yourself.

USING THE MANUAL

The manual is divided into Chapters. Each Chapter is divided into numbered Sections. Each Section consists of consecutively numbered paragraphs.

At the beginning of each numbered Section you will be referred to any illustrations which apply to the procedures in that Section. The reference numbers used in illustration captions pinpoint the pertinent Section and the Step within that Section. That is, illustration 3.2 means the illustration refers to Section 3 and Step (or paragraph) 2 within that Section.

Procedures, once described in the text, are not normally repeated. When it's necessary to refer to another Chapter, the reference will be given as Chapter and Section number. Cross references given without use of the word "Chapter" apply to Sections and/or paragraphs in the same Chapter. For example, "see Section 8" means in the same Chapter.

References to the left or right side of the vehicle assume you are sitting in the driver's seat, facing forward.

Even though we have prepared this manual with extreme care, neither the publisher nor the author can accept responsibility for any errors in, or omissions from, the information given.

➡NOTE

A *Note* provides information necessary to properly complete a procedure or information which will make the procedure easier to understand.

※ CAUTION

A *Caution* provides a special procedure or special steps which must be taken while completing the procedure where the Caution is found. Not heeding a Caution can result in damage to the assembly being worked on.

※ WARNING

A *Warning* provides a special procedure or special steps which must be taken while completing the procedure where the Warning is found. Not heeding a Warning can result in personal injury.

Introduction to the Saturn L-series

These Saturn models are available in four-door sedan or station wagon body styles. They feature transversely mounted 2.2L four-cylinder engines and 3.0L V6 engines.

All models are equipped with an electronically controlled Sequential Fuel Injection (SFI) system.

The engine transmits power to the front wheels through either a five-speed manual transaxle or a four-speed automatic transaxle via independent driveaxles.

The front suspension is a MacPherson strut design. The rear suspension employs a trailing arm, two lateral suspension arms, and shock absorbers with coil springs.

The standard power-assisted rack-and-pinion steering unit is mounted behind the engine on the front suspension subframe.

All models are equipped with power assisted front disc and rear disc or drum brakes, with an Anti-lock Brake System (ABS) available as an option.

0-6 VEHICLE IDENTIFICATION NUMBERS/BUYING PARTS

Vehicle Identification numbers

Modifications are a continuing and unpublicized process in vehicle manufacturing. Since spare parts manuals and lists are compiled on a numerical basis, the individual vehicle numbers are essential to correctly identify the component required.

VEHICLE IDENTIFICATION NUMBER (VIN)

This very important identification number is stamped on a plate attached to the dashboard inside the windshield on the driver's side of the vehicle (see illustration). It can also be found on the certification label located on the driver's side door post. The VIN also appears on the Vehicle Certificate of Title and Registration. It contains information such as where and when the vehicle was manufactured, the model year and the body style.

On the models covered by this manual the model year codes are:
Y 2000
1 2001
2 2002
3 2003
4 2004

CERTIFICATION LABEL

The certification label is attached to the driver's door post (see illustration). The plate contains the name of the manufacturer, the month and year of production, the Gross Vehicle Weight Rating (GVWR), the Gross Axle Weight Rating (GAWR) and the certification statement.

ENGINE IDENTIFICATION NUMBERS

The engine serial number can be found on the front side of the engine (see illustration).

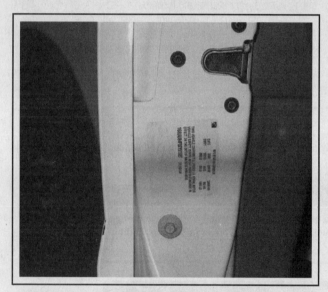

The vehicle certification label is located on the end of the driver's door

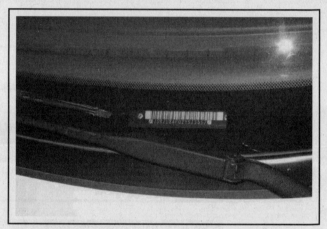

The Vehicle Identification Number (VIN) is located on a plate on top of the dash (visible through the windshield)

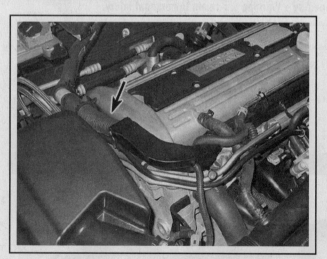

The engine ID label is attached to the valve cover on four-cylinder engines

The engine unit number and build code date can be found stamped on the side of the engine block (four-cylinder shown)

MAINTENANCE TECHNIQUES, TOOLS AND WORKING FACILITIES

Buying parts

Replacement parts are available from many sources, which generally fall into one of two categories - authorized dealer parts departments and independent retail auto parts stores. Our advice concerning these parts is as follows:

Retail auto parts stores: Good auto parts stores will stock frequently needed components which wear out relatively fast, such as clutch components, exhaust systems, brake parts, tune-up parts, etc. These stores often supply new or reconditioned parts on an exchange basis, which can save a considerable amount of money. Discount auto parts stores are often very good places to buy materials and parts needed for general vehicle maintenance such as oil, grease, filters, spark plugs, belts, touch-up paint, bulbs, etc. They also usually sell tools and general accessories, have convenient hours, charge lower prices and can often be found not far from home.

Authorized dealer parts department: This is the best source for parts which are unique to the vehicle and not generally available elsewhere (such as major engine parts, transmission parts, trim pieces, etc.).

Warranty information: If the vehicle is still covered under warranty, be sure that any replacement parts purchased - regardless of the source - do not invalidate the warranty!

To be sure of obtaining the correct parts, have engine and chassis numbers available and, if possible, take the old parts along for positive identification.

Maintenance techniques, tools and working facilities

MAINTENANCE TECHNIQUES

There are a number of techniques involved in maintenance and repair that will be referred to throughout this manual. Application of these techniques will enable the home mechanic to be more efficient, better organized and capable of performing the various tasks properly, which will ensure that the repair job is thorough and complete.

Fasteners

Fasteners are nuts, bolts, studs and screws used to hold two or more parts together. There are a few things to keep in mind when working with fasteners. Almost all of them use a locking device of some type, either a lockwasher, locknut, locking tab or thread adhesive. All threaded fasteners should be clean and straight, with undamaged threads and undamaged corners on the hex head where the wrench fits. Develop the habit of replacing all damaged nuts and bolts with new ones. Special locknuts with nylon or fiber inserts can only be used once. If they are removed, they lose their locking ability and must be replaced with new ones.

Rusted nuts and bolts should be treated with a penetrating fluid to ease removal and prevent breakage. Some mechanics use turpentine in a spout-type oil can, which works quite well. After applying the rust penetrant, let it work for a few minutes before trying to loosen the nut or bolt. Badly rusted fasteners may have to be chiseled or sawed off or removed with a special nut breaker, available at tool stores.

If a bolt or stud breaks off in an assembly, it can be drilled and removed with a special tool commonly available for this purpose. Most automotive machine shops can perform this task, as well as other repair procedures, such as the repair of threaded holes that have been stripped out.

Flat washers and lockwashers, when removed from an assembly, should always be replaced exactly as removed. Replace any damaged washers with new ones. Never use a lockwasher on any soft metal surface (such as aluminum), thin sheet metal or plastic.

Fastener sizes

For a number of reasons, automobile manufacturers are making wider and wider use of metric fasteners. Therefore, it is important to be able to tell the difference between standard (sometimes called U.S. or SAE) and metric hardware, since they cannot be interchanged.

All bolts, whether standard or metric, are sized according to diameter, thread pitch and length. For example, a standard 1/2 - 13 x 1 bolt is 1/2 inch in diameter, has 13 threads per inch and is 1 inch long. An M12 - 1.75 x 25 metric bolt is 12 mm in diameter, has a thread pitch of 1.75 mm (the distance between threads) and is 25 mm long. The two bolts are nearly identical, and easily confused, but they are not interchangeable.

In addition to the differences in diameter, thread pitch and length, metric and standard bolts can also be distinguished by examining the bolt heads. To begin with, the distance across the flats on a standard bolt head is measured in inches, while the same dimension on a metric bolt is sized in millimeters (the same is true for nuts). As a result, a standard wrench should not be used on a metric bolt and a metric wrench should not be used on a standard bolt. Also, most standard bolts have slashes radiating out from the center of the head to denote the grade or strength of the bolt, which is an indication of the amount of torque that can be applied to it. The greater the number of slashes, the greater the strength of the bolt. Grades 0 through 5 are commonly used on automobiles. Metric bolts have a property class (grade) number, rather than a slash, molded into their heads to indicate bolt strength. In this case, the higher the number, the stronger the bolt. Property class numbers 8.8, 9.8 and 10.9 are commonly used on automobiles.

Strength markings can also be used to distinguish standard hex nuts from metric hex nuts. Many standard nuts have dots stamped into one side, while metric nuts are marked with a number. The greater the number of dots, or the higher the number, the greater the strength of the nut.

Metric studs are also marked on their ends according to property class (grade). Larger studs are numbered (the same as metric bolts), while smaller studs carry a geometric code to denote grade.

It should be noted that many fasteners, especially Grades 0 through 2, have no distinguishing marks on them. When such is the case, the only way to determine whether it is standard or metric is to measure the thread pitch or compare it to a known fastener of the same size.

Standard fasteners are often referred to as SAE, as opposed to metric. However, it should be noted that SAE technically refers to a non-metric fine thread fastener only. Coarse thread non-metric fasteners are referred to as USS sizes.

Since fasteners of the same size (both standard and metric) may have different strength ratings, be sure to reinstall any bolts, studs or nuts removed from your vehicle in their original locations. Also, when replacing a fastener with a new one, make sure that the new one has a strength rating equal to or greater than the original.

0-8 MAINTENANCE TECHNIQUES, TOOLS AND WORKING FACILITIES

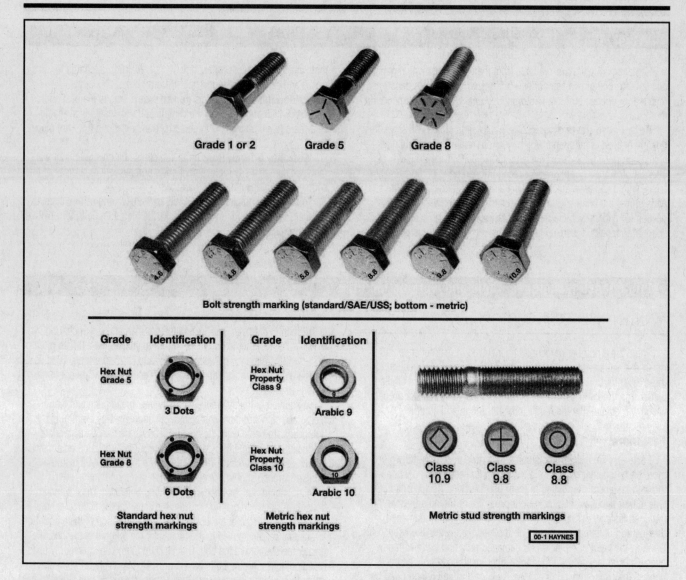

Tightening sequences and procedures

Most threaded fasteners should be tightened to a specific torque value (torque is the twisting force applied to a threaded component such as a nut or bolt). Overtightening the fastener can weaken it and cause it to break, while undertightening can cause it to eventually come loose. Bolts, screws and studs, depending on the material they are made of and their thread diameters, have specific torque values, many of which are noted in the Specifications at the end of each Chapter. Be sure to follow the torque recommendations closely. For fasteners not assigned a specific torque, a general torque value chart is presented here as a guide. These torque values are for dry (unlubricated) fasteners threaded into steel or cast iron (not aluminum). As was previously mentioned, the size and grade of a fastener determine the amount of torque that can safely be applied to it. The figures listed here are approximate for Grade 2 and Grade 3 fasteners. Higher grades can tolerate higher torque values.

Fasteners laid out in a pattern, such as cylinder head bolts, oil pan bolts, differential cover bolts, etc., must be loosened or tightened in sequence to avoid warping the component. This sequence will normally be shown in the appropriate Chapter. If a specific pattern is not given, the following procedures can be used to prevent warping.

Initially, the bolts or nuts should be assembled finger-tight only. Next, they should be tightened one full turn each, in a criss-cross or diagonal pattern. After each one has been tightened one full turn, return to the first one and tighten them all one-half turn, following the same pattern. Finally, tighten each of them one-quarter turn at a time until each fastener has been tightened to the proper torque. To loosen and remove the fasteners, the procedure would be reversed.

Component disassembly

Component disassembly should be done with care and purpose to help ensure that the parts go back together properly. Always keep track of the sequence in which parts are removed. Make note of special characteristics or marks on parts that can be installed more than one way, such as a grooved thrust washer on a shaft. It is a good idea to lay the disassembled parts out on a clean surface in the order that they were removed. It may also be helpful to make sketches or take instant photos of components before removal.

When removing fasteners from a component, keep track of their locations. Sometimes threading a bolt back in a part, or putting the washers and nut back on a stud, can prevent mix-ups later. If nuts and bolts cannot be returned to their original locations, they should be kept in a compartmented box or a series of small boxes. A cupcake or muffin tin is ideal for this purpose, since each cavity can hold the bolts and nuts from a particular area (i.e. oil pan bolts, valve cover bolts, engine mount bolts, etc.). A pan of this type is especially helpful when

MAINTENANCE TECHNIQUES, TOOLS AND WORKING FACILITIES　0-9

Metric thread sizes	Ft-lbs	Nm
M-6	6 to 9	9 to 12
M-8	14 to 21	19 to 28
M-10	28 to 40	38 to 54
M-12	50 to 71	68 to 96
M-14	80 to 140	109 to 154

Pipe thread sizes		
1/8	5 to 8	7 to 10
1/4	12 to 18	17 to 24
3/8	22 to 33	30 to 44
1/2	25 to 35	34 to 47

U.S. thread sizes		
1/4 - 20	6 to 9	9 to 12
5/16 - 18	12 to 18	17 to 24
5/16 - 24	14 to 20	19 to 27
3/8 - 16	22 to 32	30 to 43
3/8 - 24	27 to 38	37 to 51
7/16 - 14	40 to 55	55 to 74
7/16 - 20	40 to 60	55 to 81
1/2 - 13	55 to 80	75 to 108

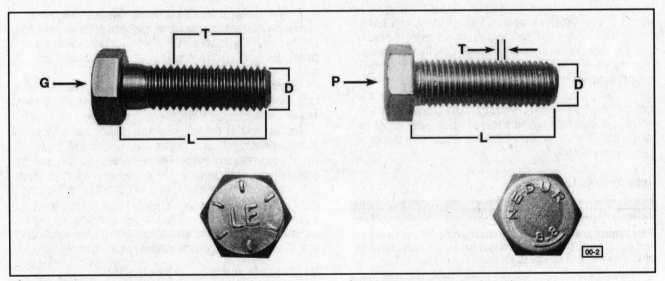

Standard (SAE and USS) bolt dimensions/grade marks
- G　Grade marks (bolt strength)
- L　Length (in inches)
- T　Thread pitch (number of threads per inch)
- D　Nominal diameter (in inches)

Metric bolt dimensions/grade marks
- P　Property class (bolt strength)
- L　Length (in millimeters)
- T　Thread pitch (distance between threads in millimeters)
- D　Diameter

working on assemblies with very small parts, such as the carburetor, alternator, valve train or interior dash and trim pieces. The cavities can be marked with paint or tape to identify the contents.

Whenever wiring looms, harnesses or connectors are separated, it is a good idea to identify the two halves with numbered pieces of masking tape so they can be easily reconnected.

Gasket sealing surfaces

Throughout any vehicle, gaskets are used to seal the mating surfaces between two parts and keep lubricants, fluids, vacuum or pressure contained in an assembly.

Many times these gaskets are coated with a liquid or paste-type gasket sealing compound before assembly. Age, heat and pressure can sometimes cause the two parts to stick together so tightly that they are very difficult to separate. Often, the assembly can be loosened by striking it with a soft-face hammer near the mating surfaces. A regular hammer can be used if a block of wood is placed between the hammer and the part. Do not hammer on cast parts or parts that could be easily damaged. With any particularly stubborn part, always recheck to make sure that every fastener has been removed.

Avoid using a screwdriver or bar to pry apart an assembly, as they can easily mar the gasket sealing surfaces of the parts, which must remain smooth. If prying is absolutely necessary, use an old broom handle, but keep in mind that extra clean up will be necessary if the wood splinters.

0-10 MAINTENANCE TECHNIQUES, TOOLS AND WORKING FACILITIES

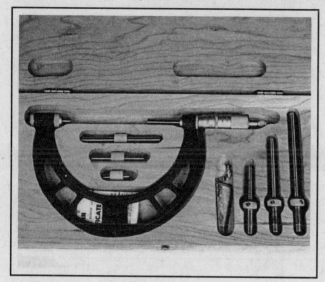

Micrometer set

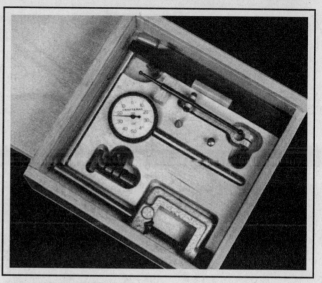

Dial indicator set

After the parts are separated, the old gasket must be carefully scraped off and the gasket surfaces cleaned. Stubborn gasket material can be soaked with rust penetrant or treated with a special chemical to soften it so it can be easily scraped off. A scraper can be fashioned from a piece of copper tubing by flattening and sharpening one end. Copper is recommended because it is usually softer than the surfaces to be scraped, which reduces the chance of gouging the part. Some gaskets can be removed with a wire brush, but regardless of the method used, the mating surfaces must be left clean and smooth. If for some reason the gasket surface is gouged, then a gasket sealer thick enough to fill scratches will have to be used during reassembly of the components. For most applications, a non-drying (or semi-drying) gasket sealer should be used.

Hose removal tips

WARNING:

If the vehicle is equipped with air conditioning, do not disconnect any of the A/C hoses without first having the system depressurized by a dealer service department or a service station.

Hose removal precautions closely parallel gasket removal precautions. Avoid scratching or gouging the surface that the hose mates against or the connection may leak. This is especially true for radiator hoses. Because of various chemical reactions, the rubber in hoses can bond itself to the metal spigot that the hose fits over. To remove a hose, first loosen the hose clamps that secure it to the spigot. Then, with slip-joint pliers, grab the hose at the clamp and rotate it around the spigot. Work it back and forth until it is completely free, then pull it off. Silicone or other lubricants will ease removal if they can be applied between the hose and the outside of the spigot. Apply the same lubricant to the inside of the hose and the outside of the spigot to simplify installation.

As a last resort (and if the hose is to be replaced with a new one anyway), the rubber can be slit with a knife and the hose peeled from the spigot. If this must be done, be careful that the metal connection is not damaged.

If a hose clamp is broken or damaged, do not reuse it. Wire-type clamps usually weaken with age, so it is a good idea to replace them with screw-type clamps whenever a hose is removed.

TOOLS

A selection of good tools is a basic requirement for anyone who plans to maintain and repair his or her own vehicle. For the owner who has few tools, the initial investment might seem high, but when compared to the spiraling costs of professional auto maintenance and repair, it is a wise one.

To help the owner decide which tools are needed to perform the tasks detailed in this manual, the following tool lists are offered: *Maintenance and minor repair, Repair/overhaul and Special.*

The newcomer to practical mechanics should start off with the *maintenance and minor repair* tool kit, which is adequate for the simpler jobs performed on a vehicle. Then, as confidence and experience grow, the owner can tackle more difficult tasks, buying additional tools as they are needed. Eventually the basic kit will be expanded into the *repair and overhaul* tool set. Over a period of time, the experienced do-it-yourselfer will assemble a tool set complete enough for most repair and overhaul procedures and will add tools from the special category when it is felt that the expense is justified by the frequency of use.

Maintenance and minor repair tool kit

The tools in this list should be considered the minimum required for performance of routine maintenance, servicing and minor repair work. We recommend the purchase of combination wrenches (box-end and open-end combined in one wrench). While more expensive than open end wrenches, they offer the advantages of both types of wrench.

Combination wrench set (1/4-inch to 1 inch or 6 mm to 19 mm)
Adjustable wrench, 8 inch
Spark plug wrench with rubber insert
Spark plug gap adjusting tool
Feeler gauge set
Brake bleeder wrench
Standard screwdriver (5/16-inch x 6 inch)
Phillips screwdriver (No. 2 x 6 inch)
Combination pliers - 6 inch
Hacksaw and assortment of blades
Tire pressure gauge
Grease gun
Oil can
Fine emery cloth

MAINTENANCE TECHNIQUES, TOOLS AND WORKING FACILITIES 0-11

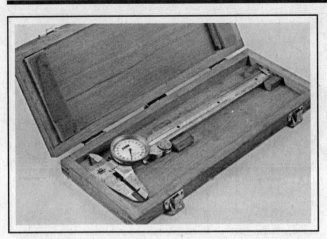
Dial caliper

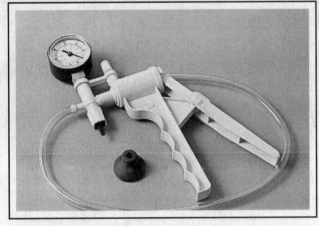
Hand-operated vacuum pump

Timing light

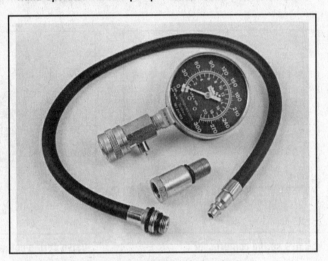
Compression gauge with spark plug hole adapter

Wire brush
Battery post and cable cleaning tool
Oil filter wrench
Funnel (medium size)
Safety goggles
Jackstands (2)
Drain pan

➥ **Note:** *If basic tune-ups are going to be part of routine maintenance, it will be necessary to purchase a good quality stroboscopic timing light and combination tachometer/dwell meter. Although they are included in the list of special tools, it is mentioned here because they are absolutely necessary for tuning most vehicles properly.*

Repair and overhaul tool set

These tools are essential for anyone who plans to perform major repairs and are in addition to those in the maintenance and minor repair tool kit. Included is a comprehensive set of sockets which, though expensive, are invaluable because of their versatility, especially when various extensions and drives are available. We recommend the 1/2-inch drive over the 3/8-inch drive. Although the larger drive is bulky and more expensive, it has the capacity of accepting a very wide range of large sockets. Ideally, however, the mechanic should have a 3/8-inch drive set and a 1/2-inch drive set.

Socket set(s)
Reversible ratchet
Extension - 10 inch
Universal joint
Torque wrench (same size drive as sockets)
Ball peen hammer - 8 ounce
Soft-face hammer (plastic/rubber)
Standard screwdriver (1/4-inch x 6 inch)
Standard screwdriver (stubby - 5/16-inch)
Phillips screwdriver (No. 3 x 8 inch)
Phillips screwdriver (stubby - No. 2)
Pliers - vise grip
Pliers - lineman's
Pliers - needle nose
Pliers - snap-ring (internal and external)
Cold chisel - 1/2-inch
Scribe
Scraper (made from flattened copper tubing)
Centerpunch
Pin punches (1/16, 1/8, 3/16-inch)
Steel rule/straightedge - 12 inch
Allen wrench set (1/8 to 3/8-inch or 4 mm to 10 mm)
A selection of files
Wire brush (large)
Jackstands (second set)
Jack (scissor or hydraulic type)

0-12 MAINTENANCE TECHNIQUES, TOOLS AND WORKING FACILITIES

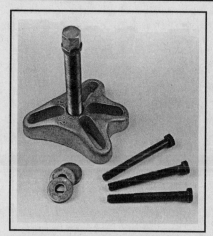
Damper/steering wheel puller

General purpose puller

Hydraulic lifter removal tool

Valve spring compressor

Valve spring compressor

Ridge reamer

➡ Note: Another tool which is often useful is an electric drill with a chuck capacity of 3/8-inch and a set of good quality drill bits.

Special tools

The tools in this list include those which are not used regularly, are expensive to buy, or which need to be used in accordance with their manufacturer's instructions. Unless these tools will be used frequently, it is not very economical to purchase many of them. A consideration would be to split the cost and use between yourself and a friend or friends. In addition, most of these tools can be obtained from a tool rental shop on a temporary basis.

This list primarily contains only those tools and instruments widely available to the public, and not those special tools produced by the vehicle manufacturer for distribution to dealer service departments. Occasionally, references to the manufacturer's special tools are included in the text of this manual. Generally, an alternative method of doing the job without the special tool is offered. However, sometimes there is no alternative to their use. Where this is the case, and the tool cannot be purchased or borrowed, the work should be turned over to the dealer service department or an automotive repair shop.

Valve spring compressor
Piston ring groove cleaning tool
Piston ring compressor
Piston ring installation tool
Cylinder compression gauge
Cylinder ridge reamer
Cylinder surfacing hone
Cylinder bore gauge
Micrometers and/or dial calipers
Hydraulic lifter removal tool
Balljoint separator
Universal-type puller
Impact screwdriver
Dial indicator set
Stroboscopic timing light (inductive pick-up)
Hand operated vacuum/pressure pump
Tachometer/dwell meter
Universal electrical multimeter
Cable hoist
Brake spring removal and installation tools
Floor jack

Buying tools

For the do-it-yourselfer who is just starting to get involved in vehicle maintenance and repair, there are a number of options available when purchasing tools. If maintenance and minor repair is the extent of the work to be done, the purchase of individual tools is satisfactory. If, on the other hand, extensive work is planned, it would be a good idea to purchase a modest tool set from one of the large retail chain stores. A set can usually be bought at a substantial savings over the individual tool prices, and they often come with a tool box. As additional tools are needed, add-on sets, individual tools and a larger tool box can be pur-

MAINTENANCE TECHNIQUES, TOOLS AND WORKING FACILITIES   0-13

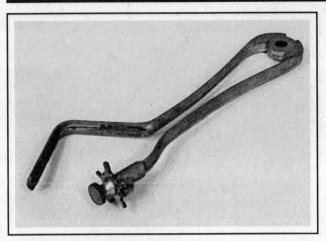
Piston ring groove cleaning tool

Ring removal/installation tool

Ring compressor

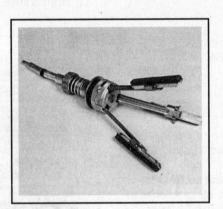
Cylinder hone

Brake hold-down spring tool

chased to expand the tool selection. Building a tool set gradually allows the cost of the tools to be spread over a longer period of time and gives the mechanic the freedom to choose only those tools that will actually be used.

Tool stores will often be the only source of some of the special tools that are needed, but regardless of where tools are bought, try to avoid cheap ones, especially when buying screwdrivers and sockets, because they won't last very long. The expense involved in replacing cheap tools will eventually be greater than the initial cost of quality tools.

Care and maintenance of tools

Good tools are expensive, so it makes sense to treat them with respect. Keep them clean and in usable condition and store them properly when not in use. Always wipe off any dirt, grease or metal chips before putting them away. Never leave tools lying around in the work area. Upon completion of a job, always check closely under the hood for tools that may have been left there so they won't get lost during a test drive.

Some tools, such as screwdrivers, pliers, wrenches and sockets, can be hung on a panel mounted on the garage or workshop wall, while others should be kept in a tool box or tray. Measuring instruments, gauges, meters, etc. must be carefully stored where they cannot be damaged by weather or impact from other tools.

When tools are used with care and stored properly, they will last a very long time. Even with the best of care, though, tools will wear out if used frequently. When a tool is damaged or worn out, replace it. Subsequent jobs will be safer and more enjoyable if you do.

HOW TO REPAIR DAMAGED THREADS

Sometimes, the internal threads of a nut or bolt hole can become stripped, usually from overtightening. Stripping threads is an all-too-common occurrence, especially when working with aluminum parts, because aluminum is so soft that it easily strips out.

Usually, external or internal threads are only partially stripped. After they've been cleaned up with a tap or die, they'll still work. Sometimes, however, threads are badly damaged. When this happens, you've got three choices:

1) Drill and tap the hole to the next suitable oversize and install a larger diameter bolt, screw or stud.
2) Drill and tap the hole to accept a threaded plug, then drill and tap the plug to the original screw size. You can also buy a plug already threaded to the original size. Then you simply drill a hole to the specified size, then run the threaded plug into the hole with a bolt and jam nut. Once the plug is fully seated, remove the jam nut and bolt.
3) The third method uses a patented thread repair kit like Heli-Coil or Slimsert. These easy-to-use kits are designed to repair damaged threads in straight-through holes and blind holes. Both are available as kits which can handle a variety of sizes and thread patterns. Drill the hole, then tap it with the special included tap. Install the Heli-Coil and the hole is back to its original diameter and thread pitch.

Regardless of which method you use, be sure to proceed calmly

0-14 MAINTENANCE TECHNIQUES, TOOLS AND WORKING FACILITIES

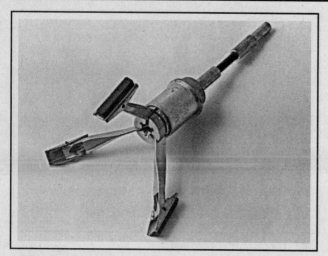
Brake cylinder hone

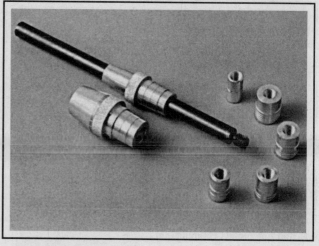
Clutch plate alignment tool

and carefully. A little impatience or carelessness during one of these relatively simple procedures can ruin your whole day's work and cost you a bundle if you wreck an expensive part.

WORKING FACILITIES

Not to be overlooked when discussing tools is the workshop. If anything more than routine maintenance is to be carried out, some sort of suitable work area is essential.

It is understood, and appreciated, that many home mechanics do not have a good workshop or garage available, and end up removing an engine or doing major repairs outside. It is recommended, however, that the overhaul or repair be completed under the cover of a roof.

A clean, flat workbench or table of comfortable working height is an absolute necessity. The workbench should be equipped with a vise that has a jaw opening of at least four inches.

As mentioned previously, some clean, dry storage space is also required for tools, as well as the lubricants, fluids, cleaning solvents, etc. which soon become necessary.

Sometimes waste oil and fluids, drained from the engine or cooling system during normal maintenance or repairs, present a disposal problem. To avoid pouring them on the ground or into a sewage system, pour the used fluids into large containers, seal them with caps and take them to an authorized disposal site or recycling center. Plastic jugs, such as old antifreeze containers, are ideal for this purpose.

Always keep a supply of old newspapers and clean rags available. Old towels are excellent for mopping up spills. Many mechanics use rolls of paper towels for most work because they are readily available

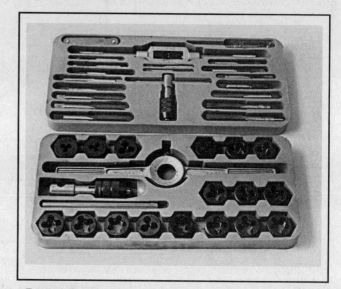
Tap and die set

and disposable. To help keep the area under the vehicle clean, a large cardboard box can be cut open and flattened to protect the garage or shop floor.

Whenever working over a painted surface, such as when leaning over a fender to service something under the hood, always cover it with an old blanket or bedspread to protect the finish. Vinyl covered pads, made especially for this purpose, are available at auto parts stores.

JACKING AND TOWING

Jacking and towing

JACKING

The jack supplied with the vehicle should only be used for raising the vehicle for changing a tire or placing jackstands under the frame.

WARNING:

Never crawl under the vehicle or start the engine when the jack is being used as the only means of support.

All vehicles are supplied with a scissors-type jack. When jacking the vehicle, it should be engaged with the notch in the rocker panel flange (see illustration).

The vehicle should be on level ground with the wheels blocked and the transmission in Park. Pry off the hub cap (if equipped) using the tapered end of the lug wrench. Loosen the lug nuts one-half turn and leave them in place until the wheel is raised off the ground.

Place the jack under the side of the vehicle in the indicated position. Use the supplied wrench to turn the jackscrew clockwise until the wheel is raised off the ground. Remove the lug nuts, pull off the wheel and install the spare.

With the beveled side in, install the lug nuts and tighten them until snug. Lower the vehicle by turning the jackscrew counterclockwise. Remove the jack and tighten the nuts in a diagonal pattern to the torque listed in the Chapter 1 Specifications. If a torque wrench is not available, have the torque checked by a service station as soon as possible. Install the hubcap by placing it in position and using the heel of your hand or a rubber mallet to seat it.

TOWING

As a general rule, the vehicle should be towed with the front (drive) wheels off the ground or, preferably, on a flat bed car carrier. If the front wheels can't be raised or a carrier isn't available, place them on a dolly. The ignition key must be in the ACC position, since the steering lock mechanism isn't strong enough to hold the front wheels straight while towing.

In emergency situations the vehicle can be towed from the front with all four wheels on the ground, provided that speeds don't exceed 35 mph and the distance is not over 50 miles. Before towing, check the transaxle fluid level (see Chapter 1). If the level is below the HOT mark on the dipstick, add fluid.

Towing equipment specifically designed for this purpose should be used and should be attached to the main structural members of the vehicle, not the bumper or brackets.

Safety is a major consideration when towing and all applicable state and local laws must be obeyed. A safety chain system must be used for all towing.

While towing, the parking brake must be released and the transmission must be in Neutral. The steering must be unlocked (ignition switch in the Off position). Remember that power steering and power brakes will not work with the engine off.

TRACTION CONTROL

On models equipped with Traction-Control system, push in the TRAC switch anytime the vehicle is on a "rolling road" tester such as a speedometer test machine or chassis dynamometer. The TRAC OFF indicator light should illuminate when the system is turned off.

The jack fits over the rocker panel flange (there are two jacking points on each side of the vehicle)

0-16 JUMP STARTING

Booster battery (jump) starting

Observe these precautions when using a booster battery to start a vehicle:

a) Before connecting the booster battery, make sure the ignition switch is in the Off position.
b) Turn off the lights, heater and other electrical loads.
c) Your eyes should be shielded. Safety goggles are a good idea.
d) Make sure the booster battery is the same voltage as the dead one in the vehicle.
e) The two vehicles MUST NOT TOUCH each other!
f) Make sure the transaxle is in Neutral (manual) or Park (automatic).
g) If the booster battery is not a maintenance-free type, remove the vent caps and lay a cloth over the vent holes.

Connect the red jumper cable to the positive (+) terminals of each battery (see illustration).

Connect one end of the black jumper cable to the negative (-) terminal of the booster battery. The other end of this cable should be connected to a good ground on the vehicle to be started, such as a bolt or bracket on the body.

Start the engine using the booster battery, then, with the engine running at idle speed, disconnect the jumper cables in the reverse order of connection.

➥Note: *If the battery has been run down or disconnected, the Powertrain Control Module (PCM) must relearn its idle and fuel mixture trim strategy for optimum drivability and performance (see Chapter 5, Section 1 for this procedure).*

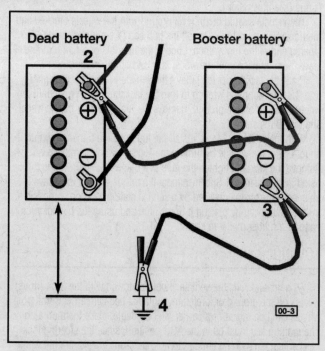

Make the booster battery cable connections in the numerical order shown (note that the negative cable of the booster battery is NOT attached to the negative terminal of the dead battery)

CONVERSION FACTORS

CONVERSION FACTORS

LENGTH (distance)
Inches (in)	X 25.4	= Millimeters (mm)	X 0.0394	= Inches (in)
Feet (ft)	X 0.305	= Meters (m)	X 3.281	= Feet (ft)
Miles	X 1.609	= Kilometers (km)	X 0.621	= Miles

VOLUME (capacity)
Cubic inches (cu in; in^3)	X 16.387	= Cubic centimeters (cc; cm^3)	X 0.061	= Cubic inches (cu in; in^3)
Imperial pints (Imp pt)	X 0.568	= Liters (l)	X 1.76	= Imperial pints (Imp pt)
Imperial quarts (Imp qt)	X 1.137	= Liters (l)	X 0.88	= Imperial quarts (Imp qt)
Imperial quarts (Imp qt)	X 1.201	= US quarts (US qt)	X 0.833	= Imperial quarts (Imp qt)
US quarts (US qt)	X 0.946	= Liters (l)	X 1.057	= US quarts (US qt)
Imperial gallons (Imp gal)	X 4.546	= Liters (l)	X 0.22	= Imperial gallons (Imp gal)
Imperial gallons (Imp gal)	X 1.201	= US gallons (US gal)	X 0.833	= Imperial gallons (Imp gal)
US gallons (US gal)	X 3.785	= Liters (l)	X 0.264	= US gallons (US gal)

MASS (weight)
Ounces (oz)	X 28.35	= Grams (g)	X 0.035	= Ounces (oz)
Pounds (lb)	X 0.454	= Kilograms (kg)	X 2.205	= Pounds (lb)

FORCE
Ounces-force (ozf; oz)	X 0.278	= Newtons (N)	X 3.6	= Ounces-force (ozf; oz)
Pounds-force (lbf; lb)	X 4.448	= Newtons (N)	X 0.225	= Pounds-force (lbf; lb)
Newtons (N)	X 0.1	= Kilograms-force (kgf; kg)	X 9.81	= Newtons (N)

PRESSURE
Pounds-force per square inch (psi; lbf/in^2; lb/in^2)	X 0.070	= Kilograms-force per square centimeter (kgf/cm^2; kg/cm^2)	X 14.223	= Pounds-force per square inch (psi; lbf/in^2; lb/in^2)
Pounds-force per square inch (psi; lbf/in^2; lb/in^2)	X 0.068	= Atmospheres (atm)	X 14.696	= Pounds-force per square inch (psi; lbf/in^2; lb/in^2)
Pounds-force per square inch (psi; lbf/in^2; lb/in^2)	X 0.069	= Bars	X 14.5	= Pounds-force per square inch (psi; lbf/in^2; lb/in^2)
Pounds-force per square inch (psi; lbf/in^2; lb/in^2)	X 6.895	= Kilopascals (kPa)	X 0.145	= Pounds-force per square inch (psi; lbf/in^2; lb/in^2)
Kilopascals (kPa)	X 0.01	= Kilograms-force per square centimeter (kgf/cm^2; kg/cm^2)	X 98.1	= Kilopascals (kPa)

TORQUE (moment of force)
Pounds-force inches (lbf in; lb in)	X 1.152	= Kilograms-force centimeter (kgf cm; kg cm)	X 0.868	= Pounds-force inches (lbf in; lb in)
Pounds-force inches (lbf in; lb in)	X 0.113	= Newton meters (Nm)	X 8.85	= Pounds-force inches (lbf in; lb in)
Pounds-force inches (lbf in; lb in)	X 0.083	= Pounds-force feet (lbf ft; lb ft)	X 12	= Pounds-force inches (lbf in; lb in)
Pounds-force feet (lbf ft; lb ft)	X 0.138	= Kilograms-force meters (kgf m; kg m)	X 7.233	= Pounds-force feet (lbf ft; lb ft)
Pounds-force feet (lbf ft; lb ft)	X 1.356	= Newton meters (Nm)	X 0.738	= Pounds-force feet (lbf ft; lb ft)
Newton meters (Nm)	X 0.102	= Kilograms-force meters (kgf m; kg m)	X 9.804	= Newton meters (Nm)

VACUUM
Inches mercury (in. Hg)	X 3.377	= Kilopascals (kPa)	X 0.2961	= Inches mercury
Inches mercury (in. Hg)	X 25.4	= Millimeters mercury (mm Hg)	X 0.0394	= Inches mercury

POWER
Horsepower (hp)	X 745.7	= Watts (W)	X 0.0013	= Horsepower (hp)

VELOCITY (speed)
Miles per hour (miles/hr; mph)	X 1.609	= Kilometers per hour (km/hr; kph)	X 0.621	= Miles per hour (miles/hr; mph)

FUEL CONSUMPTION*
Miles per gallon, Imperial (mpg)	X 0.354	= Kilometers per liter (km/l)	X 2.825	= Miles per gallon, Imperial (mpg)
Miles per gallon, US (mpg)	X 0.425	= Kilometers per liter (km/l)	X 2.352	= Miles per gallon, US (mpg)

TEMPERATURE

Degrees Fahrenheit = (°C x 1.8) + 32   Degrees Celsius (Degrees Centigrade; °C) = (°F - 32) x 0.56

*It is common practice to convert from miles per gallon (mpg) to liters/100 kilometers (l/100km), where mpg (Imperial) x l/100 km = 282 and mpg (US) x l/100 km = 235

0-18 FRACTION/DECIMAL/MILLIMETER EQUIVALENTS

FRACTION/DECIMAL/MILLIMETER EQUIVALENTS

DECIMALS to MILLIMETERS

Decimal	mm	Decimal	mm
0.001	0.0254	0.500	12.7000
0.002	0.0508	0.510	12.9540
0.003	0.0762	0.520	13.2080
0.004	0.1016	0.530	13.4620
0.005	0.1270	0.540	13.7160
0.006	0.1524	0.550	13.9700
0.007	0.1778	0.560	14.2240
0.008	0.2032	0.570	14.4780
0.009	0.2286	0.580	14.7320
0.010	0.2540	0.590	14.9860
0.020	0.5080		
0.030	0.7620		
0.040	1.0160	0.600	15.2400
0.050	1.2700	0.610	15.4940
0.060	1.5240	0.620	15.7480
0.070	1.7780	0.630	16.0020
0.080	2.0320	0.640	16.2560
0.090	2.2860	0.650	16.5100
		0.660	16.7640
0.100	2.5400	0.670	17.0180
0.110	2.7940	0.680	17.2720
0.120	3.0480	0.690	17.5260
0.130	3.3020		
0.140	3.5560		
0.150	3.8100		
0.160	4.0640	0.700	17.7800
0.170	4.3180	0.710	18.0340
0.180	4.5720	0.720	18.2880
0.190	4.8260	0.730	18.5420
		0.740	18.7960
0.200	5.0800	0.750	19.0500
0.210	5.3340	0.760	19.3040
0.220	5.5880	0.770	19.5580
0.230	5.8420	0.780	19.8120
0.240	6.0960	0.790	20.0660
0.250	6.3500		
0.260	6.6040		
0.270	6.8580	0.800	20.3200
0.280	7.1120	0.810	20.5740
0.290	7.3660	0.820	21.8280
		0.830	21.0820
0.300	7.6200	0.840	21.3360
0.310	7.8740	0.850	21.5900
0.320	8.1280	0.860	21.8440
0.330	8.3820	0.870	22.0980
0.340	8.6360	0.880	22.3520
0.350	8.8900	0.890	22.6060
0.360	9.1440		
0.370	9.3980		
0.380	9.6520		
0.390	9.9060		
		0.900	22.8600
0.400	10.1600	0.910	23.1140
0.410	10.4140	0.920	23.3680
0.420	10.6680	0.930	23.6220
0.430	10.9220	0.940	23.8760
0.440	11.1760	0.950	24.1300
0.450	11.4300	0.960	24.3840
0.460	11.6840	0.970	24.6380
0.470	11.9380	0.980	24.8920
0.480	12.1920	0.990	25.1460
0.490	12.4460	1.000	25.4000

FRACTIONS to DECIMALS to MILLIMETERS

Fraction	Decimal	mm	Fraction	Decimal	mm
1/64	0.0156	0.3969	33/64	0.5156	13.0969
1/32	0.0312	0.7938	17/32	0.5312	13.4938
3/64	0.0469	1.1906	35/64	0.5469	13.8906
1/16	0.0625	1.5875	9/16	0.5625	14.2875
5/64	0.0781	1.9844	37/64	0.5781	14.6844
3/32	0.0938	2.3812	19/32	0.5938	15.0812
7/64	0.1094	2.7781	39/64	0.6094	15.4781
1/8	0.1250	3.1750	5/8	0.6250	15.8750
9/64	0.1406	3.5719	41/64	0.6406	16.2719
5/32	0.1562	3.9688	21/32	0.6562	16.6688
11/64	0.1719	4.3656	43/64	0.6719	17.0656
3/16	0.1875	4.7625	11/16	0.6875	17.4625
13/64	0.2031	5.1594	45/64	0.7031	17.8594
7/32	0.2188	5.5562	23/32	0.7188	18.2562
15/64	0.2344	5.9531	47/64	0.7344	18.6531
1/4	0.2500	6.3500	3/4	0.7500	19.0500
17/64	0.2656	6.7469	49/64	0.7656	19.4469
9/32	0.2812	7.1438	25/32	0.7812	19.8438
19/64	0.2969	7.5406	51/64	0.7969	20.2406
5/16	0.3125	7.9375	13/16	0.8125	20.6375
21/64	0.3281	8.3344	53/64	0.8281	21.0344
11/32	0.3438	8.7312	27/32	0.8438	21.4312
23/64	0.3594	9.1281	55/64	0.8594	21.8281
3/8	0.3750	9.5250	7/8	0.8750	22.2250
25/64	0.3906	9.9219	57/64	0.8906	22.6219
13/32	0.4062	10.3188	29/32	0.9062	23.0188
27/64	0.4219	10.7156	59/64	0.9219	23.4156
7/16	0.4375	11.1125	15/16	0.9375	23.8125
29/64	0.4531	11.5094	61/64	0.9531	24.2094
15/32	0.4688	11.9062	31/32	0.9688	24.6062
31/64	0.4844	12.3031	63/64	0.9844	25.0031
1/2	0.5000	12.7000	1	1.0000	25.4000

AUTOMOTIVE CHEMICALS AND LUBRICANTS

Automotive chemicals and lubricants

A number of automotive chemicals and lubricants are available for use during vehicle maintenance and repair. They include a wide variety of products ranging from cleaning solvents and degreasers to lubricants and protective sprays for rubber, plastic and vinyl.

CLEANERS

Carburetor cleaner and choke cleaner is a strong solvent for gum, varnish and carbon. Most carburetor cleaners leave a dry-type lubricant film which will not harden or gum up. Because of this film it is not recommended for use on electrical components.

Brake system cleaner is used to remove brake dust, grease and brake fluid from the brake system, where clean surfaces are absolutely necessary. It leaves no residue and often eliminates brake squeal caused by contaminants.

Electrical cleaner removes oxidation, corrosion and carbon deposits from electrical contacts, restoring full current flow. It can also be used to clean spark plugs, carburetor jets, voltage regulators and other parts where an oil-free surface is desired.

Demoisturants remove water and moisture from electrical components such as alternators, voltage regulators, electrical connectors and fuse blocks. They are non-conductive and non-corrosive.

Degreasers are heavy-duty solvents used to remove grease from the outside of the engine and from chassis components. They can be sprayed or brushed on and, depending on the type, are rinsed off either with water or solvent.

LUBRICANTS

Motor oil is the lubricant formulated for use in engines. It normally contains a wide variety of additives to prevent corrosion and reduce foaming and wear. Motor oil comes in various weights (viscosity ratings) from 0 to 50. The recommended weight of the oil depends on the season, temperature and the demands on the engine. Light oil is used in cold climates and under light load conditions. Heavy oil is used in hot climates and where high loads are encountered. Multi-viscosity oils are designed to have characteristics of both light and heavy oils and are available in a number of weights from 5W-20 to 20W-50.

Gear oil is designed to be used in differentials, manual transmissions and other areas where high-temperature lubrication is required.

Chassis and wheel bearing grease is a heavy grease used where increased loads and friction are encountered, such as for wheel bearings, balljoints, tie-rod ends and universal joints.

High-temperature wheel bearing grease is designed to withstand the extreme temperatures encountered by wheel bearings in disc brake equipped vehicles. It usually contains molybdenum disulfide (moly), which is a dry-type lubricant.

White grease is a heavy grease for metal-to-metal applications where water is a problem. White grease stays soft under both low and high temperatures (usually from -100 to +190-degrees F), and will not wash off or dilute in the presence of water.

Assembly lube is a special extreme pressure lubricant, usually containing moly, used to lubricate high-load parts (such as main and rod bearings and cam lobes) for initial start-up of a new engine. The assembly lube lubricates the parts without being squeezed out or washed away until the engine oiling system begins to function.

Silicone lubricants are used to protect rubber, plastic, vinyl and nylon parts.

Graphite lubricants are used where oils cannot be used due to contamination problems, such as in locks. The dry graphite will lubricate metal parts while remaining uncontaminated by dirt, water, oil or acids. It is electrically conductive and will not foul electrical contacts in locks such as the ignition switch.

Moly penetrants loosen and lubricate frozen, rusted and corroded fasteners and prevent future rusting or freezing.

Heat-sink grease is a special electrically non-conductive grease that is used for mounting electronic ignition modules where it is essential that heat is transferred away from the module.

SEALANTS

RTV sealant is one of the most widely used gasket compounds. Made from silicone, RTV is air curing, it seals, bonds, waterproofs, fills surface irregularities, remains flexible, doesn't shrink, is relatively easy to remove, and is used as a supplementary sealer with almost all low and medium temperature gaskets.

Anaerobic sealant is much like RTV in that it can be used either to seal gaskets or to form gaskets by itself. It remains flexible, is solvent resistant and fills surface imperfections. The difference between an anaerobic sealant and an RTV-type sealant is in the curing. RTV cures when exposed to air, while an anaerobic sealant cures only in the absence of air. This means that an anaerobic sealant cures only after the assembly of parts, sealing them together.

Thread and pipe sealant is used for sealing hydraulic and pneumatic fittings and vacuum lines. It is usually made from a Teflon compound, and comes in a spray, a paint-on liquid and as a wrap-around tape.

CHEMICALS

Anti-seize compound prevents seizing, galling, cold welding, rust and corrosion in fasteners. High-temperature anti-seize, usually made with copper and graphite lubricants, is used for exhaust system and exhaust manifold bolts.

Anaerobic locking compounds are used to keep fasteners from vibrating or working loose and cure only after installation, in the absence of air. Medium strength locking compound is used for small nuts, bolts and screws that may be removed later. High-strength locking compound is for large nuts, bolts and studs which aren't removed on a regular basis.

Oil additives range from viscosity index improvers to chemical treatments that claim to reduce internal engine friction. It should be noted that most oil manufacturers caution against using additives with their oils.

Gas additives perform several functions, depending on their chemical makeup. They usually contain solvents that help dissolve gum and varnish that build up on carburetor, fuel injection and intake parts. They also serve to break down carbon deposits that form on the inside surfaces of the combustion chambers. Some additives contain upper cylinder lubricants for valves and piston rings, and others contain chemicals to remove condensation from the gas tank.

MISCELLANEOUS

Brake fluid is specially formulated hydraulic fluid that can withstand the heat and pressure encountered in brake systems. Care must be taken so this fluid does not come in contact with painted surfaces or plastics. An opened container should always be resealed to prevent contamination by water or dirt.

Weatherstrip adhesive is used to bond weatherstripping around doors, windows and trunk lids. It is sometimes used to attach trim pieces.

Undercoating is a petroleum-based, tar-like substance that is designed to protect metal surfaces on the underside of the vehicle from corrosion. It also acts as a sound-deadening agent by insulating the bottom of the vehicle.

Waxes and polishes are used to help protect painted and plated surfaces from the weather. Different types of paint may require the use of different types of wax and polish. Some polishes utilize a chemical or abrasive cleaner to help remove the top layer of oxidized (dull) paint on older vehicles. In recent years many non-wax polishes that contain a wide variety of chemicals such as polymers and silicones have been introduced. These non-wax polishes are usually easier to apply and last longer than conventional waxes and polishes.

0-20 SAFETY FIRST!

Safety first!

Regardless of how enthusiastic you may be about getting on with the job at hand, take the time to ensure that your safety is not jeopardized. A moment's lack of attention can result in an accident, as can failure to observe certain simple safety precautions. The possibility of an accident will always exist, and the following points should not be considered a comprehensive list of all dangers. Rather, they are intended to make you aware of the risks and to encourage a safety conscious approach to all work you carry out on your vehicle.

ESSENTIAL DOS AND DON'TS

DON'T rely on a jack when working under the vehicle. Always use approved jackstands to support the weight of the vehicle and place them under the recommended lift or support points.

DON'T attempt to loosen extremely tight fasteners (i.e. wheel lug nuts) while the vehicle is on a jack - it may fall.

DON'T start the engine without first making sure that the transmission is in Neutral (or Park where applicable) and the parking brake is set.

DON'T remove the radiator cap from a hot cooling system - let it cool or cover it with a cloth and release the pressure gradually.

DON'T attempt to drain the engine oil until you are sure it has cooled to the point that it will not burn you.

DON'T touch any part of the engine or exhaust system until it has cooled sufficiently to avoid burns.

DON'T siphon toxic liquids such as gasoline, antifreeze and brake fluid by mouth, or allow them to remain on your skin.

DON'T inhale brake lining dust - it is potentially hazardous (see *Asbestos* below).

DON'T allow spilled oil or grease to remain on the floor - wipe it up before someone slips on it.

DON'T use loose fitting wrenches or other tools which may slip and cause injury.

DON'T push on wrenches when loosening or tightening nuts or bolts. Always try to pull the wrench toward you. If the situation calls for pushing the wrench away, push with an open hand to avoid scraped knuckles if the wrench should slip.

DON'T attempt to lift a heavy component alone - get someone to help you.

DON'T rush or take unsafe shortcuts to finish a job.

DON'T allow children or animals in or around the vehicle while you are working on it.

DO wear eye protection when using power tools such as a drill, sander, bench grinder, etc. and when working under a vehicle.

DO keep loose clothing and long hair well out of the way of moving parts.

DO make sure that any hoist used has a safe working load rating adequate for the job.

DO get someone to check on you periodically when working alone on a vehicle.

DO carry out work in a logical sequence and make sure that everything is correctly assembled and tightened.

DO keep chemicals and fluids tightly capped and out of the reach of children and pets.

DO remember that your vehicle's safety affects that of yourself and others. If in doubt on any point, get professional advice.

ASBESTOS

Certain friction, insulating, sealing, and other products - such as brake linings, brake bands, clutch linings, torque converters, gaskets, etc. - may contain asbestos. Extreme care must be taken to avoid inhalation of dust from such products, since it is hazardous to health. If in doubt, assume that they do contain asbestos.

FIRE

Remember at all times that gasoline is highly flammable. Never smoke or have any kind of open flame around when working on a vehicle. But the risk does not end there. A spark caused by an electrical short circuit, by two metal surfaces contacting each other, or even by static electricity built up in your body under certain conditions, can ignite gasoline vapors, which in a confined space are highly explosive. Do not, under any circumstances, use gasoline for cleaning parts. Use an approved safety solvent.

Always disconnect the battery ground (-) cable at the battery before working on any part of the fuel system or electrical system. Never risk spilling fuel on a hot engine or exhaust component. It is strongly recommended that a fire extinguisher suitable for use on fuel and electrical fires be kept handy in the garage or workshop at all times. Never try to extinguish a fuel or electrical fire with water.

FUMES

Certain fumes are highly toxic and can quickly cause unconsciousness and even death if inhaled to any extent. Gasoline vapor falls into this category, as do the vapors from some cleaning solvents. Any draining or pouring of such volatile fluids should be done in a well ventilated area.

When using cleaning fluids and solvents, read the instructions on the container carefully. Never use materials from unmarked containers.

Never run the engine in an enclosed space, such as a garage. Exhaust fumes contain carbon monoxide, which is extremely poisonous. If you need to run the engine, always do so in the open air, or at least have the rear of the vehicle outside the work area.

If you are fortunate enough to have the use of an inspection pit, never drain or pour gasoline and never run the engine while the vehicle is over the pit. The fumes, being heavier than air, will concentrate in the pit with possibly lethal results.

THE BATTERY

Never create a spark or allow a bare light bulb near a battery. They normally give off a certain amount of hydrogen gas, which is highly explosive.

Always disconnect the battery ground (-) cable at the battery before working on the fuel or electrical systems.

If possible, loosen the filler caps or cover when charging the battery from an external source (this does not apply to sealed or maintenance-free batteries). Do not charge at an excessive rate or the battery may burst.

Take care when adding water to a non maintenance-free battery and when carrying a battery. The electrolyte, even when diluted, is very corrosive and should not be allowed to contact clothing or skin.

Always wear eye protection when cleaning the battery to prevent the caustic deposits from entering your eyes.

HOUSEHOLD CURRENT

When using an electric power tool, inspection light, etc., which operates on household current, always make sure that the tool is correctly connected to its plug and that, where necessary, it is properly grounded. Do not use such items in damp conditions and, again, do not create a spark or apply excessive heat in the vicinity of fuel or fuel vapor.

SECONDARY IGNITION SYSTEM VOLTAGE

A severe electric shock can result from touching certain parts of the ignition system (such as the spark plug wires) when the engine is running or being cranked, particularly if components are damp or the insulation is defective. In the case of an electronic ignition system, the secondary system voltage is much higher and could prove fatal.

TROUBLESHOOTING 0-21

Troubleshooting

CONTENTS

Section Symptom

Engine and performance
1. Engine will not rotate when attempting to start
2. Engine rotates but will not start
3. Engine hard to start when cold
4. Engine hard to start when hot
5. Starter motor noisy or excessively rough in engagement
6. Engine starts but stops immediately
7. Oil puddle under engine
8. Engine lopes while idling or idles erratically
9. Engine misses at idle speed
10. Engine misses throughout driving speed range
11. Engine stumbles on acceleration
12. Engine surges while holding accelerator steady
13. Engine stalls
14. Engine lacks power
15. Engine backfires
16. Pinging or knocking engine sounds during acceleration or uphill
17. Engine runs with oil pressure light on
18. Engine diesels (continues to run) after switching off

Engine electrical system
19. Battery will not hold a charge
20. Voltage warning light fails to go out
21. Voltage warning light fails to come on when key is turned on

Fuel system
22. Excessive fuel consumption
23. Fuel leakage and/or fuel odor

Cooling system
24. Overheating
25. Overcooling
26. External coolant leakage
27. Internal coolant leakage
28. Coolant loss
29. Poor coolant circulation

Clutch
30. Pedal travels to floor - no pressure or very little resistance
31. Fluid in area of master cylinder dust cover and on pedal
32. Fluid on release cylinder
33. Pedal feels spongy when depressed
34. Unable to select gears
35. Clutch slips (engine speed increases with no increase in vehicle speed)
36. Grabbing (chattering) as clutch is engaged
37. Transaxle rattling (clicking)
38. Noise in clutch area
39. Clutch pedal stays on floor
40. High pedal effort

Manual transaxle
41. Knocking noise at low speeds

Section Symptom

42. Noise most pronounced when turning
43. Clunk on acceleration or deceleration
44. Clicking noise in turns
45. Vibration
46. Noisy in neutral with engine running
47. Noisy in one particular gear
48. Noisy in all gears
49. Slips out of gear
50. Leaks lubricant
51. Locked in gear

Automatic transaxle
52. Fluid leakage
53. Transaxle fluid brown or has a burned smell
54. General shift mechanism problems
55. Engine will start in gears other than Park or Neutral
56. Transaxle slips, shifts roughly, is noisy or has no drive in forward or reverse gears

Driveaxles
57. Clicking noise in turns
58. Knock or clunk when accelerating after coasting
59. Shudder or vibration during acceleration

Brakes
60. Vehicle pulls to one side during braking
61. Noise (grinding or high-pitched squeal) when the brakes are applied
62. Brake roughness or chatter (pedal pulsates)
63. Excessive pedal effort required to stop vehicle
64. Excessive brake pedal travel
65. Dragging brakes
66. Grabbing or uneven braking action
67. Brake pedal feels spongy when depressed
68. Brake pedal travels to the floor with little resistance
69. Parking brake does not hold

Suspension and steering systems
70. Vehicle pulls to one side
71. Abnormal or excessive tire wear
72. Wheel makes a "thumping" noise
73. Shimmy, shake or vibration
74. Hard steering
75. Steering wheel does not return to center position correctly
76. Abnormal noise at the front end
77. Wander or poor steering stability
78. Erratic steering when braking
79. Excessive pitching and/or rolling around corners or during braking
80. Suspension bottoms
81. Cupped tires
82. Excessive tire wear on outside edge
83. Excessive tire wear on inside edge
84. Tire tread worn in one place
85. Excessive play or looseness in steering system
86. Rattling or clicking noise in steering gear

0-22 TROUBLESHOOTING

This section provides an easy reference guide to the more common problems which may occur during the operation of your vehicle. Various symptoms and their possible causes are grouped under headings denoting components or systems, such as Engine, Cooling system, etc. They also refer to the Chapter and/or Section that deals with the problem.

Remember that successful troubleshooting isn't a mysterious art practiced only by professional mechanics. It's simply the result of knowledge combined with an intelligent, systematic approach to a problem. Always use a process of elimination, starting with the simplest solution and working through to the most complex - and never overlook the obvious. Anyone can run the gas tank dry or leave the lights on overnight, so don't assume that you're exempt from such oversights.

Finally, always establish a clear idea why a problem has occurred and take steps to ensure that it doesn't happen again. If the electrical system fails because of a poor connection, check all other connections in the system to make sure they don't fail as well. If a particular fuse continues to blow, find out why - don't just go on replacing fuses. Remember, failure of a small component can often be indicative of potential failure or incorrect functioning of a more important component or system.

ENGINE AND PERFORMANCE

1 Engine will not rotate when attempting to start

1. Battery terminal connections loose or corroded (Chapter 1).
2. Battery discharged or faulty (Chapter 1).
3. Automatic transaxle not completely engaged in Park (Chapter 7).
4. Broken, loose or disconnected wiring in the starting circuit (Chapters 5 and 12).
5. Starter motor pinion jammed in flywheel ring gear (Chapter 5).
6. Starter solenoid faulty (Chapter 5).
7. Starter motor faulty (Chapter 5).
8. Ignition switch faulty (Chapter 12).
9. Transaxle range switch faulty (Chapter 6).
10. Starter pinion or driveplate teeth worn or broken (Chapter 5).

2 Engine rotates but will not start

1. Fuel tank empty.
2. Battery discharged (engine rotates slowly) (Chapter 5).
3. Battery terminal connections loose or corroded (Chapter 1).
4. Leaking fuel injector(s), fuel pump, pressure regulator, etc. (Chapter 4).
5. Fuel not reaching fuel injection system (Chapter 4).
6. Ignition components damp or damaged (Chapter 5).
7. Worn, faulty or incorrectly gapped spark plugs (Chapter 1).
8. Broken, loose or disconnected wires at the ignition coil(s) or faulty coil(s) (Chapter 5).

3 Engine hard to start when cold

1. Battery discharged or low (Chapter 1).
2. Fuel system malfunctioning (Chapter 4).
3. Emissions or engine control system malfunctioning (Chapter 6).

4 Engine hard to start when hot

1. Air filter clogged (Chapter 1).
2. Fuel not reaching the fuel injection system (Chapter 4).
3. Corroded battery connections, especially ground (Chapter 1).
4. Emissions or engine control system malfunctioning (Chapter 6).

5 Starter motor noisy or excessively rough in engagement

1. Pinion or driveplate gear teeth worn or broken (Chapter 5).
2. Starter motor mounting bolts loose or missing (Chapter 5).

6 Engine starts but stops immediately

1. Loose or faulty electrical connections at coil pack or alternator (Chapter 5).
2. Insufficient fuel reaching the fuel injectors (Chapter 4).
3. Vacuum leak at the gasket between the intake manifold/plenum and throttle body (Chapters 1 and 4).
4. Restricted exhaust system (most likely the catalytic converter) (Chapters 4 and 6).

7 Oil puddle under engine

1. Oil pan gasket and/or oil pan drain bolt seal leaking (Chapters 1 and 2).
2. Oil pressure sending unit leaking (Chapter 2).
3. Valve cover gaskets leaking (Chapter 2).
4. Engine oil seals leaking (Chapter 2).

8 Engine lopes while idling or idles erratically

1. Vacuum leakage (Chapter 4).
2. Leaking EGR valve or plugged PCV valve (Chapter 6).
3. Air filter clogged (Chapter 1).
4. Fuel pump not delivering sufficient fuel to the fuel injection system (Chapter 4).
5. Leaking head gasket (Chapter 2).
6. Camshaft lobes worn (Chapter 2).

9 Engine misses at idle speed

1. Spark plugs worn or not gapped properly (Chapter 1).
2. Faulty spark plug wires (Chapter 1).
3. Vacuum leaks (Chapters 1 and 4).
4. Uneven or low compression (Chapter 2C).

10 Engine misses throughout driving speed range

1. Fuel filter clogged and/or impurities in the fuel system (Chapters 1 and 4).
2. Low fuel output at the injector (Chapter 4).
3. Faulty or incorrectly gapped spark plugs (Chapter 1).
4. Leaking spark plug wires (Chapter 1).
5. Faulty emission system components (Chapter 6).
6. Low or uneven cylinder compression pressures (Chapter 2).
7. Weak or faulty ignition system (Chapter 5).
8. Vacuum leak in fuel injection system, intake manifold or vacuum hoses (Chapter 4).

11 Engine stumbles on acceleration

1. Spark plugs fouled (Chapter 1).
2. Fuel injection system malfunctioning (Chapter 4).
3. Fuel filter clogged (Chapter 1).
4. Intake manifold air leak (Chapter 4).

12 Engine surges while holding accelerator steady

1. Intake air leak (Chapter 4).
2. Fuel pump faulty (Chapter 4).
3. Defective Throttle Position (TP) sensor (Chapter 6).
4. Defective ECM (Chapter 6).

13 Engine stalls

1. Idle speed incorrect (Chapters 1 and 4).
2. Fuel filter clogged and/or water and impurities in the fuel system (Chapters 1 and 4).
3. Ignition components damp or damaged (Chapter 5).
4. Faulty emissions system components (Chapter 6).
5. Faulty or incorrectly gapped spark plugs (Chapter 1).
6. Faulty spark plug wires (Chapter 1).
7. Vacuum leak in the intake manifold or vacuum hoses (Chapter 4).

14 Engine lacks power

1. Faulty or incorrectly gapped spark plugs (Chapter 1).
2. Restricted exhaust system (most likely the catalytic converter (Chapters 4 and 6).
3. Fuel injection system malfunctioning (Chapter 4).
4. Faulty coil(s) (Chapter 5).
5. Brakes binding (Chapter 1).
6. Automatic transaxle fluid level incorrect (Chapter 1).
7. Fuel filter clogged and/or impurities in the fuel system (Chapter 1).
8. Emission control system not functioning properly (Chapter 6).
9. Low or uneven cylinder compression pressures (Chapter 2).

15 Engine backfires

1. Emissions system not functioning properly (Chapter 6).
2. Fuel injection system malfunctioning (Chapter 4).
3. Vacuum leak at fuel injectors, intake manifold or vacuum hoses (Chapter 4).
4. Valves sticking (Chapter 2).

16 Pinging or knocking engine sounds during acceleration or uphill

1. Incorrect grade of fuel.
2. Fuel injection system malfunctioning Chapter 4).
3. Improper or damaged spark plugs or wires (Chapter 1).
4. Worn or damaged ignition components (Chapter 5).
5. Faulty emissions system (Chapter 6).
6. Vacuum leak (Chapter 4).

17 Engine runs with oil pressure light on

1. Low oil level (Chapter 1).
2. Short in wiring circuit (Chapter 12).
3. Faulty oil pressure sender (Chapter 2).
4. Oil viscosity too low or oil diluted.
5. Worn engine bearings and/or oil pump (Chapter 2).

18 Engine diesels (continues to run) after switching off

1. Excessive engine operating temperature (Chapter 3).
2. Excessive carbon deposits on valves and pistons.

ENGINE ELECTRICAL SYSTEM

19 Battery will not hold a charge

1. Alternator drivebelt defective or not adjusted properly (Chapter 1).
2. Battery terminals loose or corroded (Chapter 1).
3. Alternator not charging properly (Chapter 5).
4. Loose, broken or faulty wiring in the charging circuit (Chapter 5).
5. Short in vehicle wiring (Chapters 5 and 12).
6. Internally defective battery (Chapters 1 and 5).

20 Voltage warning light fails to go out

1. Faulty alternator or charging circuit (Chapter 5).
2. Alternator drivebelt defective or out of adjustment (Chapter 1).
3. Alternator voltage regulator inoperative (Chapter 5).

21 Voltage warning light fails to come on when key is turned on

1. Warning light bulb defective (Chapter 12).
2. Fault in the printed circuit, dash wiring or bulb holder (Chapter 12).

FUEL SYSTEM

22 Excessive fuel consumption

1. Dirty or clogged air filter element (Chapter 1).
2. Emissions system not functioning properly (Chapter 6).
3. Fuel injection system malfunctioning (Chapter 4).
4. Low tire pressure or incorrect tire size (Chapter 1).

23 Fuel leakage and/or fuel odor

1. Leak in a fuel feed or vent line (Chapter 4).
2. Tank overfilled.
3. Evaporative emissions control canister defective (Chapters 1 and 6).
4. Fuel injector seals faulty (Chapter 4).

COOLING SYSTEM

24 Overheating

1. Insufficient coolant in system (Chapter 1).
2. Water pump drivebelt defective or out of adjustment (Chapter 1).
3. Radiator core blocked or grille restricted (Chapter 3).
4. Thermostat faulty (Chapter 3).
5. Electric cooling fan blades broken or cracked (Chapter 3).
6. Radiator cap not maintaining proper pressure (Chapter 3).

25 Overcooling

Incorrect (opening temperature too low) or faulty thermostat (Chapter 3).

TROUBLESHOOTING

26 External coolant leakage

1. Deteriorated/damaged hoses or loose clamps (Chapters 1 and 3)
2. Water pump seal defective (Chapters 1 and 3).
3. Leakage from radiator core (Chapter 3).
4. Engine drain or water jacket core plugs leaking (Chapter 2).

27 Internal coolant leakage

1. Leaking cylinder head gasket (Chapter 2).
2. Cracked cylinder bore or cylinder head (Chapter 2).

28 Coolant loss

1. Too much coolant in system (Chapter 1).
2. Coolant boiling away because of overheating (Chapter 3).
3. Internal or external leakage (Chapter 3).
4. Faulty radiator cap (Chapter 3).

29 Poor coolant circulation

1. Inoperative water pump (Chapter 3).
2. Restriction in cooling system (Chapters 1 and 3).
3. Water pump drivebelt defective or out of adjustment (Chapter 1).
4. Thermostat sticking (Chapter 3).

CLUTCH

30 Pedal travels to floor - no pressure or very little resistance

1. Master or release cylinder faulty (Chapter 8).
2. Hose/pipe burst or leaking (Chapter 8).
3. Connections leaking (Chapter 8).
4. No fluid in reservoir (Chapter 8).
5. If fluid level in reservoir rises as pedal is depressed, master cylinder center valve seal is faulty (Chapter 8).
6. If there is fluid on dust seal at master cylinder, piston primary seal is leaking (Chapter 8).
7. Broken release bearing or fork (Chapter 8).
8. Faulty pressure plate diaphragm spring (Chapter 8).

31 Fluid in area of master cylinder dust cover and on pedal

Rear seal failure in master cylinder (Chapter 8).

32 Fluid on release cylinder

Release cylinder plunger seal faulty (Chapter 8).

33 Pedal feels spongy when depressed

Air in system (Chapter 8).

34 Unable to select gears

1. Faulty transaxle (Chapter 7).
2. Faulty clutch disc or pressure plate (Chapter 8).
3. Faulty release lever or release bearing (Chapter 8).
4. Faulty shift lever assembly or control cables (Chapter 8).

35 Clutch slips (engine speed increases with no increase in vehicle speed)

1. Clutch plate worn (Chapter 8).
2. Clutch plate is oil soaked by leaking rear main seal (Chapters 2 and 8).
3. Clutch plate not seated (Chapter 8).
4. Warped pressure plate or flywheel (Chapter 8).
5. Weak diaphragm springs (Chapter 8).
6. Clutch plate overheated. Allow to cool.

36 Grabbing (chattering) as clutch is engaged

1. Oil on clutch plate lining, burned or glazed facings (Chapter 8).
2. Worn or loose engine or transaxle mounts (Chapter 2).
3. Worn splines on clutch plate hub (Chapter 8).
4. Warped pressure plate or flywheel (Chapter 8).
5. Burned or smeared resin on flywheel or pressure plate (Chapter 8).

37 Transaxle rattling (clicking)

1. Release lever loose (Chapter 8).
2. Clutch plate damper spring failure (Chapter 8).

38 Noise in clutch area

1. Fork shaft improperly installed (Chapter 8).
2. Faulty bearing (Chapter 8).

39 Clutch pedal stays on floor

1. Clutch master cylinder piston binding in bore (Chapter 8).
2. Broken release bearing or fork (Chapter 8).

40 High pedal effort

1. Piston binding in bore (Chapter 8).
2. Pressure plate faulty (Chapter 8).
3. Incorrect size master or release cylinder (Chapter 8).

MANUAL TRANSAXLE

41 Knocking noise at low speeds

1. Worn driveaxle constant velocity (CV) joints (Chapter 8).
2. Worn side gear shaft counterbore in differential case (Chapter 7A).*

42 Noise most pronounced when turning

Differential gear noise (Chapter 7A).*

43 Clunk on acceleration or deceleration

1. Loose engine or transaxle mounts (Chapter 2).
2. Worn differential pinion shaft in case.*
3. Worn side gear shaft counterbore in differential case (Chapter 7A).*
4. Worn or damaged driveaxle inboard CV joints (Chapter 8).

44 Clicking noise in turns

Worn or damaged outboard CV joint (Chapter 8).

45 Vibration

1. Rough wheel bearing (Chapter 10).
2. Damaged driveaxle (Chapter 8).
3. Out-of-round tires (Chapter 1).
4. Tire out of balance (Chapters 1 and 10).
5. Worn CV joint (Chapter 8).

46 Noisy in neutral with engine running

1. Damaged input gear bearing (Chapter 7A).*
2. Damaged clutch release bearing (Chapter 8).

47 Noisy in one particular gear

1. Damaged or worn constant mesh gears (Chapter 7A).*
2. Damaged or worn synchronizers (Chapter 7A).*
3. Bent reverse fork (Chapter 7A).*
4. Damaged fourth speed gear or output gear (Chapter 7A).*
5. Worn or damaged reverse idler gear or idler bushing (Chapter 7A).*

48 Noisy in all gears

1. Insufficient lubricant (Chapter 7A).
2. Damaged or worn bearings (Chapter 7A).*
3. Worn or damaged input gear shaft and/or output gear shaft (Chapter 7A).*

49 Slips out of gear

1. Worn or improperly adjusted linkage (Chapter 7A).
2. Transaxle loose on engine (Chapter 7A).
3. Shift linkage does not work freely, binds (Chapter 7A).
4. Input gear bearing retainer broken or loose (Chapter 7A).*
5. Worn shift fork (Chapter 7A).*

50 Leaks lubricant

1. Side gear shaft seals worn (Chapter 7).
2. Excessive amount of lubricant in transaxle (Chapters 1 and 7A).
3. Loose or broken input gear shaft bearing retainer (Chapter 7A).*
4. Input gear bearing retainer O-ring and/or lip seal damaged (Chapter 7A).*

51 Locked in gear

Lock pin or interlock pin missing (Chapter 7A).*

* Although the corrective action necessary to remedy the symptoms described is beyond the scope of this manual, the above information should be helpful in isolating the cause of the condition so that the owner can communicate clearly with a professional mechanic.

AUTOMATIC TRANSAXLE

➡ Note: Due to the complexity of the automatic transaxle, it's difficult for the home mechanic to properly diagnose and service this component. For problems other than the following, the vehicle should be taken to a dealer service department or a transmission shop.

52 Fluid leakage

1. Automatic transmission fluid is a deep red color. Fluid leaks should not be confused with engine oil, which can easily be blown by airflow to the transaxle.

2. To pinpoint a leak, first remove all built-up dirt and grime from the transaxle housing with degreasing agents and/or steam cleaning. Drive the vehicle at low speeds so air flow will not blow the leak far from its source. Raise the vehicle and determine where the leak is coming from. Common areas of leakage are:
   a) *Fluid pan*
   b) *Fill plug (Chapter 1)*
   c) *Fluid cooler lines (Chapter 7)*
   d) *Vehicle Speed Sensor (Chapter 6)*

53 Transaxle fluid brown or has a burned smell

Transaxle overheated. Change fluid (Chapter 1).

54 General shift mechanism problems

1. Chapter 7 deals with checking and adjusting the shift linkage on automatic transaxles. Common problems which may be attributed to a poorly adjusted linkage are:
   a) *Engine starting in gears other than Park or Neutral.*
   b) *Indicator on shifter pointing to a gear other than the one actually being used.*
   c) *Vehicle moves when in Park.*
2. Refer to Chapter 7 for the shift linkage adjustment procedure.

55 Engine will start in gears other than Park or Neutral

Transmission range switch malfunctioning (Chapter 6).

56 Transaxle slips, shifts roughly, is noisy or has no drive in forward or reverse gears

There are many probable causes for the above problems, but the home mechanic should be concerned with only one possibility - fluid level. Before taking the vehicle to a repair shop, check the level and condition of the fluid as described in Chapter 1.

Correct the fluid level as necessary or change the fluid and filter if needed. If the problem persists, have a professional diagnose the probable cause.

Driveaxles

57 Clicking noise in turns

Worn or damaged outer CV joint. Check for cut or damaged boots (Chapter 1). Repair as necessary (Chapter 8).

58 Knock or clunk when accelerating after coasting

Worn or damaged CV joint. Check for cut or damaged boots (Chapter 1). Repair as necessary (Chapter 8).

59 Shudder or vibration during acceleration

1. Worn or damaged CV joints. Repair or replace as necessary (Chapter 8).
2. Sticking inner joint assembly. Correct or replace as necessary (Chapter 8).

Brakes

➡ Note: Before assuming that a brake problem exists, make sure . . .

a) The tires are in good condition and properly inflated (Chapter 1).
b) The front end alignment is correct (Chapter 10).
c) The vehicle isn't loaded with weight in an unequal manner.

60 Vehicle pulls to one side during braking

1. Incorrect tire pressures (Chapter 1).
2. Front end out of alignment (have the front end aligned).
3. Unmatched tires on same axle.
4. Restricted brake lines or hoses (Chapter 9).
5. Sticking caliper or wheel cylinder piston (Chapter 9).
6. Loose suspension parts (Chapter 10).
7. Contaminated brake pad or shoe material (Chapter 9).

61 Noise (grinding or high-pitched squeal) when the brakes are applied

1. Disc brake pads worn out. Replace pads with new ones immediately (Chapter 9).
2. Drum brake shoes worn out. Replace the shoes immediately (Chapter 9).

62 Brake roughness or chatter (pedal pulsates)

1. Excessive brake disc lateral runout or brake drum out-of-round (Chapter 9).
2. Parallelism of disc not within specifications (Chapter 9).
3. Uneven pad wear caused by caliper not sliding due to improper clearance or dirt (Chapter 9).
4. Defective brake disc (Chapter 9).

63 Excessive pedal effort required to stop vehicle

1. Malfunctioning power brake booster (Chapter 9).
2. Partial system failure (Chapter 9).
3. Excessively worn pads (Chapter 9).
4. One or more caliper or wheel cylinder pistons seized or sticking (Chapter 9).
5. Brake pads contaminated with oil or grease (Chapter 9).
6. New pads or shoes installed and not yet seated. It will take a while for the new material to seat.

64 Excessive brake pedal travel

1. Partial brake system failure (Chapter 9).
2. Insufficient fluid in master cylinder (Chapters 1 and 9).
3. Air trapped in system (Chapter 9).
4. Faulty master cylinder (Chapter 9).

65 Dragging brakes

1. Master cylinder pistons not returning correctly (Chapter 9).
2. Restricted brake lines or hoses (Chapters 1 and 9).
3. Incorrect parking brake adjustment (Chapter 9).
4. Defective brake calipers (Chapter 9).

66 Grabbing or uneven braking action

1. Malfunction of proportioning valve (Chapter 9).
2. Malfunction of power brake booster unit (Chapter 9).
3. Binding brake pedal mechanism (Chapter 9).
4. Contaminated brake linings (Chapter 9).

67 Brake pedal feels spongy when depressed

1. Air in hydraulic lines (Chapter 9).
2. Master cylinder mounting bolts loose (Chapter 9).
3. Master cylinder defective (Chapter 9).

68 Brake pedal travels to the floor with little resistance

Little or no fluid in the master cylinder reservoir caused by leaking caliper, or loose, damaged or disconnected brake lines (Chapter 9).

69 Parking brake does not hold

Parking brake cables improperly adjusted (Chapter 9).

SUSPENSION AND STEERING SYSTEMS

➡ Note: Before attempting to diagnose the suspension and steering systems, perform the following preliminary checks:

a) Check the tire pressures and look for uneven wear.
b) Check the steering universal joints or coupling from the column to the steering gear for loose fasteners and wear.
c) Check the front and rear suspension and the steering gear assembly for loose and damaged parts.
d) Look for out-of-round or out-of-balance tires, bent rims and loose and/or rough wheel bearings.

70 Vehicle pulls to one side

1. Mismatched or uneven tires (Chapter 10).
2. Broken or sagging springs (Chapter 10).
3. Wheel alignment incorrect (Chapter 10).
4. Front brakes dragging (Chapter 9).

71 Abnormal or excessive tire wear

1. Front wheel alignment incorrect (Chapter 10).
2. Sagging or broken springs (Chapter 10).
3. Tire out-of-balance (Chapter 10).
4. Worn strut or shock absorber (Chapter 10).
5. Overloaded vehicle.
6. Tires not rotated regularly.

72 Wheel makes a "thumping" noise

1. Blister or bump on tire (Chapter 1).
2. Improper strut or shock absorber action (Chapter 10).

73 Shimmy, shake or vibration

1. Tire or wheel out-of-balance or out-of-round (Chapter 10).
2. Loose or worn wheel bearings (Chapter 10).
3. Worn tie-rod ends (Chapter 10).
4. Worn balljoints (Chapter 10).

5 Excessive wheel runout (Chapter 10).
6 Blister or bump on tire (Chapter 1).

74 Hard steering

1 Lack of lubrication at balljoints, tie-rod ends and steering gear assembly (Chapter 10).
2 Front wheel alignment incorrect (Chapter 10).
3 Low tire pressure (Chapter 1).

75 Steering wheel does not return to center position correctly

1 Lack of lubrication at balljoints and tie-rod ends (Chapters 1 and 10).
2 Binding in steering column (Chapter 10).
3 Defective rack-and-pinion assembly (Chapter 10).
4 Front wheel alignment problem (Chapter 10).

76 Abnormal noise at the front end

1 Lack of lubrication at balljoints and tie-rod ends (Chapter 1).
2 Loose upper strut mount (Chapter 10).
3 Worn tie-rod ends (Chapter 10).
4 Loose stabilizer bar (Chapter 10).
5 Loose wheel lug nuts (Chapter 1).
6 Loose suspension bolts (Chapter 10).

77 Wander or poor steering stability

1 Mismatched or uneven tires (Chapter 10).
2 Lack of lubrication at balljoints or tie-rod ends (Chapters 1 and 10).
3 Worn struts or shock absorbers (Chapter 10).
4 Loose stabilizer bar (Chapter 10).
5 Broken or sagging springs (Chapter 10).
6 Front wheel alignment incorrect.
7 Loose steering gear mounting fasteners (Chapter 10).

78 Erratic steering when braking

1 Wheel bearings worn (Chapter 10).
2 Broken or sagging springs (Chapter 10).
3 Leaking caliper (Chapter 9).
4 Warped brake discs (Chapter 9).
5 Worn steering gear clamp bushing (Chapter 10).
6 Wheel alignment incorrect.

79 Excessive pitching and/or rolling around corners or during braking

1 Loose stabilizer bar (Chapter 10).
2 Worn struts/shock absorbers or mounts (Chapter 10).
3 Broken or sagging springs (Chapter 10).
4 Overloaded vehicle.

80 Suspension bottoms

1 Overloaded vehicle.
2 Worn struts or shock absorbers (Chapter 10).
3 Incorrect, broken or sagging springs (Chapter 10).

81 Cupped tires

1 Front wheel alignment incorrect (Chapter 10).
2 Worn struts or shock absorbers (Chapter 10).
3 Wheel bearings worn (Chapter 10).
4 Excessive tire or wheel runout (Chapter 10).
5 Worn balljoints (Chapter 10).

82 Excessive tire wear on outside edge

1 Inflation pressures incorrect (Chapter 1).
2 Excessive speed in turns.
3 Wheel alignment incorrect (excessive toe-in or positive camber). Have professionally aligned.
4 Suspension arm bent or twisted (Chapter 10).

83 Excessive tire wear on inside edge

1 Inflation pressures incorrect (Chapter 1).
2 Wheel alignment incorrect (toe-out or excessive negative camber). Have professionally aligned.
3 Loose or damaged steering components (Chapter 10).

84 Tire tread worn in one place

1 Tires out-of-balance.
2 Damaged or buckled wheel. Inspect and replace if necessary.
3 Defective tire (Chapter 1).

85 Excessive play or looseness in steering system

1 Wheel bearings worn (Chapter 10).
2 Tie-rod end loose or worn (Chapter 10).
3 Steering gear loose (Chapter 10).

86 Rattling or clicking noise in steering gear

1 Steering gear mounting bolts loose (Chapter 10).
2 Steering gear defective (Chapter 10).

0-28 TROUBLESHOOTING

Notes

Section

1 Maintenance schedule
2 Introduction
3 Tune-up general information
4 Fluid level checks
5 Tire and tire pressure checks
6 Engine oil and filter change
7 Windshield wiper blade inspection and replacement
8 Battery check, maintenance and charging
9 Cooling system check
10 Tire rotation
11 Seat belt check
12 Brake check
13 Interior ventilation filter - replacement
14 Underhood hose check and replacement
15 Steering, suspension and driveaxle boot check
16 Exhaust system check
17 Fuel system check
18 Drivebelt check and replacement
19 Air filter check and replacement
20 Cooling system servicing (draining, flushing and refilling)
21 Brake fluid change
22 Automatic transaxle fluid change
23 Manual transaxle lubricant level check and change
24 Fuel filter replacement
25 Spark plug check and replacement

1

TUNE-UP AND ROUTINE MAINTENANCE

1-2 TUNE-UP AND ROUTINE MAINTENANCE

1 Maintenance schedule

The maintenance intervals in this manual are provided with the assumption that you, not the dealer, will be doing the work. These are the minimum maintenance intervals recommended by the factory for vehicles that are driven daily. If you wish to keep your vehicle in peak condition at all times, you may wish to perform some of these procedures even more often. Because frequent maintenance enhances the efficiency, performance and resale value of your car, we encourage you to do so. If you drive in dusty areas, tow a trailer, idle or drive at low speeds for extended periods or drive for short distances (less than four miles) in below freezing temperatures, shorter intervals are also recommended.

When your vehicle is new, it should be serviced by a factory authorized dealer service department to protect the factory warranty. In many cases, the initial maintenance check is done at no cost to the owner.

EVERY 250 MILES (400 KM) OR WEEKLY, WHICHEVER COMES FIRST

Check the engine oil level (Section 4)
Check the engine coolant level (Section 4)
Check the brake and clutch fluid level (Section 4)
Check the windshield washer fluid level (Section 4)
Check the power steering fluid level (Section 4)
Check the tires and tire pressures (Section 5)

EVERY 3000 MILES (4800 KM) OR 3 MONTHS, WHICHEVER COMES FIRST

All items listed above plus:
Change the engine oil and oil filter (Section 6)

EVERY 6000 MILES OR 6 MONTHS, WHICHEVER COMES FIRST

All items listed above plus:
Inspect (and replace, if necessary) the windshield wiper blades (Section 7)
Check and service the battery (Section 8)
Check the cooling system (Section 9)
Rotate the tires (Section 10)
Check the seat belts (Section 11)
Inspect the brake system (Section 12)

EVERY 15,000 MILES (24,000 KM) OR 12 MONTHS, WHICHEVER COMES FIRST

All items listed above plus:
Replace the interior ventilation filter (Section 13)
Check all underhood hoses (Section 14)
Inspect the suspension, steering components and driveaxle boots (Section 15)*
Check the exhaust system (Section 16)
Check the fuel system (Section 17)
Check the engine drivebelts (Section 18)

EVERY 30,000 MILES (30,000 KM) OR 24 MONTHS, WHICHEVER COMES FIRST

All items listed above plus:
Check (and replace, if necessary) the air filter (Section 19)*
Service the cooling system (drain, flush and refill) (Section 20)
Change the brake fluid (Section 21)
Replace the automatic transaxle fluid (Section 22)**

EVERY 60,000 MILES (96,000 KM) OR 48 MONTHS, WHICHEVER COMES FIRST

Replace the manual transaxle lubricant (Section 23)

EVERY 100,000 MILES (166,000 KM)

Replace the fuel filter (Section 24)
Replace the spark plugs (Section 25)
Replace the timing belt (V6 engines) (see Chapter 2B)

This item is affected by "severe" operating conditions as described below. If your vehicle is operated under "severe" conditions, perform all maintenance indicated with an asterisk (*) at 3000 mile/3 month intervals. Severe conditions are indicated if you mainly operate your vehicle under one or more of the following conditions:

Operating in dusty areas
Towing a trailer
Idling for extended periods and/or low speed operation

** If operated under one or more of the following conditions, change the or automatic transaxle fluid lubricant every 30,000 miles:

In heavy city traffic where the outside temperature regularly reaches 90-degrees F (32-degrees C) or higher
In hilly or mountainous terrain
Frequent towing of a trailer

TUNE-UP AND ROUTINE MAINTENANCE 1-3

Engine compartment layout (four-cylinder engine)

1. Brake/clutch fluid reservoir
2. Air filter housing
3. Coolant expansion tank
4. Engine oil filler cap
5. Battery
6. Power steering fluid reservoir
7. Engine oil dipstick
8. Oil filter cover
9. Underhood fuse/relay block
10. Windshield washer fluid reservoir

1-4 TUNE-UP AND ROUTINE MAINTENANCE

Engine compartment layout (V6 engine)

1. Brake fluid reservoir
2. Underhood fuse/relay block
3. Air filter housing
4. Coolant expansion tank
5. Engine oil dipstick
6. Engine oil filler cap
7. Power steering fluid reservoir
8. Battery
9. Windshield washer fluid reservoir

TUNE-UP AND ROUTINE MAINTENANCE 1-5

2 Introduction

This Chapter is designed to help the home mechanic maintain the Saturn with the goals of maximum performance, economy, safety and reliability in mind.

Included is a master maintenance schedule, followed by procedures dealing specifically with each item on the schedule. Visual checks, adjustments, component replacement and other helpful items are included. Refer to the accompanying illustrations of the engine compartment and the underside of the vehicle for the locations of various components.

Servicing the vehicle, in accordance with the mileage/time maintenance schedule and the step-by-step procedures will result in a planned maintenance program that should produce a long and reliable service life. Keep in mind that it is a comprehensive plan, so maintaining some items but not others at the specified intervals will not produce the same results.

As you service the vehicle, you will discover that many of the procedures can - and should - be grouped together because of the nature of the particular procedure you're performing or because of the close proximity of two otherwise unrelated components to one another.

For example, if the vehicle is raised for chassis lubrication, you should inspect the exhaust, suspension, steering and fuel systems while you're under the vehicle. When you're rotating the tires, it makes good sense to check the brakes since the wheels are already removed. Finally, let's suppose you have to borrow or rent a torque wrench. Even if you only need it to tighten the spark plugs, you might as well check the torque of as many critical fasteners as time allows.

The first step in this maintenance program is to prepare yourself before the actual work begins. Read through all the procedures you're planning to do, then gather up all the parts and tools needed. If it looks like you might run into problems during a particular job, seek advice from a mechanic or an experienced do-it-yourselfer.

OWNER'S MANUAL AND VECI LABEL INFORMATION

Your vehicle owner's manual was written for your year and model and contains very specific information on component locations, specifications, fuse ratings, part numbers, etc. The Owner's Manual is an important resource for the do-it-yourselfer to have; if one was not supplied with your vehicle, it can generally be ordered from a dealer parts department.

Among other important information, the Vehicle Emissions Control Information (VECI) label contains specifications and procedures for applicable tune-up adjustments and, in some instances, spark plugs (see Chapter 6 for more information on the VECI label). The information on this label is the exact maintenance data recommended by the manufacturer. This data often varies by intended operating altitude, local emissions regulations, month of manufacture, etc.

This Chapter contains procedural details, safety information and more ambitious maintenance intervals than you might find in manufacturer's literature. However, you may also find procedures or specifications in your Owner's Manual or VECI label that differ with what's printed here. In these cases, the Owner's Manual or VECI label can be considered correct, since it is specific to your particular vehicle.

3 Tune-up general information

The term tune-up is used in this manual to represent a combination of individual operations rather than one specific procedure.

If, from the time the vehicle is new, the routine maintenance schedule is followed closely and frequent checks are made of fluid levels and high wear items, as suggested throughout this manual, the engine will be kept in relatively good running condition and the need for additional work will be minimized.

More likely than not, however, there will be times when the engine is running poorly due to lack of regular maintenance. This is even more likely if a used vehicle, which has not received regular and frequent maintenance checks, is purchased. In such cases, an engine tune-up will be needed outside of the regular routine maintenance intervals.

The first step in any tune-up or diagnostic procedure to help correct a poor running engine is a cylinder compression check. A compression check (see Chapter 2C) will help determine the condition of internal engine components and should be used as a guide for tune-up and repair procedures. If, for instance, a compression check indicates serious internal engine wear, a conventional tune-up will not improve the performance of the engine and would be a waste of time and money. Because of its importance, the compression check should be done by someone with the right equipment and the knowledge to use it properly.

The following procedures are those most often needed to bring a generally poor running engine back into a proper state of tune.

MINOR TUNE-UP

Check all engine related fluids (Section 4)
Clean, inspect and test the battery (Section 8)
Check the cooling system (Section 9)
Check all underhood hoses (Section 14)
Check the fuel system (Section 17)
Check the air filter (Section 19)

MAJOR TUNE-UP

All items listed under Minor tune-up, plus . . .
Check the drivebelt (Section 18)
Replace the air filter (Section 19)
Replace the fuel filter (Section 24)
Replace the spark plugs (Section 25)

4 Fluid level checks (every 250 miles [400 km] or weekly)

1 Fluids are an essential part of the lubrication, cooling, brake and windshield washer systems. Because the fluids gradually become depleted and/or contaminated during normal operation of the vehicle, they must be periodically replenished. See *Recommended lubricants and fluids* at the end of this Chapter before adding fluid to any of the following components.

➟Note: **The vehicle must be on level ground when fluid levels are checked.**

1-6 TUNE-UP AND ROUTINE MAINTENANCE

ENGINE OIL

▸ **Refer to illustrations 4.2a, 4.2b, 4.4 and 4.6**

2 The oil level is checked with a dipstick, which is attached to the engine block (see illustrations). The dipstick extends through a metal tube down into the oil pan.

3 The oil level should be checked before the vehicle has been driven, or about 5 minutes after the engine has been shut off. If the oil is checked immediately after driving the vehicle, some of the oil will remain in the upper part of the engine, resulting in an inaccurate reading on the dipstick.

4 Pull the dipstick out of the tube and wipe all the oil from the end with a clean rag or paper towel. Insert the clean dipstick all the way back into the tube and pull it out again. Note the oil at the end of the dipstick. At its highest point, the level should be between the MIN and MAX marks on the dipstick (see illustration).

5 It takes one quart of oil to raise the level from the MIN mark to the MAX mark on the dipstick. Do not allow the level to drop below the MIN mark or oil starvation may cause engine damage. Conversely, overfilling the engine (adding oil above the MAX mark) may cause oil fouled spark plugs, oil leaks or oil seal failures. Maintaining the oil level above the MAX mark can cause excessive oil consumption.

6 To add oil, remove the filler cap from the valve cover (see illustration). After adding oil, wait a few minutes to allow the level to stabilize, then pull out the dipstick and check the level again. Add more oil if required. Install the filler cap and tighten it by hand only.

7 Checking the oil level is an important preventive maintenance step. A consistently low oil level indicates oil leakage through damaged seals, defective gaskets or past worn rings or valve guides. If the oil looks milky in color or has water droplets in it, the cylinder head gasket(s) may be blown or the head(s) or block may be cracked. The engine should be checked immediately. The condition of the oil should also be checked. Whenever you check the oil level, slide your thumb and index finger up the dipstick before wiping off the oil. If you see small dirt or metal particles clinging to the dipstick, the oil should be changed (see Section 6).

ENGINE COOLANT

▸ **Refer to illustrations 4.8 and 4.9**

※ WARNING:

Do not allow antifreeze to come in contact with your skin or painted surfaces of the vehicle. Flush contaminated areas immediately with plenty of water. Don't store new coolant or leave old coolant lying around where it's accessible to children or pets - they're attracted by its sweet smell. Ingestion of even a small amount of coolant can be fatal! Wipe up garage floor and drip pan spills immediately. Keep antifreeze containers covered and repair cooling system leaks as soon as they're noticed.

8 All vehicles covered by this manual are equipped with a pressurized coolant recovery system. A plastic expansion tank located at the front of the engine compartment is connected by a hose to the radiator (see illustration). As the engine heats up during operation, the expanding coolant fills the tank.

9 The coolant level in the tank should be checked regularly.

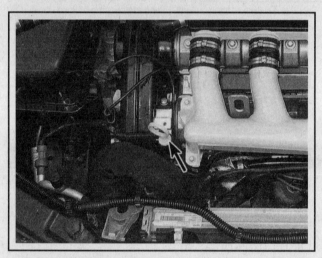

4.2a The oil dipstick is located at the right front corner of the engine on V6 engines

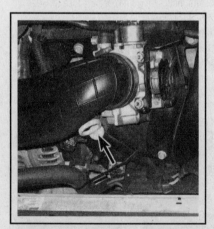

4.2b On four-cylinder engines the oil dipstick is located on the forward side of the engine, next to the throttle body

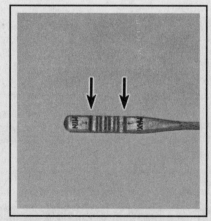

4.4 The oil level should be in the safe range - if it's below the MIN or ADD mark, add enough oil to bring it up to or near the MAX or FULL mark

4.6 The oil filler cap is located on the valve cover - always make sure the area around the opening is clean before unscrewing the cap to prevent dirt from contaminating the engine (four-cylinder engine shown)

TUNE-UP AND ROUTINE MAINTENANCE

4.8 The cooling system expansion tank is located at the left side of the engine compartment

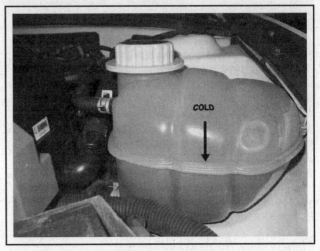

4.9 When the engine is cold, the engine coolant level should be at the COLD mark (expansion tank seam)

WARNING:

Do not remove the expansion tank cap to check the coolant level when the engine is warm! The level in the tank varies with the temperature of the engine.

When the engine is cold, the coolant level should be at the COLD mark on the reservoir (see illustration). If it isn't then remove the cap from the tank and add a 50/50 mixture of DEX-COOL antifreeze and water.

10 Drive the vehicle, let the engine cool completely then recheck the coolant level. Don't use rust inhibitors or additives. If only a small amount of coolant is required to bring the system up to the proper level, water can be used. However, repeated additions of water will dilute the antifreeze and water solution. In order to maintain the proper ratio of antifreeze and water, always top up the coolant level with the correct mixture. An empty plastic milk jug or bleach bottle makes an excellent container for mixing coolant.

11 If the coolant level drops consistently, there may be a leak in the system. Inspect the radiator, hoses, filler cap, drain plugs and water pump (see Section 9). If no leaks are noted, have the expansion tank cap pressure tested by a service station.

12 If you have to remove the expansion tank cap wait until the engine has cooled completely, then wrap a thick cloth around the cap and unscrew it slowly, stopping if you hear a hissing noise. If coolant or steam escapes, let the engine cool down longer, then remove the cap.

13 Check the condition of the coolant as well. It should be relatively clear. If it's brown or rust colored, the system should be drained, flushed and refilled. Even if the coolant appears to be normal, the corrosion inhibitors wear out, so it must be replaced at the specified intervals.

BRAKE AND CLUTCH FLUID

▶ Refer to illustration 4.15

14 The brake master cylinder is mounted on the front of the power booster unit in the engine compartment. The hydraulic clutch master cylinder used on manual transaxle vehicles is located next to the brake master cylinder.

15 The brake master cylinder and the clutch master cylinder share a common reservoir. To check the fluid level of either system, simply look at the MAX and MIN marks on the brake fluid reservoir (see illustration).

16 If the level is low, wipe the top of the reservoir cover with a clean rag to prevent contamination of the brake system before lifting the cover.

17 Add only the specified brake fluid to the reservoir (refer to *Recommended lubricants and fluids* at the end of this Chapter or to your owner's manual). Mixing different types of brake fluid can damage the system. Fill the brake master cylinder reservoir only to the MAX line.

WARNING:

Use caution when filling the reservoir - brake fluid can harm your eyes and damage painted surfaces. Do not use brake fluid that is more than one year old or has been left open. Brake fluid absorbs moisture from the air. Excess moisture can cause a dangerous loss of braking.

4.15 The brake fluid level should be kept between the MIN and MAX marks on the translucent plastic reservoir; the same reservoir contains the clutch fluid and is connected to the clutch master cylinder by a hose

18 While the reservoir cap is removed, inspect the master cylinder reservoir for contamination. If deposits, dirt particles or water droplets are present, the system should be drained and refilled.

1-8 TUNE-UP AND ROUTINE MAINTENANCE

19 After filling the reservoir to the proper level, make sure the cap is properly seated to prevent fluid leakage.

20 The fluid in the brake master cylinder will drop slightly as the brake pads at each wheel wear down during normal operation. If the master cylinder requires repeated replenishing to keep it at the proper level, this is an indication of leakage in the brake or clutch system, which should be corrected immediately. If the brake system shows an indication of leakage check all brake lines and connections, along with the calipers, wheel cylinders and booster (see Section 12 for more information). If the hydraulic clutch system shows an indication of leakage check all clutch lines and connections, along with the clutch release cylinder (see Chapter 8 for more information).

21 If, upon checking the brake or clutch master cylinder fluid level, you discover the reservoir empty or nearly empty, the systems should be checked, repaired and bled (see Chapters 8 and 9).

WINDSHIELD WASHER FLUID

▶ Refer to illustration 4.22

22 Fluid for the windshield washer system is stored in a plastic reservoir located at the left front of the engine compartment (see illustration).

23 In milder climates, plain water can be used in the reservoir, but it should be kept no more than 2/3 full to allow for expansion if the water freezes. In colder climates, use windshield washer system antifreeze, available at any auto parts store, to lower the freezing point of the fluid. Mix the antifreeze with water in accordance with the manufacturer's directions on the container.

CAUTION:

Do not use cooling system antifreeze - it will damage the vehicle's paint.

POWER STEERING FLUID

▶ Refer to illustrations 4.25a, 4.25b and 4.26

24 Check the power steering fluid level periodically to avoid steering system problems, such as damage to the pump.

CAUTION:

DO NOT hold the steering wheel against either stop (extreme left or right turn) for more than five seconds. If you do, the power steering pump could be damaged.

25 The fluid reservoir for the power steering system is mounted to the power steering pump at the left front corner of the engine on four-cylinder models, or separately on the radiator fan shroud on V6 models (see illustrations).

26 Four-cylinder engine models have a translucent reservoir; the level can be checked without removing the reservoir cap. The fluid level should be kept at the COLD mark when the engine is cool (see illustration). On V6 engine models, the fluid level can be checked by removing the dipstick. With the engine cool, the fluid level should be kept between the MIN and MAX marks on the dipstick.

27 Add small amounts of fluid until the level is correct.

CAUTION:

Do not overfill the reservoir. If too much fluid is added, remove the excess with a clean syringe or suction pump.

28 If the reservoir requires frequent fluid additions, all power steering hoses, hose connections, the power steering pump and the steering gear assembly should be carefully checked for leaks.

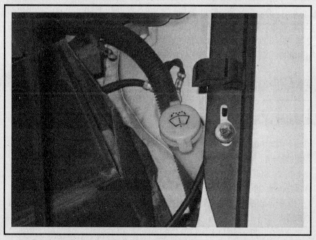

4.22 The windshield washer fluid reservoir is located in the left front corner of the engine compartment

4.25a The power steering fluid reservoir is located at the left side of the engine on four-cylinder models

4.25b The power steering fluid reservoir is located on the radiator fan shroud on V6 models

4.26 With the engine cool, the fluid level should be at the COLD mark

TUNE-UP AND ROUTINE MAINTENANCE

5 Tire and tire pressure checks (every 250 miles [400 km] or weekly)

▶ Refer to illustrations 5.2, 5.3, 5.4a, 5.4b and 5.8

1 Periodic inspection of the tires may spare you the inconvenience of being stranded with a flat tire. It can also provide you with vital information regarding possible problems in the steering and suspension systems before major damage occurs.

2 The original tires on this vehicle are equipped with 1/2-inch wide bands that will appear when tread depth reaches 1/16-inch, at which point they can be considered worn out. Tread wear can be monitored with a simple, inexpensive device known as a tread depth indicator (see illustration).

3 Note any abnormal tread wear (see illustration). Tread pattern irregularities such as cupping, flat spots and more wear on one side than the other are indications of front end alignment and/or balance problems. If any of these conditions are noted, take the vehicle to a tire shop or service station to correct the problem.

4 Look closely for cuts, punctures and embedded nails or tacks. Sometimes a tire will hold air pressure for a short time or leak down very slowly after a nail has embedded itself in the tread. If a slow leak persists, check the valve stem core to make sure it is tight (see illustration). Examine the tread for an object that may have embedded itself in the tire or for a "plug" that may have begun to leak (radial tire punctures are repaired with a plug that is installed in a puncture). If a puncture is suspected, it can be easily verified by spraying a solution of soapy water onto the puncture area (see illustration). The soapy solution will bubble if there is a leak. Unless the puncture is unusually large, a tire shop or service station can usually repair the tire.

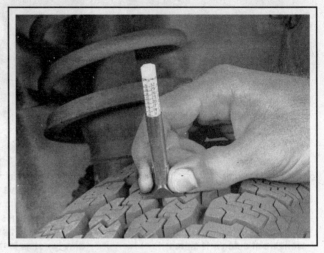

5.2 A tire tread depth indicator should be used to monitor tire wear - they are available at auto parts stores and service stations and cost very little

5 Carefully inspect the inner sidewall of each tire for evidence of brake fluid leakage. If you see any, inspect the brakes immediately.

6 Correct air pressure adds miles to the life span of the tires, improves mileage and enhances overall ride quality. Tire pressure cannot be accurately estimated by looking at a tire, especially if it's a

UNDERINFLATION

INCORRECT TOE-IN OR EXTREME CAMBER

CUPPING

Cupping may be caused by:
- Underinflation and/or mechanical irregularities such as out-of-balance condition of wheel and/or tire, and bent or damaged wheel.
- Loose or worn steering tie-rod or steering idler arm.
- Loose, damaged or worn front suspension parts.

OVERINFLATION

FEATHERING DUE TO MISALIGNMENT

5.3 This chart will help you determine the condition of your tires, the probable cause(s) of abnormal wear and the corrective action necessary

1-10 TUNE-UP AND ROUTINE MAINTENANCE

5.4a If a tire loses air on a steady basis, check the valve core first to make sure it's snug (special inexpensive wrenches are commonly available at auto parts stores)

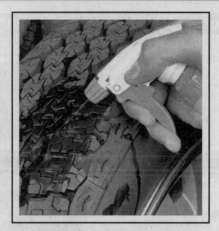

5.4b If the valve core is tight, raise the corner of the vehicle with the low tire and spray a soapy water solution onto the tread as the tire is turned slowly - slow leaks will cause small bubbles to appear

5.8 To extend the life of your tires, check the air pressure at least once a week with an accurate gauge (don't forget the spare!)

radial. A tire pressure gauge is essential. Keep an accurate gauge in the glove compartment. The pressure gauges attached to the nozzles of air hoses at gas stations are often inaccurate.

7 Always check tire pressure when the tires are cold. Cold, in this case, means the vehicle has not been driven over a mile in the three hours preceding a tire pressure check. A pressure rise of four to eight pounds is not uncommon once the tires are warm.

8 Unscrew the valve cap protruding from the wheel or hubcap and push the gauge firmly onto the valve stem (see illustration). Note the reading on the gauge and compare the figure to the recommended tire pressure shown on the tire placard on the driver's side door. Be sure to reinstall the valve cap to keep dirt and moisture out of the valve stem mechanism. Check all four tires and, if necessary, add enough air to bring them up to the recommended pressure.

9 Don't forget to keep the spare tire inflated to the specified pressure (refer to the pressure molded into the tire sidewall).

6 Engine oil and filter change (every 3000 miles [5000 km] or 3 months)

▶ Refer to illustrations 6.7, 6.11, 6.12a, 6.12b and 6.19

➡Note: These vehicles are equipped with an oil life indicator system that illuminates a light or message on the instrument panel when the system deems it necessary to change the oil. A number of factors are taken into consideration to determine when the oil should be considered "worn out." Generally, this system will allow the vehicle to accumulate more miles between oil changes than the traditional 3000 mile interval, but we believe that frequent oil changes are "cheap insurance" and will prolong engine life. If you do decide not to change your oil every 3000 miles and rely on the oil life indicator instead, make sure you don't exceed 10,000 miles before the oil is changed, regardless of what the oil life indicator shows.

1 Frequent oil changes are the most important preventive maintenance procedures that can be done by the home mechanic. As engine oil ages, it becomes diluted and contaminated, which leads to premature engine wear.

6.7 Use a proper size box-end wrench or socket to remove the oil drain plug and avoid rounding it off

6.11 Unscrew the cap to access the oil filter

6.12a Remove the filter cartridge . . .

TUNE-UP AND ROUTINE MAINTENANCE

6.12b ... then separate the element from the cap

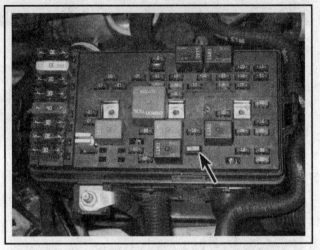

6.19 The reset button for the Change Engine Oil light is located on the engine compartment fuse/relay block

2 Make sure that you have all the necessary tools before you begin this procedure. You should also have plenty of rags or newspapers handy for mopping up oil spills.

3 Access to the oil drain plug and filter will be improved if the vehicle can be lifted on a hoist, driven onto ramps or supported by jackstands.

✲✲ WARNING:

Do not work under a vehicle supported only by a jack - always use jackstands!

4 If you haven't changed the oil on this vehicle before, get under it and locate the oil drain plug and the oil filter. The exhaust components will be warm as you work, so note how they are routed to avoid touching them when you are under the vehicle.

5 Start the engine and allow it to reach normal operating temperature - oil and sludge will flow out more easily when warm. If new oil, a filter or tools are needed, use the vehicle to go get them and warm up the engine/oil at the same time. Park on a level surface and shut off the engine when it's warmed up. Remove the oil filler cap from the valve cover.

6 Raise the vehicle and support it on jackstands. Make sure it is safely supported!

7 Being careful not to touch the hot exhaust components, position a drain pan under the plug in the bottom of the engine, then remove the plug (see illustration). It's a good idea to wear a rubber glove while unscrewing the plug the final few turns to avoid being scalded by hot oil.

8 It may be necessary to move the drain pan slightly as oil flow slows to a trickle. Inspect the old oil for the presence of metal particles.

9 After all the oil has drained, wipe off the drain plug with a clean rag. Any small metal particles clinging to the plug would immediately contaminate the new oil.

10 Clean the area around the drain plug opening, reinstall the plug and tighten it securely, but don't strip the threads.

11 Move the drain pan into position under the oil filter. On four-cylinder engines, the canister-type oil filter is located at the front left side of the engine and is accessible from the top of the vehicle (see illustration). On V6 models, the canister-type filter is also located at the front of the engine but is accessed from under the vehicle.

12 Unscrew the oil filter cap and withdraw it, together with the element (see illustrations).

13 Use a clean rag to remove all oil, dirt and sludge from the oil filter housing and cap.

14 Install a new O-ring seal in the groove on the retaining cap, then install the new element in the cap and insert them both in the filter housing. Screw on the cap and tighten it securely.

15 Remove all tools and materials from under the vehicle, being careful not to spill the oil in the drain pan, then lower the vehicle.

16 Add new oil to the engine through the oil filler cap. Use a funnel to prevent oil from spilling onto the top of the engine. Pour four quarts of fresh oil into the engine. Wait a few minutes to allow the oil to drain into the pan, then check the level on the dipstick (see Section 4 if necessary). If the oil level is in the OK range, install the filler cap.

17 Start the engine and run it for about a minute. While the engine is running, look under the vehicle and check for leaks at the oil pan drain plug and around the oil filter. If either one is leaking, stop the engine and tighten the plug or filter slightly.

18 Wait a few minutes, then recheck the level on the dipstick. Add oil as necessary to bring the level into the OK range.

19 Be sure to reset the Change Engine Oil light. With the ignition key turned to the RUN position, open the cover on the underhood fuse/relay block and press and hold the reset button for five seconds (see illustration). This should reset the system.

20 During the first few trips after an oil change, make it a point to check frequently for leaks and proper oil level.

21 The old oil drained from the engine cannot be reused in its present state and should be disposed of. Check with your local auto parts store, disposal facility or environmental agency to see if they will accept the oil for recycling. After the oil has cooled it can be drained into a container (capped plastic jugs, topped bottles, milk cartons, etc.) for transport to one of these disposal sites. Don't dispose of the oil by pouring it on the ground or down a drain!

1-12 TUNE-UP AND ROUTINE MAINTENANCE

7 Windshield wiper blade inspection and replacement (every 6000 miles [10,000 km] or 6 months)

▶ Refer to illustrations 7.4a and 7.4b

1 The windshield wiper and blade assembly should be inspected periodically for damage, loose components and cracked or worn blade elements.

2 Road film can build up on the wiper blades and affect their efficiency, so they should be washed regularly with a mild detergent solution.

3 If the wiper blade elements are cracked, worn or warped, or no longer clean adequately, they should be replaced with new ones.

4 Lift the arm assembly away from the glass for clearance, pry up on the release lever, then slide the wiper blade assembly out of the hook in the end of the arm (see illustrations).

5 Attach the new wiper to the arm. Connection can be confirmed by an audible click.

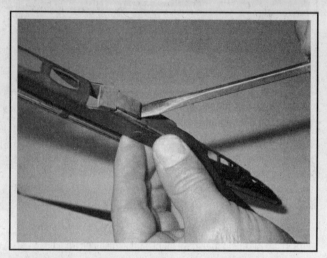

7.4a To release the blade holder, pry up on the release tab . . .

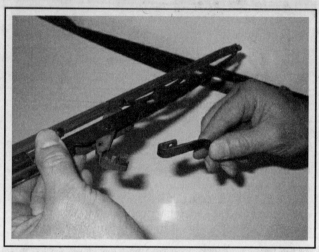

7.4b . . . and pull the wiper blade in the direction of the windshield to separate it from the arm

8 Battery check, maintenance and charging (every 6000 miles [10,000 km] or 6 months)

▶ Refer to illustrations 8.1, 8.5, 8.7a, 8.7b and 8.7c

⁂ WARNING:

Certain precautions must be followed when checking and servicing the battery. Hydrogen gas, which is highly flammable, is always present in the battery cells, so keep lighted tobacco and all other open flames and sparks away from the battery. The electrolyte inside the battery is actually dilute sulfuric acid, which will cause injury if splashed on your skin or in your eyes. It will also ruin clothes and painted surfaces. When removing the battery cables, always detach the negative cable first and hook it up last!

1 A routine preventive maintenance program for the battery in your vehicle is the only way to ensure quick and reliable starts. But before performing any battery maintenance, make sure that you have the proper equipment necessary to work safely around the battery (see illustration).

2 There are also several precautions that should be taken whenever battery maintenance is performed. Before servicing the battery, always turn the engine and all accessories off and disconnect the cable from the negative terminal of the battery.

3 The battery produces hydrogen gas, which is both flammable and explosive. Never create a spark, smoke or light a match around the battery. Always charge the battery in a ventilated area.

4 Electrolyte contains poisonous and corrosive sulfuric acid. Do not allow it to get in your eyes, on your skin or on your clothes. Never ingest it. Wear protective safety glasses when working near the battery. Keep children away from the battery.

5 Note the external condition of the battery. If the positive terminal and cable clamp on your vehicle's battery is equipped with a rubber protector, make sure that it's not torn or damaged. It should completely cover the terminal. Look for any corroded or loose connections, cracks in the case or cover or loose hold-down clamps. Also check the entire length of each cable for cracks and frayed conductors (see illustration).

6 If corrosion, which looks like white, fluffy deposits is evident, particularly around the terminals, the battery should be removed for cleaning. Loosen the cable bolts with a wrench, being careful to remove the ground cable first, and slide them off the terminals. Then disconnect the hold-down clamp bolt and nut, remove the clamp and lift the battery from the engine compartment.

7 Clean the cable ends thoroughly with a battery brush or a terminal cleaner and a solution of warm water and baking soda. Wash the terminals and the side of the battery case with the same solution but make sure that the solution doesn't get into the battery. When cleaning the cables, terminals and battery case, wear safety goggles and rubber gloves to prevent any solution from coming in contact with your eyes or hands. Wear old clothes too - even diluted, sulfuric acid splashed onto clothes will burn holes in them. If the terminals have been corroded, clean them up with a terminal cleaner (see illustrations). Thoroughly wash all cleaned areas with plain water.

TUNE-UP AND ROUTINE MAINTENANCE

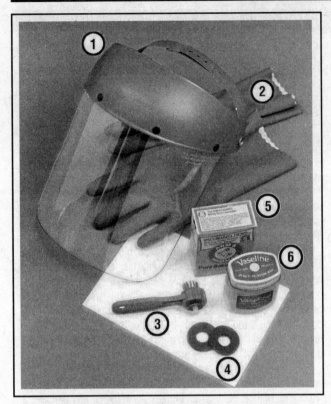

8.1 Tools and materials required for battery maintenance
1 **Face shield/safety goggles** - When removing corrosion with a brush, the acidic particles can easily fly up into your eyes
2 **Rubber gloves** - Another safety item to consider when servicing the battery - remember that's acid inside the battery!
3 **Battery terminal/cable cleaner** - This wire brush cleaning tool will remove all traces of corrosion from the battery posts and cable clamps
4 **Treated felt washers** - Placing one of these on each post, directly under the cable clamps, will help prevent corrosion
5 **Baking soda** - A solution of baking soda and water can be used to neutralize corrosion
6 **Petroleum jelly** - A layer of this on the battery posts will help prevent corrosion

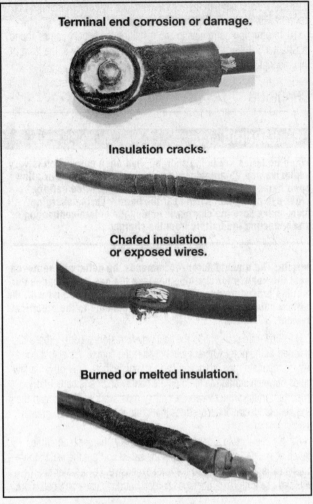

8.5 Typical battery cable problems

8 Make sure that the battery tray is in good condition and the hold-down clamp bolts are tight. If the battery is removed from the tray, make sure no parts remain in the bottom of the tray when the battery is reinstalled. When reinstalling the hold-down clamp bolts, do not overtighten them.

8.7a A tool like this one (available at auto parts stores) is used to clean the side-terminal type battery-cable contact area

8.7b Use the brush side of the tool to finish the job

8.7c Regardless of the type of tool used on the battery and cables, a clean, shiny surface should be the result

1-14 TUNE-UP AND ROUTINE MAINTENANCE

9 Any metal parts of the vehicle damaged by corrosion should be covered with a zinc-based primer, then painted.

10 Information on removing and installing the battery can be found in Chapter 5. Information on jump starting can be found at the front of this manual.

CHARGING

WARNING:

When batteries are being charged, hydrogen gas, which is very explosive and flammable, is produced. Do not smoke or allow open flames near a charging or a recently charged battery. Wear eye protection when near the battery during charging. Also, make sure the charger is unplugged before connecting or disconnecting the battery from the charger.

→Note: The manufacturer recommends the battery be removed from the vehicle for charging because the gas that escapes during this procedure can damage the paint. Fast charging with the battery cables connected can result in damage to the electrical system.

11 Slow-rate charging is the best way to restore a battery that's discharged to the point where it will not start the engine. It's also a good way to maintain the battery charge in a vehicle that's only driven a few miles between starts. Maintaining the battery charge is particularly important in the winter when the battery must work harder to start the engine and electrical accessories that drain the battery are in greater use.

12 It's best to use a one or two-amp battery charger (sometimes called a "trickle" charger). They are the safest and put the least strain on the battery. They are also the least expensive. For a faster charge, you can use a higher amperage charger, but don't use one rated more than 1/10th the amp/hour rating of the battery. Rapid boost charges that claim to restore the power of the battery in one to two hours are hardest on the battery and can damage batteries not in good condition. This type of charging should only be used in emergency situations.

13 The average time necessary to charge a battery should be listed in the instructions that come with the charger. As a general rule, a trickle charger will charge a battery in 12 to 16 hours.

14 Remove all the cell caps (if equipped) and cover the holes with a clean cloth to prevent spattering electrolyte. Disconnect the negative battery cable and hook the battery charger cable clamps up to the battery posts (positive to positive, negative to negative), then plug in the charger. Make sure it is set at 12-volts if it has a selector switch.

15 If you're using a charger with a rate higher than two amps, check the battery regularly during charging to make sure it doesn't overheat. If you're using a trickle charger, you can safely let the battery charge overnight after you've checked it regularly for the first couple of hours.

16 If the battery has removable cell caps, measure the specific gravity with a hydrometer every hour during the last few hours of the charging cycle. Hydrometers are available inexpensively from auto parts stores - follow the instructions that come with the hydrometer. Consider the battery charged when there's no change in the specific gravity reading for two hours and the electrolyte in the cells is gassing (bubbling) freely. The specific gravity reading from each cell should be very close to the others. If not, the battery probably has a bad cell(s).

17 Some batteries with sealed tops have built-in hydrometers on the top that indicate the state of charge by the color displayed in the hydrometer window. Normally, a bright-colored hydrometer indicates a full charge and a dark hydrometer indicates the battery still needs charging.

18 If the battery has a sealed top and no built-in hydrometer, you can hook up a digital voltmeter across the battery terminals to check the charge. A fully charged battery should read 12.5 volts or higher.

19 Further information on the battery and jump-starting can be found in Chapter 5 and at the front of this manual.

9 Cooling system check (every 6000 miles [10,000 km] or 6 months)

▶ Refer to illustration 9.4

1 Many major engine failures can be caused by a faulty cooling system.

2 The engine must be cold for the cooling system check, so perform the following procedure before the vehicle is driven for the day or after it has been shut off for at least three hours.

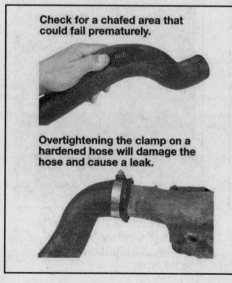

Check for a chafed area that could fail prematurely.

Overtightening the clamp on a hardened hose will damage the hose and cause a leak.

Check for a soft area indicating the hose has deteriorated inside.

Check each hose for swelling and oil-soaked ends. Cracks and breaks can be located by squeezing the hose.

9.4 Hoses, like drivebelts, have a habit of failing at the worst possible time - to prevent the inconvenience of a blown radiator or heater hose, inspect them carefully as shown here

TUNE-UP AND ROUTINE MAINTENANCE

3 Remove the pressure relief cap from the expansion tank. Clean the cap thoroughly, inside and out, with clean water. The presence of rust or corrosion in the expansion tank means the coolant should be changed (see Section 20). The coolant inside the expansion tank should be relatively clean and transparent. If it's rust colored, drain the system and refill it with new coolant.

4 Carefully check the radiator hoses and the smaller diameter heater hoses (see illustrations in Chapter 3). Inspect each coolant hose along its entire length, replacing any hose which is cracked, swollen or deteriorated (see illustration). Cracks will show up better if the hose is squeezed. Pay close attention to hose clamps that secure the hoses to cooling system components. Hose clamps can pinch and puncture hoses, resulting in coolant leaks.

5 Make sure that all hose connections are tight. A leak in the cooling system will usually show up as white or rust colored deposits on the area adjoining the leak. If wire-type clamps are used on the hoses, it may be a good idea to replace them with screw-type clamps.

6 Clean the front of the radiator and air conditioning condenser with compressed air, if available, or a soft brush. Remove all bugs, leaves, etc. embedded in the radiator fins. Be extremely careful not to damage the cooling fins or cut your fingers on them.

7 If the coolant level has been dropping consistently and no leaks are detectable, have the expansion tank cap and cooling system pressure checked at a service station.

10 Tire rotation (every 6000 miles [10,000 km] or 6 months)

▶ Refer to illustration 10.2

1 The tires should be rotated at the specified intervals and whenever uneven wear is noticed. Since the vehicle will be raised and the tires removed anyway, check the brakes also (see Section 12).

2 Radial tires must be rotated in a specific pattern (see illustration). If your vehicle has a compact spare tire, don't include it in the rotation pattern.

3 Refer to the information in *Jacking and towing* at the front of this manual for the proper procedure to follow when raising the vehicle and changing a tire. If the brakes must be checked, don't apply the parking brake as stated.

4 The vehicle must be raised on a hoist or supported on jackstands to get all four wheels off the ground. Make sure the vehicle is safely supported!

5 After the rotation procedure is finished, check and adjust the tire pressures as necessary and be sure to check the lug nut tightness.

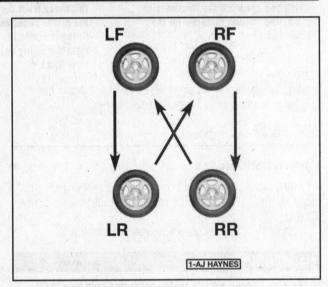

10.2 Four-tire rotation pattern

11 Seat belt check (every 6000 miles [10,000 km] or 6 months)

1 Check seat belts, buckles, latch plates and guide loops for obvious damage and signs of wear.

2 See if the seat belt reminder light comes on when the key is turned to the Run or Start position. A chime should also sound.

3 The seat belts are designed to lock up during a sudden stop or impact, yet allow free movement during normal driving. Make sure the retractors return the belt against your chest while driving and rewind the belt fully when the buckle is unlatched.

4 If any of the above checks reveal problems with the seat belt system, replace parts as necessary.

12 Brake check (every 6000 miles [10,000 km] or 6 months)

※※ WARNING:
The dust created by the brake system is harmful to your health. Never blow it out with compressed air and don't inhale any of it. An approved filtering mask should be worn when working on the brakes. Do not, under any circumstances, use petroleum-based solvents to clean brake parts. Use brake system cleaner only!

➡ **Note: For detailed photographs of the brake system, refer to Chapter 9.**

1 In addition to the specified intervals, the brakes should be inspected every time the wheels are removed or whenever a defect is suspected.

2 Any of the following symptoms could indicate a potential brake system defect: The vehicle pulls to one side when the brake pedal is depressed; the brakes make squealing or dragging noises when

1-16 TUNE-UP AND ROUTINE MAINTENANCE

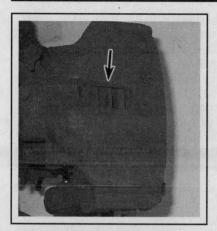

12.6 You will find an inspection hole like this in each caliper through which you can view the thickness of remaining friction material for the inner pad

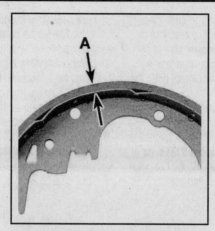

12.14 If the lining is bonded to the brake shoe, measure the lining thickness from the outer surface to the metal shoe, as shown here; if the lining is riveted to the shoe, measure from the lining outer surface to the rivet head

12.16 Carefully peel back the wheel cylinder boot and check for leaking fluid, indicating that the cylinder must be replaced

applied; brake pedal travel is excessive; the pedal pulsates; or brake fluid leaks, usually onto the inside of the tire or wheel.

DISC BRAKES

♦ Refer to illustration 12.6

3 Disc brakes can be visually checked without removing any parts except the wheels. Remove the hub caps (if applicable) and loosen the wheel lug nuts a quarter turn each.

4 Raise the vehicle and place it securely on jackstands.

※ WARNING:

Never work under a vehicle that is supported only by a jack!

5 Remove the wheels. Now visible is the disc brake caliper which contains the pads. There is an outer brake pad and an inner pad. Both must be checked for wear.

6 Measure the thickness of the outer pad at each end of the caliper and the inner pad through the inspection hole in the caliper body (see illustration). Compare the measurement with the limit given in this Chapter's Specifications; if any brake pad thickness is less than specified, then all brake pads must be replaced (see Chapter 9).

7 If you're in doubt as to the exact pad thickness or quality, remove them for measurement and further inspection (see Chapter 9).

8 Check the disc for score marks, wear and burned spots. If any of these conditions exist, the disc should be removed for servicing or replacement (see Chapter 9).

9 Before installing the wheels, check all the brake lines and hoses for damage, wear, deformation, cracks, corrosion, leakage, bends and twists, particularly in the vicinity of the rubber hoses and calipers.

10 Install the wheels, lower the vehicle and tighten the wheel lug nuts to the torque given in this Chapter's Specifications.

DRUM BRAKES

♦ Refer to illustrations 12.14 and 12.16

11 On models with rear drum brakes, make sure the parking brake is off then tap on the outside of the drum with a rubber mallet to loosen it.

12 Remove the brake drums. If the drum still won't come off, refer to Chapter 9

13 With the drums removed, carefully clean the brake assembly with brake system cleaner.

※ WARNING:

Don't blow the dust out with compressed air and don't inhale any of it (it is harmful to your health).

14 Note the thickness of the lining material on both front and rear brake shoes (see illustration). Compare the measurement with the limit given in this Chapter's Specifications; if any lining thickness is less than specified, then all of the brake shoes must be replaced (see Chapter 9). The shoes should also be replaced if they're cracked, glazed (shiny areas), or covered with brake fluid.

15 Make sure all the brake assembly springs are connected and in good condition.

16 Check the brake components for signs of fluid leakage. With your finger or a small screwdriver, carefully pry back the rubber cups on the wheel cylinder located at the top of the brake shoes (see illustration). Any leakage here is an indication that the wheel cylinders should be replaced immediately (see Chapter 9). Also, check all hoses and connections for signs of leakage.

17 Wipe the inside of the drum with a clean rag and denatured alcohol or brake cleaner. Again, be careful not to breathe the dangerous brake dust.

18 Check the inside of the drum for cracks, score marks, deep scratches and "hard spots" which will appear as small discolored areas. If imperfections cannot be removed with fine emery cloth, the drum must be taken to an automotive machine shop for resurfacing.

19 Repeat the procedure for the remaining wheel. If the inspection reveals that all parts are in good condition, reinstall the brake drums, install the wheels and lower the vehicle to the ground.

BRAKE BOOSTER CHECK

20 Sit in the driver's seat and perform the following sequence of tests.

TUNE-UP AND ROUTINE MAINTENANCE 1-17

21 With the brake fully depressed, start the engine - the pedal should move down a little when the engine starts.
22 With the engine running, depress the brake pedal several times - the travel distance should not change.
23 Depress the brake, stop the engine and hold the pedal in for about 30 seconds - the pedal should neither sink nor rise.

24 Restart the engine, run it for about a minute and turn it off. Then firmly depress the brake several times - the pedal travel should decrease with each application.
25 If your brakes do not operate as described, the brake booster has failed. Refer to Chapter 9 for the replacement procedure.

13 Interior ventilation filter - replacement (every 12,000 miles [24,000 km] or 12 months)

▶ Refer to illustrations 13.2, 13.3 and 13.4

1 These models are equipped with an air filtering element for the interior ventilation system. The access panel is located in the cowl.
2 Pull up on the hood-to-cowl rubber weatherseal (see illustration).
3 Open the cover for access to the filter (see illustration).
4 Release the filter tabs at each end and remove the filter (see illustration).
5 Installation is the reverse of the removal procedure.

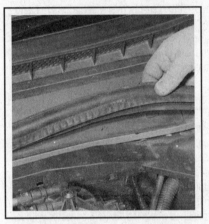

13.2 Pull back the rubber seal . . .

13.3 . . . open the filter door . . .

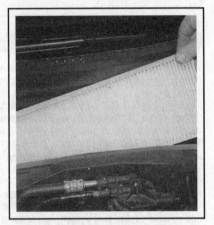

13.4 . . . release the tabs securing the filter element, then remove it from the housing

14 Underhood hose check and replacement (every 15,000 miles [24,000 km] or 12 months)

※ WARNING:

Replacement of air conditioning hoses must be left to a dealer service department or air conditioning shop that has the equipment to depressurize the system safely. Never remove air conditioning components or hoses until the system has been depressurized.

GENERAL

1 High temperatures under the hood can cause deterioration of the rubber and plastic hoses used for engine, accessory and emission systems operation. Periodic inspection should be made for cracks, loose clamps, material hardening and leaks.
2 Information specific to the cooling system hoses can be found in Section 9.
3 Most (but not all) hoses are secured to the fittings with clamps. Where clamps are used, check to be sure they haven't lost their tension, allowing the hose to leak. If clamps aren't used, make sure the hose has not expanded and/or hardened where it slips over the fitting, allowing it to leak.

PCV SYSTEM HOSE

4 To reduce hydrocarbon emissions, crankcase blow-by gas is vented through the PCV valve in the rocker arm cover to the intake manifold via a rubber hose on most models. The blow-by gases mix with incoming air in the intake manifold before being burned in the combustion chambers.
5 Check the PCV hose for cracks, leaks and other damage. Disconnect it from the valve cover and the intake manifold and check the inside for obstructions. If it's clogged, clean it out with solvent.

VACUUM HOSES

6 It's quite common for vacuum hoses, especially those in the emissions system, to be color coded or identified by colored stripes molded into them. Various systems require hoses with different wall thickness, collapse resistance and temperature resistance. When replacing hoses, be sure the new ones are made of the same material.
7 Often the only effective way to check a hose is to remove it completely from the vehicle. If more than one hose is removed, be sure to label the hoses and fittings to ensure correct installation.

8 When checking vacuum hoses, be sure to include any plastic T-fittings in the check. Inspect the fittings for cracks and the hose where it fits over each fitting for distortion, which could cause leakage.

9 A small piece of vacuum hose (1/4-inch inside diameter) can be used as a stethoscope to detect vacuum leaks. Hold one end of the hose to your ear and probe around vacuum hoses and fittings, listening for the "hissing" sound characteristic of a vacuum leak.

WARNING:

When probing with the vacuum hose stethoscope, be careful not to come into contact with moving engine components such as drivebelts, the cooling fan, etc.

FUEL HOSE

WARNING:

Gasoline is flammable, so take extra precautions when you work on any part of the fuel system. Don't smoke or allow open flames or bare light bulbs near the work area, and don't work in a garage where a gas-type appliance (such as a water heater or clothes dryer) is present. Since fuel is carcinogenic, wear latex gloves when there's a possibility of being exposed to fuel, and, if you spill any fuel on your skin, rinse it off immediately with soap and water. Mop up any spills immediately and do not store fuel-soaked rags where they could ignite. The fuel system is under constant pressure, so, if any fuel lines are to be disconnected, the fuel pressure in the system must be relieved first (see Chapter 4 for more information). When you perform any kind of work on the fuel system, wear safety glasses and have a Class B type fire extinguisher on hand.

10 The fuel lines are usually under pressure, so if any fuel lines are to be disconnected be prepared to catch spilled fuel.

WARNING:

Your vehicle is equipped with fuel injection and you must relieve the fuel system pressure before servicing the fuel lines. Refer to Chapter 4 for the fuel system pressure relief procedure.

11 Check all flexible fuel lines for deterioration and chafing. Check especially for cracks in areas where the hose bends and just before fittings, such as where a hose attaches to the fuel pump, fuel filter and fuel injection unit.

12 When replacing a hose, use only hose that is specifically designed for your fuel injection system.

13 Spring-type clamps are sometimes used on fuel return or vapor lines. These clamps often lose their tension over a period of time, and can be "sprung" during removal. Replace all spring-type clamps with screw clamps whenever a hose is replaced. Some fuel lines use spring-lock type couplings, which require a special tool to disconnect. See Chapter 4 for more information on this type of coupling.

METAL LINES

14 Sections of metal line are often used for fuel line between the fuel pump and the fuel injection unit. Check carefully to make sure the line isn't bent, crimped or cracked.

15 If a section of metal fuel line must be replaced, use seamless steel tubing only, since copper and aluminum tubing do not have the strength necessary to withstand vibration caused by the engine.

16 Check the metal brake lines where they enter the master cylinder and brake proportioning unit (if used) for cracks in the lines and loose fittings. Any sign of brake fluid leakage calls for an immediate thorough inspection of the brake system.

15 Steering, suspension and driveaxle boot check (every 15,000 miles [24,000 km] or 12 months)

▶ Refer to illustrations 15.4, 15.10, 15.11 and 15.14

➡ Note: For detailed illustrations of the steering and suspension components, refer to Chapter 10.

WITH THE WHEELS ON THE GROUND

1 With the vehicle stopped and the front wheels pointed straight ahead, rock the steering wheel gently back and forth. If freeplay is excessive, a front wheel bearing, steering shaft universal joint or lower arm balljoint is worn or the steering gear is out of adjustment or broken. Refer to Chapter 10 for the appropriate repair procedure.

2 Other symptoms, such as excessive vehicle body movement over rough roads, swaying (leaning) around corners and binding as the steering wheel is turned, may indicate faulty steering and/or suspension components.

3 Check the shock absorbers by pushing down and releasing the vehicle several times at each corner. If the vehicle does not come back to a level position within one or two bounces, the shocks/struts are worn and must be replaced. When bouncing the vehicle up and down, listen for squeaks and noises from the suspension components.

4 Check the struts and shock absorbers for evidence of fluid leakage (see illustration). A light film of fluid is no cause for concern. Make

15.4 Check the shocks for leakage at the indicated area

sure that any fluid noted is from the struts/shocks and not from some other source. If leakage is noted, replace the struts/shocks as a set.

5 Check the struts and shocks to be sure they are securely

TUNE-UP AND ROUTINE MAINTENANCE

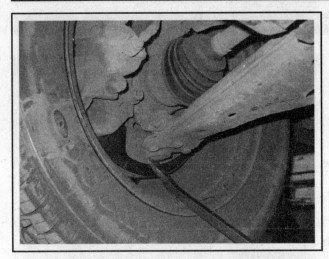

15.10 To check a balljoint for wear, try to pry the control arm up and down to make sure there is no play in the balljoint (if there is, replace it)

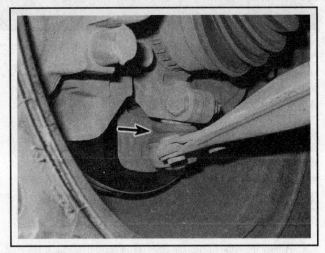

15.11 Check the balljoint boot for damage also

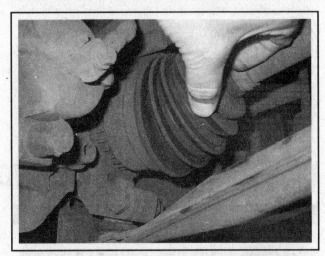

15.14 Flex the driveaxle boots by hand to check for cracks and/or leaking grease

mounted and undamaged. Check the upper mounts for damage and wear. If damage or wear is noted, replace the shocks as a set (front and rear).

6 If the shocks must be replaced, refer to Chapter 10 for the procedure.

UNDER THE VEHICLE

7 Raise the vehicle with a floor jack and support it securely on jackstands. See *Jacking and towing* at the front of this book for the proper jacking points.

8 Check the tires for irregular wear patterns and proper inflation. See Section 5 in this Chapter for information regarding tire wear and Chapter 10 for information on wheel bearing replacement.

9 Inspect the universal joint between the steering shaft and the steering gear housing. Check the steering gear housing for lubricant leakage. Make sure that the dust seals and boots are not damaged and that the boot clamps are not loose. Check the steering linkage for looseness or damage. Check the tie-rod ends for excessive play. Look for loose bolts, broken or disconnected parts and deteriorated rubber bushings on all suspension and steering components. While an assistant turns the steering wheel from side to side, check the steering components for free movement, chafing and binding. If the steering components do not seem to be reacting with the movement of the steering wheel, try to determine where the slack is located.

10 Check the balljoints for wear by trying to move each control arm up and down with a prybar (see illustration) to ensure that its balljoint has no play. If any balljoint does have play, replace it. See Chapter 10 for the balljoint replacement procedure.

11 Inspect the balljoint boots for damage and leaking grease (see illustration). Replace the balljoints with new ones if they are damaged (see Chapter 10).

12 At the rear of the vehicle, inspect the suspension arm bushings for deterioration. Additional information on suspension components can be found in Chapter 10.

DRIVEAXLE BOOT CHECK

➡ **Note:** *For detailed illustrations of the driveaxles, refer to Chapter 8.*

13 The driveaxle boots are very important because they prevent dirt, water and foreign material from entering and damaging the constant velocity (CV) joints. Oil and grease can cause the boot material to deteriorate prematurely, so it's a good idea to wash the boots with soap and water. Because it constantly pivots back and forth following the steering action of the front hub, the outer CV boot wears out sooner and should be inspected regularly.

14 Inspect the boots for tears and cracks as well as loose clamps (see illustration). If there is any evidence of cracks or leaking lubricant, they must be replaced as described in Chapter 8.

16 Exhaust system check (every 15,000 miles [24,000 km] or 12 months)

♦ Refer to illustration 16.2

1 With the engine cold (at least three hours after the vehicle has been driven), check the complete exhaust system from the engine to the end of the tailpipe. Ideally, the inspection should be done with the vehicle on a hoist to permit unrestricted access. If a hoist isn't available, raise the vehicle and support it securely on jackstands.

2 Check the exhaust pipes and connections for evidence of leaks, severe corrosion and damage. Make sure that all brackets and hangers are in good condition and tight (see illustration).

3 At the same time, inspect the underside of the body for holes, corrosion, open seams, etc. which may allow exhaust gases to enter the passenger compartment. Seal all body openings with silicone or body putty.

4 Rattles and other noises can often be traced to the exhaust system, especially the mounts and hangers. Try to move the pipes, muffler and catalytic converter. If the components can come in contact with the body or suspension parts, secure the exhaust system with new mounts.

5 Check the running condition of the engine by inspecting inside the end of the tailpipe. The exhaust deposits here are an indication of engine state-of-tune. If the pipe is black and sooty or coated with white deposits, the engine may need a tune-up, including a thorough fuel system inspection and adjustment.

16.2 Be sure to check each exhaust system rubber hanger for damage

17 Fuel system check (every 15,000 miles [24,000 km] or 12 months)

※ WARNING:

Gasoline is flammable, so take extra precautions when you work on any part of the fuel system. Don't smoke or allow open flames or bare light bulbs near the work area, and don't work in a garage where a gas-type appliance (such as a water heater or clothes dryer) is present. Since fuel is carcinogenic, wear latex gloves when there's a possibility of being exposed to fuel, and, if you spill any fuel on your skin, rinse it off immediately with soap and water. Mop up any spills immediately and do not store fuel-soaked rags where they could ignite. When you perform any kind of work on the fuel system, wear safety glasses and have a Class B type fire extinguisher on hand. The fuel system is under constant pressure, so, before any lines are disconnected, the fuel system pressure must be relieved (see Chapter 4).

1 If you smell gasoline while driving or after the vehicle has been sitting in the sun, inspect the fuel system immediately.

2 Remove the fuel filler cap and inspect if for damage and corrosion. The gasket should have an unbroken sealing imprint. If the gasket is damaged or corroded, install a new cap.

3 Inspect the fuel feed line for cracks. Make sure that the connections between the fuel lines and the fuel injection system and between the fuel lines and the in-line fuel filter are tight.

※ WARNING:

Your vehicle is fuel injected, so you must relieve the fuel system pressure before servicing fuel system components. The fuel system pressure relief procedure is outlined in Chapter 4.

4 Since some components of the fuel system - the fuel tank and part of the fuel feed line, for example - are underneath the vehicle, they can be inspected more easily with the vehicle raised on a hoist. If that's not possible, raise the vehicle and support it on jackstands.

5 With the vehicle raised and safely supported, inspect the gas tank and filler neck for punctures, cracks and other damage. The connection between the filler neck and the tank is particularly critical. Sometimes a rubber filler neck will leak because of loose clamps or deteriorated rubber. Inspect all fuel tank mounting brackets and straps to be sure that the tank is securely attached to the vehicle.

※ WARNING:

Do not, under any circumstances, try to repair a fuel tank (except rubber components). A welding torch or any open flame can easily cause fuel vapors inside the tank to explode.

6 Carefully check all rubber hoses and metal lines leading away from the fuel tank. Check for loose connections, deteriorated hoses, crimped lines and other damage. Repair or replace damaged sections as necessary (see Chapter 4).

18 Drivebelt check and replacement (every 15,000 miles [24,000 km] or 12 months)

ACCESSORY DRIVEBELT

1 A single serpentine drivebelt is located at the front of the engine and plays an important role in the overall operation of the engine and its components. Due to its function and material make up, the belt is prone to wear and should be periodically inspected. The serpentine belt drives the alternator and air conditioning compressor. Although the belt should be inspected at the recommended intervals, replacement may not be necessary for more than 100,000 miles.

CHECK

▶ Refer to illustration 18.3

2 With the engine stopped, inspect the full length of the drivebelt for cracks and separation of the belt plies. It will be necessary to turn the engine (using a wrench or socket and bar on the crankshaft pulley bolt) in order to move the belt from the pulleys so that the belt can be inspected thoroughly. Twist the belt between the pulleys so that both sides can be viewed. Also check for fraying, and glazing which gives the belt a shiny appearance. Check the pulleys for nicks, cracks, distortion and corrosion.

3 Note that it is not unusual for a ribbed belt to exhibit small cracks in the edges of the belt ribs, and unless these are extensive or very deep, belt replacement is not essential (see illustration).

REPLACEMENT

▶ Refer to illustration 18.5

4 To remove the drivebelt, loosen the right front wheel lug nuts, then raise the front of the vehicle and support it on jackstands. Remove the right front wheel and remove the lower splash shield from the underbody.

5 Note how the drivebelt is routed, then remove the belt from the pulleys. If you're working on a V6 engine, remove the air filter housing (see Chapter 4), right motor mount (see Chapter 2B) and use a wrench on the tensioner center bolt and turn the tensioner clockwise to release the drivebelt tension. If you're working on a four-cylinder engine, insert a 3/8-inch drive ratchet or breaker bar into the tensioner hole and pull the handle counterclockwise to release the drivebelt tension (see illustration).

6 Fit the new drivebelt onto the crankshaft, alternator, power steering pump, and air conditioning compressor pulleys, as applicable, then turn the release the tensioner and locate the drivebelt on the pulley. Make sure that the drivebelt is correctly seated in all of the pulley grooves, then release the tensioner.

7 Install the lower splash shield and wheel, then lower the car to the ground. Tighten the lug nuts to the torque listed in this Chapter's Specifications.

TENSIONER REPLACEMENT

▶ Refer to illustration 18.9

8 Remove the drivebelt as described previously.

9 On four-cylinder models, remove the bolt securing the tensioner to the engine block. On V6 models, remove the two bolts securing the tensioner, then detach the tensioner from the engine (see illustration).

10 Installation is the reverse of removal. Be sure to tighten the tensioners bolt(s) to the torque listed in this Chapter's Specifications.

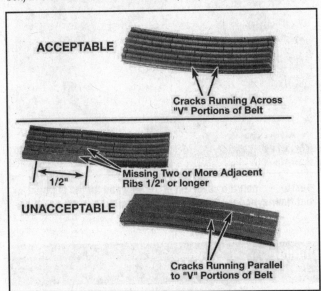

18.3 Here are some of the more common problems associated with drivebelts (check the belts very carefully to prevent an untimely breakdown)

18.5 Rotate the tensioner arm to relieve belt tension

18.9 On V6 models, remove the tensioner mounting bolts, then remove the tensioner

1-22 TUNE-UP AND ROUTINE MAINTENANCE

19 Air filter check and replacement (every 30,000 miles [48,000 km] or 24 months)

▶ Refer to illustrations 19.1a and 19.1b

1 The air filter is located inside a housing at the right (passenger's) side of the engine compartment. To remove the air filter, loosen the clamp securing the inlet tube to the air filter cover, release the clamps that secure the two halves of the air cleaner housing together, then separate the cover halves and remove the air filter element (see illustrations).

2 Inspect the outer surface of the filter element. If it is dirty, replace it. If it is only moderately dusty, it can be reused by blowing it clean from the back to the front surface with compressed air. Because it is a pleated paper type filter, it cannot be washed or oiled. If it cannot be cleaned satisfactorily with compressed air, discard and replace it. While the cover is off, be careful not to drop anything down into the housing.

CAUTION:
Never drive the vehicle with the air cleaner removed. Excessive engine wear could result and backfiring could even cause a fire under the hood.

3 Wipe out the inside of the air cleaner housing.
4 Place the new filter into the air cleaner housing, making sure it seats properly.
5 Installation of the housing is the reverse of removal.

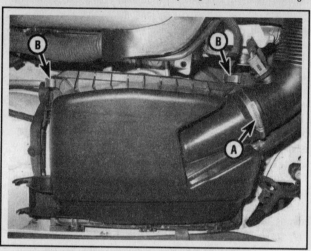

19.1a Loosen the intake hose clamp (A), then unlatch these clips (B) . . .

19.1b . . . pull the cover out of the way and lift the element out (four-cylinder engine shown, V6 similar)

20 Cooling system servicing (draining, flushing and refilling) (every 30,000 miles [48,000 km] or 24 months)

WARNING:
Do not allow antifreeze to come in contact with your skin or painted surfaces of the vehicle. Rinse off spills immediately with plenty of water. Antifreeze is highly toxic if ingested. Never leave antifreeze lying around in an open container or in puddles on the floor; children and pets are attracted by its sweet smell and may drink it. Check with local authorities on disposing of used anti-freeze. Many communities have collection centers that will see that antifreeze is disposed of safely.

➡ Note: Non-toxic antifreeze is now manufactured and available at local auto parts stores, but even this type should be disposed of properly.

DRAINING

▶ Refer to illustration 20.3

1 Periodically, the cooling system should be drained, flushed and refilled to replenish the antifreeze mixture and prevent formation of rust and corrosion, which can impair the performance of the cooling system and cause engine damage. When the cooling system is serviced, all hoses and the expansion tank cap should be checked and replaced if necessary.

2 Apply the parking brake and block the wheels. Raise the front of the vehicle and support it securely on jackstands, then remove the under-vehicle splash shield.

WARNING:
If the vehicle has just been driven, wait several hours to allow the engine to cool down before beginning this procedure.

3 Move a large container under the radiator drain to catch the coolant. The coolant can be drained either by detaching the lower radiator hose from the radiator or by turning the knob on the radiator drain valve (see illustration). Remove the cap from the coolant expansion tank and allow the coolant to drain.

4 While the coolant is draining, check the condition of the radiator hoses, heater hoses and clamps (refer to Section 9 if necessary).

5 Replace any damaged clamps or hoses.

TUNE-UP AND ROUTINE MAINTENANCE

FLUSHING

6 Fill the cooling system with clean water, following the *Refilling* procedure (see Step 12).

7 Start the engine and allow it to reach normal operating temperature, then rev up the engine a few times.

8 Turn the engine off and allow it to cool completely, then drain the system as described earlier.

9 Repeat Steps 6 through 8 until the water being drained is free of contaminants.

10 In severe cases of contamination or clogging of the radiator, remove the radiator (see Chapter 3) and have a radiator repair facility clean and repair it if necessary.

11 Many deposits can be removed by the chemical action of a cleaner available at auto parts stores. Follow the procedure outlined in the manufacturer's instructions.

➡ Note: **When the coolant is regularly drained and the system refilled with the correct antifreeze/water mixture, there should be no need to use chemical cleaners or descalers.**

REFILLING

12 Close and tighten the radiator drain.

13 Place the heater temperature control in the maximum heat position.

14 Slowly add new coolant (a 50/50 mixture of water and antifreeze) to the expansion tank until the level is at the COLD mark on the expansion tank.

15 Leave the expansion tank cap off and run the engine in a well-ventilated area until the thermostat opens (coolant will begin flowing through the radiator and the upper radiator hose will become hot).

20.3 **The radiator drain fitting is located at the bottom of the radiator - before opening the valve, push a short length of rubber hose onto the plastic fitting to prevent the coolant from splashing**

16 Turn the engine off and let it cool. Add more coolant mixture to bring the level to the COLD mark on the expansion tank.

17 Squeeze the upper radiator hose to expel air, then add more coolant mixture if necessary. Replace the expansion tank cap.

18 Start the engine, allow it to reach normal operating temperature and check for leaks. Also, set the heater and blower controls to the maximum setting and check to see that the heater output from the air ducts is warm. This is a good indication that all air has been purged from the cooling system.

21 Brake fluid change (every 30,000 miles [48,000 km] or 24 months)

✳✳ WARNING:

Brake fluid can harm your eyes and damage painted surfaces, so use extreme caution when handling or pouring it. Do not use brake fluid that has been standing open or is more than one year old. Brake fluid absorbs moisture from the air. Excess moisture can cause a dangerous loss of braking effectiveness.

1 At the specified intervals, the brake fluid should be drained and replaced. Since the brake fluid may drip or splash when pouring it, place plenty of rags around the master cylinder to protect any surrounding painted surfaces.

2 Before beginning work, purchase the specified brake fluid (see *Recommended lubricants and fluids* at the end of this Chapter).

3 Remove the cap from the master cylinder reservoir.

4 Using a hand suction pump or similar device, withdraw the fluid from the master cylinder reservoir.

5 Add new fluid to the master cylinder until it rises to the base of the filler neck.

6 Bleed the brake system as described in Chapter 9 at all four brakes until new and uncontaminated fluid is expelled from the bleeder screw. Be sure to maintain the fluid level in the master cylinder as you perform the bleeding process. If you allow the master cylinder to run dry, air will enter the system.

7 Refill the master cylinder with fluid and check the operation of the brakes. The pedal should feel solid when depressed, with no sponginess.

✳✳ WARNING:

Do not operate the vehicle if you are in doubt about the effectiveness of the brake system.

1-24 TUNE-UP AND ROUTINE MAINTENANCE

22 Automatic transaxle fluid change (see *Maintenance schedule* for service intervals)

♦ **Refer to illustrations 22.10a, 22.10b, 22.12, 22.17 and 22.19**

1 At the specified time intervals, the transaxle fluid should be drained and replaced. Since the fluid will remain hot long after driving, perform this procedure only after everything has cooled down completely.

2 Before beginning work, purchase the specified transaxle fluid (see *Recommended lubricants and fluids* at the end of this Chapter) and a new filter.

3 Other tools necessary for this job include jackstands to support the vehicle in a raised position, a drain pan capable of holding several quarts, newspapers and clean rags.

4 Raise and support the vehicle on jackstands.

5 With a drain pan in place, remove the front and side transaxle pan mounting bolts.

6 Loosen the rear pan bolts one turn.

7 Carefully pry the transaxle pan loose with a screwdriver, allowing the fluid to drain.

8 Remove the remaining bolts, pan and gasket. Carefully clean the gasket surface of the transaxle to remove all traces of the old gasket and sealant.

9 Drain the fluid from the transaxle pan, clean the pan with solvent and dry it with compressed air. Be careful not to lose the magnet.

10 Remove the filter and pry out the seal (see illustrations).

11 Push a new filter seal fully into its bore, then install the new filter.

12 Make sure the gasket surface on the transaxle pan is clean, then install the new gasket (see illustration). Put the pan in place against the transaxle and install the bolts. Working around the pan, tighten each bolt a little at a time until the final torque figure is reached.

13 Lower the vehicle and add the specified amount of automatic transmission fluid through the vent/fill cap and check the fluid level (see below).

14 Check under the vehicle for leaks during the first few trips.

FLUID LEVEL CHECK

15 The manufacturer states that routine checks of the automatic transaxle fluid are not necessary; this procedure should only be used when refilling the transaxle after the fluid has been drained, unless an obvious leak has been detected. Low fluid level can lead to slipping or loss of drive, while overfilling can cause foaming and loss of fluid.

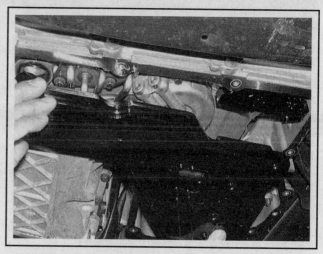

22.10a Pull the transaxle filter straight down and out of the transaxle - there are no fasteners

✱✱ WARNING:

This procedure is potentially dangerous and is best left to a professional shop with a safe lifting apparatus. The vehicle must be kept level while being safely raised high enough for access to the check plug on the transaxle.

16 With the vehicle raised and safely supported, start the engine, then move the shift lever through all the gear ranges, ending in Park.

➡**Note:** Incorrect fluid level readings will result if the vehicle has just been driven at high speeds for an extended period, in hot weather in city traffic, or if it has been pulling a trailer. If any of these conditions apply, wait until the fluid has cooled (about 30 minutes).

17 Remove the vent/fill cap (see illustration 22.19). With the engine running and the transaxle at normal operating temperature (having idled for 3 to 5 minutes), locate the check plug on the transaxle. The check

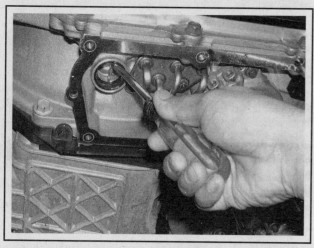

22.10b Pry out the old seal, being careful not to damage the aluminum housing

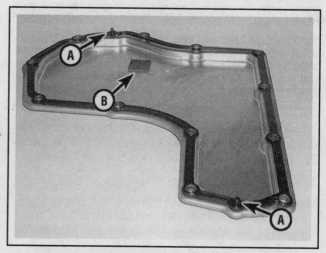

22.12 Place the gasket on the pan, aligning the plastic pins (A) with the holes in the pan for alignment (they also help align the pan to the transaxle) - make sure the magnet (B) is clean and in place before installing the pan

TUNE-UP AND ROUTINE MAINTENANCE 1-25

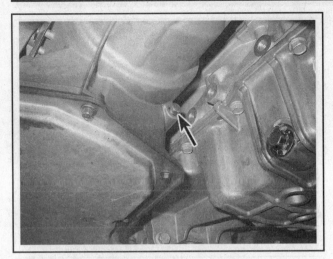

22.17 Location of the transaxle fluid check plug

22.19 Location of the vent/fill plug on the automatic transaxle

plug is located near the pan, adjacent to the engine oil drain plug (see illustration).

18 Place a container under the check plug and remove it. Observe the fluid as it drips into the pan, indicating correct fluid level.

19 The fluid level should be at the bottom of the check hole. If fluid pours out excessively, the transaxle may have been overfilled. Double-check to make sure the vehicle is level. If no fluid drips from the check hole, add small amounts of fluid through the vent/fill cap at the top of the transaxle until the level is at the bottom of the check hole (see illustration). A long-necked funnel will be necessary to add fluid.

20 The condition of the fluid should also be checked along with the level. If the fluid in the drain pan is a dark reddish-brown color, or if the fluid has a burned smell, the fluid should be changed (see above). If you're in doubt about the condition of the fluid, purchase some new fluid and compare the two for color and smell.

21 Be sure to install the check plug and tighten it securely when you're done.

23 Manual transaxle lubricant level check and change (see *Maintenance schedule* for service intervals)

CHECK

1 Most manual transaxles do not have a dipstick. To check the fluid level, raise the vehicle and support it securely on jackstands. On the front side of the transaxle housing you will see a plug. Remove it. If the lubricant level is correct, it should be up to the lower edge of the hole.

2 If the transaxle needs more lubricant (if the level is not up to the hole), use a syringe or a gear oil pump to add more. Stop filling the transaxle when the lubricant begins to run out the hole.

3 Install the plug and tighten it securely. Drive the vehicle a short distance, then check for leaks.

CHANGE

4 Raise the vehicle and support it securely on jackstands.
5 Move a drain pan, rags, newspapers and wrenches under the transaxle.
6 Remove the transaxle fill plug (if equipped) on the front of the case and the drain plug at the bottom of the case, then allow the lubricant to drain into the pan.
7 After the lubricant has drained completely, reinstall the drain plug and tighten it securely.
8 Using a hand pump, syringe or funnel, fill the transaxle with the specified lubricant until it is level with the lower edge of the filler hole. Reinstall the fill plug and tighten it securely.
9 Lower the vehicle.
10 Drive the vehicle for a short distance, then check the drain and fill plugs (if equipped) for leakage.

24 Fuel filter replacement (see *Maintenance schedule* for service intervals)

♦ Refer to illustration 24.4

WARNING:

Gasoline is extremely flammable, so take extra precautions when you work on any part of the fuel system. Don't smoke or allow open flames or bare light bulbs near the work area, and don't work in a garage where a gas-type appliance (such as a water heater or clothes dryer) is present. Since fuel is carcinogenic, wear latex gloves when there's a possibility of being exposed to fuel, and, if you spill any fuel on your skin, rinse it off immediately with soap and water. Mop up any spills immediately and do not store fuel-soaked rags where they could ignite. When you perform any kind of work on the fuel system, wear safety glasses and have a Class B type fire extinguisher on hand.

1 The fuel filter is mounted under the vehicle on the right side, in front of the gas tank.

1-26 TUNE-UP AND ROUTINE MAINTENANCE

2 Relieve the fuel system pressure (see Chapter 4).

3 If necessary, raise the vehicle and support it securely on jackstands. Inspect the fittings at both ends of the filter to see if they're clean. If more than a light coating of dust is present, clean the fittings before proceeding.

4 Disconnect the fuel lines at the fuel filter (see illustration). Detach the lines, one at a time; be prepared for fuel spillage.

5 After the lines are detached, check the fittings for damage and distortion. If they were damaged in any way during removal, new ones must be used when the lines are reattached to the new filter.

6 Remove the fuel filter from the mounting clamp, while noting the direction the fuel filter is installed.

7 Install the new filter in the same direction. Carefully push each hose onto the filter until it's seated against the collar on the fitting, then install the clips. Make sure the clips are securely attached to the hose fittings - if they come off, the hoses could back off the filter and a fire could result!

8 Start the engine and check for fuel leaks.

24.4 Use a screwdriver and depress the clip to release the fuel line (A) then squeeze the two tabs to release the line on the other side (B)

25 Spark plug check and replacement (see *Maintenance schedule* for service intervals)

♦ Refer to illustrations 25.3, 25.6a, 25.6b, 25.8, 25.10a and 25.10b

WARNING:
Because of the very high voltage generated by the ignition system (as much as 40,000 volts), use extreme care when you're servicing ignition components such as the ignition coil pack and spark plugs.

1 Disconnect the cable from the negative battery terminal (see Chapter 5, Section 1).

2 Remove the ignition coil pack(s) (see Chapter 5).

3 In most cases, the tools necessary for spark plug replacement include a spark plug socket which fits onto a ratchet (spark plug sockets are padded inside to prevent damage to the porcelain insulators on the new plugs), various extensions and a gap gauge to check and adjust the gaps on the new plugs (see illustration). A torque wrench should be used to tighten the new plugs.

4 The best approach when replacing the spark plugs is to purchase the new ones in advance, adjust them to the proper gap and replace the plugs one at a time. When buying the new spark plugs, be sure to obtain the correct plug type for your particular engine. This information can be found in the Specifications Section at the end of this Chapter or in your Owner's manual.

5 Allow the engine to cool completely before attempting to remove any of the plugs. These engines are equipped with aluminum cylinder heads, which can be damaged if the spark plugs are removed when the engine is hot. While you are waiting for the engine to cool, check the new plugs for defects and adjust the gaps.

6 The gap is checked by inserting the proper-thickness gauge between the electrodes at the tip of the plug (see illustration). The gap between the electrodes should be the same as the one specified on the Emissions Control Information label or in this Chapter's Specifications. The gauge should just slide between the electrodes with a slight amount of drag. If the gap is incorrect, use the adjuster on the gauge body to bend the curved side electrode slightly until the proper gap is obtained (see illustration). If the side electrode is not exactly over the center electrode, bend it with the adjuster until it is. Check for cracks in the porcelain insulator (if any are found, the plug should not be used).

➡ Note: We recommend using a tapered thickness gauge when checking platinum or iridium-type spark plugs. Other types of gauges may scrape the thin coating from the electrodes, thus dramatically shortening the life of the plugs. However, if dual-electrode spark plugs are used, a wire-type gauge will have to be used.

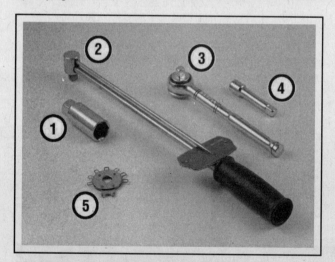

25.3 Tools required for changing spark plugs

1 **Spark plug socket** - This will have special padding inside to protect the spark plug porcelain insulator
2 **Torque wrench** - Although not mandatory, use of this tool is the best way to ensure that the plugs are tightened properly
3 **Ratchet** - Standard hand tool to fit the plug socket
4 **Extension** - Depending on model and accessories, you may need special extensions and universal joints to reach one or more of the plugs
5 **Spark plug gap gauge** - This gauge for checking the gap comes in a variety of styles. Make sure the gap for your engine is included

TUNE-UP AND ROUTINE MAINTENANCE

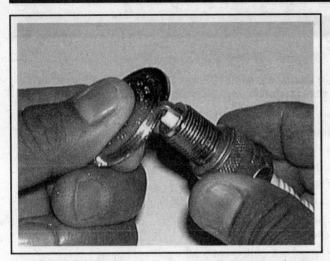

25.6a Spark plug manufacturers recommend using a tapered thickness gauge when checking the gap - slide the thin side into the gap and turn it until the gauge just fills the gap, then read the thickness on the gauge - do not force the tool into the gap or use the tapered portion to widen a gap

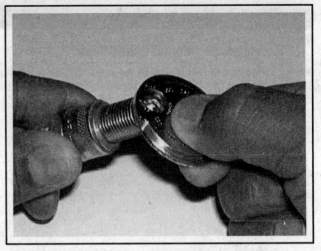

25.6b To change the gap, bend the side electrode only, using the adjuster hole in the tool, and be very careful not to crack or chip the porcelain insulator surrounding the center electrode

7 If compressed air is available, use it to blow any dirt or foreign material away from the spark plug hole. The idea here is to eliminate the possibility of debris falling into the cylinder as the spark plug is removed.

8 Place the spark plug socket over the plug and remove it from the engine by turning it in a counterclockwise direction (see illustration).

9 Compare the spark plug to those shown in the photos located on the inside back cover to get an indication of the general running condition of the engine.

10 Apply a small amount of anti-seize compound to the spark plug threads (see illustration). Install one of the new plugs into the hole until you can no longer turn it with your fingers, then tighten it with a torque wrench (if available) or the ratchet. It is a good idea to slip a short length of rubber hose over the end of the plug to use as a tool to thread it into place (see illustration). The hose will grip the plug well enough to turn it, but will start to slip if the plug begins to cross-thread in the hole - this will prevent damaged threads and the accompanying repair costs.

11 Repeat the procedure for the remaining spark plugs.

12 After replacing all the plugs, install the ignition coil pack(s) (see Chapter 5).

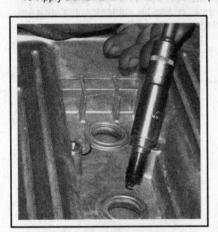

25.8 Use a ratchet and extension to remove the spark plugs

25.10a Apply a thin coat of anti-seize compound to the spark plug threads

25.10b A length of snug-fitting rubber hose will save time and prevent damaged threads when installing the spark plugs

1-28 TUNE-UP AND ROUTINE MAINTENANCE

Specifications

Recommended lubricants and fluids

→ **Note:** Listed here are manufacturer recommendations at the time this manual was written. Manufacturers occasionally upgrade their fluid and lubricant specifications, so check with your local auto parts store for current recommendations.

Engine oil	
Type	API "certified for gasoline engines"
Viscosity	SAE 5W-30
Fuel	Unleaded gasoline, 87 Octane minimum
Automatic transaxle fluid	DEXRON ® III
Manual transaxle lubricant	Manual transaxle fluid (Saturn part # 21018899)
Brake fluid	DOT 3 brake fluid
Clutch fluid	DOT 3 brake fluid
Engine coolant	50/50 mixture of DEX-COOL® and distilled water
Power steering system	Power steering fluid (Saturn part # 21007583)

Capacities*

Engine oil (including filter)	
Four-cylinder engine	5.0 quarts (4.7L)
V6 engine	5.0 quarts (4.7L)
Coolant	
Four-cylinder engine	7.4 quarts (7.0L)
V6 engine	
2002 and earlier models	7.8 quarts (7.4L)
2003 and later models	8.5 quarts (8.0L)
Automatic transaxle (dry fill)	6.9 quarts (6.5L)

→ **Note:** Since this is a dry-fill specification, the amount required during a routine fluid change will be substantially less. The best way to determine the amount of fluid to add during a routine fluid change is to measure the amount drained. Begin the refill procedure by initially adding 1/3rd of the amount drained. Then, with the engine running, add 1/2-pint at a time (cycling the shifter through each gear position between additions) until the level is correct. It is important to not overfill the transaxle (see Section 22).

Manual transaxle	2.0 quarts (1.9L)

*All capacities approximate. Add as necessary to bring up to appropriate level.

Ignition system

Spark plug type and gap	
Type	
Four-cylinder engine	AC41-981 or equivalent
V6 engine	Bosch FLR9LTE or equivalent
Gap	
Four-cylinder engine	0.045 inch (1.1 mm)
V6 engine	0.040 inch (1.0 mm)
Engine firing order	
Four-cylinder engine	1-3-4-2
V6 engine	1-2-3-4-5-6

Brakes

Disc brake pad lining thickness (minimum)	1/8 inch
Drum brake shoe lining thickness (minimum)	1/16 inch
Parking brake adjustment	3 to 5 clicks

TUNE-UP AND ROUTINE MAINTENANCE

Torque specifications	Ft-lbs (unless otherwise indicated)	Nm
Engine oil drain plug		
Four-cylinder engine	18	25
V6 engine	19	25
Automatic transaxle fluid pan bolts	108 in-lbs	12
Manual transaxle drain plug	37	50
Spark plugs		
Four-cylinder engine	15	20
V6 engine	19	25
Drivebelt tensioner bolts		
Four-cylinder engine	37	50
V6 engine	30	40
Wheel lug nuts	92	125

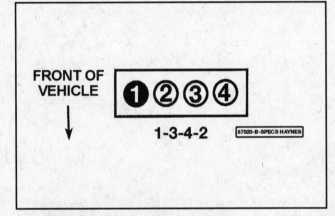

Cylinder locations (four-cylinder engines)

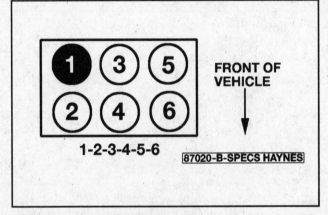

Cylinder locations (V6 engine)

1-30 TUNE-UP AND ROUTINE MAINTENANCE

Notes

Section

1 General information
2 Repair operations possible with the engine in the vehicle
3 Top Dead Center (TDC) for number one piston - locating
4 Valve cover - removal and installatio
5 Intake manifold - removal and installation
6 Exhaust manifold - removal and installation
7 Engine front cover - removal and installation
8 Timing chain and sprockets - removal, inspection and installation
9 Balance shaft chain and balance shafts - removal, inspection and installation
10 Crankshaft pulley and front oil seal - removal and installation
11 Camshafts and hydraulic lash adjusters - removal, inspection and installation
12 Cylinder head - removal and installation
13 Oil pan - removal and installation
14 Oil pump - removal, inspection and installation
15 Flywheel/driveplate - removal and installation
16 Rear main oil seal - replacement
17 Powertrain mounts - check and replacement

Reference to other Chapters

CHECK ENGINE light on - See Chapter 6
Compression check - See Chapter 2C
Drivebelt check, adjustment and replacement - See Chapter 1
Engine oil and filter change - See Chapter 1
Engine overhaul - general information - See Chapter 2C
Engine - removal and installation - See Chapter 2C
Spark plug replacement - See Chapter 1
Water pump - removal and installation - See Chapter 3

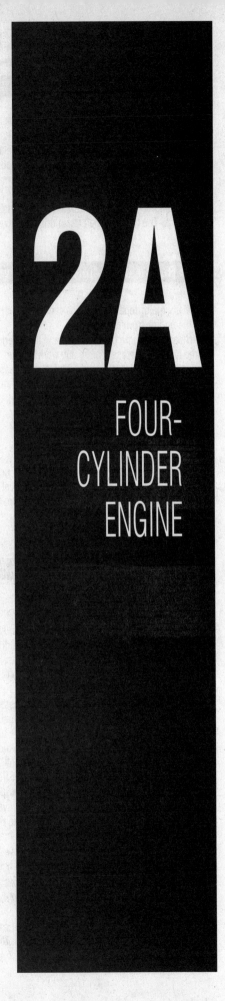

2A
FOUR-
CYLINDER
ENGINE

2A-2 FOUR-CYLINDER ENGINE

1 General information

This Part of Chapter 2 is devoted to in-vehicle repair procedures for the 2.2L DOHC (Double Overhead Camshaft), engine as well as procedures such as timing chain and sprocket(s), balance shaft chain and balance shafts and oil pan removal. All information concerning engine removal and installation and engine block overhaul can be found in Part C of this Chapter.

This engine is equipped with a single timing chain to drive the camshafts. The balance shaft chain drives the two balance shafts and the water pump sprocket. The balance shaft chain is mounted directly behind the camshaft timing chain. The camshaft timing chain and balance shaft chain procedure can be performed with the engine in the vehicle.

The Specifications included in this Part of Chapter 2 apply only to the procedures contained in this Part. Information concerning engine removal and overhaul or replacement can be found in Chapter 2, Part C.

2 Repair operations possible with the engine in the vehicle

Many major repair operations can be accomplished without removing the engine from the vehicle.

Clean the engine compartment and the exterior of the engine with some type of degreaser before any work is done. It will make the job easier and help keep dirt out of the internal areas of the engine.

Depending on the components involved, it may be helpful to remove the hood to improve access to the engine as repairs are performed (refer to Chapter 11 if necessary). Cover the fenders to prevent damage to the paint. Special pads are available, but an old bedspread or blanket will also work.

If vacuum, exhaust, oil or coolant leaks develop, indicating a need for gasket or seal replacement, the repairs can generally be made with the engine in the vehicle. The intake and exhaust manifold gaskets, oil pan gasket, crankshaft oil seals and cylinder head gasket are all accessible with the engine in place.

Exterior engine components, such as the intake and exhaust manifolds, the oil pan, the oil pump, the water pump, the starter motor, the alternator and the fuel system components can be removed for repair with the engine in place.

Since the cylinder head can be removed without pulling the engine, camshaft and valve component servicing can also be accomplished with the engine in the vehicle. Replacement of the timing chain, balance shaft chain and sprockets is also possible with the engine in the vehicle. Balance shaft removal, however, will require removal of the engine.

In extreme cases caused by a lack of necessary equipment, repair or replacement of piston rings, pistons, connecting rods and rod bearings is possible with the engine in the vehicle. However, this practice is not recommended because of the cleaning and preparation work that must be done to the components involved.

3 Top Dead Center (TDC) for number one piston - locating

▶ Refer to illustration 3.5

1 Top Dead Center (TDC) is the highest point in the cylinder that each piston reaches as it travels up-and-down during crankshaft rotation. Each piston reaches TDC on the compression stroke and again on the exhaust stroke, but TDC generally refers to piston position on the compression stroke.

2 Positioning the piston(s) at TDC is an essential part of certain other repair procedures discussed in this manual.

3 Before beginning this procedure, be sure to place the transmission in Neutral and apply the parking brake or block the rear wheels. Remove the spark plugs (see Chapter 1). Disable the ignition system by disconnecting the wiring harness connector from the ignition module (four cylinder models) or from each coil-on-plug assembly (V6 models) (see Chapter 5). Also disable the fuel system by unplugging the electrical connector in the wiring harness to the fuel injectors.

4 In order to bring any piston to TDC, the crankshaft must be turned using one of the methods outlined below. When looking at the front of the engine, normal crankshaft rotation is clockwise.

a) The preferred method is to turn the crankshaft with a socket and ratchet attached to the bolt threaded into the front of the crankshaft.

b) A remote starter switch, which may save some time, can also be used. Follow the instructions included with the switch. Once the piston is close to TDC, use a socket and ratchet as described in the previous paragraph.

c) If an assistant is available to turn the ignition switch to the Start position in short bursts, you can get the piston close to TDC without a remote starter switch. Make sure your assistant is out of the vehicle, away from the ignition switch, then use a socket and ratchet as described in Paragraph a) to complete the procedure.

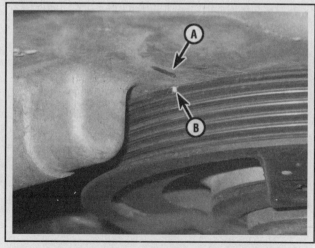

3.5 Timing marks - align the pointer on the engine front cover (A) with the notch in the crankshaft pulley (B)

FOUR CYLINDER ENGINE 2A-3

5 Insert a compression gauge into the number one cylinder spark plug hole. Turn the crankshaft (see Step 4 above) until compression registers on the gauge, then turn it slowly until the TDC mark on the timing chain cover is aligned with the notch on the crankshaft pulley (see illustration).

6 After the number one piston has been positioned at TDC on the compression stroke, TDC for any of the remaining pistons can be located by turning the crankshaft and following the firing order.

a) *On four-cylinder engines, divide the crankshaft pulley into two equal sections with chalk marks at each point, each indicating 180-degrees of crankshaft rotation. Rotating the engine past TDC no. 1 to the next mark will place the engine at TDC for cylinder no. 3.*

b) *On V6 engines, divide the crankshaft pulley into three equal sections with chalk marks at each point, each indicating 120-degrees of crankshaft rotation. Rotating the engine past TDC no. 1 to the next mark will place the engine at TDC for cylinder no. 2.*

4 Valve covers - removal and installation

REMOVAL

◆ **Refer to illustrations 4.2, 4.3a, 4.3b, 4.4, 4.5 and 4.6**

1 Disconnect the cable from the negative battery terminal (see Chapter 5, Section 1). Remove the ignition coil assembly from the valve cover (see Chapter 5).

2 Detach the PCV hose from the valve cover (see illustration).

3 Detach the wiring harness and the fuel line bracket from the timing chain end of the valve cover (see illustrations) and position the assembly away from the valve cover.

4 Remove the nuts and detach the fuel injector harness from the mounting studs (see illustration).

5 Remove the bolts and detach the ground strap and coolant tube

4.2 Squeeze the clamp and detach the PCV hose from the valve cover

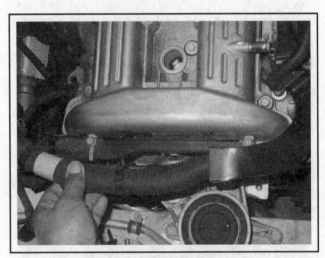

4.3a Working on the timing chain end of the engine, disconnect the wiring harness from the bracket

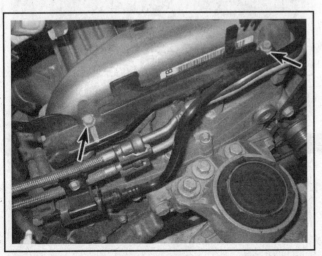

4.3b Remove the fuel line bracket bolts and position the assembly off to the side

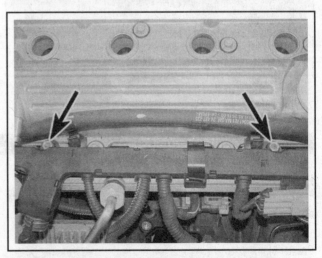

4.4 Remove these two nuts and detach the fuel injector wiring harness for access to the valve cover front mounting bolts

2A-4 FOUR-CYLINDER ENGINE

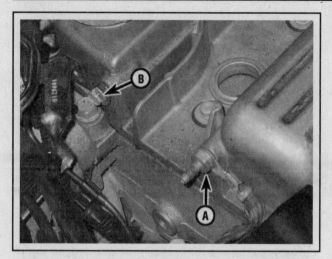

4.5 Remove the ground strap bolt (A) and the coolant tube bracket bolt (B)

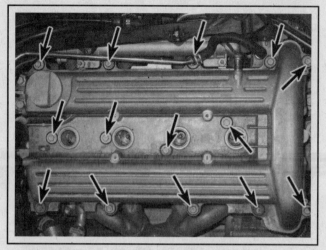

4.6 Location of the valve cover mounting bolts

bracket from the valve cover (see illustration).

6 Remove the valve cover bolts (see illustration) then lift the valve cover off. Tap gently with a soft-face hammer, if necessary, to break the gasket seal.

INSTALLATION

▶ Refer to illustrations 4.8 and 4.9

7 Clean the gasket surfaces on the intake manifold, cylinder head and valve cover. Use a shop rag, lacquer thinner or acetone to wipe off all residue and gasket material from the sealing surfaces.

8 Insert a new valve cover gasket into the grooved recess in the valve cover. Make sure the gasket is positioned properly in the groove (see illustration).

9 Install new O-rings and spark plug seals in the valve cover (see illustration).

10 The remainder of installation is the reverse of the removal Steps. Tighten the valve cover bolts evenly, starting with the center bolts and working out, to the torque listed in this Chapter's Specifications.

11 Reconnect the battery (see Chapter 5, Section 1).

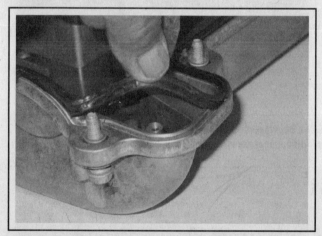

4.8 Install the gasket into the grooved recess in the valve cover

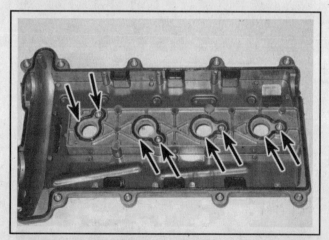

4.9 Be sure to change all the spark plug seals and O-rings in the valve cover

5 Intake manifold - removal and installation

REMOVAL

▶ Refer to illustration 5.6

1 Disconnect the cable from the negative battery terminal (see Chapter 5).

2 Remove the throttle body (see Chapter 4).

3 Disconnect any electrical connectors that would interfere with manifold removal. Open the wiring harness clips and detach all wiring harnesses from the manifold.

4 Disconnect any vacuum hoses connected to the manifold (this will vary by year). Mark the hoses, if necessary, to ensure correct reassembly.

5 Remove the dipstick tube mounting bolt and position the dipstick to the side.

6 Remove the intake manifold mounting bolts and nuts (see illustration).

7 Lift the intake manifold from the engine compartment.

FOUR CYLINDER ENGINE 2A-5

INSTALLATION

8 Install a new gasket, if necessary.

→ **Note:** *The intake manifold gasket does not need to be replaced unless it has become damaged during the removal process. Make sure the mating surfaces of the manifold and cylinder head are clean.*

9 Install the manifold over the studs on the cylinder head. Install the bolts and nuts and tighten them finger-tight.

10 Tighten the bolts to the torque listed in this Chapter's Specifications, starting with the center bolts and working towards the ends.

11 The remainder of installation is the reverse of the removal steps.

12 Reconnect the battery (see Chapter 5, Section 1).

13 Run the engine and check for vacuum leaks.

5.6 Location of the intake manifold mounting bolts

6 Exhaust manifold - removal and installation

※ WARNING:

The engine must be completely cool before beginning this procedure.

REMOVAL

◆ **Refer to illustrations 6.2, 6.3, 6.5 and 6.7**

1 Disconnect the cable from the negative battery terminal (see Chapter 5).

2 If equipped, remove the AIR valve bracket nut, then remove the bolts and disconnect the AIR pipe from the exhaust manifold (see illustration).

→ **Note:** *It isn't necessary to detach the hose from the AIR valve.*

3 Remove the exhaust manifold heat shield (see illustration).

4 Raise the vehicle and support it on jackstands.

5 Detach the exhaust pipe from the manifold (see illustration).

6.2 Location of the AIR pipe and bracket fasteners

6.3 Location of the heat shield mounting bolts

6.5 Exhaust pipe-to-manifold nuts

2A-6 FOUR-CYLINDER ENGINE

6 Follow the lead from the oxygen sensor up to its electrical connector, then unplug the connector. Also detach the lead from its retaining clip.

7 Remove the exhaust manifold mounting nuts and detach the manifold from the cylinder head (see illustration).

INSTALLATION

8 Using a scraper, thoroughly clean the mating surfaces on the cylinder head, manifold and exhaust pipe. Remove the residue with a solvent such as acetone or lacquer thinner.

9 Check that the mating surfaces are perfectly flat and not damaged in any way. A warped or damaged manifold may require machining or, if severe enough, replacement. Install the new gasket to the cylinder head studs and place the manifold on the cylinder head. Tighten the nuts evenly, working from the center outwards, to the torque listed in this Chapter's Specifications.

10 Connect the exhaust pipe to the manifold and tighten the nuts evenly to the torque listed in this Chapter's Specifications.

11 The remainder of installation is the reverse of the removal steps.

6.7 Exhaust manifold mounting nuts

12 Reconnect the battery (see Chapter 5, Section 1).
13 Run the engine and check for exhaust leaks.

7 Engine front cover - removal and installation

REMOVAL

▶ Refer to illustrations 7.7a and 7.7b

1 Disconnect the cable from the negative battery terminal (see Chapter 5).
2 Drain the engine oil (see Chapter 1).
3 Remove the drivebelt (see Chapter 1).
4 Remove the drivebelt tensioner from the front cover.
5 Remove the crankshaft pulley (see Section 10).
6 Loosen the right front wheel lug nuts, then raise the front of the vehicle and support it securely on jackstands. Remove the right front wheel.
7 Loosen the engine cover fasteners gradually and evenly, then remove the fasteners (see illustrations).

➡ Note: Draw a sketch of the engine cover and cover fasteners. Identify the location of all bolts for installation in their original locations.

8 Remove the water pump bolt from the engine front cover (see illustration 7.7b).
9 Remove the front cover.
10 Remove the engine cover-to-block gasket.

INSTALLATION

11 Inspect and clean all sealing surfaces of the engine front cover and the block.

※ **CAUTION:**

Be very careful when scraping on aluminum engine parts. Aluminum is soft and gouges easily. Severely gouged parts may require replacement.

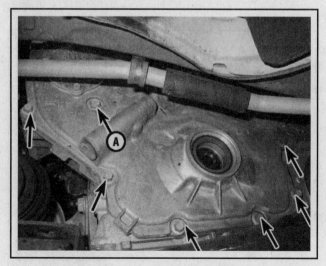

7.7a The engine cover mounting bolts can be accessed from below . . .

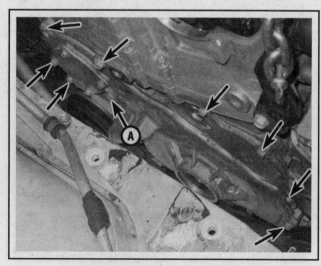

7.7b . . . and from above the engine compartment - don't forget the front cover/water pump bolt (A)

FOUR CYLINDER ENGINE 2A-7

12 If necessary, replace the crankshaft front oil seal in the front cover (see Section 10).
13 Install the front cover gasket on the engine block.

> **CAUTION:**
> The engine cover gasket is reusable. Make sure the gasket has not been damaged. Install a new gasket if necessary.

14 Install the front cover and cover fasteners. Make sure the hub on the inner rotor is aligned with the flats on the crankshaft and the engine cover fasteners are in their original locations. Tighten the fasteners by hand until the cover is contacting the block around its entire periphery.
15 Install the long water pump bolt.
16 Tighten the bolts to the torque listed in this Chapter's Specifications.
17 Install the drivebelt and tensioner. Tighten the drivebelt tensioner to the torque listed in this Chapter's Specifications.
18 Install the crankshaft pulley (see Section 10).
19 Reinstall the remaining parts in the reverse order of removal.
20 Fill the crankcase with the recommended oil (see Chapter 1).
21 Reconnect the battery (see Chapter 5, Section 1).
22 Start the engine and check for leaks. Check all fluid levels.

8 Timing chain and sprockets - removal, inspection and installation

REMOVAL

▸ Refer to illustrations 8.8, 8.9, 8.10, 8.11a, 8.11b, 8.12, 8.14 and 8.15

1 Disconnect the cable from the negative battery terminal (see Chapter 5).
2 Set the engine to TDC for cylinder number one (see Section 3).
3 Drain the engine oil (see Chapter 1).
4 Remove the drivebelt (see Chapter 1).
5 Remove the drivebelt tensioner from the front cover.
6 Remove the engine front cover (see Section 7).
7 Remove the valve cover (see Section 4).
8 Remove the timing chain tensioner (see illustration).
9 Remove the upper timing chain guide (see illustration).
10 Remove the exhaust camshaft sprocket bolt (see illustration). Be sure to discard the bolt and install a new bolt on reassembly.
11 Remove the adjustable timing chain guide (see illustrations).
12 Unscrew the access bolt and remove the fixed timing chain guide upper mounting bolt (see illustration).

8.8 Remove the timing chain tensioner from the cylinder head

8.9 Remove the upper timing chain guide mounting bolts

8.10 Use a wrench on the hex drive on the camshaft to prevent the camshaft from turning while loosening the sprocket bolt

8.11a Remove the adjustable timing chain guide mounting bolt . . .

2A-8 FOUR-CYLINDER ENGINE

8.11b ... then lift the guide out through the top of the cylinder head

8.12 Access plug for the fixed timing chain guide upper mounting bolt

8.14 Use a wrench on the hex drive on the camshaft when loosening the camshaft sprocket bolt

8.15 Carefully remove the timing chain and the intake camshaft sprocket through the top of the cylinder head

15 Remove the timing chain through the top of the cylinder head (see illustration).

16 Remove the timing chain drive sprocket and slide the timing chain oiling nozzle off the engine block.

INSPECTION

17 Clean all parts with clean solvent and dry with compressed air, if available.

18 Inspect the chain tensioner for excessive wear or other damage. Be sure to drain all the oil out of the chain tensioner if it is to be reused.

19 Inspect the timing chain guides for deep grooves, excessive wear, or other damage.

20 Inspect the timing chain for excessive wear or damage.

21 Inspect the crankshaft and camshaft sprockets for chipped or broken teeth, excessive wear, or damage.

22 Replace any component that is in questionable condition.

INSTALLATION

▶ **Refer to illustrations 8.23, 8.25, 8.29, 8.33a, 8.33b, 8.33c, 8.33d and 8.33e**

23 If the crankshaft has been rotated during this procedure, make

13 Remove the fixed timing chain guide lower mounting bolt and lift the guide from the engine block.

14 Remove the intake camshaft sprocket bolt (see illustration). Be sure to discard the bolt and install a new bolt on reassembly.

8.23 The round dot (alignment mark) on the sprocket should be in the 5 o'clock position. When installing the chain, one of the silver plated links must be aligned with this dot

8.25 The copper link must align with the INT on the intake camshaft and the silver link with the EXH on the exhaust camshaft

FOUR CYLINDER ENGINE 2A-9

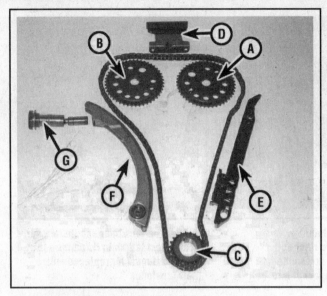

8.29 Timing chain component details

- A Intake camshaft sprocket
- B Exhaust camshaft sprocket
- C Crankshaft sprocket
- D Upper timing chain guide
- E Fixed timing chain guide
- F Adjustable timing chain guide
- G Timing chain tensioner

sure the number one piston is at the top of it's stroke (TDC) (see Section 3). The timing mark (round dot) should point to 5 o'clock position on the crankshaft sprocket (see illustration).

24 Install the intake camshaft sprocket onto the camshaft. Be sure to install a new bolt. Tighten the intake camshaft sprocket bolt lightly, finger tight at this time.

※※ CAUTION:

Do not turn the camshaft more than 1/2 turn to avoid any valve/piston contact. The camshafts should be positioned correctly before the timing chain is installed.

25 Install the timing chain by lowering it from the top through the opening. Be sure the timing chain drops down around both sides of the cylinder block bosses. Be sure the bright colored link (copper) on the chain is aligned with the INT designation on the camshaft sprocket (see illustration).

➡ **Note: The copper link will be installed at the intake camshaft sprocket (front) while the silver links will be installed at the crankshaft sprocket and the exhaust camshaft sprocket (rear).**

26 Drape the timing chain over the crankshaft sprocket and engage the plated link (silver) on the chain with the crankshaft sprocket timing mark located in the 5 o'clock position (see illustration 8.23).

27 Install the adjustable timing chain guide. Install the bolts and tighten them to the torque listed in this Chapter's Specifications.

28 Install the exhaust camshaft sprocket onto the camshaft. Be sure to install a new bolt. Be sure the plated link (silver) on the chain is aligned with the EXH designation on the camshaft sprocket (see illustration 8.25). Tighten the exhaust camshaft sprocket bolt lightly, finger tight at this time.

29 Install the fixed timing chain guide (see illustration). Tighten the bolts to the torque listed in this Chapter's Specifications.

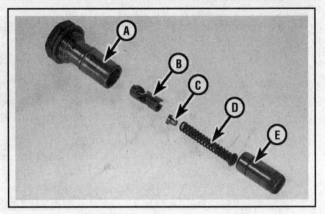

8.33a Timing tensioner details

- A Timing chain tensioner body
- B Ratchet cylinder
- C Spring adjuster
- D Spring
- E Piston

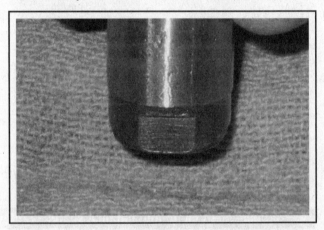

8.33b Install the piston with the flats locked into the jaws of the vise

30 Install the upper timing chain guide (see illustration 8.9). Tighten the bolts to the torque listed in this Chapter's Specifications.

31 Install a 24 mm wrench onto the intake camshaft hex as a back-up, and torque the intake camshaft bolt to the torque listed in this Chapter's Specifications.

32 Install a 24 mm wrench onto the exhaust camshaft hex as a back-up, and torque the exhaust camshaft bolt to the torque listed in this Chapter's Specifications.

33 Install the timing chain tensioner. The timing chain tensioner must be installed in its compressed state. Follow the steps to correctly compress the tensioner.

※※ CAUTION:

The timing chain tensioner must be installed in the compressed state. Do not install a tensioner in its released state. Damage to the tensioner and timing chain will occur.

a) Completely disassemble the tensioner and drain all the oil (see illustration). Inspect the tensioner body, the piston and all components for scoring or damage. If necessary, replace the tensioner with a new one.

b) Install the tensioner piston into the vise with the flats seated in the jaws of the vice (see illustration).

2A-10 FOUR-CYLINDER ENGINE

8.33c Align the groove in the ratchet cylinder with the pin in the piston

8.33d Using a flat-bladed screwdriver, drive the ratchet cylinder down to the bottom and rotate it clockwise to lock it into position

8.33e The tensioner should measure the correct length in its compressed state or it must be replaced with a new tensioner

c) Install the ratchet cylinder into the piston, aligning the groove with the locating pin (see illustration).

d) Drive the ratchet cylinder into the piston with a flat-bladed screwdriver. Rotate the ratchet cylinder clockwise when it reaches the bottom (see illustration). The ratchet cylinder should be locked into position.

e) The tensioner must measure 2.83 inches (72 mm) from end-to-end (see illustration).

34 Install the timing chain oiling nozzle. Tighten the bolt to the torque listed in this Chapter's Specifications.

35 Apply a small amount of RTV sealant to the threads and install the timing chain guide access plug. Tighten the bolt to the torque listed in this Chapter's Specifications.

36 Install the valve cover (see Section 4).
37 Install the engine front cover (see Section 7).
38 The remainder of installation is the reverse of the removal Steps.
39 Reconnect the battery (see Chapter 5, Section 1).
40 Run the engine and check for oil or coolant leaks.

9 Balance shaft chain and balance shafts - removal, inspection and installation

→**Note:** This procedure covers removal of the balance shaft chain and balance shafts, but take note that the shafts themselves can only be removed from the engine block after the engine has been removed from the vehicle. If there is a problem with the balance shafts that does warrant their removal, the engine would have to be removed anyway, since replacement of the balance shaft bushings is a job that must be left to an automotive machine shop. If you're just removing or replacing the chain, ignore the steps that don't apply.

REMOVAL

▶ Refer to illustrations 9.5, 9.6 and 9.9

1 Disconnect the cable from the negative battery terminal (see Chapter 5).
2 Drain the engine oil (see Chapter 1).
3 Remove the timing chain, timing chain guides and sprockets (see Section 8).
4 Check to make sure the engine is positioned at TDC for cylinder number 1 (see Section 3).

⁂ CAUTION:

Do not rotate the engine to find TDC number 1 when the timing chain is removed unless the engine has been rotated accidentally. If the engine is not positioned at TDC number 1, the camshafts must be removed to prevent damage to the valves (see Section 11).

9.5 Location of the balance shaft chain tensioner mounting bolts

5 Remove the balance shaft chain tensioner (see illustration).
6 Remove the adjustable balance shaft chain guide (see illustration).

FOUR CYLINDER ENGINE   2A-11

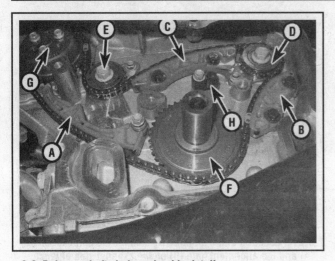

9.6 Balance shaft chain and guide details

- A Adjustable balance shaft chain guide
- B Small balance shaft chain guide
- C Upper balance shaft chain guide
- D Intake side (front) balance shaft sprocket
- E Exhaust side (rear) balance shaft sprocket
- F Crankshaft/balance shaft sprocket
- G Water pump sprocket
- H Timing chain oiling nozzle

7 Remove the small balance shaft chain guide (see illustration 9.6).
8 Remove the upper balance shaft chain guide (see illustration 9.6).
9 Remove the balance shaft drive chain (see illustration).

➡ **Note:** *To aid in removal, gather all the slack in the chain between the water pump sprocket and the crankshaft sprocket.*

10 If you're removing the balance shafts (engine removed from the vehicle), remove the balance shaft retainer bolts.
11 Remove the balance shafts from the engine block.

✱ CAUTION:

Mark each balance shaft to insure correct reassembly. The balance shafts are not interchangeable. Do not install the balance shaft into the wrong bore or extreme engine vibration will occur.

9.18 The timing mark (round dot) should point to the 6 o'clock position (approximately)

9.9 Balance shaft sprocket/chain alignment marks (A) and retainer bolts (B)

INSPECTION

12 Clean all parts with clean solvent and dry with compressed air, if available.
13 Inspect the chain tensioners for excessive wear or other damage.
14 Inspect the balance shaft chain guides for deep grooves, excessive wear, or other damage.
15 Inspect the balance shaft chain for excessive wear or damage.
16 Inspect the crankshaft and water pump sprockets for chipped or broken teeth, excessive wear, or damage.
17 Replace any component that is damaged.

INSTALLATION

▶ **Refer to illustrations 9.18, 9.20, 9.21a, 9.21b, 9.26a, 9.26b and 9.28**

18 Before installing the balance shaft chain, make sure the crankshaft timing mark (round dot) is pointing towards the 6 o'clock position (see illustration).

✱ CAUTION:

Do not rotate the engine to find TDC number 1 after the timing chain has been removed unless the engine has been rotated accidentally. If the engine is not positioned at TDC number 1, the camshafts must be removed to prevent damage to the valves (see Section 11).

19 Install the balance shafts into the bores and tighten the balance shaft retainer bolts to the torque listed in this Chapter's Specifications.
20 Align the balance shaft sprockets before installing the balance shaft chain. Starting with the intake side balance shaft, place the alignment arrow pointing up, then temporarily install a drill bit into the alignment hole and the sprocket teeth to lock the balance shaft sprocket in place (see illustration).
21 Now position the exhaust side (rear) balance shaft sprocket with the arrow pointing down and aligned with the cutout in the retainer, then install a drill bit into the alignment hole to hold the sprocket (see illustrations).

2A-12 FOUR-CYLINDER ENGINE

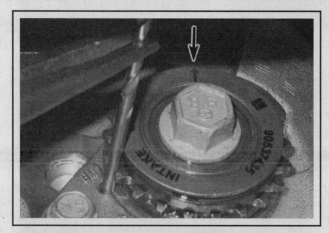

9.20 With the arrow on the intake side balance shaft sprocket pointing up (and aligned with the cutout on the balance shaft retainer, not visible in this photo, but similar to the one shown in illustration 9.21a), install a drill bit into the hole to lock the sprocket in place

9.21a Location of the alignment notch for the sprocket arrow (A) and the alignment hole (B) on the exhaust side balance shaft sprocket

9.21b Install the drill bit into the exhaust balance shaft retainer to lock it into position

22 Install the balance shaft chain onto the balance shaft/crankshaft sprocket and the balance shafts. Align the colored links with the alignment marks on each sprocket. Position the copper-colored link onto the intake side balance shaft, aligning the mark with the colored link at approximately the 12 o'clock position (see illustration 9.9).

➥**Note:** *The copper link will be installed at the intake balance shaft sprocket (front) while the silver links will be installed at the crankshaft sprocket and the exhaust balance shaft sprocket (rear).*

23 Working clockwise, position the second colored link (silver) on the crankshaft/balance shaft sprocket, aligning the mark on the sprocket with the colored link at the 6 o'clock position (see illustration 9.18).

24 Finally, pass the chain over the water pump sprocket, under the exhaust balance shaft sprocket and into position. Align the third colored link (silver) on the exhaust balance shaft sprocket, aligning the mark on the sprocket with the colored link at the 6 o'clock position.

25 Install the balance shaft chain guides (see illustration 9.6). Tighten the bolts to the torque listed in this Chapter's Specifications.

26 Reset the balance shaft chain tensioner. Turn the tensioner plunger 90-degrees in the bore and compress the tensioner plunger (see illustration). Rotate the plunger back to the original position at 12 o'clock and install a paper clip through the hole in the body into the plunger (see illustration).

27 Install the balance shaft chain tensioner and torque the bolts to the Specifications listed in this Chapter.

28 Remove the drill bit to release the plunger (see illustration).

29 Recheck all the balance shaft chain timing marks.

30 Install the timing chain (see Section 8) and all components removed previously.

31 Reconnect the battery (see Chapter 5, Section 1).

9.26a Rotate the plunger 90-degrees, align the holes in the body and piston . . .

9.26b . . . then install a drill bit to retain the piston in the locked position

9.28 After the tensioner is installed and the bolts tightened, remove the drill bit

FOUR CYLINDER ENGINE 2A-13

10 Crankshaft pulley and front oil seal - removal and installation

REMOVAL

▶ Refer to illustrations 10.4, 10.5 and 10.7

1 Disconnect the cable from the negative battery terminal (see Chapter 5).
2 Remove the drivebelt (see Chapter 1).
3 Raise the vehicle and support it securely on jackstands.
4 Remove the splash shield from below the engine compartment (see illustration).
5 Use a breaker bar and socket to remove the crankshaft pulley center bolt (see illustration). Discard the bolt and obtain a new one for installation.

➡ **Note:** It will be necessary to lock the pulley in position using a strap wrench or a large pin spanner. Be sure to wrap a length of old drivebelt around the pulley if you are using a strap wrench.

6 Slide the puller off the nose of the crankshaft. If the pulley is stuck, use a puller that bolts to the three threaded holes in the pulley hub. Additionally, a spacer, such as a deep socket that just fit into the hole in the pulley and bears on the crankshaft, will be required to avoid damage to the crankshaft.
7 Use a seal puller to remove the crankshaft front oil seal (see illustration). A screwdriver may be used instead, if the tip is wrapped with tape to avoid scratching the crankshaft.
8 Clean the seal bore and check it for nicks or gouges. Also examine the area of the hub that rides in the seal for signs of abnormal wear or scoring. For many popular engines, repair sleeves are available to restore a smooth finish to the sealing surface. Check with your auto parts store.

INSTALLATION

▶ Refer to illustration 10.9

9 Coat the lip of the new seal with clean engine oil and drive it into the bore with a seal driver or a socket slightly smaller in diameter than the seal (see illustration). The open side of the seal faces into the engine.
10 Using clean engine oil, lubricate the sealing surface of the hub. Install the crankshaft pulley/damper with a special installation tool, available at most auto parts stores. Do not use a hammer to install the pulley/damper. Install a new center bolt and tighten it to the torque listed in this Chapter's Specifications.

➡ **Note: You must use a new pulley bolt.**

11 The remainder of the installation is the reverse of the removal procedure.
12 Reconnect the battery (see Chapter 5, Section 1).

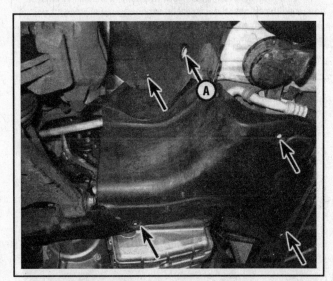

10.4 Remove the splash shield mounting bolts - one bolt can be removed through the access hole (A) in the inner fender splash shield

10.5 A large pin spanner can be used to prevent the pulley from rotating

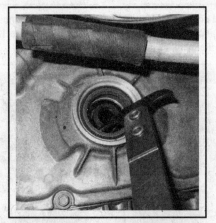

10.7 Use a seal puller to remove the old crankshaft seal, taking care not to damage the crankshaft or the seal bore in the cover

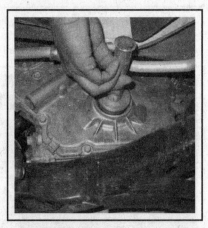

10.9 Driving the new front cover seal in with a seal driver

2A-14 FOUR-CYLINDER ENGINE

11 Camshafts and hydraulic lash adjusters - removal, inspection and installation

Note: This is a difficult procedure, involving special tools. Read through the entire Section and obtain the necessary tools before beginning the procedure.

REMOVAL

▶ Refer to illustrations 11.5a and 11.5b

1 Disconnect the cable from the negative battery terminal (see Chapter 5).
2 Remove the valve cover (see Section 4).
3 Set the engine to TDC for cylinder number one (see Section 3), then turn the crankshaft counterclockwise until the engine is set at 60-degrees before TDC. At this point, the diamond-shaped hole on the intake camshaft should be in the 12 o'clock position.

CAUTION:

Do not remove the camshafts with the engine at TDC number 1 or the valves and pistons will be damaged.

4 Remove the upper timing chain guide (see Section 8).
5 Install a special tool to secure the camshaft sprockets in position (see illustrations). This camshaft locking tool (jig) can be purchased through a dealership parts department or at specialty automotive tool suppliers.
6 Remove the camshaft sprocket bolts and slide the camshaft sprockets forward, then tighten the wingnuts to hold the sprockets securely.

Intake camshaft

▶ Refer to illustrations 11.7a, 11.7b, 11.8, 11.9 and 11.10

7 Each camshaft cap is marked with a number indicating its position (see illustrations). Loosen each bearing cap nut slowly and evenly, allowing the camshaft to lift from the cylinder head, parallel to the surface of the cylinder head.

11.5a Install a camshaft locking tool to hold the sprockets and timing chain in place - make sure the camshaft sprockets are locked properly and the tool is bolted to the cylinder head

11.5b The diamond-shaped hole on the intake camshaft should be in the 12 o'clock position

11.7a The camshaft bearing cap designations are stamped onto each cap

11.7b Make sure the arrow faces the timing chain end of the engine

FOUR CYLINDER ENGINE 2A-15

11.8 Remove each rocker arm . . .

11.9 . . . and store them in an organized manner so they can be returned to their original locations

11.10 Pull the lash adjusters from their bores in the head and store them along with their corresponding rocker arms

✳✳ CAUTION:

The caps must be installed in their original locations. Keep all parts from each camshaft together; never mix parts from one camshaft with those for another.

8 Remove the rocker arms (see illustration).
9 Place the rocker arms in a suitable container, in order, so they can be reinstalled in their original positions (see illustration).
10 Remove the hydraulic lash adjusters from their bores in the cylinder head (see illustration). Store these with their corresponding rocker arms so they can be reinstalled in their original locations.

Exhaust camshaft

11 Mark the exhaust bearing caps in the original positions and remove them from the cylinder head. Each camshaft cap is designated with a number (see illustrations 11.7a and 11.7b). Loosen each bearing cap nut slowly and evenly, allowing the camshaft to lift from the cylinder head, parallel to the surface of the cylinder head.

✳✳ CAUTION:

The camshaft bearing caps are numbered to identify the locations of the caps. The caps must be installed in their original locations. Keep all parts from each camshaft together; never mix parts from one camshaft with those for another.

12 Mark the positions of the rocker arms so they can be reinstalled in their original locations, then remove the rocker arms.
13 Place the rocker arms in a suitable container so they can be separated and identified (see illustration 11.9).
14 Lift the hydraulic lash adjusters from their bores in the cylinder head. Identify and separate the adjusters so they can be reinstalled in their original locations (see illustration 11.10).

INSPECTION

▶ Refer to illustrations 11.15, 11.18, 11.19, 11.20, 11.21 and 11.22

15 Check each hydraulic lash adjuster for excessive wear, scoring, pitting, or an out-of-round condition (see illustration). Replace as necessary.

16 Measure the outside diameter of each adjuster at the top and bottom of the adjuster. Then take a second set of measurements at a right angle to the first. If any measurement is significantly different from the others, the adjuster is tapered or out of round and must be replaced. If the necessary equipment is available, measure the diameter of the lash adjuster and the inside diameter of the corresponding cylinder head bore. Subtract the diameter of the lash adjuster from the bore diameter to obtain the oil clearance. Compare the measurements obtained to those given in this Chapter's Specifications. If the adjusters or the cylinder head bores are excessively worn, new adjusters or a new cylinder head, or both, may be required. If the valve train is noisy, particularly if the noise persists after a cold start, you can suspect a faulty lash adjuster.

17 Inspect the rocker arms for signs of wear or damage. The areas of wear are the tip that contacts the valve stem, the socket that contacts the lash adjuster and the roller that contacts the camshaft (see illustration 11.15).

18 Examine the camshaft lobes for scoring, pitting, galling (wear due to rubbing), and evidence of overheating (blue, discolored areas).

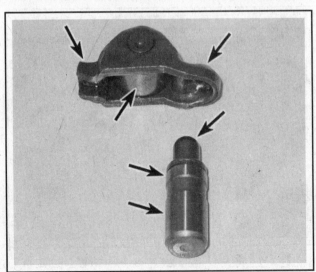

11.15 Check the rocker arms and lash adjusters for wear at the indicated points

11.18 Check the cam lobes for pitting, excessive wear, and scoring. If scoring is excessive, as shown here, replace the camshaft

11.19 Measure each camshaft lobe height with a micrometer

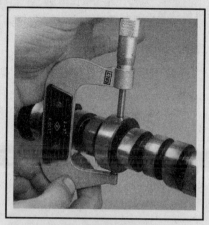

11.20 Measure each journal diameter with a micrometer. If any journal is less than the specified minimum, replace the camshaft

Look for flaking of the hardened surface layer of each lobe (see illustration). If any such wear is evident, replace the camshaft.

19 Measure the lobe height of each cam lobe on the intake camshaft, and record your measurements (see illustration). Compare the measurements for excessive variation; if the lobe heights vary more than 0.005 inch (0.125 mm), replace the camshaft. Compare the lobe height measurements on the exhaust camshaft and follow the same procedure. Do not compare intake camshaft lobe heights with exhaust camshaft lobe heights, as they are different. Only compare intake lobes with intake lobes and exhaust lobes with exhaust lobes.

20 Inspect the camshaft bearing journals and the cylinder head bearing surfaces for pitting or excessive wear. If any such wear is evident, replace the component concerned. Using a micrometer, measure the diameter of each camshaft bearing journal at several points (see illustration). If the diameter of any journal is less than specified, replace the camshaft.

21 To check the bearing journal oil clearance, remove the rocker arms and hydraulic lash adjusters (if not already done), use a suitable solvent and a clean lint-free rag to clean all bearing surfaces, then install the camshafts and bearing caps with a piece of Plastigage across each journal (see illustration). Tighten the bearing cap bolts to the specified torque. Don't rotate the camshafts.

22 Remove the bearing caps and measure the width of the flattened Plastigage with the Plastigage scale (see illustration). Scrape off the Plastigage with your fingernail or the edge of a credit card. Don't scratch or nick the journals or bearing caps.

23 If the oil clearance of any bearing is worn beyond the specified service limit, install a new camshaft and repeat the check. If the clearance is still excessive, replace the cylinder head.

24 To check camshaft endplay, remove the hydraulic lash adjusters, clean the bearing surfaces carefully, and install the camshafts and bearing caps. Tighten the bearing cap bolts to the specified torque, then measure the endplay using a dial indicator mounted on the cylinder head so that its tip bears on the camshaft end.

25 Lightly but firmly tap the camshaft fully toward the gauge, zero the gauge, then tap the camshaft fully away from the gauge and note the gauge reading. If the measured endplay is at or beyond the specified service limit, install a new camshaft thrust cap and repeat the check. If the clearance is still excessive, the camshaft or the cylinder head must be replaced.

11.21 Lay a strip of Plastigage on each camshaft journal, in line with the camshaft

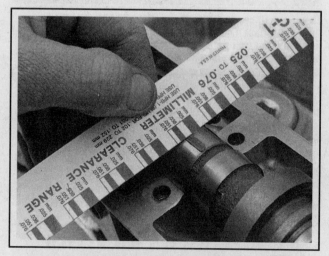

11.22 Compare the width of the crushed Plastigage to the scale on the package to determine the journal oil clearance

FOUR CYLINDER ENGINE 2A-17

INSTALLATION

▶ **Refer to illustration 11.29**

26 Lubricate the rocker arms and hydraulic lash adjusters with engine assembly lubricant or fresh engine oil. Install the adjusters into their original bores, then install the rocker arms in their correct locations.

27 Lubricate the camshafts with camshaft installation lubricant and install them in their correct locations. Position the camshafts with the slots in the end of the camshafts positioned as shown in illustration 11.29, aligning them with the slots in the camshaft sprockets.

28 Install the camshaft bearing caps in their correct locations, except for the front end and rear end bearing caps on each camshaft. Install the cap bolts and tighten by hand until snug. Tighten the bolts in four to five steps, starting with the center cap and working to the outside caps, to the torque listed in this Chapter's Specifications.

29 Slide the camshaft sprockets and timing chain along the guide pins toward the camshafts. Rotate the camshafts with an open-end wrench on the hex drive on each camshaft until the slots are aligned with the projections on the sprockets (see illustration). Install new bolts and tighten the camshaft sprockets to the torque listed in this Chapter's Specifications (see Section 8).

30 Remove the camshaft locking tool from the cylinder head. Then install the front and rear camshaft caps and tighten them to the torque listed in this Chapter's Specifications. Note that the rear cap on the

11.29 Camshaft and timing sprocket alignment details

intake camshaft is equipped with larger bolts and requires a different torque.

31 Install the upper timing chain guide (see Section 8). Rotate the engine by hand two revolutions - if you feel any resistance, stop and find out why.

32 The remainder of installation is the reverse of removal.

33 Reconnect the battery (see Chapter 5, Section 1).

12 Cylinder head - removal and installation

❈❈ CAUTION:

The engine must be completely cool when the head is removed. Failure to allow the engine to cool off could result in head warpage.

REMOVAL

1 Disconnect the cable from the negative battery terminal (see Chapter 5).
2 Wait until the engine is completely cool, then drain the cooling system (see Chapter 1).
3 Remove the drivebelt (see Chapter 1) and the drivebelt tensioner.
4 Remove the exhaust manifold (see Section 6).
5 Remove the intake manifold (see Section 5).
6 Remove the timing chain (see Section 8).
7 Label and disconnect the electrical connectors from the cylinder head that will interfere with removal. Use tape and mark each connector to insure correct reassembly.
8 Remove the cylinder head bolts and discard them, following the reverse of the tightening sequence (see illustration 12.16). Loosen the bolts in sequence 1/4-turn at a time. If the head is to be completely overhauled, refer to Section 11 for removal of the camshafts, rocker arms and hydraulic lash adjusters.
9 Use a prybar at the corners of the head-to-block mating surface to break the gasket seal. Do not pry between the cylinder head and engine block in the gasket sealing area.

10 Lift the cylinder head off the engine. If resistance is felt, place a wood block against the end and strike the wood block with a hammer. Store the cylinder head on wood blocks to prevent damage to the gasket sealing surfaces.

11 Remove the old cylinder head gasket. Before removing, note the correct orientation of the gasket for correct installation.

INSTALLATION

▶ **Refer to illustration 12.16**

12 The mating surfaces of the cylinder heads and block must be perfectly clean when the heads are installed. Use a gasket scraper to remove all traces of carbon and old gasket material, then clean the mating surfaces with lacquer thinner or acetone. If there's oil on the mating surfaces when the cylinder heads are installed, the gaskets may not seal correctly and leaks may develop. When working on the engine block, cover the open areas of the engine with shop rags to keep debris out during repair and reassembly. Use a vacuum cleaner to remove any debris that falls into the cylinders.

13 Check the engine block and cylinder head mating surfaces for nicks, deep scratches and other damage.

14 Use a tap of the correct size to chase the threads in the cylinder head bolt holes. Dirt, corrosion, sealant and damaged threads will affect torque readings.

15 Make sure the new gasket is located on the dowels in the block.

16 Carefully position the cylinder head on the engine block without disturbing the gasket. Install new cylinder head bolts and, following the

2A-18 FOUR-CYLINDER ENGINE

recommended sequence (see illustration), tighten the bolts to the torque listed in this Chapter's Specifications. All the main cylinder head bolts (numbers 1 through 10) are tightened in the first Step and second Step. The four smaller bolts located on the front of the cylinder head are the only ones tightened in the third Step. Mark a stripe on each of the main cylinder head bolts to help keep track of the bolts that have been tightened the additional 155-degrees.

➥ **Note: The method used for the head bolt tightening procedure is referred to as a "torque-angle" method. A special torque angle gauge (available at most auto parts stores) is available to attach to a breaker bar and socket for better accuracy during the tightening procedure.**

17 Install the timing chain (see Section 8).
18 Install the exhaust manifold (see Section 6).
19 Install the intake manifold (see Section 5).
20 The remaining installation steps are the reverse of removal.
21 Reconnect the battery (see Chapter 5, Section 1).
22 Change the engine oil and filter (Chapter 1), then start the engine and check carefully for oil and coolant leaks.

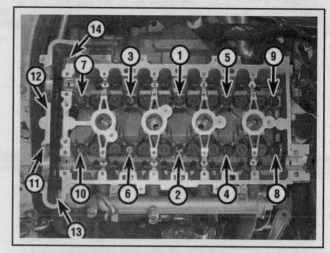

12.16 Cylinder head bolt tightening sequence

13 Oil pan - removal and installation

REMOVAL

1 Drain the engine oil (see Chapter 1).
2 Loosen the right-front wheel lug nuts, raise the front of the vehicle and support it securely on jackstands. Remove the right front wheel.
3 Remove the splash shield from below the right side of the engine compartment (see illustration 10.4).
4 Remove the lower air conditioning compressor mounting bolt (see Chapter 3). Loosen, but don't remove, the other compressor mounting bolts.
5 Remove the dipstick an the dipstick tube (the tube is bolted to the intake manifold).
6 Remove the oil pan bolts. Follow the reverse of the tightening sequence (see illustration 13.10).
7 Carefully remove the oil pan from the lower crankcase.

※ CAUTION:

If the oil pan is difficult to separate from the lower crankcase, use a rubber mallet or a block of wood and a hammer to jar it loose. If it's stubborn and still won't come off, pry carefully on casting protrusions (not the mating surfaces!).

INSTALLATION

▸ **Refer to illustration 13.10**

8 Using a gasket scraper, thoroughly clean all old gasket material from the lower crankcase and oil pan. Remove residue and oil film with a solvent such as acetone or lacquer thinner.
9 Apply a 2 mm bead of RTV sealant to the perimeter of the oil

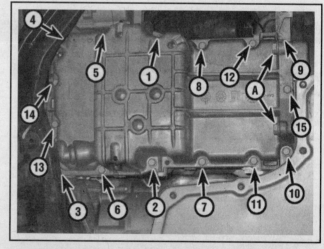

13.10 Oil pan bolt tightening sequence. Tighten the pan-to-transaxle bolts (A) until they're snug (but not too tight), then tighten the pan to block bolts in numerical order to the torque listed in this Chapter's Specifications, then tighten bolts (A) to the torque listed in this Chapter's Specifications

pan, inboard of the bolt holes, and around the oil suction port. Allow the sealant to set-up before installing the oil pan to the engine (but be sure to install the pan in the time given by the sealant manufacturer).

10 Install the oil pan and bolts (see illustration). Follow the correct torque sequence and tighten the bolts to the torque listed in this Chapter's Specifications.

11 The remaining installation is the reverse of removal. Be sure to tighten the wheel lug nuts to the torque listed in the Chapter 1 Specifications.

FOUR CYLINDER ENGINE 2A-19

14 Oil pump - removal, inspection and installation

REMOVAL

♦ Refer to illustrations 14.5a and 14.5b

1 Drain the engine oil (see Chapter 1).
2 Remove the drivebelt (see Chapter 1).
3 Loosen the right-front wheel lug nuts, raise the front of the vehicle and support it securely on jackstands. Remove the right front wheel.
4 Remove the engine front cover (see Section 7).
5 Working on the backside of the engine cover, loosen the oil pump cover screws a little at a time until they're all loose (see illustrations). When all of the screws are loose, remove the cover.

INSPECTION

♦ Refer to illustrations 14.8a, 14.8b, 14.8c and 14.10

6 Note any identification marks on the rotors and withdraw the rotors from the pump body. If no marks can be seen, use a permanent marker and make your own to ensure that they will be installed correctly.
7 Thoroughly clean and dry the components.
8 Inspect the rotors for obvious wear or damage. If either rotor, the pump body or the cover is scored or damaged, the complete oil pump assembly must be replaced. Also check the inner-to-outer rotor tip clearance, the outer rotor-to-housing clearance, and the rotor-to-cover side clearance (see illustrations).
9 If the oil pump components are in acceptable condition, dip the rotors in clean engine oil and install them into the pump body with any identification marks positioned as noted during disassembly.
10 Remove the oil pressure relief valve components from the oil pump body. Thoroughly clean and dry the components. Inspect the components for obvious wear or damage. Install them in the correct order (see illustration).

INSTALLATION

11 Install the rotors into the housing with the hub of the inner rotor facing the engine front cover. The inner rotor hub must be installed cor-

14.5a Location of the oil pump cover mounting screws

14.5b Lift the oil pump cover from the oil pump assembly

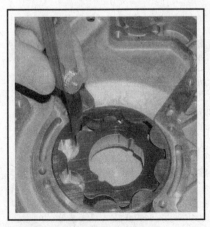

14.8a Using a feeler gauge to check the inner-to-outer rotor tip clearance . . .

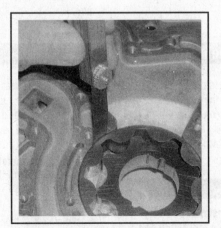

14.8b . . . and the outer rotor-to-housing clearance

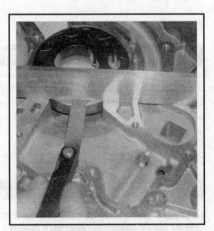

14.8c Use a straightedge and a feeler gauge to check the rotor-to-cover clearance

2A-20 FOUR-CYLINDER ENGINE

rectly or the engine front cover will not fasten properly.

12 Install the oil pump cover and screws and tighten by hand until snug. Then tighten the screws gradually and evenly to the torque listed in this Chapter's Specifications.

13 Install the engine front cover (see Section 7).

14 Refer to Chapter 1 and fill the engine with fresh engine oil. Install a new oil filter. Refill the cooling system.

15 Start the engine and check for leaks.

16 Run the engine and make sure oil pressure comes up to normal quickly. If it doesn't, stop the engine and find out the cause. Severe engine damage can result from running an engine with insufficient oil pressure!

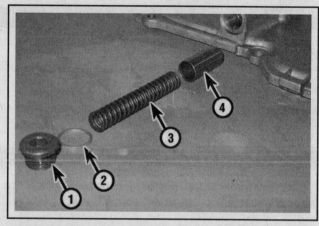

14.10 Oil pressure relief valve component details

1   Oil pressure relief valve plug
2   Washer
3   Spring
4   Piston

15 Flywheel/driveplate - removal and installation

REMOVAL

1 Raise the vehicle and support it securely on jackstands, then refer to Chapter 7 and remove the transaxle. If it's leaking, now would be a very good time to replace the front pump seal/O-ring (automatic transaxle only).

2 If you're working on a manual transaxle equipped vehicles, remove the pressure plate and clutch disc (see Chapter 8). Now is a good time to check/replace the clutch components.

3 Use a center punch or paint to make alignment marks on the flywheel/driveplate and crankshaft to ensure correct alignment during reinstallation.

4 Remove the bolts that secure the flywheel/driveplate to the crankshaft. If the crankshaft turns, wedge a screwdriver in the ring gear teeth to jam the flywheel.

5 Remove the flywheel/driveplate from the crankshaft. Since the flywheel is fairly heavy, be sure to support it while removing the last bolt. Automatic transmission equipped vehicles have a spacer between the crankshaft and the driveplate.

INSTALLATION

6 Clean the flywheel to remove grease and oil. Inspect the surface for cracks, rivet grooves, burned areas and score marks. Light scoring can be removed with emery cloth. Check for cracked and broken ring gear teeth. Lay the flywheel on a flat surface and use a straightedge to check for warpage.

7 Clean and inspect the mating surfaces of the flywheel/driveplate and the crankshaft. If the crankshaft rear seal is leaking, replace it before reinstalling the flywheel/driveplate (see Section 16).

8 Position the flywheel/driveplate against the crankshaft. Be sure to align the marks made during removal. Note that some engines have an alignment dowel or staggered bolt holes to ensure correct installation. Before installing the bolts, apply thread locking compound to the threads.

9 Wedge a screwdriver in the ring gear teeth to keep the flywheel/driveplate from turning and tighten the bolts to the torque listed in this Chapter's Specifications. Work up to the final torque in three or four steps.

10 The remainder of installation is the reverse of the removal procedure.

16 Rear main oil seal - replacement

1 The one-piece rear main oil seal is pressed into engine block and the crankcase reinforcement section. Remove the transaxle (see Chapter 7), the clutch components, if equipped (see Chapter 8) and the flywheel (see Section 15).

2 Pry out the old seal with a special seal removal tool or a flat-blade screwdriver.

✳✳✳ CAUTION:

To prevent an oil leak after the new seal is installed, be very careful not to scratch or otherwise damage the crankshaft sealing surface or the bore in the engine block.

3 Clean the crankshaft and seal bore in the block thoroughly and de-grease these areas by wiping them with a rag soaked in lacquer thinner or acetone. Lubricate the lip of the new seal and the outer diameter of the crankshaft with engine oil.

4 Position the new seal onto the crankshaft. Make sure the edges of the new oil seal are not rolled over.

➡**Note: When installing the new seal, if so marked, the words THIS SIDE OUT on the seal must face out, toward the rear of the engine.**

Use a special rear main oil seal installation tool or a socket with the exact diameter of the seal to drive the seal in place. Make sure the seal is not off-set; it must be flush along the entire circumference of the engine block and the crankcase reinforcement section.

5 The remainder of installation is the reverse of removal.

FOUR CYLINDER ENGINE 2A-21

17 Powertrain mounts - check and replacement

CHECK

1 Engine mounts seldom require attention, but broken or deteriorated mounts should be replaced immediately or the added strain placed on the driveline components may cause damage or wear.

2 During the check, the engine must be raised slightly to remove the weight from the mounts.

3 Raise the vehicle and support it securely on jackstands, then position a jack under the engine oil pan. Place a large block of wood between the jack head and the oil pan, then carefully raise the engine just enough to take the weight off the mounts.

※ WARNING:

DO NOT place any part of your body under the engine when it's supported only by a jack!

4 Check the mounts to see if the rubber is cracked, hardened or separated from the bushing in the center of the mount.

5 Check for relative movement between the mount and the engine or frame (use a large screwdriver or prybar to attempt to move the mounts). If movement is noted, lower the engine and tighten the mount fasteners.

REPLACEMENT

▸ Refer to illustration 16.9a, 16.9b, 16.9c, 16.9d and 16.9e

6 Disconnect the negative battery cable from the battery (see Chapter 5).

7 Raise the vehicle and support it securely on jackstands.

8 Place a large block of wood between the jack head and the oil pan, then carefully raise the engine just enough to take the weight off the mounts.

※ CAUTION:

Do not disconnect more than one mount at a time unless the engine will be removed from the vehicle.

9 Remove the engine mount through-bolt/nuts and detach the mount from the chassis bracket (see illustrations).

10 Remove the nuts holding the mount to the engine bracket.

11 Installation is the reverse of removal. Use thread-locking compound on the mount bolts and be sure to tighten them securely.

12 Reconnect the battery (see Chapter 5, Section 1).

17.9a Location of the passenger side engine mount upper mounting bolts

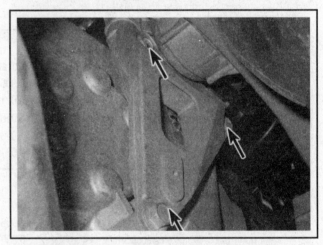

17.9b Location of the transaxle mount bracket, as seen from below - the transaxle mount can be accessed from above the transaxle

17.9c Location of the through-bolt on the front engine mount

17.9d Location of the rear engine mount through-bolt

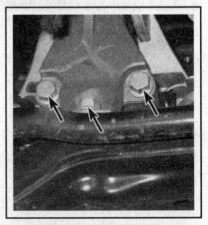

17.9e The rear engine mount bracket can be accessed from below

2A-22 FOUR-CYLINDER ENGINE

Specifications

General
Firing order	1-3-4-2
Compression ratio	10:1
Compression pressure	See Chapter 2C
Bore	3.385 to 3.386 inches (85.9 to 86.0 mm)
Stroke	Not available
Displacement	134 cubic inches (2.2 liters)
Oil pressure	See Chapter 2C

Timing chain tensioner
Timing chain tensioner compressed length	2.83 inches (72.0 mm)

Hydraulic lash adjuster
Lash adjuster bore diameter	0.4730 to 0.4739 inch (12.013 to 12.037 mm)
Lash adjuster diameter	0.4723 to 0.4728 inch (11.986 to 12.000 mm)
Lash adjuster-to-bore clearance	0.0005 to 0.0020 inch (0.013 to 0.051 mm)

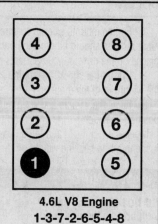

4.6L V8 Engine
1-3-7-2-6-5-4-8

Cylinder locations and firing order

Camshafts
Lobe lift (intake and exhaust)	Not available
Allowable lobe lift variation	0.005 inch (0.125 mm)
Endplay	0.0016 to 0.0057 inch (0.040 to 0.144 mm)
Journal diameter (all)	1.0604 to 1.0614 inches (26.935 to 26.960 mm)
Bearing inside diameter (all)	1.0630 to 1.0638 inches (27.00 to 27.021 mm)
Journal-to-bearing (oil) clearance	0.0015 to 0.0034 inch (0.040 to 0.086 mm)

Oil pump
Outer rotor-to-oil pump housing clearance limit	0.011 inch (0.277 mm)
Inner rotor-to-outer rotor tip clearance limit	0.006 inch (0.150 mm)
Rotor-to-cover side clearance limit	0.005 inch (0.128 mm)

Torque specifications

	Ft-lbs (unless otherwise indicated)	Nm
Camshaft sprocket bolts*		
Step 1	63	85
Step 2	Tighten an additional 30-degrees	
Camshaft bearing cap bolts		
Intake camshaft rear cap bolts	19	25
All other camshaft cap bolts	89 in-lbs	10
Crankshaft pulley bolt*		
Step 1	74	100
Step 2	Tighten an additional 75-degrees	
Cylinder head bolts*		
Step 1 - Main bolts (1 through 10)	22	30
Step 2 - Main bolts (1 through 10)	Tighten an additional 155-degrees	
Step 3 - Front bolts (11 through 14)	25	35

FOUR CYLINDER ENGINE 2A-23

Torque specifications

	Ft-lbs (unless otherwise indicated)	Nm
Drivebelt tensioner bolt	37	50
Flywheel/driveplate bolts		
Step 1	39	53
Step 2	Tighten an additional 25-degrees	
Exhaust manifold-to-cylinder head nuts	13	18
Exhaust manifold heat shield bolts	18	25
Exhaust pipe-to-manifold nuts	22	30
Engine front cover perimeter bolts	18	25
Engine front cover water pump bolt	18	25
Intake manifold bolts/nuts	89 in-lbs	10
Oil pump cover-to-engine front cover screws	53 in-lbs	6
Oil pump pressure relief valve plug	30	40
Oil pan-to-crankcase reinforcement bolts	18	25
Oil pan-to-transaxle bolts	26	35
Balance shaft chain tensioner	89 in-lbs	10
Balance shaft chain guides		
Adjustable balance shaft chain guide bolts	89 in-lbs	10
Small balance shaft chain guide bolts	89 in-lbs	10
Upper balance shaft guide bolts	89 in-lbs	10
Balance shaft retainer bolts	89 in-lbs	10
Timing chain tensioner	55	75
Timing chain guides		
Adjustable timing chain guide bolts	89 in-lbs	10
Fixed timing chain guide bolts	89 in-lbs	10
Upper timing chain guide bolts	89 in-lbs	10
Timing chain oiling nozzle bolt	89 in-lbs	10
Timing chain guide access hole plug	66	90
Valve cover bolts	89 in-lbs	10
Valve cover ground strap bolt	89 in-lbs	10
Water pump bolts	18	25
Water pump drain bolt	15	20

*Bolt(s) must be replaced.

2A-24 FOUR-CYLINDER ENGINE

Notes

Section

1 General information
2 Repair operations possible with the engine in the vehicle
3 Top Dead Center (TDC) for number one piston - locating
4 Valve cover(s) - removal and installation
5 Intake manifold - removal and installation
6 Exhaust manifolds - removal and installation
7 Timing belt cover - removal and installation
8 Timing belt and sprockets - removal, inspection and installation
9 Crankshaft pulley and front oil seal - removal and installation
10 Camshafts and followers - removal, inspection and installation
11 Cylinder heads - removal and installation
12 Oil pan - removal and installation
13 Oil pump - removal, inspection and installation
14 Driveplate - removal and installation
15 Rear main oil seal - replacement
16 Engine mounts - check and replacement

Reference to other Chapters

CHECK ENGINE light on - See Chapter 6

2B

V6 ENGINE

2B-2 V6 ENGINE

1 General information

This Part of Chapter 2 is devoted to in-vehicle repair procedures for the 3.0L Double Overhead Cam (DOHC) V6 engines. This engine is equipped with two camshafts per cylinder head. All four camshaft sprockets are driven by one timing belt. This engine has aluminum cylinder heads, an iron block and two intake valves and two exhaust valves per cylinder. All information concerning engine removal and installation can be found in Part C of this Chapter.

These engines are an interference design. In the event the timing belt breaks, the pistons will interfere with the valves and cause damage. The timing belt, tensioner and cylinder heads can be removed with the engine in the vehicle. It will be necessary to obtain special tools to perform many repair operations on this engine.

The Specifications included in this Part of Chapter 2 apply only to the procedures contained in this Part. Part C of Chapter 2 contains the Specifications necessary for certain engine rebuilding procedures.

2 Repair operations possible with the engine in the vehicle

Many major repair operations can be accomplished without removing the engine from the vehicle.

If possible, clean the engine compartment and the exterior of the engine with some type of pressure washer before any work is started. It will make the job easier and help keep dirt out of the internal areas of the engine.

It may help to remove the hood to improve access to the engine as repairs are performed (refer to Chapter 11 if necessary).

If vacuum, exhaust, oil or coolant leaks develop, indicating a need for gasket or seal replacement, the repairs can generally be made with the engine in the vehicle. The intake and exhaust manifold gaskets, timing cover gasket, oil pan gasket, crankshaft oil seals and cylinder head gaskets are all accessible with the engine in place.

Exterior engine components, such as the intake and exhaust manifolds, the oil pan, the water pump, the starter motor, the alternator and the fuel system components can be removed for repair with the engine in place.

Since the cylinder heads can be removed without pulling the engine, valve component servicing can also be accomplished with the engine in the vehicle. Replacement of the timing belt and camshaft sprockets and oil pump is also possible with the engine in the vehicle.

In extreme cases caused by a lack of necessary equipment, repair or replacement of piston rings, pistons, connecting rods and rod bearings is also possible with the engine in the vehicle. However, this practice is not recommended because of the cleaning and preparation work that must be done to the components involved.

3 Top Dead Center (TDC) for number one piston - locating

▶ Refer to illustrations 3.1a and 3.1b

Refer to Chapter 2, Part A for the TDC locating procedure, with the following exception:

a) To verify TDC number 1, remove the engine timing belt cover (see Section 7) and make sure the marks on the sprockets align with the marks on the engine/sprocket covers (see illustrations). At this point number one cylinder is at TDC on the compression stroke.

3.1a When the engine is at TDC on the compression stroke for cylinder number one, the camshaft sprocket marks will be aligned with the cutouts on the rear timing belt covers - (rear cylinder head shown)

3.1b Camshaft sprocket alignment marks - front cylinder head

V6 ENGINE 2B-3

4 Valve covers - removal and installation

4.11a Install the rubber valve cover gasket into the groove in the valve cover . . .

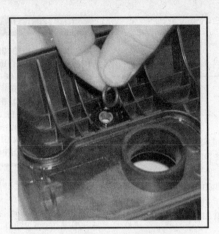
4.11b . . . and install new O-rings around the valve cover bolt holes

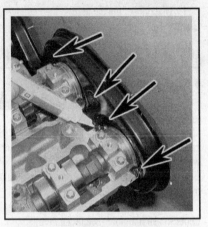
4.12a Apply RTV sealant to the camshaft front bearing caps at the cylinder head . . .

REMOVAL

1 Disconnect the cable from the negative battery terminal (see Chapter 5).
2 Remove the ignition coil pack(s) (see Chapter 5).
3 Remove the intake manifold runners and the upper intake manifold (see Section 5).

Front valve cover

4 Remove the oil filler cap.
5 Remove the lifting bracket mounting bolts and the bracket assembly.
6 Loosen the valve cover bolts and remove the valve cover. If it's stuck, tap it with a hammer and block of wood.

➡Note: The valve cover contains eight rubber O-rings that may stick to the surface of the cylinder head. Be sure to collect each O-ring and install it in the original location.

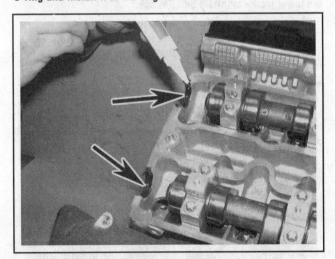
4.12b . . . and to the semi-circular cutouts on the rear of the cylinder head

Rear valve cover

7 Remove the knock sensor harness connector and position the harness off to the side (see Chapter 6).
8 Remove the lifting bracket mounting bolts and the bracket assembly.
9 Loosen the valve cover bolts and remove the cover. If it's stuck, tap it with a hammer and block of wood.

➡Note: The valve cover contains eight rubber O-rings that may stick to the surface of the cylinder head. Be sure to collect each O-ring and install it in the original location.

INSTALLATION

▶ Refer to illustrations 4.11a, 4.11b, 4.12a and 4.12b

10 The mating surfaces of each cylinder head and valve cover must be perfectly clean when the valve covers are installed. Remove all traces of sealant, and clean the mating surfaces with lacquer thinner or acetone. If there's old sealant or oil on the mating surfaces when the valve cover is installed, oil leaks may develop.
11 Install the rubber valve cover gasket into the groove in the valve cover (see illustration). Also, install the rubber O-rings into the grooves around the valve cover bolts (see illustration).
12 Apply a bead of RTV sealant into the corners where the front camshaft bearing cap meets the cylinder head (see illustration), and around the housing seals at the rear (see illustration).
13 Carefully position the valve cover on the cylinder head and install the bolts.

➡Note: Install the covers within five minutes of applying the RTV sealant. Tighten the bolts to the torque listed in this Chapter's Specifications.

14 The remaining installation steps are the reverse of removal.
15 Reconnect the battery (see Chapter 5, Section 1).
16 Start the engine and check for oil leaks as the engine warms up.

2B-4 V6 ENGINE

5 Intake manifold - removal and installation

1 Disconnect the cable from the negative battery terminal (see Chapter 5).

UPPER INTAKE MANIFOLD (PLENUM)

▶ Refer to illustrations 5.4 and 5.5

2 Remove the air filter housing and the intake air duct (see Chapter 4).

3 Disconnect the throttle body solenoid, the intake manifold runner control solenoid and any other electrical connector that will interfere with the intake manifold removal.

4 Remove the vacuum hoses from the intake manifold runner control vacuum actuator and the vacuum solenoid (see illustration).

5 Remove the clamps from the intake manifold runners at the upper intake manifold (see illustration).

6 Disconnect the EGR pipe from the upper intake manifold (see Chapter 6).

7 Remove the upper intake manifold mounting bolts and lift the manifold from the engine. Clamp off, then disconnect the coolant hose from the throttle body after the upper intake manifold has been slightly lifted.

8 Carefully set the upper intake manifold in place.

9 Install the upper intake manifold bolts. Tighten the bolts to the torque listed in this Chapter's Specifications.

10 Install the intake manifold runners at the intake manifold. Tighten the clamp bolts to the torque listed in this Chapter's Specifications.

11 The remaining installation steps are the reverse of removal. Check the coolant level, adding as necessary (see Chapter 1).

12 Reconnect the battery (see Chapter 5, Section 1), then start the engine and check carefully for leaks.

LOWER INTAKE MANIFOLD

13 Remove the upper intake manifold and the intake manifold runners (see Steps 1 through 7).

14 Depressurize the fuel system, then remove the fuel rail and injectors (see Chapter 4).

15 Remove the lower intake manifold mounting bolts.

16 Remove the lower intake manifold spacer from the engine block.

17 Remove the O-rings from the lower intake manifold spacer.

18 Clean the surface of the spacer and the engine block.

✹✹ CAUTION:

The mating surfaces of the cylinder heads, engine block and intake manifold must be perfectly clean. Gasket removal solvents in aerosol cans are available at most auto parts stores and may be helpful when removing old gasket material that's stuck to the cylinder heads and intake manifold. Since the cylinder heads are aluminum and the lower intake manifold spacer is aluminum, aggressive scraping can cause damage! Be sure to follow directions printed on the container, and use only a plastic-tipped scraper, not a metal one.

19 Clean the mating surfaces with lacquer thinner or acetone. If there's old sealant or oil on the mating surfaces when the lower intake manifold is installed, oil or vacuum leaks may develop.

20 When working on the cylinder heads and engine block, cover the open engine areas with shop rags to keep debris out of the engine. Use a vacuum cleaner to remove any material that falls into the intake ports in the cylinder heads.

21 Use a tap of the correct size to chase the threads in the bolt holes, then use compressed air (if available) to remove the debris from the holes.

✹✹ WARNING:

Wear safety glasses or a face shield to protect your eyes when using compressed air!

22 Install the new O-rings into the spacer.

23 Install the spacer onto the engine block.

24 Carefully set the lower intake manifold in place. Don't disturb the spacer and don't move the lower intake manifold fore-and-aft after it contacts the spacer on the engine block.

25 Install the lower intake manifold bolts and tighten them to the torque listed in this Chapter's Specifications. Start with the center bolts and work to the outside and tighten the bolts evenly, working up to the final torque in several steps.

➡ Note: Apply a small amount of thread locking compound to the lower intake manifold bolt threads.

26 The remaining installation steps are the reverse of removal. Check the coolant level, adding as necessary (see Chapter 1).

27 Reconnect the battery (see Chapter 5, Section 1), then start the engine and check carefully for leaks.

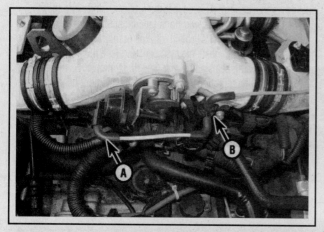

5.4 Remove the vacuum hoses from the IMRC vacuum actuator (A) and the IMRC vacuum solenoid (B)

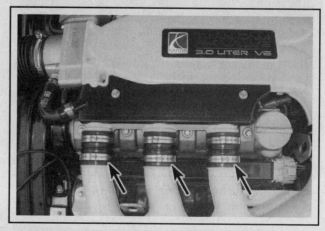

5.5 Remove the intake manifold runner clamps

V6 ENGINE   2B-5

6 Exhaust manifolds - removal and installation

⁂ WARNING:

Wait until the engine is completely cool before beginning this procedure.

REMOVAL

1 Disconnect the cable from the negative battery terminal (see Chapter 5).
2 Raise the vehicle and support it securely on jackstands.
3 Remove the splash shield from below the engine compartment.
4 Working under the vehicle, apply penetrating oil to the exhaust pipe-to-manifold studs and nuts (they're usually corroded or rusty).

Front exhaust manifold

5 Drain the coolant (see Chapter 1).
6 Remove the oxygen sensor from the exhaust manifold (see Chapter 6).
7 Install an engine support fixture (see Section 7) and remove the transaxle mount through-bolt.
8 Raise the engine slightly to access the coolant extension housing.
9 Remove the oil dipstick and dipstick tube.
10 Remove the coolant extension housing (see Chapter 3).
11 Remove the power steering pipe bracket.
12 Remove the front exhaust pipe assembly (see Chapter 4).
13 Remove the oil filter housing assembly.
14 Remove the mounting nuts from the exhaust manifold, then detach the manifold from the cylinder head.

Rear exhaust manifold

15 Disconnect the EGR pipe from the exhaust manifold (see Chapter 6).
16 Remove the oxygen sensor from the exhaust manifold (see Chapter 6).
17 Remove the rear exhaust pipe assembly from the vehicle (see Chapter 6, Section 19).
18 Remove the mounting nuts from the exhaust manifold, then detach the manifold from the cylinder head.

INSTALLATION

19 Check the exhaust manifolds for cracks. Make sure the bolt threads are clean and undamaged. The exhaust manifold and cylinder head mating surfaces must be clean before the exhaust manifolds are reinstalled - use a gasket scraper to remove all carbon deposits.
20 Position a new gasket in place and slip the exhaust manifold over the studs on the cylinder head. Install the mounting nuts.
21 When tightening the mounting nuts, work from the center outwards, alternating between top and bottom rows. Tighten the bolts to the torque listed in this Chapter's Specifications.
22 The remaining installation steps are the reverse of removal. When reconnecting the EGR tube to the exhaust manifold, use a slight amount of anti-seize compound on the threads.
23 If the front exhaust manifold was removed, refill the cooling system (see Chapter 1).
24 Reconnect the battery (see Chapter 5, Section 1).
25 Start the engine and check for leaks.

7 Timing belt cover - removal and installation

REMOVAL

▶ Refer to illustrations 7.4 and 7.6

1 Disconnect the cable from the negative battery terminal (see Chapter 5).
2 Drain the engine oil and remove the oil filter (see Chapter 1).
3 Raise the vehicle and support it securely on jackstands.
4 Remove the inner fender splash shield from the right side of the vehicle (see illustration).
5 Loosen the water pump pulley and power steering pump pulley bolts. Leave the pulleys and the drivebelt on the vehicle at this time.
6 Install an engine support fixture (see illustration).

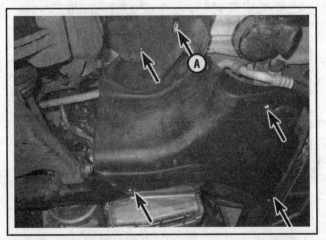

7.4 Remove the splash shield mounting bolts - one bolt can be removed through the access hole (A) in the inner fender splash shield

7.6 Install an engine support fixture to support the engine (four-cylinder engine shown, V6 similar)

2B-6 V6 ENGINE

7 Remove the engine mount and the drivebelt (see Chapter 1).
8 Remove the front crankshaft pulley and oil seal (see Section 9).
9 Remove the engine mount bracket from the timing belt cover.
10 Remove the water pump pulley.
11 Remove the power steering pump pulley (see Chapter 10).
12 Remove the drivebelt tensioner.
13 Disconnect the harness connectors at the timing belt cover and position the wiring harness off to the side.
14 Remove the timing belt cover mounting bolts.
15 Remove the timing belt cover from the engine.

INSTALLATION

16 Clean the timing belt cover, engine block and inspect the timing belt cover for cracks and damage.

➡ **Note: If the timing belt cover is damaged or chipped, replace it with a new timing belt cover.**

17 Install the timing belt cover to the engine. Tighten the timing belt cover bolts to the torque listed in this Chapter's Specifications.
18 Install the remaining parts in the reverse order of removal.
19 Reconnect the battery (see Chapter 5, Section 1)
20 Add the proper type and quantity of engine oil and coolant (see Chapter 1). Run the engine and check for leaks.

8 Timing belt and sprockets - removal, inspection and installation

➡ **Note 1:** Because this is an "interference" engine design, if the belt has broken, there will be damage to the valves and/or pistons and will require removal of the cylinder heads.

➡ **Note 2:** The timing belt replacement procedure requires several special tools. The special tools are sold as a kit (J-42069) that includes two camshaft locking tools, a crankshaft locking tool, a crankshaft wedge and a camshaft timing belt gauge. Consult with a specialty tool distributor or dealership parts department for availability.

➡ **Note 3:** These engines are equipped with two different types of timing belt tensioners. Engines before number 578511 are equipped with the original type tensioner while engines after 578512 are equipped with the updated tensioner. Early tensioners are combined with the upper timing belt guide and tensioner as a complete assembly. Later tensioners are equipped with separate upper timing belt guides. The timing belt installation will differ slightly depending upon the type of tensioner. Check the engine identification number on the engine block to determine the type of timing belt tensioner equipped on the engine.

REMOVAL

▶ **Refer to illustrations 8.4, 8.5a, 8.5b, 8.6a, 8.6b, 8.7a, 8.7b, 8.8a and 8.8b**

1 Disconnect the cable from the negative battery terminal (see Chapter 5).
2 Remove the timing belt cover (see Section 7).
3 Set the engine to TDC for cylinder number one (see Section 3), then turn the crankshaft counterclockwise until the engine is set at 60-degrees before TDC. Remove the spark plugs (see Chapter 1).

✳✳ CAUTION:

The initial position of the crankshaft at 60-degrees BTDC allows the tool to be rotated for proper number 1 TDC alignment. The camshaft timing marks will align after the crankshaft locking tool has been installed and rotated into position.

4 Install the crankshaft locking tool onto the crankshaft (see illustration).
5 Rotate the crankshaft clockwise approximately 60-degrees until the locking tool lever arm rests against the water pump pulley flange (see illustrations). Lock the tool in this position. The camshaft sprock-

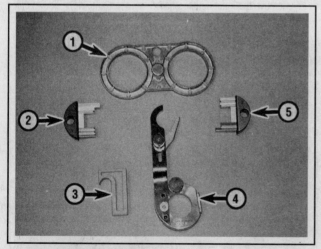

8.4 Timing belt tool kit details
1 Timing belt gauge
2 Camshaft sprocket locking tool
3 Timing belt wedge
4 Crankshaft locking tool
5 Camshaft sprocket locking tool

8.5a Install the crankshaft locking tool over the crankshaft sprocket flange, then turn the crankshaft clockwise until the tool contacts the water pump . . .

V6 ENGINE 2B-7

8.5b ... then lock the tool in place by moving the lever arm against the water pump flange and tightening the thumbscrew

ets should align correctly (see illustrations 3.1a and 3.1b). The engine is now positioned exactly at TDC number 1.

6 Install the special tools that lock the camshaft sprockets (see illustrations).

7 Loosen the timing belt tensioner (see illustrations).
8 Loosen the upper timing belt guide pulley (see illustration) and the lower timing belt guide pulley (see illustration) and rotate them to release belt tension.
9 Remove the timing belt.
10 The camshaft sprockets can be removed at this point, if they are damaged, or to replace the oil seals. Remove the keys from the shafts so they don't fall out and get lost.

❊❊❊ CAUTION:
Don't allow the camshaft(s) to turn.

11 If it's worn or damaged, or if you're replacing the crankshaft front oil seal, the crankshaft sprocket can now be removed (see Section 9). If it won't come off by hand, carefully pry it off. Also remove the timing belt guide(s), noting how they are installed.

INSPECTION

12 Inspect the sprocket teeth for wear and damage. Check the timing belt for any cracks or oil residue. Also check the camshaft for excessive endplay (see Section 10). Check the timing belt tensioner for smooth operation. Replace any worn parts with new ones.

8.6a Install the camshaft sprocket locking tool between the two camshaft sprockets ...

8.6b ... and push the tool into position - the TOP mark must be up. Repeat this on the other set of camshafts

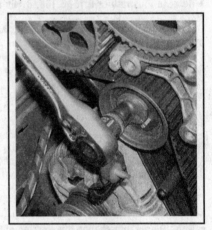

8.7a Loosen the timing belt tensioner bolt

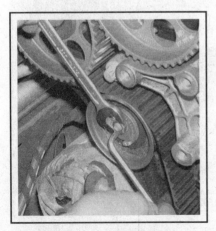

8.7b Rotate the tensioner using an Allen key

8.8a Loosen the upper timing belt guide pulley ...

8.8b ... and the lower timing belt guide pulley mounting bolts

2B-8 V6 ENGINE

8.15 Use a wrench on the hex drive on the camshaft to prevent the camshaft from turning

13 Now that the timing belt is removed, inspect the water pump (see Chapter 3).

INSTALLATION

◆ Refer to illustrations 8.15, 8.16a, 8.16b and 8.16c

14 Remove all dirt and oil from the timing belt area. Clean the teeth of the sprockets with lacquer thinner.
15 If any of the timing belt sprockets were removed, install them now with their keys and tighten the bolts to the torque listed in this Chapter's Specifications (see illustration).
16 Recheck the position of the timing marks (see illustrations).
17 Remove the crankshaft locking tool.

Engines before 578511

◆ Refer to illustrations 8.18, 8.19, 8.20, 8.21a, 8.21b, 8.23, 8.25, 8.27 and 8.36

18 Mark the position of the upper and lower timing belt guide pulleys. Place the mark at the point farthest from the bolt (see illustration).

➡ Note: Do not interchange the position of the two timing belt guide pulleys. Each timing belt guide pulley is designed with different shim thicknesses and bolt colors.

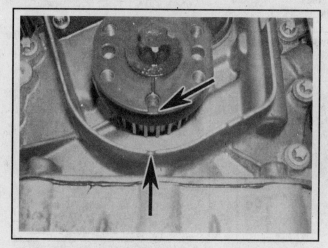

8.16c Align the crankshaft sprocket timing mark with the notch in the oil pump cover

8.16a Location of the number 1 and 2 camshaft timing marks (rear cylinder head)

8.16b Location of the number 3 and 4 camshaft timing marks (front cylinder head)

19 Install the timing belt over the number 3 and 4 camshaft sprockets on the front cylinder head (see illustration). Align the marks on the timing belt with the marks on the timing belt rear cover.

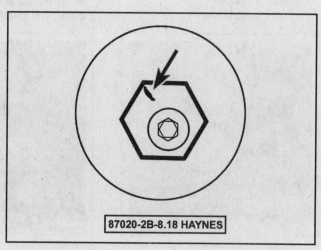

8.18 Mark the hex of the guide pulley at the point farthest from the bolt

V6 ENGINE 2B-9

8.19 Align the timing belt paint marks with the camshaft sprocket marks and the notches on the rear covers (number 3 and 4 on front cylinder head)

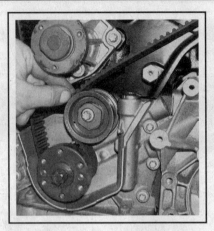

8.20 Route the timing belt over the lower timing belt guide pulley

8.21a Install the timing belt around the crankshaft sprocket with the timing belt mark aligned with the sprocket mark

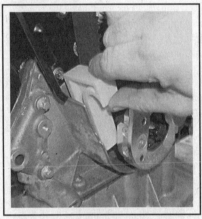

8.21b Install the timing belt wedge between the belt and the oil pump cover flange

8.23 Align the timing belt paint marks with the camshaft sprocket marks and the notches on the rear covers (number 1 and 2)

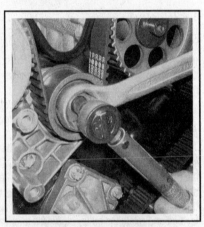

8.25 Tighten the guide bolt slightly, allowing it to move for adjustment purposes

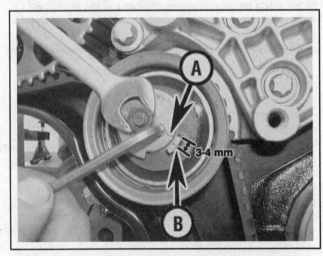

8.27 Adjust the tensioner center alignment mark (A) 1/8-inch (3 mm) above the mark on the spring loaded idler on the tensioner (B)

20 Route the timing belt over the lower timing belt guide pulley (see illustration).

21 Install the timing belt around the crankshaft sprocket. Align the mark on the timing belt with the sprocket mark (see illustration). Install the timing belt wedge to prevent the timing belt from moving out of position (see illustration).

22 Install the timing belt around the timing belt tensioner.

23 Install the timing belt over the number 1 and 2 camshaft sprockets on the rear cylinder head (see illustration). Align the marks on the timing belt with the marks on the timing belt rear cover.

24 Install the timing belt under the upper timing belt guide.

25 Tighten the upper timing belt guide pulley with enough resistance to allow the pulley to move slightly (see illustration).

26 Reinstall the crankshaft locking tool (see Steps 4 and 5).

27 Adjust the tensioner center alignment mark 1/8 inch (3 mm) above the mark on the spring loaded idler on the tensioner (see illustration).

28 Remove the timing belt wedge.

29 Rotate the upper timing belt guide pulley so the mark made on the hex of the pulley is at the 10 o'clock position, then tighten the pulley bolt slightly. Do not fully torque the bolt at this time. The number 1

8.32 Make sure the timing gauge marks are aligned with the sprocket marks

8.36 Use a wrench to hold the guide pulley stationary while tightening the pulley bolt

and 2 (rear) camshaft sprocket marks should be very close to the number 1 TDC marks.

30 Rotate the lower timing belt guide pulley to the 11 o'clock position and tighten the pulley bolt slightly. The number 3 and 4 (front) camshaft sprocket marks should be very close to the number 1 TDC marks.

31 Remove the camshaft locking tools from the camshaft sprockets.

32 Install the timing belt gauge over number 3 and 4 camshaft sprockets (front) (see illustration).

33 Using hand pressure, pull the timing belt between the crankshaft sprocket and the timing belt tensioner. This will remove any slack in the timing belt. The alignment marks on the timing belt gauge should not move more than 1 mm or the width of the timing mark. If the timing belt gauge marks are incorrect, loosen the lower timing belt guide pulley bolt and rotate the pulley clockwise to allow the timing marks to retard slightly. Recheck the marks by pulling on the timing belt between the crankshaft sprocket and the timing belt tensioner.

34 Remove the crankshaft locking tool.

✲✲ CAUTION:

Do not rotate the engine counterclockwise or the timing belt slack between the crankshaft sprocket and the tensioner will be altered causing timing belt adjustment problems.

35 Rotate the engine 1-3/4 turns clockwise and install the crankshaft locking tool. Continue turning the crankshaft until number 1 TDC is located. Lock the tool against the water pump flange.

36 Adjust the lower timing belt guide pulley by rotating the pulley counterclockwise until the timing marks on the number 3 and 4 camshaft sprockets (front) are aligned with the timing belt gauge marks. Use a wrench to hold the guide pulley stationary while tightening the pulley bolt (see illustration). Torque the lower timing belt guide pulley bolt to 15 ft-lbs (temporary torque specification).

37 Remove the crankshaft locking tool.

✲✲ CAUTION:

Do not rotate the engine counterclockwise or the timing belt slack between the crankshaft sprocket and the tensioner will be altered causing timing belt adjustment problems.

38 Rotate the engine 1-3/4 turns clockwise and install the crankshaft locking tool. Continue turning the crankshaft until number 1 TDC is located. Lock the tool against the water pump flange.

39 Recheck the camshaft sprocket marks on camshaft sprockets number 3 and 4 and realign them if necessary. If they are aligned, continue with the next step.

40 Install the timing belt gauge over number 1 and 2 camshaft sprockets (rear) (see illustration 8.32).

41 Install the camshaft sprocket locking tool onto the number 3 and 4 camshaft sprockets (see illustration 8.6c).

42 Adjust the upper timing belt guide pulley by rotating the pulley counterclockwise until the timing marks on the number 1 and 2 camshaft sprockets (rear) are aligned with the timing belt gauge marks. Use a wrench to hold the guide pulley stationary while tightening the pulley bolt. Torque the lower timing belt guide pulley bolt to 15 ft-lbs (temporary torque specification).

43 Remove the crankshaft locking tool and the camshaft sprocket locking tool.

✲✲ CAUTION:

Do not rotate the engine counterclockwise or the timing belt slack between the crankshaft sprocket and the tensioner will be altered causing timing belt adjustment problems.

44 Rotate the engine 1-3/4 turns clockwise and install the crankshaft locking tool. Continue turning the crankshaft until number 1 TDC is located. Lock the tool against the water pump flange.

45 Recheck the camshaft sprocket marks on camshaft sprocket numbers 1 and 2 and realign them if necessary. If they are aligned, continue with the next step.

46 Tighten the upper and lower timing belt guide pulleys to the torque listed in this Chapter's Specifications.

47 Adjust the tensioner center alignment mark 1/8-inch (3 mm) above the mark on the spring loaded idler on the tensioner (see illustration 8.27). Torque the timing belt tensioner bolt to the torque listed in this Chapter's Specifications.

48 Recheck all the alignment marks. Repeat the procedure if the alignment marks are not correct.

49 Install the timing belt cover (see Section 7).

50 Installation is the reverse of removal.

V6 ENGINE 2B-11

Engines after 578512

51 With the camshaft locking tools installed onto the camshafts sprockets, check the distance between the camshaft timing marks using the green test belt included in the timing belt tool kit (J-42069).

52 Install the timing belt over the number 1 and 2 camshaft sprockets on the rear cylinder head (see illustration 8.23). Align the timing belt marks with the sprocket timing marks and install the timing belt.

53 Install the timing belt around the timing belt tensioner.

54 Install the timing belt under the upper timing belt guide pulley (see illustration 8.25).

55 Install the timing belt over the number 3 and 4 camshaft sprockets on the front cylinder head (see illustration 8.19). Align the timing belt marks with the sprocket timing marks and install the timing belt.

56 Install the timing belt around the crankshaft sprocket. Align the mark on the timing belt with the sprocket mark (see illustration 8.21a). Install the timing belt wedge to prevent the timing belt from moving out of position (see illustration 8.21b).

57 Rotate the crankshaft counterclockwise to gather the belt slack on the lower timing belt guide pulley side of the engine. Route the timing belt over the lower timing belt guide pulley (see illustration 8.20).

58 Remove the timing belt wedge.

59 Install the crankshaft locking tool (see Steps 4 and 5).

60 Remove the camshaft sprocket locking tools.

61 Install the timing belt gauge over number 3 and 4 camshaft sprockets (front) (see illustration 8.32).

62 Install the camshaft sprocket locking tool onto the 3 and 4 camshaft sprockets (front) if the marks are aligned.

63 Rotate the number 1 camshaft sprocket timing mark counterclockwise to remove the timing belt slack from the camshaft sprockets. The timing marks should be slightly retarded.

64 To tighten the timing belt tension around the camshaft sprockets, rotate the upper timing belt guide pulley counterclockwise until the timing marks are aligned.

65 Install the camshaft sprocket locking tool onto the 1 and 2 camshaft sprockets (rear) if the marks are aligned.

66 Tighten the upper timing belt guide pulley to the torque listed in this Chapter's Specifications.

67 Adjust the tensioner center alignment mark 1/8-inch (3 mm) above the mark on the spring loaded idler on the tensioner (see illustration 8.27).

68 Remove the crankshaft locking tool and the camshaft sprocket locking tool.

✱✱ CAUTION:

Do not rotate the engine counterclockwise or the timing belt slack between the crankshaft sprocket and the tensioner will be altered causing timing belt adjustment problems.

69 Rotate the engine 1-3/4 turns clockwise and install the crankshaft locking tool. Continue turning the crankshaft until number 1 TDC is located. Lock the tool against the water pump flange.

70 Recheck the camshaft sprocket marks and realign them if necessary. If they are aligned, continue with the next step.

71 Adjust the tensioner center alignment mark 1/8-inch (3 mm) above the mark on the spring loaded idler on the tensioner (see illustration 8.27). Tighten the timing belt tensioner bolt to the torque listed in this Chapter's Specifications.

72 Recheck the camshaft sprocket timing marks and the crankshaft sprocket marks.

73 Install the timing belt cover (see Section 7).

74 Reconnect the battery (see Chapter 5, Section 1).

75 The remainder of installation is the reverse of removal.

9 Crankshaft pulley and front oil seal - removal and installation

REMOVAL

▶ Refer to illustrations 9.7, 9.8a and 9.8b

1 Disconnect the cable from the negative battery terminal (see Chapter 5).

2 Loosen the right front wheel lug nuts. Raise the vehicle and support it securely on jackstands. Remove the wheel

3 Remove the inner fender splash shield from the passenger side of the vehicle (see illustration 7.4).

4 Remove the crankshaft pulley bolts using the pressure from the drivebelt to prevent the engine from rotating.

➡ Note: If the engine rotates during bolt removal, lock the flywheel using a special tool to prevent engine rotation.

5 Remove the drivebelt (see Chapter 1).

6 Remove the timing belt cover (see Section 7) and the timing belt (see Section 8).

7 Use a pin spanner (see illustration) to hold the crankshaft sprocket stationary, then remove the crankshaft center bolt.

9.7 Here's a homemade pin spanner being used to immobilize the crankshaft sprocket

2B-12 V6 ENGINE

9.8a Remove the crankshaft sprocket . . .

9.8b . . . and the hub from the crankshaft

8 Remove the crankshaft sprocket and the seal hub from the crankshaft (see illustrations).

9 Use a seal puller to remove the crankshaft front oil seal. A screwdriver may be used instead, if the tip is wrapped with tape to avoid scratching the crankshaft.

10 Clean the seal bore and check it for nicks or gouges. Also examine the area of the hub that rides in the seal for signs of abnormal wear or scoring.

INSTALLATION

11 Coat the lip of the new seal with clean engine oil and drive it into the bore with a seal driver or a deep socket slightly smaller in diameter than the seal. The open side of the seal faces into the engine.

12 Lubricate the oil seal contact surface of the crankshaft seal hub (see illustration 9.8b) with clean engine oil.

13 Install the crankshaft sprocket. Install the center bolt and tighten it to the torque listed in this Chapter's Specifications.

14 Reconnect the battery (see Chapter 5, Section 1).

15 The remainder of installation is the reverse of removal.

10 Camshafts and cam followers - removal, inspection and installation

REMOVAL

1 Disconnect the cable from the negative battery terminal (see Chapter 5).
2 Set the engine to 60-degrees before TDC.

※ CAUTION:

Do not remove the camshafts with the engine at TDC number 1 or the valves could be damaged.

3 Remove the valve cover(s) (see Section 4).
4 Remove the timing belt (see Section 8).
5 Remove the camshaft sprockets (see illustration 8.15). Mark each camshaft sprocket in its original position to insure correct reassembly. Discard the bolts.

→Note: Mark the position of the camshaft dowels to the camshaft sprockets. This will insure accurate reassembly of the camshaft sprockets later.

6 Mark the bearing caps in the original positions and remove them from the cylinder head. Each camshaft cap is designated with a letter and a number.

※ CAUTION:

The caps must be installed in their original locations. Keep all parts from each camshaft together; never mix parts from one camshaft with those for another. Loosen each bearing cap nut slowly and evenly, starting from the center bearing caps and working out, allowing the camshaft to lift from the cylinder head parallel to the surface of the cylinder head.

Front cylinder head camshafts and followers

♦ Refer to illustrations 10.8 and 10.9

7 The camshaft bearing caps on the front cylinder head are designated with an R plus the number of the cap. Mark down on paper the location of each camshaft bearing cap to insure correct reassembly. If no marks are present, or they are hard to see, make your own - the bearing caps must be replaced in their original positions.

8 Remove the camshaft from the cylinder head (see illustration). Be sure to remove the camshaft seal with the camshaft.

9 Obtain twelve small, clean containers, and number them 1 to 12. Remove each cam follower and place them in the containers (see illustration). Do not interchange the cam followers, or the rate of wear will be much increased. Store the followers right-side up to prevent the oil from draining out of the hydraulic lash adjuster mechanisms.

V6 ENGINE 2B-13

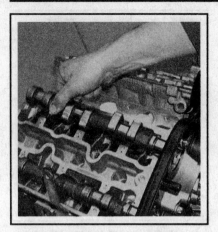

10.8 Lift the camshaft from the cylinder head

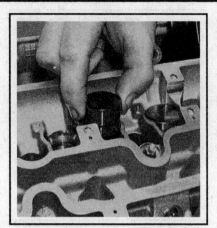

10.9 Lift the cam follower from the cylinder head bore. A magnet or suction cup may be required to extract them

10.10 The letter and number designations are located on the same side of the camshaft housing

Rear cylinder head camshafts and followers

▶ Refer to illustration 10.10

10 The camshaft bearing caps on the rear cylinder head are designated with an L plus the number of the cap (see illustration). Mark down on paper the location of each camshaft bearing cap to insure correct reassembly. If no marks are present, or they are hard to see, make your own - the bearing caps must be reinstalled in their original positions.

11 Remove the camshaft from the cylinder head (see illustration 10.8). Be sure to remove the camshaft seal with the camshaft.

12 Obtain twelve small, clean containers, and number them 1 to 12. Remove each cam follower and place them in the containers (see illustration 10.9). Do not interchange the cam followers, or the rate of wear will be much increased. Store the followers right-side up to prevent the oil from draining out of the hydraulic lash adjuster mechanisms.

INSPECTION

▶ Refer to illustrations 10.14, 10.15, 10.16, 10.18, 10.19a and 10.19b

13 With the camshafts and cam followers removed, check each for signs of obvious wear (scoring, pitting, etc) and for roundness and replace if necessary.

14 Measure the outside diameter of each cam follower - take measurements at the top and bottom of each cam follower, then a second set at right-angles to the first; if any measurement is significantly different from the others, the cam follower is tapered or out-of round and must be replaced (see illustration). If the necessary equipment is available, measure the inside diameter of the corresponding cylinder head bore. Check the specifications listed in this Chapter; if the cam followers or the cylinder head bores are excessively worn, new cam followers and/or a new cylinder head may be required.

15 Visually examine the camshaft lobes for score marks, pitting, galling (wear due to rubbing) and evidence of overheating (blue, discolored areas) (see illustration). Look for flaking away of the hardened surface layer of each lobe. If any such signs are evident, replace the component concerned.

16 Measure the lobe height of each cam lobe on the intake camshaft, and record your measurements (see illustration). Compare the measurements for excessive variation; if the lobe heights vary more than 0.005 inch (0.125 mm), replace the camshaft. Compare the lobe height measurements on the exhaust camshaft and follow the same procedure. Do not compare intake camshaft lobe heights with exhaust camshaft lobe heights, as they are different. Only compare intake lobes with intake lobes and exhaust lobes with exhaust lobes.

10.14 Measure the cam follower outside diameter at several points

10.15 Check the cam lobes for pitting, excessive wear, and scoring. If scoring is excessive, as shown here, replace the camshaft

10.16 Measure the camshaft lobe height (greatest dimension) with a micrometer

2B-14 V6 ENGINE

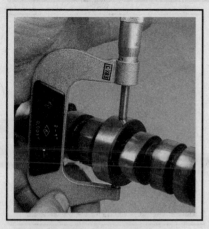

10.18 Measure each journal diameter with a micrometer. If any journal is less than the specified minimum, replace the camshaft

10.19a Lay a strip of Plastigage on each camshaft journal, in line with the camshaft

10.19b Compare the width of the crushed Plastigage to the scale on the package to determine the journal oil clearance

17 Examine the camshaft bearing journals and the cylinder head bearing surfaces for signs of obvious wear or pitting. If any such signs are evident, consult an automotive machine shop for advice. Also check that the bearing oilways in the cylinder head are clear.

18 Using a micrometer, measure the diameter of each journal at several points (see illustration). If the diameter of any one journal is less than the specified value, replace the camshaft.

19 To check the bearing journal running clearance, remove the cam followers, use a suitable solvent and a clean lint-free rag to clean carefully all bearing surfaces, then install the camshafts and bearing caps with a strand of Plastigage across each journal (see illustration). Tighten the bearing cap bolts in the given sequence (see Installation) to the specified torque setting (do not rotate the camshafts), then remove the bearing caps and use the scale provided to measure the width of the compressed strands (see illustration). Scrape off the Plastigage with your fingernail or the edge of a credit card - don't scratch or nick the journals or bearing caps.

20 If the running clearance of any bearing is found to be worn to beyond the specified service limits, install a new camshaft and repeat the check; if the clearance is still excessive, the cylinder head must be replaced.

21 To check camshaft endplay, remove the cam followers, clean the bearing surfaces carefully and install the camshafts and bearing caps. Tighten the bearing cap bolts to the specified torque, then measure the endplay using a dial indicator mounted on the cylinder head so that its tip bears on the camshaft right-hand end.

22 Tap the camshaft fully towards the gauge, zero the gauge, then tap the camshaft fully away from the gauge and note the gauge reading. If the endplay measured is found to be at or beyond the specified service limit, install a new camshaft and repeat the check; if the clearance is still excessive, the cylinder head must be replaced.

INSTALLATION

23 Make sure the engine is positioned at 60-degrees before TDC.

CAUTION:

Do not install the camshafts with the engine at TDC number 1 or the valves and pistons will be damaged.

24 On reassembly, liberally oil the cylinder head cam follower bores and the cam followers. Carefully install the cam followers in the cylinder head, ensuring that each cam follower is replaced into its original bore, and is the correct way up.

25 It is highly recommended that new camshaft oil seals are installed, as a precaution against later failure. The new seals are installed with the camshaft and bearing cap installation.

26 Liberally oil the camshaft bearings (not the caps) and lobes.

Front cylinder head camshafts

27 Ensuring that each camshaft is in its original location, install the camshafts, locating the pin on the front of the exhaust camshaft at 12 o'clock and the pin on the intake camshaft at 7 o'clock. This position will allow the least spring loaded tension on the valves and camshaft on initial installation.

Rear cylinder head camshafts

28 Ensuring that each camshaft is in its original location, install the camshafts, locating the pin on the front of the exhaust camshaft at 1 o'clock and the pin in the intake camshaft at 11 o'clock. This position will allow the least spring loaded tension on the valves and camshaft on initial installation.

Both cylinder head camshafts

▶ Refer to illustrations 10.31a and 10.31b

29 Apply a small amount of anaerobic sealant between the camshaft cap and the cylinder head on the front caps. This will prevent oil leaks.

30 Apply a film of camshaft installation lube to the journals and lobes, then install each of the camshaft bearing caps to its previously-noted position, so that its numbered sides are aligned (see illustration 10.10).

31 Ensuring that each cap is kept square to the cylinder head as it is tightened down, tighten the camshaft bearing cap bolts slowly and by one turn at a time, in the proper sequence, until each cap touches the cylinder head (see illustrations). Tighten the camshaft caps to the torque listed in this Chapter's Specifications.

32 Wipe off all surplus sealant, so that none is left to find its way into any oilways. Follow the sealant manufacturer's instructions as to the time needed for curing.

33 Install the sprockets to the camshafts (see Section 8).

V6 ENGINE 2B-15

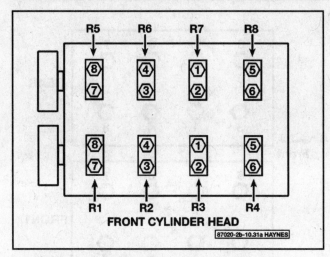

10.31a Camshaft bearing cap marks and tightening sequence for the front cylinder head

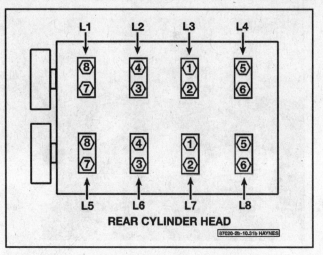

10.31b Camshaft bearing cap marks and tightening sequence for the rear cylinder head

34 The remainder of the reassembly procedure, including replacement of the timing belt and setting the valve timing, is as described in Section 8.

35 The remainder of installation is the reverse of removal.
36 Reconnect the battery (see Chapter 5, Section 1).

11 Cylinder heads - removal and installation

❋ CAUTION:

The engine must be completely cool when the cylinder heads are removed. Failure to allow the engine to cool off could result in cylinder head warpage.

➡ Note: Cylinder head removal is a difficult and time-consuming job requiring several special tools; read through the procedure and obtain the necessary tools before beginning.

REMOVAL

1 Disconnect the cable from the negative battery terminal (see Chapter 5).
2 Drain the cooling system (see Chapter 1).
3 Remove the valve covers (see Section 4).
4 Remove the upper and lower intake manifolds and the intake manifold spacer (see Section 5).
5 Remove the coolant crossover housing (see Chapter 3).
6 Remove the exhaust manifolds (see Section 6).
7 Remove the timing belt (see Section 8).
8 Remove the upper radiator hose and the coolant extension housing (see Chapter 3).
9 If the cylinder head is to be completely overhauled, refer to Section 10 for removal of the camshafts.
10 Following the reverse of the tightening sequence (see illustration 11.18b), use a breaker bar to remove the cylinder head bolts. Loosen the bolts in sequence 1/4-turn at a time.
11 Use a pry bar at the corners of the cylinder head-to-engine block mating surface to break the cylinder head gasket seal. Do not pry between the cylinder head and engine block in the gasket sealing area.

12 Lift the cylinder head(s) off the engine. If the head is stuck, place a wood block against the end and strike the wood block with a hammer.

❋ CAUTION:

The cylinder heads are aluminum; store them on wood blocks to prevent damage to the gasket sealing surfaces.

13 Remove the cylinder head gasket(s). Before removing, note which gasket goes on which side (they are different and cannot be interchanged).

INSTALLATION

▸ Refer to illustrations 11.17 and 11.18

❋ CAUTION:

New cylinder head bolts must be used for reassembly. Failure to use new bolts may result in cylinder head gasket leakage and engine damage.

14 The mating surfaces of the cylinder heads and engine block must be perfectly clean when the cylinder heads are installed. Use a gasket scraper to remove all traces of carbon and old gasket material, then clean the mating surfaces with lacquer thinner or acetone. If there's oil on the mating surfaces when the cylinder heads are installed, the gaskets may not seal correctly and leaks may develop. When working on the engine block, cover the open areas of the engine with shop rags to keep debris out during repair and reassembly. Use a vacuum cleaner to remove any debris that falls into the cylinders.

2B-16 V6 ENGINE

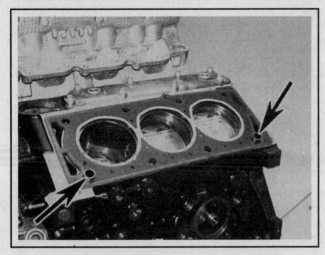

11.17 Install the head gasket over the locating dowels

※ **CAUTION:**

Do not use abrasive wheels or sharp metal scrapers on the heads or block surface. Use a plastic scraper and chemical gasket remover, or the head gasket surfaces could have future leaks.

15 Check the engine block and cylinder head mating surfaces for nicks, deep scratches and other damage.
16 Use a tap of the correct size to chase the threads in the cylinder head bolt holes. Dirt, corrosion, sealant and damaged threads will affect torque readings.
17 Make sure the part numbers on the head gaskets face UP. Install the head gasket, locating the dowels in the alignment holes in the gaskets (see illustration).
18 Carefully position the cylinder heads on the engine block without disturbing the gaskets. Install the NEW cylinder head bolts. The cylin-

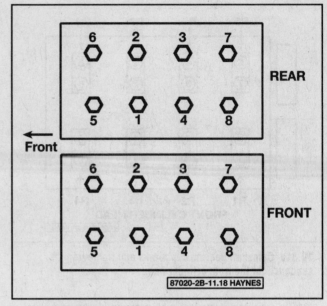

11.18 Cylinder head bolt tightening sequence

der head bolts are torque-to-yield design and they cannot be reused. Following the recommended sequence (see illustration), tighten the cylinder head bolts, in five steps, to the torque listed in this Chapter's Specifications.

➡ **Note: The method used for the cylinder head bolt tightening procedure is referred to as "torque-angle" or "torque-to-yield" method. Follow the procedure exactly. Tighten the bolts in Step 1 using a torque wrench, then use a breaker bar and a torque-angle adapter for Steps 2 through 5.**

19 The remaining installation steps are the reverse of removal.
20 Change the engine oil and filter (Chapter 1), then start the engine and check carefully for oil and coolant leaks.
21 Reconnect the battery (see Chapter 5, Section 1).

12 Oil pan - removal and installation

REMOVAL

1 Disconnect the negative battery cable from the battery (see Chapter 5).
2 Refer to Chapter 1 and drain the engine oil and remove the oil filter.
3 Raise the vehicle and support it securely on jackstands.
4 Remove the bracket assembly from the front section of the oil pan.
5 Remove the bolts from the transaxle bellhousing brace at the oil pan.
6 Remove the oil level dipstick.
7 Remove the oil pan mounting bolts.
8 Carefully separate the oil pan from the engine block. Don't pry between the engine block and oil pan or damage to the sealing surfaces may result and oil leaks could develop. Instead, dislodge the oil pan with a large rubber mallet or a wood block and a hammer.

INSTALLATION

▸ **Refer to illustrations 12.12a and 12.12b**

9 Use a gasket scraper or putty knife to remove all traces of old gasket material and sealant from the pan and engine block.

※ **CAUTION:**

Be careful not to gouge the oil pan or block, or oil leaks could develop later.

10 Clean the mating surfaces with lacquer thinner or acetone. Make sure the bolt holes in the engine block are clean.
11 Install special alignment dowels (tool J-44715) into the designated holes (one forward and one rear) of the engine block reinforcement. These special tools allow the oil pan to remain aligned, without sliding around.
12 Apply a 2 mm bead of RTV sealant to the flange of the oil

V6 ENGINE 2B-17

12.12a Apply a small amount of RTV sealant at the front and rear seams, where the oil pump meets the engine block . . .

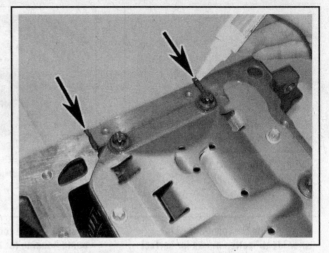

12.12b . . . and the rear seal retainer meets the engine block

pan, 3 mm from the inside edge. Apply a small amount of RTV sealant at the front and rear seams on the engine block, where the rear seal retainer and the oil pump meet the engine block (see illustrations).

13 Carefully position the oil pan against the engine block, then install the front bracket assembly and tighten the bolts finger tight. Install the bolts into the bellhousing brace/oil pan and tighten them finger tight. Apply thread locking compound to the threads of the oil pan bolts, then install them. Remove the alignment dowels and tighten the bolts to the torque listed in this Chapter's Specifications. Start at the center of the oil pan and work out toward the ends in a spiral pattern.

14 The remaining steps are the reverse of removal.

✼✼ CAUTION:

Don't forget to refill the engine with oil before starting it (see Chapter 1).

15 Reconnect the battery (see Chapter 5, Section 1).
16 Start the engine and check carefully for oil leaks at the oil pan. Drive the vehicle and check again.

13 Oil pump - removal, inspection and installation

REMOVAL

1 Raise the front of the vehicle and support it securely on jackstands.

13.12 Remove the oil pump cover from the oil pump housing

2 Disconnect the cable from the negative terminal of the battery (see Chapter 5, Section 1).
3 Drain the engine oil (see Chapter 1).
4 Remove the drivebelt (see Chapter 1).
5 Remove the timing belt cover (see Section 7) and the timing belt (see Section 8).
6 Remove the rear timing belt cover.
7 Remove the air conditioning compressor (see Chapter 3) and tie it out of the way with a piece of wire. Don't disconnect the refrigerant lines from the compressor.
8 Remove the alternator (see Chapter 5).
9 Remove the oil pan-to-oil pump mounting bolts (see Section 12).
10 Remove the crankshaft sprocket (see Section 9).
11 Remove the oil pump mounting bolts and separate the pump from the engine block.

INSPECTION

▶ **Refer to illustrations 13.12, 13.13a, 13.13b, 13.17a, 13.17b, 13.17c and 13.17d**

12 Working on the backside of the oil pump, loosen the oil pump cover mounting bolts gradually and evenly. Loosen each bolt in several steps. Remove the oil pump cover (see illustration).

2B-18 V6 ENGINE

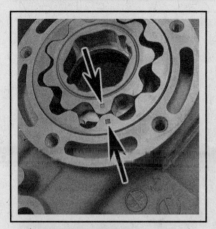

13.13a Note the locations of the rotor identification marks for correct reassembly - the rotor marks must face the same direction

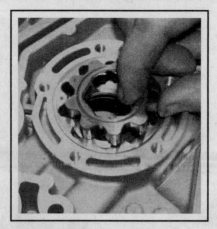

13.13b Lift the rotors from the oil pump housing for inspection

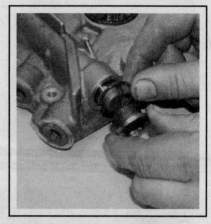

13.17a Remove the oil pressure relief valve components from the oil pump housing for inspection

13 Note any identification marks on the rotors and withdraw the rotors from the pump body (see illustrations).

14 Thoroughly clean and dry the components.

15 Inspect the rotors for obvious wear or damage. If either rotor, the pump body or the cover is scored or damaged, the complete oil pump assembly must be replaced.

16 If the oil pump components are in acceptable condition, dip the rotors in clean engine oil and install them into the pump body with any identification marks positioned as noted during disassembly.

17 Remove the oil pressure relief valve components from the oil pump body. Thoroughly clean and dry the components. Inspect the components for obvious wear or damage. Install them in the correct order (see illustrations).

INSTALLATION

▸ **Refer to illustrations 13.18 and 13.20**

18 Using a seal driver, install a new seal into the bore of the pump. Install a new gasket onto the engine block (see illustration). Coat the oil pump side of the gasket with anaerobic sealant. Be careful to apply a thin layer of sealant and keep any sealant from entering the oil passageways.

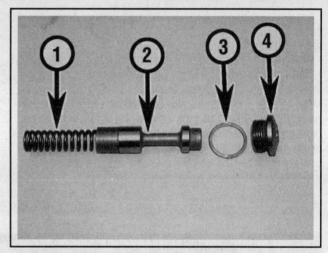

13.17b Oil pressure relief valve components

| 1 | Spring | 3 | Washer |
| 2 | Plunger | 4 | Plug |

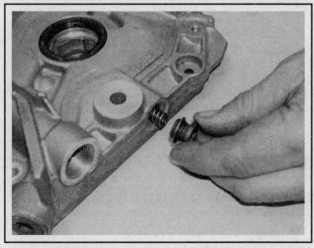

13.17c Remove the oil pressure control valve components from the oil pump housing

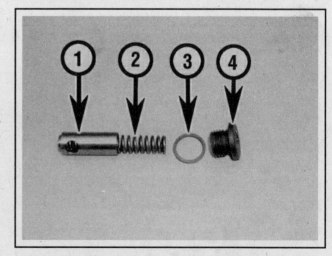

13.17d Oil control valve components

| 1 | Plunger | 3 | Sealing washer |
| 2 | Spring | 4 | Plug |

V6 ENGINE 2B-19

13.18 Install the oil pump housing gasket - do not allow sealant to leak into the oil passageways

13.20 Install the oil pump housing onto the engine block

19 Prime the oil pump prior to installation. Pour clean oil into the pick-up port and turn the pump by hand.

20 Install the oil pump to the engine (see illustration). Apply a threadlocking compound to the oil pump bolts and tighten the bolts to the torque listed in this Chapter's Specifications.

➡Note: The oil pump mounting bolts must be retorqued after the alternator bracket has been installed onto the oil pump/engine block.

21 The remainder of installation is the reverse of removal procedure.
22 Reconnect the battery (see Chapter 5, Section 1).
23 Fill the engine with the correct type and quantity of oil. Start the engine and check for leaks.

14 Driveplate - removal and installation

This procedure is essentially the same as for the four cylinder engine. Refer to Part A and follow the procedure outlined there. However, use bolt torque listed in this Chapter's Specifications.

15 Rear main oil seal - replacement

This procedure is essentially the same as for the four cylinder engine. Refer to Part A and follow the procedure outlined there.

16 Engine mounts - check and replacement

This procedure is essentially the same as for the four cylinder engine. Refer to part A and follow the procedure outlined there.

2B-20 V6 ENGINE

Specifications

General

Displacement	182 cubic inches (3.0 liters)
Bore and stroke	3.44 x 3.40 inches (86.0 x 85.0 mm)
Cylinder numbers (front to rear)	
Front cylinder head	2-4-6
Rear cylinder head	1-3-5
Firing order	1-2-3-4-5-6

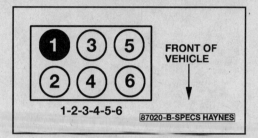

Cylinder location and firing order

Camshaft

Lobe lift (intake and exhaust)	Not available
Allowable lobe lift variation	0.005 inch (0.125 mm)
Lifter diameter	1.2976 to 1.2982 inch (32.959 to 32.975 mm)
Lifter clearance	0.0010 to 0.0026 inch (0.025 to 0.066 mm)
Endplay	0.0016 to 0.0057 inch (0.040 to 0.144 mm)
Journal diameter (all)	1.099 to 1.100 inches (27.939 to 27.960 mm)
Journal-to-bearing (oil) clearance	
Standard	0.0015 to 0.0020 inch (0.038 to 0.050 mm)
Service limit	0.002 inch maximum (0.05 mm)

Torque specifications

	Ft-lbs (unless otherwise indicated)	Nm
Camshaft sprocket bolts*		
Step 1	37	50
Step 2	Tighten an additional 60-degrees	
Step 3	Tighten an additional 15-degrees	
Camshaft cap bolts	71 in-lbs	8
Drive belt tensioner bolts	18	25
Crankshaft sprocket bolt		
Step 1	184	250
Step 2	Tighten an additional 45-degrees	
Step 3	Tighten an additional 15-degrees	
Cylinder head bolts*		
Step 1	18	25
Step 2	Tighten an additional 90-degrees	
Step 3	Tighten an additional 90-degrees	
Step 4	Tighten an additional 90-degrees	
Step 5	Tighten an additional 15-degrees	
Crankshaft pulley bolts	15	20
Driveplate bolts		
Step 1	48	65
Step 2	Tighten an additional 30-degrees	
Step 3	Tighten an additional 15-degrees	
Engine lifting bracket bolts	71 in-lbs	8
Exhaust manifold-to-cylinder head nuts	15	20
Exhaust pipe-to-exhaust manifold nuts	25	30
Engine mount bracket bolts	41	55

V6 ENGINE 2B-21

Torque specifications (continued)

	Ft-lbs (unless otherwise indicated)	Nm
Intake manifold		
Upper intake manifold bolts/nut	71 in-lbs	8
Intake manifold runner clamp bolts	71 in-lbs	8
Lower intake manifold bolts	15	20
Oil pan-to-engine block bolts		
Front bracket-to-oil pan bolts	30	40
Transaxle bellhousing brace-to-oil pan bolts	48	65
Oil pan bolts	132 in-lbs	15
Oil pump-to-engine block mounting bolts	80 in-lbs	8
Oil pick-up tube bolts	71 in-lbs	8
Power steering pump pulley bolts	71 in-lbs	8
Timing belt cover bolts	71 in-lbs	8
Timing belt upper guide pulley bolt	30	40
Timing belt lower guide pulley bolt	30	40
Timing belt tensioner bolt	15	20
Valve cover bolts	71 in-lbs	8
Water pump pulley bolts	71 in-lbs	8

* *Bolt(s) must be replaced.*

Notes

2C
GENERAL ENGINE OVERHAUL PROCEDURES

Section
1. General information - engine overhaul
2. Oil pressure check
3. Cylinder compression check
4. Vacuum gauge diagnostic checks
5. Engine rebuilding alternatives
6. Engine removal - methods and precautions
7. Engine - removal and installation
8. Engine overhaul - disassembly sequence
9. Pistons and connecting rods - removal and installation
10. Crankshaft - removal and installation
11. Engine overhaul - reassembly sequence
12. Initial start-up and break-in after overhaul

Reference to other Chapters
CHECK ENGINE light on – See Chapter 6

2C-2 GENERAL ENGINE OVERHAUL PROCEDURES

1 General information - engine overhaul

▶ **Refer to illustrations 1.1, 1.2, 1.3, 1.4, 1.5 and 1.6**

Included in this portion of Chapter 2 are general information and diagnostic testing procedures for determining the overall mechanical condition of your engine.

The information ranges from advice concerning preparation for an overhaul and the purchase of replacement parts and/or components to detailed, step-by-step procedures covering removal and installation.

The following Sections have been written to help you determine whether your engine needs to be overhauled and how to remove and install it once you've determined it needs to be rebuilt. For information concerning in-vehicle engine repair, see Chapter 2A or 2B.

The Specifications included in this Part are general in nature and include only those necessary for testing the oil pressure and checking the engine compression. Refer to Chapter 2A or 2B for additional engine Specifications.

It's not always easy to determine when, or if, an engine should be completely overhauled, because a number of factors must be considered.

High mileage is not necessarily an indication that an overhaul is needed, while low mileage doesn't preclude the need for an overhaul. Frequency of servicing is probably the most important consideration. An engine that's had regular and frequent oil and filter changes, as well as other required maintenance, will most likely give many thousands of miles of reliable service. Conversely, a neglected engine may require an overhaul very early in its service life.

Excessive oil consumption is an indication that piston rings, valve seals and/or valve guides are in need of attention. Make sure that oil leaks aren't responsible before deciding that the rings and/or guides are bad. Perform a cylinder compression check to determine the extent of the work required (see Section 3). Also check the vacuum readings under various conditions (see Section 4).

Check the oil pressure with a gauge installed in place of the oil pressure sending unit and compare it to this Chapter's Specifications (see Section 2). If it's extremely low, the bearings and/or oil pump are probably worn out.

Loss of power, rough running, knocking or metallic engine noises, excessive valve train noise and high fuel consumption rates may also point to the need for an overhaul, especially if they're all present at the same time. If a complete tune-up doesn't remedy the situation, major mechanical work is the only solution.

An engine overhaul involves restoring the internal parts to the specifications of a new engine. During an overhaul, the piston rings are replaced and the cylinder walls are reconditioned (rebored and/or honed) (see illustrations 1.1 and 1.2). If a rebore is done by an automotive machine shop, new oversize pistons will also be installed. The

1.1 An engine block being bored. An engine rebuilder will use special machinery to recondition the cylinder bores

1.2 If the cylinders are bored, the machine shop will normally hone the engine on a machine like this

GENERAL ENGINE OVERHAUL PROCEDURES   2C-3

1.3 A crankshaft having a main bearing journal ground

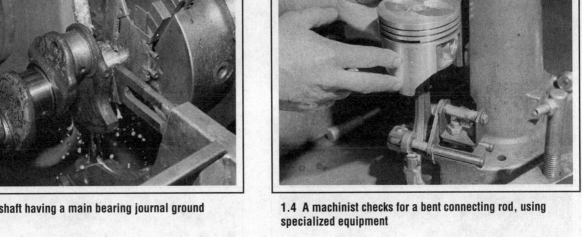

1.4 A machinist checks for a bent connecting rod, using specialized equipment

main bearings, connecting rod bearings and camshaft bearings are generally replaced with new ones and, if necessary, the crankshaft may be reground to restore the journals (see illustration 1.3). Generally, the valves are serviced as well, since they're usually in less-than-perfect condition at this point. While the engine is being overhauled, other components, such as the distributor, starter and alternator, can be rebuilt as well. The end result should be similar to a new engine that will give many trouble free miles.

→**Note: Critical cooling system components such as the hoses, drivebelts, thermostat and water pump should be replaced with new parts when an engine is overhauled. The radiator should be checked carefully to ensure that it isn't clogged or leaking (see Chapter 3). If you purchase a rebuilt engine or short block, some rebuilders will not warranty their engines unless the radiator has been professionally flushed. Also, we don't recommend overhauling the oil pump - always install a new one when an engine is rebuilt.**

Overhauling the internal components on today's engines is a difficult and time-consuming task which requires a significant amount of specialty tools and is best left to a professional engine rebuilder (see illustrations 1.4, 1.5 and 1.6). A competent engine rebuilder will handle the inspection of your old parts and offer advice concerning the reconditioning or replacement of the original engine; never purchase parts or have machine work done on other components until the block has been thoroughly inspected by a professional machine shop. As a general rule, time is the primary cost of an overhaul, especially since the vehicle may be tied up for a minimum of two weeks or more. Be aware that some engine builders only have the capability to rebuild the engine you bring them while other rebuilders have a large inventory of rebuilt exchange engines in stock. Also be aware that many machine shops could take as much as two weeks time to completely rebuild your engine depending on shop workload. Sometimes it makes more sense to simply exchange your engine for another engine that's already rebuilt to save time.

1.5 A bore gauge being used to check the main bearing bore

1.6 Uneven piston wear like this indicates a bent connecting rod

2C-4 GENERAL ENGINE OVERHAUL PROCEDURES

2 Oil pressure check

Refer to illustrations 2.2 and 2.3

1 Low engine oil pressure can be a sign of an engine in need of rebuilding. A "low oil pressure" indicator (often called an "idiot light") is not a test of the oiling system. Such indicators only come on when the oil pressure is dangerously low. Even a factory oil pressure gauge in the instrument panel is only a relative indication, although much better for driver information than a warning light. A better test is with a mechanical (not electrical) oil pressure gauge.

2 Locate the oil pressure sending unit on the engine block:
  a) On four-cylinder engines, the oil pressure sending unit is located on the front left side of the engine block near the oil filter assembly (see illustration).
  b) On V6 engines, the oil pressure sending unit is located on the oil pump housing on the timing belt side of the engine, near the oil pan.

3 Unscrew and remove the oil pressure sending unit and screw in the hose for your oil pressure gauge (see illustration). If necessary, install an adapter fitting. Use Teflon tape or thread sealant on the threads of the adapter and/or the fitting on the end of your gauge's hose.

4 Connect an accurate tachometer to the engine, according to the tachometer manufacturer's instructions.

5 Check the oil pressure with the engine running (normal operating temperature) at the specified engine speed, and compare it to this Chapter's Specifications. If it's extremely low, the bearings and/or oil pump are probably worn out.

2.2 On four-cylinder engines, the oil pressure sending unit is located on the front left side of the engine block near the oil filter

2.3 Remove the oil pressure sending unit and install an oil pressure gauge

3 Cylinder compression check

Refer to illustration 3.6

1 A compression check will tell you what mechanical condition the upper end of your engine (pistons, rings, valves, head gaskets) is in. Specifically, it can tell you if the compression is down due to leakage caused by worn piston rings, defective valves and seats or a blown head gasket.

→ **Note: The engine must be at normal operating temperature and the battery must be fully charged for this check.**

2 Begin by cleaning the area around the spark plugs before you remove them (compressed air should be used, if available). The idea is to prevent dirt from getting into the cylinders as the compression check is being done.

3 Remove all of the spark plugs from the engine (see Chapter 1).

4 Block the throttle wide open.

5 Disable the ignition system by unplugging the electrical connector(s) from the coil pack(s) (see Chapter 5). Also disable the fuel system by unplugging the electrical connector in the harness to the fuel injectors (see Chapter 4) or by removing the fuel pump relay.

6 Install a compression gauge in the spark plug hole (see illustration).

3.6 Use a compression gauge with a threaded fitting for the spark plug hole, not the type that requires hand pressure to maintain the seal

GENERAL ENGINE OVERHAUL PROCEDURES 2C-5

7 Crank the engine over at least seven compression strokes and watch the gauge. The compression should build up quickly in a healthy engine. Low compression on the first stroke, followed by gradually increasing pressure on successive strokes, indicates worn piston rings. A low compression reading on the first stroke, which doesn't build up during successive strokes, indicates leaking valves or a blown head gasket (a cracked head could also be the cause). Deposits on the undersides of the valve heads can also cause low compression. Record the highest gauge reading obtained.

8 Repeat the procedure for the remaining cylinders and compare the results to this Chapter's Specifications.

9 Add some engine oil (about three squirts from a plunger-type oil can) to each cylinder, through the spark plug hole, and repeat the test.

10 If the compression increases after the oil is added, the piston rings are definitely worn. If the compression doesn't increase significantly, the leakage is occurring at the valves or head gasket. Leakage past the valves may be caused by burned valve seats and/or faces or warped, cracked or bent valves.

11 If two adjacent cylinders have equally low compression, there's a strong possibility that the head gasket between them is blown. The appearance of coolant in the combustion chambers or the crankcase would verify this condition.

12 If one cylinder is slightly lower than the others, and the engine has a slightly rough idle, a worn lobe on the camshaft could be the cause.

13 If the compression is unusually high, the combustion chambers are probably coated with carbon deposits. If that's the case, the cylinder head(s) should be removed and decarbonized.

14 If compression is way down or varies greatly between cylinders, it would be a good idea to have a leak-down test performed by an automotive repair shop. This test will pinpoint exactly where the leakage is occurring and how severe it is.

4 Vacuum gauge diagnostic checks

♦ Refer to illustrations 4.4 and 4.6

A vacuum gauge provides inexpensive but valuable information about what is going on in the engine. You can check for worn rings or cylinder walls, leaking head or intake manifold gaskets, incorrect carburetor adjustments, restricted exhaust, stuck or burned valves, weak valve springs, improper ignition or valve timing and ignition problems.

Unfortunately, vacuum gauge readings are easy to misinterpret, so they should be used in conjunction with other tests to confirm the diagnosis.

Both the absolute readings and the rate of needle movement are important for accurate interpretation. Most gauges measure vacuum in inches of mercury (in-Hg). The following references to vacuum assume the diagnosis is being performed at sea level. As elevation increases (or atmospheric pressure decreases), the reading will decrease. For every 1,000 foot increase in elevation above approximately 2,000 feet, the gauge readings will decrease about one inch of mercury.

Connect the vacuum gauge directly to the intake manifold vacuum, not to ported (throttle body) vacuum (see illustration). Be sure no hoses are left disconnected during the test or false readings will result.

Before you begin the test, allow the engine to warm up completely. Block the wheels and set the parking brake. With the transaxle in Park, start the engine and allow it to run at normal idle speed.

WARNING:

Keep your hands and the vacuum gauge clear of the fans.

Read the vacuum gauge; an average, healthy engine should normally produce about 17 to 22 in-Hg with a fairly steady needle (see illustration). Refer to the following vacuum gauge readings and what they indicate about the engine's condition:

1 A low steady reading usually indicates a leaking gasket between the intake manifold and cylinder head(s) or throttle body, a leaky vacuum hose, late ignition timing or incorrect camshaft timing. Check ignition timing with a timing light and eliminate all other possible causes, utilizing the tests provided in this Chapter before you remove the timing chain cover to check the timing marks.

2 If the reading is three to eight inches below normal and it fluctuates at that low reading, suspect an intake manifold gasket leak at an intake port or a faulty fuel injector.

3 If the needle has regular drops of about two-to-four inches at a steady rate, the valves are probably leaking. Perform a compression check or leak-down test to confirm this.

4 An irregular drop or down-flick of the needle can be caused by a sticking valve or an ignition misfire. Perform a compression check or leak-down test and read the spark plugs.

5 A rapid vibration of about four in-Hg vibration at idle combined with exhaust smoke indicates worn valve guides. Perform a leak-down test to confirm this. If the rapid vibration occurs with an increase in engine speed, check for a leaking intake manifold gasket or head gasket, weak valve springs, burned valves or ignition misfire.

6 A slight fluctuation, say one inch up and down, may mean ignition problems. Check all the usual tune-up items and, if necessary, run the engine on an ignition analyzer.

4.4 A simple vacuum gauge can be handy in diagnosing engine condition and performance

2C-6 GENERAL ENGINE OVERHAUL PROCEDURES

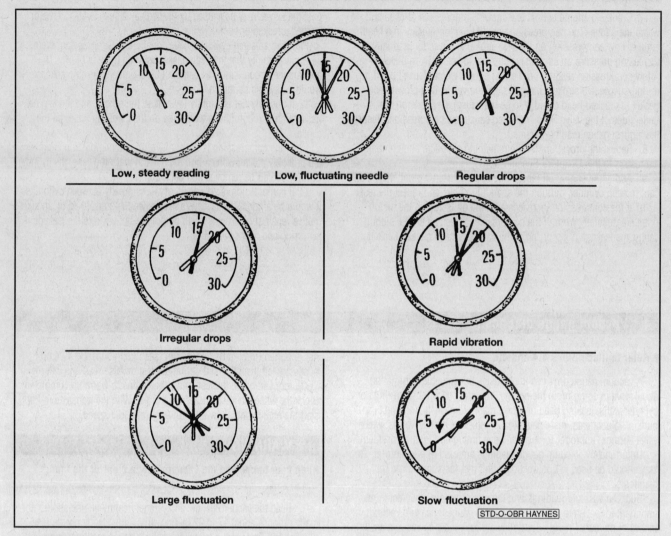

4.6 Typical vacuum gauge readings

7 If there is a large fluctuation, perform a compression or leakdown test to look for a weak or dead cylinder or a blown head gasket.

8 If the needle moves slowly through a wide range, check for a clogged PCV system, incorrect idle fuel mixture, throttle body or intake manifold gasket leaks.

9 Check for a slow return after revving the engine by quickly snapping the throttle open until the engine reaches about 2,500 rpm and let it shut. Normally the reading should drop to near zero, rise above normal idle reading (about 5 in-Hg over) and return to the previous idle reading. If the vacuum returns slowly and doesn't peak when the throttle is snapped shut, the rings may be worn. If there is a long delay, look for a restricted exhaust system (often the muffler or catalytic converter). An easy way to check this is to temporarily disconnect the exhaust ahead of the suspected part and redo the test.

5 Engine rebuilding alternatives

The do-it-yourselfer is faced with a number of options when purchasing a rebuilt engine. The major considerations are cost, warranty, parts availability and the time required for the rebuilder to complete the project. The decision to replace the engine block, piston/connecting rod assemblies and crankshaft depends on the final inspection results of your engine. Only then can you make a cost effective decision whether to have your engine overhauled or simply purchase an exchange engine for your vehicle.

Some of the rebuilding alternatives include:

Individual parts - If the inspection procedures reveal that the engine block and most engine components are in reusable condition, purchasing individual parts and having a rebuilder rebuild your engine may be the most economical alternative. The block, crankshaft and piston/connecting rod assemblies should all be inspected carefully by a machine shop first.

Short block - A short block consists of an engine block with a crankshaft and piston/connecting rod assemblies already installed. All new bearings are incorporated and all clearances will be correct. The existing camshafts, valve train components, cylinder head and external parts can be bolted to the short block with little or no machine shop work necessary.

Long block - A long block consists of a short block plus an oil

GENERAL ENGINE OVERHAUL PROCEDURES 2C-7

pump, oil pan, cylinder head, valve cover, camshaft and valve train components, timing sprockets and chain or gears and timing cover. All components are installed with new bearings, seals and gaskets incorporated throughout. The installation of manifolds and external parts is all that's necessary.

Low mileage used engines - Some companies now offer low mileage used engines which is a very cost effective way to get your vehicle up and running again. These engines often come from vehicles which have been in totaled in accidents or come from other countries which have a higher vehicle turn over rate. A low mileage used engine also usually has a similar warranty like the newly remanufactured engines.

Give careful thought to which alternative is best for you and discuss the situation with local automotive machine shops, auto parts dealers and experienced rebuilders before ordering or purchasing replacement parts.

6 Engine removal - methods and precautions

♦ Refer to illustrations 6.1, 6.2, 6.3 and 6.4

If you've decided that an engine must be removed for overhaul or major repair work, several preliminary steps should be taken. Read all removal and installation procedures carefully prior to committing to this job.

Locating a suitable place to work is extremely important. Adequate work space, along with storage space for the vehicle, will be needed. If a shop or garage isn't available, at the very least a flat, level, clean work surface made of concrete or asphalt is required.

Cleaning the engine compartment and engine before beginning the removal procedure will help keep tools clean and organized (see illustrations 6.1 and 6.2).

An engine hoist will also be necessary. Make sure the hoist is rated in excess of the combined weight of the engine and transaxle. Safety is of primary importance, considering the potential hazards involved in removing the engine from the vehicle.

A vehicle hoist will be necessary for engine removal on four-cylinder engines with a manual transaxle, since on these models the engine/transaxle and subframe assembly must be lowered from the engine compartment, then the vehicle is raised and the powertrain unit is removed from under the vehicle. If the necessary equipment is not available, the engine will have to be removed by a qualified automotive repair facility.

If you're a novice at engine removal, get at least one helper. One person cannot easily do all the things you need to do to remove a big

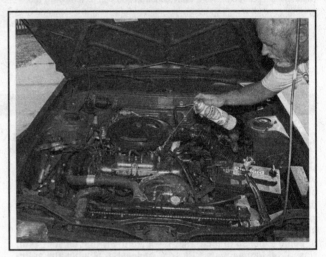

6.1 After tightly wrapping water-vulnerable components, use a spray cleaner on everything, with particular concentration on the greasiest areas, usually around the valve cover and lower edges of the block. If one section dries out, apply more cleaner

6.2 Depending on how dirty the engine is, let the cleaner soak in according to the directions and hose off the grime and cleaner. Get the rinse water down into every area you can get at; then dry important components with a hair dryer or paper towels

6.3 Get an engine stand sturdy enough to firmly support the engine while you're working on it. Stay away from three-wheeled models: they have a tendency to tip over more easily, so get a four-wheeled unit.

2C-8 GENERAL ENGINE OVERHAUL PROCEDURES

heavy engine and transaxle assembly from the engine compartment. Also helpful is to seek advice and assistance from someone who's experienced in engine removal.

Plan the operation ahead of time. Arrange for or obtain all of the tools and equipment you'll need prior to beginning the job (see illustrations 6.3 and 6.4). Some of the equipment necessary to perform engine removal and installation safely and with relative ease are (in addition to a vehicle hoist and an engine hoist) a heavy duty floor jack (preferably fitted with a transaxle jack head adapter), complete sets of wrenches and sockets as described in the front of this manual, wooden blocks, plenty of rags and cleaning solvent for mopping up spilled oil, coolant and gasoline.

Plan for the vehicle to be out of use for quite a while. A machine shop can do the work that is beyond the scope of the home mechanic. Machine shops often have a busy schedule, so before removing the engine, consult the shop for an estimate of how long it will take to rebuild or repair the components that may need work.

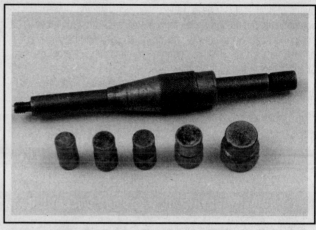

6.4 A clutch alignment tool will be necessary if you're working on a model with a manual transaxle

7 Engine - removal and installation

❊ WARNING 1:

Gasoline is extremely flammable, so take extra precautions when you work on any part of the fuel system. Don't smoke or allow open flames or bare light bulbs near the work area, and don't work in a garage where a gas-type appliance (such as a water heater or clothes dryer) is present. Since gasoline is carcinogenic, wear fuel-resistant gloves when there's a possibility of being exposed to fuel, and, if you spill any fuel on your skin, rinse it off immediately with soap and water. Mop up any spills immediately and do not store fuel-soaked rags where they could ignite. The fuel system is under constant pressure, so, if any fuel lines are to be disconnected, the fuel pressure in the system must be relieved first (see Chapter 4 for more information). When you perform any kind of work on the fuel system, wear safety glasses and have a Class B type fire extinguisher on hand.

❊ WARNING 2:

The engine must be completely cool before beginning this procedure.

FOUR-CYLINDER MODELS WITH AN AUTOMATIC TRANSAXLE AND ALL V6 MODELS

▶ Refer to illustrations 7.9, 7.25, 7.26, 7.33a, 7.33b and 7.33c

Removal

1 Disconnect the cable from the negative battery terminal (see Chapter 5).
2 Relieve the fuel system pressure (see Chapter 4).
3 Disconnect the fuel lines from the fuel rail (see Chapter 4).
4 Remove the fender splash shields and the hood (see Chapter 11). Cover the fenders and cowl using special pads. An old bedspread or blanket will also work.
5 Remove the accessory drivebelt (see Chapter 1).
6 Remove the air filter housing (see Chapter 4).

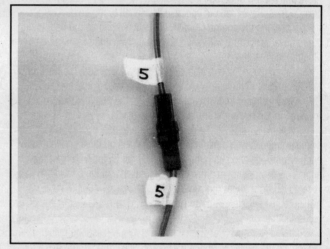

7.9 Label both ends of each wire and hose before disconnecting it

7 Disconnect the accelerator cable, the cruise control cable, if equipped, and bracket from the engine and position them aside (see Chapter 4).
8 Remove the battery and the battery tray (see Chapter 5).
9 Clearly label and disconnect all vacuum lines, emissions hoses, wiring harness connectors and fuel lines. Masking tape and/or a touch up paint applicator work well for marking items (see illustration). Take instant photos or sketch the locations of components and brackets.
10 Disconnect the electrical connectors from the PCM (see Chapter 6).
11 Detach the positive cables and the electrical connectors from the engine compartment fuse/relay box (see Chapter 5) and the ground cable from the vehicle.
12 Detach any other electrical connectors between the engine and the vehicle.
13 Loosen the front wheel lug nuts, then raise the vehicle and secure it on jackstands. Remove the front wheels and tires.

GENERAL ENGINE OVERHAUL PROCEDURES 2C-9

7.25 Remove the torque converter bolts through the bellhousing cutout after the starter is removed

7.26 The engine/transaxle brace is bolted to the engine block and the transaxle housing

→Note: *Keep in mind that during this procedure you'll have to adjust the height of the vehicle to perform certain operations.*

14 Detach the heat shields, exhaust brackets and the exhaust pipes from the exhaust manifold(s) (see Chapter 4).
15 Drain the cooling system (see Chapter 1).
16 Drain the engine oil (see Chapter 1).
17 Detach the lower radiator hose from the engine (see Chapter 3).
18 Lower the vehicle and detach the heater hoses at the firewall (see Chapter 3).
19 Remove the upper radiator hose (see Chapter 3).
20 Remove the cooling fan(s) and shroud(s) (see Chapter 3).
21 Remove the radiator (see Chapter 3).
22 On V6 models, disconnect the Transaxle Control Module (TCM) harness at the module (see Chapter 7B) and at the harness connector next to the brake master cylinder. Remove the TCM harness from the engine compartment.
23 Raise the vehicle and secure it on jackstands.
24 Remove the starter (see Chapter 5).
25 Working through the starter/transaxle bellhousing cutout, remove the torque converter mounting bolts (see illustration). Separate the torque converter from the driveplate.

→Note: *Four-cylinder models are equipped with three torque converter bolts. V6 models are equipped with six of them, but* you only have to remove the three bolts that are recessed into the driveplate.

26 Remove the transaxle-to-engine brace (see illustration).
27 Disconnect the transaxle range switch connector (see Chapter 6).
28 Remove the alternator (see Chapter 5).
29 Remove the air conditioning compressor (see Chapter 3) but do not disconnect the air conditioning lines. Use wire to tie the compressor to the frame. It is not necessary to discharge the refrigerant from the system.
30 Disconnect the power steering pump return line and hose from the power steering pump (see Chapter 10).
31 Remove the power steering fluid reservoir and position it off to the side (see Chapter 10).

→Note: *The power steering fluid reservoir is an integral component of the power steering pump on four-cylinder models. On V6 models, the reservoir is separate and is mounted to the radiator support.*

32 Disconnect the shift cable(s) from the transaxle (see Chapter 7B). Also disconnect any wiring harness connectors from the transaxle and cable brackets from the engine.
33 Roll the engine hoist into position and attach it to the lifting brackets with a couple pieces of heavy-duty chain (see illustrations). Take up the slack in the sling or chain, but don't lift the engine.

7.33a Attach the chain to the engine lifting bracket on the front of the cylinder head

7.33b Attach the other end of the chain to the engine lifting bracket on the rear of the cylinder head

7.33c Raise the engine using the engine hoist, just enough to take the weight off the mounts

2C-10 GENERAL ENGINE OVERHAUL PROCEDURES

> **WARNING:**
> DO NOT place any part of your body under the engine when it's supported only by a hoist or other lifting device.

34 Position a floor jack under the transaxle. Be sure to place a block of wood between the floor jack head and the transaxle.

35 Remove the through-bolts from the forward and rear engine mounts (see Chapters 2A or 2B). The transaxle must be raised slightly to allow the engine to angle away from the bellhousing.

36 Remove the passenger side engine mount and bracket (see Chapter 2A).

37 Remove the lower and upper transaxle-to-engine mounting bolts.

38 Recheck to be sure nothing is still connecting the engine or vehicle. Disconnect anything still remaining.

39 Raise the engine slightly and inspect it thoroughly once more to make sure that nothing is still attached, then slowly raise the engine out of the engine compartment. Check carefully to make sure nothing is hanging up. Raise the transaxle slightly to allow the engine to angle away from the transaxle as the engine is lifted from the engine compartment.

40 Remove the driveplate (see Chapter 2A or 2B) and mount the engine on an engine stand (see illustration 6.3).

41 Inspect the engine and transaxle mounts (see the "Engine mount" Section in Chapter 2A or 2B). If they're worn or damaged, replace them.

Installation

42 Install the driveplate (see Chapter 2A or 2B).

43 Carefully lower the engine into the engine compartment and reattach it to the engine mount (see Chapter 2A or 2B).

44 Guide the torque converter into the crankshaft following the procedure outlined in Chapter 7B. Install the transaxle-to-engine bolts and tighten them securely.

> **CAUTION:**
> DO NOT use the bolts to force the transaxle and engine together!

45 Install the engine-to-transaxle brace and tighten the bolts securely.

46 Reinstall the remaining components in the reverse order of removal.

47 Reconnect the battery (see Chapter 5, Section 1).

48 Add coolant, oil and transaxle fluid as needed.

49 Run the engine and check for leaks and proper operation of all accessories, then install the hood and test drive the vehicle.

50 Have the air conditioning system re-charged and leak tested, if it was discharged.

FOUR-CYLINDER MODELS WITH MANUAL TRANSAXLES

▶ Refer to illustration 7.87

➡**Note 1:** Engine removal on these models is a difficult job, especially for the do-it-yourself mechanic working at home. Because of the vehicle's design, the manufacturer states that the engine and transaxle have to be removed as a unit from the bottom of the vehicle, not the top. With a floor jack and jackstands, the vehicle can't be raised high enough or supported safely enough for the engine/transaxle assembly to slide out from underneath. The manufacturer recommends that removal of the engine transaxle assembly only be performed with the use of a frame-contact type vehicle hoist.

➡**Note 2:** Read through the entire Section before beginning this procedure. The engine and transaxle are removed as a unit from below, then separated outside the vehicle.

Removal

51 Park the vehicle on a frame-contact type vehicle hoist, then engage the arms of the hoist with the jacking points of the vehicle. Raise the hoist arms until they contact the vehicle, but not so much that the wheels come off the ground.

52 Disconnect the cable from the negative battery terminal (see Chapter 5).

53 Relieve the fuel system pressure (see Chapter 4).

54 Disconnect the fuel lines from the fuel rail (see Chapter 4).

55 Remove the fender splash shields and the hood (see Chapter 11). Cover the fenders and cowl using special pads. An old bedspread or blanket will also work.

56 Remove the accessory drivebelt (see Chapter 1).

57 Remove the air filter housing (see Chapter 4).

58 Disconnect the accelerator cable, the cruise control cable, if equipped, and bracket from the engine and position them aside (see Chapter 4).

59 Remove the battery and the battery tray (see Chapter 5).

60 Clearly label and disconnect all vacuum lines, emissions hoses, wiring harness connectors and fuel lines. Masking tape and/or a touch up paint applicator work well for marking items (see illustration 7.9). Take instant photos or sketch the locations of components and brackets.

61 Disconnect the electrical connectors from the PCM (see Chapter 6).

62 Detach the positive cables and the electrical connectors from the engine compartment fuse/relay box (see Chapter 5) and the ground cable from the vehicle.

63 Detach any other electrical connectors between the engine and the vehicle.

64 Loosen the front wheel lug nuts, then raise the vehicle. Remove the front wheels and tires.

➡**Note:** Keep in mind that during this procedure you'll have to adjust the height of the vehicle to perform certain operations.

65 Detach the heat shields, exhaust brackets and the exhaust pipes from the exhaust manifold(s) (see Chapter 4).

66 Remove the exhaust pipe from the exhaust manifold to the catalytic converter (see Chapter 6).

67 Drain the cooling system (see Chapter 1).

68 Drain the engine oil (see Chapter 1).

69 Detach the lower radiator hose from the engine (see Chapter 3).

70 Lower the vehicle and detach the heater hoses at the firewall (see Chapter 3).

71 Remove the upper radiator hose (see Chapter 3).

72 Remove the cooling fan(s) and shroud(s) (see Chapter 3).

73 Remove the radiator (see Chapter 3).

74 Remove the clutch release cylinder and hydraulic line (see Chapter 8).

GENERAL ENGINE OVERHAUL PROCEDURES 2C-11

75 Remove the starter (see Chapter 5).

76 Disconnect the control shaft lever from the transaxle (see Chapter 7A). Also disconnect any wiring harness connectors from the transaxle.

77 Remove the front and rear engine/transaxle mount through-bolts (see Chapter 2A).

78 Remove the air conditioning compressor (see Chapter 3). Use wire to tie the compressor to a bracket or other component mounted on the uni-body structure (not the subframe). It is not necessary to discharge the refrigerant from the system.

79 Disconnect the power steering pump return line and hose from the power steering pump (see Chapter 10).

80 Remove the power steering fluid reservoir and position it off to the side (see Chapter 10).

➞Note: *The power steering fluid reservoir is an integral component of the power steering pump on four-cylinder models.*

81 Disconnect the stabilizer bar links (see Chapter 10).

82 Disconnect the balljoints from the lower control arms (see Chapter 10).

83 Disconnect the tie rod ends from the steering knuckles (see Chapter 10).

84 Disconnect the power steering fluid lines from the power steering gear (see Chapter 10).

➞Note: *The power steering gear will remain on the subframe assembly when it is lowered from the vehicle.*

85 Disconnect the intermediate shaft from the power steering gear (see Chapter 10).

86 Unplug the downstream oxygen sensor electrical connector.

87 Roll the hoist into position and attach the sling or chain to it (see illustration). Tighten the bolts securely. Take up the slack until there is slight tension on the hoist. Remember that the transaxle end of the engine will be heavier, so position the chain on the hoist so it balances the engine and the transaxle level with the vehicle.

➞Note 1: *Depending on the design of the engine hoist, it may be helpful to position the hoist from the side of the vehicle, so that when the engine/transaxle assembly is lowered, it will fit between the legs of the hoist.*

➞Note 2: *The sling or chain must be long enough to allow the engine hoist to lower the engine/transaxle assembly to the ground, without letting the hoist arm contact the vehicle.*

88 Remove the right-side engine mount and the transaxle mount on the left side.

89 Recheck to be sure nothing is still connecting the engine or transaxle to the vehicle. Disconnect and label anything still remaining.

90 Remove the subframe mounting bolts (see Chapter 10). Do not separate the subframe from the engine/transaxle assembly at this time. The engine/transaxle assembly and subframe assembly will be lowered from the vehicle as a complete unit.

91 Position four jackstands under the subframe, then slowly lower the engine/transaxle and subframe assembly onto the jackstands.

92 Once the powertrain/subframe assembly is on the jackstands, disconnect the engine lifting hoist and raise the vehicle until it clears the powertrain.

93 Reconnect the chain or sling to support the engine and transaxle, then unbolt the subframe.

94 Raise the engine/transaxle assembly off of the subframe, then

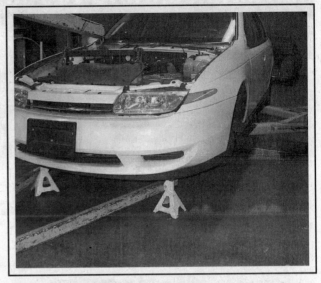

7.87 The engine/transaxle and subframe assembly can be lowered onto jackstands using the engine hoist. After the subframe assembly is resting safely on the jackstands, the engine hoist can be removed and the vehicle can be raised up

support the engine with blocks of wood or another floor jack, while leaving the sling or chain attached. Support the transaxle with another floor jack, preferably one with a transaxle jack head adapter. At this point the transaxle can be unbolted and removed from the engine. Be very careful to ensure that the components are supported securely so they won't topple off their supports during disconnection.

95 Reconnect the lifting chain to the engine, then raise the engine and attach it to an engine stand.

Installation

> ※※ **WARNING:**
>
> The manufacturer recommends replacing the subframe bolts, cage nuts and steering gear mounting fasteners with new ones whenever they are removed.

96 Installation is the reverse of removal, noting the following points:

a) Check the engine/transaxle mounts. If they're worn or damaged, replace them.

b) Attach the transaxle to the engine following the procedure described in Chapter 7.

c) When installing the subframe, tighten the subframe mounting bolts to the torque listed in Chapter 10 Specifications. Note the locations of the various size bolts. Replace all the subframe bolts and nuts with new ones.

d) Add coolant, oil, power steering and transaxle fluids as needed (see Chapter 1).

e) Reconnect the battery (see Chapter 5, Section 1).

f) Run the engine and check for proper operation and leaks. Shut off the engine and recheck fluid levels.

ENGINE BEARING ANALYSIS

Debris

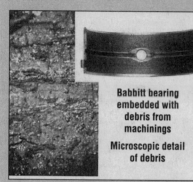

Babbitt bearing embedded with debris from machinings

Microscopic detail of debris

Microscopic detail of gouges

Overplated copper alloy bearing gouged by cast iron debris

Aluminum bearing embedded with glass beads

Microscopic detail of glass beads

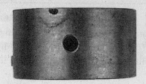

Damaged lining caused by dirt left on the bearing back

Misassembly

Result of a lower half assembled as an upper - blocking the oil flow

Excessive oil clearance is indicated by a short contact arc

Polished and oil-stained backs are a result of a poor fit in the housing bore

Result of a wrong, reversed, or shifted cap

Overloading

Damage from excessive idling which resulted in an oil film unable to support the load imposed

Damaged upper connecting rod bearings caused by engine lugging; the lower main bearings (not shown) were similarly affected

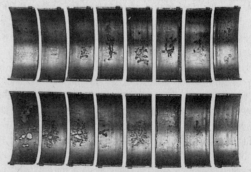

The damage shown in these upper and lower connecting rod bearings was caused by engine operation at a higher-than-rated speed under load

Misalignment

A warped crankshaft caused this pattern of severe wear in the center, diminishing toward the ends

A poorly finished crankshaft caused the equally spaced scoring shown

A tapered housing bore caused the damage along one edge of this pair

A bent connecting rod led to the damage in the "V" pattern

Lubrication

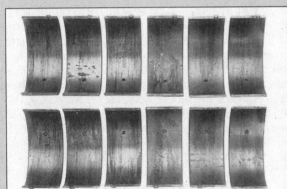
Result of dry start: The bearings on the left, farthest from the oil pump, show more damage

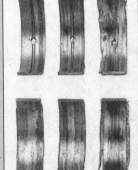
Result of a low oil supply or oil starvation

Severe wear as a result of inadequate oil clearance

Corrosion

Microscopic detail of corrosion

Corrosion is an acid attack on the bearing lining generally caused by inadequate maintenance, extremely hot or cold operation, or interior oils or fuels

Microscopic detail of cavitation

Example of cavitation - a surface erosion caused by pressure changes in the oil film

Damage from excessive thrust or insufficient axial clearance

Bearing affected by oil dilution caused by excessive blow-by or a rich mixture

© 1986 Federal-Mogul Corporation
Copy and photographs courtesy of Federal Mogul Corporation

2C-14　GENERAL ENGINE OVERHAUL PROCEDURES

8　Engine overhaul - disassembly sequence

1 It's much easier to remove the external components if the engine is mounted on a portable engine stand. A stand can often be rented quite cheaply from an equipment rental yard. Before the engine is mounted on a stand, the flywheel/driveplate should be removed from the engine.

2 If a stand isn't available, it's possible to remove the external engine components with it blocked up on the floor. Be extra careful not to tip or drop the engine when working without a stand.

3 If you're going to obtain a rebuilt engine, all external components must come off first, to be transferred to the replacement engine. These components include:

Clutch and flywheel (models with manual transaxle)
Driveplate (models with automatic transaxle)
Ignition system components
Emissions-related components
Engine mounts and mount brackets
Engine rear cover (spacer plate between flywheel/driveplate and engine block)
Intake/exhaust manifolds
Fuel injection components
Oil filter
Ignition coil pack(s) and spark plugs
Thermostat and housing assembly
Water pump

➡ **Note:** *When removing the external components from the engine, pay close attention to details that may be helpful or important during installation. Note the installed position of gaskets, seals, spacers, pins, brackets, washers, bolts and other small items.*

4 If you're going to obtain a short block (assembled engine block, crankshaft, pistons and connecting rods), then remove the timing chain or belt, cylinder head(s), oil pan, oil pump pick-up tube, oil pump and water pump from your engine so that you can turn in your old short block to the rebuilder as a core. See *Engine rebuilding alternatives* for additional information regarding the different possibilities to be considered.

9　Pistons and connecting rods - removal and installation

REMOVAL

♦ **Refer to illustrations 9.1, 9.3 and 9.4**

➡ **Note:** *Prior to removing the piston/connecting rod assemblies, remove the cylinder head and oil pan (see Chapter 2A).*

1 Use your fingernail to feel if a ridge has formed at the upper limit of ring travel (about 1/4-inch down from the top of each cylinder). If carbon deposits or cylinder wear have produced ridges, they must be completely removed with a special tool (see illustration). Follow the manufacturer's instructions provided with the tool. Failure to remove the ridges before attempting to remove the piston/connecting rod assemblies may result in piston breakage.

2 After the cylinder ridges have been removed, turn the engine so the crankshaft is facing up.

3 Before the main bearing cap assembly and connecting rods are removed, check the connecting rod endplay with feeler gauges. Slide them between the first connecting rod and the crankshaft throw until the play is removed (see illustration). Repeat this procedure for each connecting rod. The endplay is equal to the thickness of the feeler gauge(s). Check with an automotive machine shop for the endplay service limit (a typical endplay limit should measure between 0.005 to 0.015 inch [0.127 to 0.369 mm]). If the play exceeds the service limit, new connecting rods will be required. If new rods (or a new crankshaft) are installed, the endplay may fall under the minimum allowable. If it does, the rods will have to be machined to restore it. If necessary, consult an automotive machine shop for advice.

4 Check the connecting rods and caps for identification marks. If they aren't plainly marked, use paint or marker to clearly identify each rod and cap (1, 2, 3, etc., depending on the cylinder they're associated with) (see illustration).

9.1 Before you try to remove the pistons, use a ridge reamer to remove the raised material (ridge) from the top of the cylinders

9.3 Checking the connecting rod endplay (side clearance)

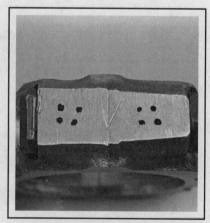

9.4 If the connecting rods and caps are not marked, use permanent ink to mark the caps to the rods by cylinder number (for example, this would be the No. 4 connecting rod)

GENERAL ENGINE OVERHAUL PROCEDURES 2C-15

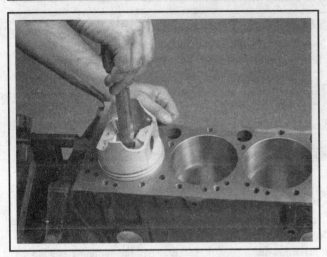

9.13 Install the piston ring into the cylinder then push it down into position using a piston so the ring will be square in the cylinder

9.14 With the ring square in the cylinder, measure the ring end gap with a feeler gauge

5 Remove the connecting rod cap bolts from the number one connecting rod.

➡ **Note: New connecting rod cap bolts must be used when reassembling the engine, but save the old bolts - they'll be required for the bearing oil clearance check during reassembly.**

6 Remove the number one connecting rod cap and bearing insert. Don't drop the bearing insert out of the cap.

7 Remove the bearing insert and push the connecting rod/piston assembly out through the top of the engine. Use a wooden dowel to push on the connecting rod. If resistance is felt, double-check to make sure that all of the ridge was removed from the cylinder.

8 Repeat the procedure for the remaining cylinders.

9 After removal, reassemble the connecting rod caps and bearing inserts in their respective connecting rods and install the cap bolts finger tight. Leaving the old bearing inserts in place until reassembly will help prevent the connecting rod bearing surfaces from being accidentally nicked or gouged.

10 The pistons and connecting rods are now ready for inspection and overhaul at an automotive machine shop.

PISTON RING INSTALLATION

▸ **Refer to illustrations 9.13, 9.14, 9.15, 9.19a, 9.19b and 9.22**

11 Before installing the new piston rings, the ring end gaps must be checked. It's assumed that the piston ring side clearance has been checked and verified correct.

12 Lay out the piston/connecting rod assemblies and the new ring sets so the ring sets will be matched with the same piston and cylinder during the end gap measurement and engine assembly.

13 Insert the top (number one) ring into the first cylinder and square it up with the cylinder walls by pushing it in with the top of the piston (see illustration). The ring should be near the bottom of the cylinder, at the lower limit of ring travel.

14 To measure the end gap, slip feeler gauges between the ends of the ring until a gauge equal to the gap width is found (see illustration). The feeler gauge should slide between the ring ends with a slight amount of drag. A typical ring gap should fall between 0.010 and 0.020 inch [0.25 to 0.50 mm] for compression rings and up to 0.030 inch [0.76 mm] for the oil ring steel rails. If the gap is larger or smaller than specified, double-check to make sure you have the correct rings before proceeding.

15 If the gap is too small, it must be enlarged or the ring ends may come in contact with each other during engine operation, which can cause serious damage to the engine. If necessary, increase the end gaps by filing the ring ends very carefully with a fine file. Mount the file in a vise equipped with soft jaws, slip the ring over the file with the ends contacting the file face and slowly move the ring to remove material from the ends. When performing this operation, file only by pushing the ring from the outside end of the file towards the vise (see illustration).

16 Excess end gap isn't critical unless it's greater than 0.040 inch (1.01 mm). Again, double-check to make sure you have the correct ring type.

17 Repeat the procedure for each ring that will be installed in the first cylinder and for each ring in the remaining cylinders. Remember to keep rings, pistons and cylinders matched up.

18 Once the ring end gaps have been checked/corrected, the rings can be installed on the pistons.

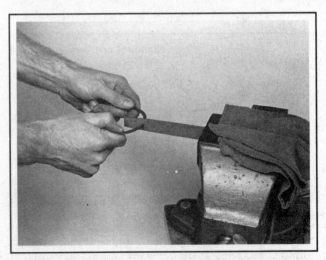

9.15 If the ring end gap is too small, clamp a file in a vise as shown and file the piston ring ends - be sure to remove all raised material

2C-16 GENERAL ENGINE OVERHAUL PROCEDURES

9.19a Installing the spacer/expander in the oil ring groove

9.19b DO NOT use a piston ring installation tool when installing the oil control side rails

9.22 Use a piston ring installation tool to install the number 2 and the number 1 (top) rings - be sure the directional mark on the piston ring(s) is facing toward the top of the piston

19 The oil control ring (lowest one on the piston) is usually installed first. It's composed of three separate components. Slip the spacer/expander into the groove (see illustration). If an anti-rotation tang is used, make sure it's inserted into the drilled hole in the ring groove. Next, install the upper side rail in the same manner (see illustration). Don't use a piston ring installation tool on the oil ring side rails, as they may be damaged. Instead, place one end of the side rail into the groove between the spacer/expander and the ring land, hold it firmly in place and slide a finger around the piston while pushing the rail into the groove. Finally, install the lower side rail.

20 After the three oil ring components have been installed, check to make sure that both the upper and lower side rails can be rotated smoothly inside the ring grooves.

21 The number two (middle) ring is installed next. It's usually stamped with a mark which must face up, toward the top of the piston. Do not mix up the top and middle rings, as they have different cross-sections.

➡ **Note:** Always follow the instructions printed on the ring package or box - different manufacturers may require different approaches.

22 Use a piston ring installation tool and make sure the identification mark is facing the top of the piston, then slip the ring into the middle groove on the piston (see illustration). Don't expand the ring any more than necessary to slide it over the piston.

23 Install the number one (top) ring in the same manner. Make sure the mark is facing up. Be careful not to confuse the number one and number two rings.

24 Repeat the procedure for the remaining pistons and rings.

INSTALLATION

25 Before installing the piston/connecting rod assemblies, the cylinder walls must be perfectly clean, the top edge of each cylinder bore must be chamfered, and the crankshaft must be in place.

26 Remove the cap from the end of the number one connecting rod (refer to the marks made during removal). Remove the original bearing inserts and wipe the bearing surfaces of the connecting rod and cap with a clean, lint-free cloth. They must be kept spotlessly clean.

Connecting rod bearing oil clearance check

▸ Refer to illustrations 9.30, 9.35, 9.37 and 9.41

27 Clean the back side of the new upper bearing insert, then lay it in place in the connecting rod.

28 Make sure the tab on the bearing fits into the recess in the rod. Don't hammer the bearing insert into place and be very careful not to nick or gouge the bearing face. Don't lubricate the bearing at this time.

29 Clean the back side of the other bearing insert and install it in the rod cap. Again, make sure the tab on the bearing fits into the recess in the cap, and don't apply any lubricant. It's critically important that the mating surfaces of the bearing and connecting rod are perfectly clean and oil free when they're assembled.

30 Position the piston ring gaps at the specified intervals around the piston as shown (see illustration).

31 Lubricate the piston and rings with clean engine oil and attach a piston ring compressor to the piston. Leave the skirt protruding about 1/4-inch to guide the piston into the cylinder. The rings must be compressed until they're flush with the piston.

32 Rotate the crankshaft until the number one connecting rod journal is at BDC (bottom dead center) and apply a liberal coat of engine oil to the cylinder walls.

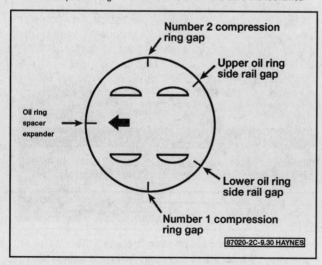

9.30 Position the piston ring end gaps as shown

GENERAL ENGINE OVERHAUL PROCEDURES 2C-17

9.35 Use a plastic or wooden hammer handle to push the piston into the cylinder

9.37 Place Plastigage on each connecting rod bearing journal, parallel to the crankshaft centerline

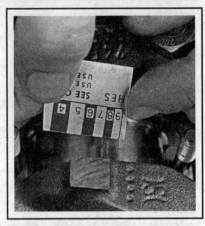

9.41 Use the scale on the Plastigage package to determine the bearing oil clearance - be sure to measure the widest part of the Plastigage and use the correct scale; it comes with both standard and metric scales

33 With the arrow on top of the piston facing the front (timing belt end or timing chain) of the engine, gently insert the piston/connecting rod assembly into the number one cylinder bore and rest the bottom edge of the ring compressor on the engine block. Install the pistons with the cavity mark(s) or arrow facing toward the timing belt or timing chain end of the engine.

34 Tap the top edge of the ring compressor to make sure it's contacting the block around its entire circumference.

35 Gently tap on the top of the piston with the end of a wooden or plastic hammer handle (see illustration) while guiding the end of the connecting rod into place on the crankshaft journal (a pair of wooden dowels would be helpful for this). The piston rings may try to pop out of the ring compressor just before entering the cylinder bore, so keep some downward pressure on the ring compressor. Work slowly, and if any resistance is felt as the piston enters the cylinder, stop immediately. Find out what's hanging up and fix it before proceeding. Do not, for any reason, force the piston into the cylinder - you might break a ring and/or the piston.

36 Once the piston/connecting rod assembly is installed, the connecting rod bearing oil clearance must be checked before the rod cap is permanently installed.

37 Cut a piece of the appropriate size Plastigage slightly shorter than the width of the connecting rod bearing and lay it in place on the number one connecting rod journal, parallel with the journal axis (see illustration).

38 Clean the connecting rod cap bearing face and install the rod cap. Make sure the mating mark on the cap is on the same side as the mark on the connecting rod (see illustration 9.4).

39 Install the old rod bolts, at this time, and tighten them to the torque listed in this Chapter's Specifications.

➥Note: *Use a thin-wall socket to avoid erroneous torque readings that can result if the socket is wedged between the rod cap and the bolt. If the socket tends to wedge itself between the fastener and the cap, lift up on it slightly until it no longer contacts the cap. DO NOT rotate the crankshaft at any time during this operation.*

40 Remove the fasteners and detach the rod cap, being very careful not to disturb the Plastigage. Discard the cap bolts at this time as they cannot be reused.

➥Note: *You MUST use new connecting rod bolts.*

41 Compare the width of the crushed Plastigage to the scale printed on the Plastigage envelope to obtain the oil clearance (see illustration). The connecting rod oil clearance is usually about 0.001 to 0.002 inch. Consult an automotive machine shop for the clearance specified for the rod bearings on your engine.

42 If the clearance is not as specified, the bearing inserts may be the wrong size (which means different ones will be required). Before deciding that different inserts are needed, make sure that no dirt or oil was between the bearing inserts and the connecting rod or cap when the clearance was measured. Also, recheck the journal diameter. If the Plastigage was wider at one end than the other, the journal may be tapered. If the clearance still exceeds the limit specified, the bearing will have to be replaced with an undersize bearing.

✻✻✻ CAUTION:

When installing a new crankshaft always use a standard size bearing.

Final installation

43 Carefully scrape all traces of the Plastigage material off the rod journal and/or bearing face. Be very careful not to scratch the bearing - use your fingernail or the edge of a plastic card.

44 Make sure the bearing faces are perfectly clean, then apply a uniform layer of clean moly-base grease or engine assembly lube to both of them. You'll have to push the piston into the cylinder to expose the face of the bearing insert in the connecting rod.

45 Slide the connecting rod back into place on the journal, install the rod cap, install the new bolts and tighten them to the torque listed in this Chapter's Specifications.

✻✻✻ CAUTION:

Install new connecting rod cap bolts. Do NOT reuse old bolts - they have stretched and cannot be reused (see Step 5).

46 Repeat the entire procedure for the remaining pistons/connecting rods.

47 The important points to remember are:
 a) *Keep the back sides of the bearing inserts and the insides of the connecting rods and caps perfectly clean when assembling them.*

2C-18 GENERAL ENGINE OVERHAUL PROCEDURES

b) Make sure you have the correct piston/rod assembly for each cylinder.
c) The arrow or mark on the piston must face the front (timing chain on four-cylinder engines or timing belt on V6 engines) of the engine.
d) Lubricate the cylinder walls liberally with clean oil.
e) Lubricate the bearing faces when installing the rod caps after the oil clearance has been checked.

48 After all the piston/connecting rod assemblies have been correctly installed, rotate the crankshaft a number of times by hand to check for any obvious binding.

49 As a final step, check the connecting rod endplay, as described in Step 3. If it was correct before disassembly and the original crankshaft and rods were reinstalled, it should still be correct. If new rods or a new crankshaft were installed, the endplay may be inadequate. If so, the rods will have to be removed and taken to an automotive machine shop for resizing.

10 Crankshaft - removal and installation

REMOVAL

▸ Refer to illustrations 10.1 and 10.3

➡ Note: The crankshaft can be removed only after the engine has been removed from the vehicle. It's assumed that the flywheel or driveplate, crankshaft pulley, timing belt or timing chain, oil pan, oil pump body, oil filter and piston/connecting rod assemblies have already been removed. The rear main oil seal retainer must be unbolted and separated from the block before proceeding with crankshaft removal.

1 Before the crankshaft is removed, measure the endplay. Mount a dial indicator with the indicator in line with the crankshaft and just touching the end of the crankshaft as shown (see illustration).

2 Pry the crankshaft all the way to the rear and zero the dial indicator. Next, pry the crankshaft to the front as far as possible and check the reading on the dial indicator. The distance traveled is the endplay. A typical crankshaft endplay will fall between 0.003 to 0.010 inch (0.076 to 0.254 mm). If it is greater than that, check the crankshaft thrust surfaces for wear after it's removed. If no wear is evident, new main bearings should correct the endplay.

3 If a dial indicator isn't available, feeler gauges can be used. Gently pry the crankshaft all the way to the front of the engine. Slip feeler gauges between the crankshaft and the front face of the thrust bearing or washer to determine the clearance (see illustration).

4 Loosen the lower crankcase perimeter bolts and the lower crankcase bolts (four-cylinder engine) or the main bearing/bridge bolts (V6 engines) 1/4-turn at a time each, until they can be removed by hand. Follow the reverse of the tightening sequence (see illustrations 10.19a and 10.19b).

※ CAUTION:

The lower crankcase bolts (four-cylinder engine) and main bearing/bridge bolts (V6 engine) must be replaced with new ones upon installation. Save the old bolts, however, as they will be used for the main bearing oil clearance check.

5 Remove the lower crankcase (four-cylinder) or main bearing bridge (V6). If you're working on a V6, proceed to remove the main bearing caps. Try not to drop the bearing inserts if they come out with the lower crankcase or main bearing caps.

➡ Note: If you're working on a V6 engine, note the orientation of the main bearing caps. They should have markings on them indicating their installed positions and an arrow pointing towards the timing belt end of the engine. If there are no marks, be sure to make your own before removing the caps.

6 Carefully lift the crankshaft out of the engine. It may be a good idea to have an assistant available, since the crankshaft is quite heavy and awkward to handle. With the bearing inserts in place inside the engine block and main bearing caps and bridge or the lower crankcase, reinstall the caps and bridge or lower crankcase onto the engine block and tighten the bolts finger tight. On V6 engines, make sure the caps are in the exact order in which they were removed, with the arrow pointing toward the front (timing belt end) of the engine.

10.1 Checking crankshaft endplay with a dial indicator

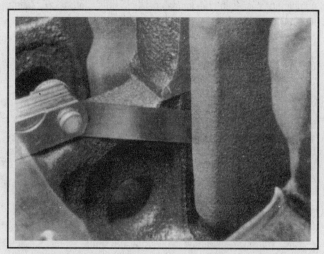

10.3 Checking the crankshaft endplay with feeler gauges at the thrust bearing journal

GENERAL ENGINE OVERHAUL PROCEDURES   2C-19

INSTALLATION

7 Crankshaft installation is the first step in engine reassembly. It's assumed at this point that the engine block and crankshaft have been cleaned, inspected and repaired or reconditioned.

8 Position the engine block with the bottom facing up.

9 Remove the mounting bolts and lift off the lower crankcase or bearing bridge and main bearing caps.

10 If they're still in place, remove the original bearing inserts from the block and from the main bearing caps or lower crankcase. Wipe the bearing surfaces of the block and bearing cap/lower crankcase saddle with a clean, lint-free cloth. They must be kept spotlessly clean. This is critical for determining the correct bearing oil clearance.

MAIN BEARING OIL CLEARANCE CHECK

▶ Refer to illustrations 10.17, 10.19a, 10.19b and 10.21

11 Without mixing them up, clean the back sides of the new upper main bearing inserts (with grooves and oil holes) and lay one in each main bearing saddle in the engine block. Each upper bearing (engine block) has an oil groove and oil hole in it.

❋❋ CAUTION:

The oil holes in the block must line up with the oil holes in the engine block inserts.

The thrust washer or thrust bearing insert must be installed in the correct location.

➡ **Note: The thrust bearing on the four-cylinder engine is located on the engine block number 2 journal. The thrust washers on the V6 engine are located on the 4th journal in the main bearing cap (lower) and engine block (upper).**

Clean the back sides of the lower main bearing inserts and lay them in the corresponding location in the lower crankcase saddles (four-cylinder) or main bearing caps (V6). Make sure the tab on the bearing insert fits into the recess in the block or main bearing caps.

❋❋ CAUTION:

Do not hammer the bearing insert into place and don't nick or gouge the bearing faces. DO NOT apply any lubrication at this time.

10.17 Place the Plastigage onto the crankshaft bearing journal as shown

12 Clean the faces of the bearing inserts in the block and the crankshaft main bearing journals with a clean, lint-free cloth.

13 Check or clean the oil holes in the crankshaft, as any dirt here can go only one way - straight through the new bearings.

14 Once you're certain the crankshaft is clean, carefully lay it in position in the cylinder block.

15 Before the crankshaft can be permanently installed, the main bearing oil clearance must be checked.

16 Cut several strips of the appropriate size of Plastigage. They must be slightly shorter than the width of the main bearing journal.

17 Place one piece on each crankshaft main bearing journal, parallel with the journal axis as shown (see illustration).

18 Clean the faces of the bearing inserts in the lower crankcase or main bearing caps. Hold the bearing inserts in place and install the lower crankcase or caps onto the crankshaft and cylinder block. DO NOT disturb the Plastigage.

19 Apply clean engine oil to all bolt threads prior to installation, then install all bolts finger-tight. Tighten the lower crankcase or bridge/bearing cap bolts in the sequence shown (see illustrations) progressing in steps, to the torque listed in this Chapter's Specifications. DO NOT rotate the crankshaft at any time during this operation.

20 Remove the bolts in the reverse order of the tightening sequence and carefully lift the lower crankcase or main bearing caps straight up and off the block. Do not disturb the Plastigage or rotate the crankshaft.

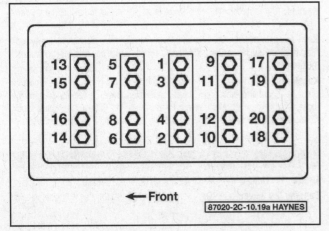

10.19a Lower crankcase bolt tightening sequence (four-cylinder engines)

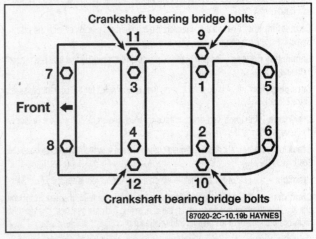

10.19b Main bearing cap bolt and crankshaft bearing bridge bolt tightening sequence on V6 engines

GLOSSARY

B

Backlash - The amount of play between two parts. Usually refers to how much one gear can be moved back and forth without moving gear with which it's meshed.

Bearing Caps - The caps held in place by nuts or bolts which, in turn, hold the bearing surface. This space is for lubricating oil to enter.

Bearing clearance - The amount of space left between shaft and bearing surface. This space is for lubricating oil to enter.

Bearing crush - The additional height which is purposely manufactured into each bearing half to ensure complete contact of the bearing back with the housing bore when the engine is assembled.

Bearing knock - The noise created by movement of a part in a loose or worn bearing.

Blueprinting - Dismantling an engine and reassembling it to EXACT specifications.

Bore - An engine cylinder, or any cylindrical hole; also used to describe the process of enlarging or accurately refinishing a hole with a cutting tool, as to bore an engine cylinder. The bore size is the diameter of the hole.

Boring - Renewing the cylinders by cutting them out to a specified size. A boring bar is used to make the cut.

Bottom end - A term which refers collectively to the engine block, crankshaft, main bearings and the big ends of the connecting rods.

Break-in - The period of operation between installation of new or rebuilt parts and time in which parts are worn to the correct fit. Driving at reduced and varying speed for a specified mileage to permit parts to wear to the correct fit.

Bushing - A one-piece sleeve placed in a bore to serve as a bearing surface for shaft, piston pin, etc. Usually replaceable.

C

Camshaft - The shaft in the engine, on which a series of lobes are located for operating the valve mechanisms. The camshaft is driven by gears or sprockets and a timing chain. Usually referred to simply as the cam.

Carbon - Hard, or soft, black deposits found in combustion chamber, on plugs, under rings, on and under valve heads.

Cast iron - An alloy of iron and more than two percent carbon, used for engine blocks and heads because it's relatively inexpensive and easy to mold into complex shapes.

Chamfer - To bevel across (or a bevel on) the sharp edge of an object.

Chase - To repair damaged threads with a tap or die.

Combustion chamber - The space between the piston and the cylinder head, with the piston at top dead center, in which air-fuel mixture is burned.

Compression ratio - The relationship between cylinder volume (clearance volume) when the piston is at top dead center and cylinder volume when the piston is at bottom dead center.

Connecting rod - The rod that connects the crank on the crankshaft with the piston. Sometimes called a con rod.

Connecting rod cap - The part of the connecting rod assembly that attaches the rod to the crankpin.

Core plug - Soft metal plug used to plug the casting holes for the coolant passages in the block.

Crankcase - The lower part of the engine in which the crankshaft rotates; includes the lower section of the cylinder block and the oil pan.

Crank kit - A reground or reconditioned crankshaft and new main and connecting rod bearings.

Crankpin - The part of a crankshaft to which a connecting rod is attached.

Crankshaft - The main rotating member, or shaft, running the length of the crankcase, with offset throws to which the connecting rods are attached; changes the reciprocating motion of the pistons into rotating motion.

Cylinder sleeve - A replaceable sleeve, or liner, pressed into the cylinder block to form the cylinder bore.

D

Deburring - Removing the burrs (rough edges or areas) from a bearing.

Deglazer - A tool, rotated by an electric motor, used to remove glaze from cylinder walls so a new set of rings will seat.

E

Endplay - The amount of lengthwise movement between two parts. As applied to a crankshaft, the distance that the crankshaft can move forward and back in the cylinder block.

F

Face - A machinist's term that refers to removing metal from the end of a shaft or the face of a larger part, such as a flywheel.

Fatigue - A breakdown of material through a large number of loading and unloading cycles. The first signs are cracks followed shortly by breaks.

Feeler gauge - A thin strip of hardened steel, ground to an exact thickness, used to check clearances between parts.

Free height - The unloaded length or height of a spring.

Freeplay - The looseness in a linkage, or an assembly of parts, between the initial application of force and actual movement. Usually perceived as slop or slight delay.

Freeze plug - See Core plug.

G

Gallery - A large passage in the block that forms a reservoir for engine oil pressure.

Glaze - The very smooth, glassy finish that develops on cylinder walls while an engine is in service.

H

Heli-Coil - A rethreading device used when threads are worn or damaged. The device is installed in a retapped hole to reduce the thread size to the original size.

I

Installed height - The spring's measured length or height, as installed on the cylinder head. Installed height is measured from the spring seat to the underside of the spring retainer.

J

Journal - The surface of a rotating shaft which turns in a bearing.

K

Keeper - The split lock that holds the valve spring retainer in position on the valve stem.

Key - A small piece of metal inserted into matching grooves machined into two parts fitted together - such as a gear pressed onto a shaft - which prevents slippage between the two parts.

Knock - The heavy metallic engine sound, produced in the combustion chamber as a result of abnormal combustion - usually detonation. Knock is usually caused by a loose or worn bearing. Also referred to as detonation, pinging and spark knock. Connecting rod or main bearing knocks are created by too much oil clearance or insufficient lubrication.

L

Lands - The portions of metal between the piston ring grooves.

Lapping the valves - Grinding a valve face and its seat together with lapping compound.

Lash - The amount of free motion in a gear train, between gears, or in a mechanical assembly, that occurs before movement can begin. Usually refers to the lash in a valve train.

Lifter - The part that rides against the cam to transfer motion to the rest of the valve train.

M

Machining - The process of using a machine to remove metal from a metal part.
Main bearings - The plain, or babbitt, bearings that support the crankshaft.
Main bearing caps - The cast iron caps, bolted to the bottom of the block, that support the main bearings.

O

O.D. - Outside diameter.
Oil gallery - A pipe or drilled passageway in the engine used to carry engine oil from one area to another.
Oil ring - The lower ring, or rings, of a piston; designed to prevent excessive amounts of oil from working up the cylinder walls and into the combustion chamber. Also called an oil-control ring.
Oil seal - A seal which keeps oil from leaking out of a compartment. Usually refers to a dynamic seal around a rotating shaft or other moving part.
O-ring - A type of sealing ring made of a special rubberlike material; in use, the O-ring is compressed into a groove to provide the sealing action.
Overhaul - To completely disassemble a unit, clean and inspect all parts, reassemble it with the original or new parts and make all adjustments necessary for proper operation.

P

Pilot bearing - A small bearing installed in the center of the flywheel (or the rear end of the crankshaft) to support the front end of the input shaft of the transmission.
Pip mark - A little dot or indentation which indicates the top side of a compression ring.
Piston - The cylindrical part, attached to the connecting rod, that moves up and down in the cylinder as the crankshaft rotates. When the fuel charge is fired, the piston transfers the force of the explosion to the connecting rod, then to the crankshaft.
Piston pin (or wrist pin) - The cylindrical and usually hollow steel pin that passes through the piston. The piston pin fastens the piston to the upper end of the connecting rod.
Piston ring - The split ring fitted to the groove in a piston. The ring contacts the sides of the ring groove and also rubs against the cylinder wall, thus sealing space between piston and wall. There are two types of rings: Compression rings seal the compression pressure in the combustion chamber; oil rings scrape excessive oil off the cylinder wall.
Piston ring groove - The slots or grooves cut in piston heads to hold piston rings in position.
Piston skirt - The portion of the piston below the rings and the piston pin hole.
Plastigage - A thin strip of plastic thread, available in different sizes, used for measuring clearances. For example, a strip of plastigage is laid across a bearing journal and mashed as parts are assembled. Then parts are disassembled and the width of the strip is measured to determine clearance between journal and bearing. Commonly used to measure crankshaft main-bearing and connecting rod bearing clearances.
Press-fit - A tight fit between two parts that requires pressure to force the parts together. Also referred to as drive, or force, fit.
Prussian blue - A blue pigment; in solution, useful in determining the area of contact between two surfaces. Prussian blue is commonly used to determine the width and location of the contact area between the valve face and the valve seat.

R

Race (bearing) - The inner or outer ring that provides a contact surface for balls or rollers in bearing.
Ream - To size, enlarge or smooth a hole by using a round cutting tool with fluted edges.

Ring job - The process of reconditioning the cylinders and installing new rings.
Runout - Wobble. The amount a shaft rotates out-of-true.

S

Saddle - The upper main bearing seat.
Scored - Scratched or grooved, as a cylinder wall may be scored by abrasive particles moved up and down by the piston rings.
Scuffing - A type of wear in which there's a transfer of material between parts moving against each other; shows up as pits or grooves in the mating surfaces.
Seat - The surface upon which another part rests or seats. For example, the valve seat is the matched surface upon which the valve face rests. Also used to refer to wearing into a good fit; for example, piston rings seat after a few miles of driving.
Short block - An engine block complete with crankshaft and piston and, usually, camshaft assemblies.
Static balance - The balance of an object while it's stationary.
Step - The wear on the lower portion of a ring land caused by excessive side and back-clearance. The height of the step indicates the ring's extra side clearance and the length of the step projecting from the back wall of the groove represents the ring's back clearance.
Stroke - The distance the piston moves when traveling from top dead center to bottom dead center, or from bottom dead center to top dead center.
Stud - A metal rod with threads on both ends.

T

Tang - A lip on the end of a plain bearing used to align the bearing during assembly.
Tap - To cut threads in a hole. Also refers to the fluted tool used to cut threads.
Taper - A gradual reduction in the width of a shaft or hole; in an engine cylinder, taper usually takes the form of uneven wear, more pronounced at the top than at the bottom.
Throws - The offset portions of the crankshaft to which the connecting rods are affixed.
Thrust bearing - The main bearing that has thrust faces to prevent excessive end-play, or forward and backward movement of the crankshaft.
Thrust washer - A bronze or hardened steel washer placed between two moving parts. The washer prevents longitudinal movement and provides a bearing surface for thrust surfaces of parts.
Tolerance - The amount of variation permitted from an exact size of measurement. Actual amount from smallest acceptable dimension to largest acceptable dimension.

U

Umbrella - An oil deflector placed near the valve tip to throw oil from the valve stem area.
Undercut - A machined groove below the normal surface.
Undersize bearings - Smaller diameter bearings used with re-ground crankshaft journals.

V

Valve grinding - Refacing a valve in a valve-refacing machine.
Valve train - The valve-operating mechanism of an engine; includes all components from the camshaft to the valve.
Vibration damper - A cylindrical weight attached to the front of the crankshaft to minimize torsional vibration (the twist-untwist actions of the crankshaft caused by the cylinder firing impulses). Also called a harmonic balancer.

W

Water jacket - The spaces around the cylinders, between the inner and outer shells of the cylinder block or head, through which coolant circulates.
Web - A supporting structure across a cavity.
Woodruff key - A key with a radiused backside (viewed from the side).

2C-22 GENERAL ENGINE OVERHAUL PROCEDURES

10.21 Use the scale on the Plastigage package to determine the bearing oil clearance - be sure to measure the widest part of the Plastigage and use the correct scale; it comes with both standard and metric scales

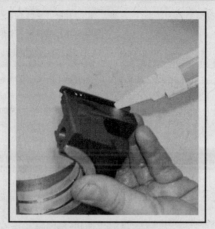

10.27a Before installing the rear main bearing cap on V6 engines, fill the side grooves with RTV sealant . . .

10.27b . . . then install the cap

If the main bearing caps (V6 models) are difficult to remove, tap them gently from side-to-side with a soft-face hammer to loosen it.

21 Compare the width of the crushed Plastigage on each journal to the scale printed on the Plastigage envelope to determine the main bearing oil clearance (see illustration). Check with an automotive machine shop for the oil clearance for your engine.

22 If the clearance is not as specified, the bearing inserts may be the wrong size (which means different ones will be required). Before deciding if different inserts are needed, make sure that no dirt or oil was between the bearing inserts and the caps or block when the clearance was measured. If the Plastigage was wider at one end than the other, the crankshaft journal may be tapered. If the clearance still exceeds the limit specified, the bearing insert(s) will have to be replaced with an undersize bearing insert(s).

❊❊ CAUTION:

When installing a new crankshaft always install a standard bearing insert set.

23 Carefully scrape all traces of the Plastigage material off the main bearing journals and/or the bearing insert faces. Be sure to remove all residue from the oil holes. Use your fingernail or the edge of a plastic card - don't nick or scratch the bearing faces.

FINAL INSTALLATION

▶ Refer to illustrations 10.27a, 10.27b, 10.29a, 10.29b, 10.29c, 10.29d and 10.30

24 Carefully lift the crankshaft out of the cylinder block.

25 Clean the bearing insert faces in the cylinder block, then apply a thin, uniform layer of moly-base grease or engine assembly lube to each of the bearing surfaces. Be sure to coat the thrust faces as well as the journal face of the thrust bearing.

26 Make sure the crankshaft journals are clean, then lay the crankshaft back in place in the cylinder block.

27 Clean the bearing insert faces and apply the same lubricant to them. Clean the engine block and the mating surface of the lower

10.29a After backing out the bearing bridge adjusting sleeves and tightening the bridge/bearing cap bolts to the proper torque, tighten the sleeves to the torque listed in this Chapter's Specifications . . .

crankcase or the bearing caps thoroughly. The surfaces must be free of oil residue. Install the lower crankcase (four-cylinder engine) or the bearing caps and bridge (V6 engine).

➡ Note: On V6 engines, before installing the rear main bearing cap, fill the side grooves of the cap with RTV sealant (see illustrations).

28 Prior to installation, apply clean engine oil to all bolt threads, wiping off any excess, then install all bolts finger-tight.

❊❊ CAUTION:

Remember, new bolts must be used.

29 Tighten the bolts to the torque listed in this Chapter's Specifications following the correct torque sequence (see illustrations 10.19a and 10.19b).

➡ Note: On V6 engines, it will be necessary to adjust the height of the crankshaft bearing bridge adjusting sleeves. With the bridge/bearing cap bolts loose, unscrew the sleeves until they are not contacting the engine block. Tighten the main bearing cap bolts (in the correct sequence) to the torque listed in this

GENERAL ENGINE OVERHAUL PROCEDURES 2C-23

10.29b ... then install the crankshaft outer bearing bridge bolts ...

10.29c ... and tighten them to the torque listed in this Chapter's Specifications

10.29d Inject RTV sealant down each of the rear main bearing cap grooves until it emerges through the joints

Chapter's Specifications, then apply thread locking compound to the sleeves and tighten them to the torque listed in this Chapter's Specifications (see illustration). Finally, install and tighten the outer crankshaft bearing bridge bolts to the torque listed in this Chapter's Specifications (see illustrations). Once all of the bridge/bearing cap bolts have been properly tightened, inject sealant down into the grooves in the rear main bearing cap (see illustration).

30 On four-cylinder engines, install the lower crankcase perimeter bolts and tighten them to the torque listed in this Chapter's Specifications (see illustration).

31 Recheck the crankshaft endplay with a feeler gauge or a dial indicator. The endplay should be correct if the crankshaft thrust faces aren't worn or damaged and if new bearings have been installed.

32 Rotate the crankshaft a number of times by hand to check for any obvious binding. It should rotate with a running torque of 50 in-lbs or less. If the running torque is too high, correct the problem at this time.

33 Install the new rear main oil seal (see Chapter 2A).

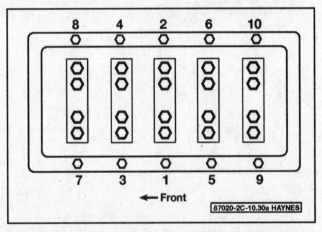

10.30 Lower crankcase perimeter bolt tightening sequence on the four-cylinder engine

11 Engine overhaul - reassembly sequence

1 Before beginning engine reassembly, make sure you have all the necessary new parts, gaskets and seals as well as the following items on hand:

   Common hand tools
   A 1/2-inch drive torque wrench
   New engine oil
   Gasket sealant
   Thread locking compound

2 If you obtained a short block it will be necessary to install the cylinder head, the oil pump and pick-up tube, the oil pan, the water pump, the timing belt or chain and timing cover, and the valve cover (see Chapter 2A or 2B). In order to save time and avoid problems, the external components must be installed in the following general order:

   Thermostat and housing cover
   Water pump
   Intake and exhaust manifolds
   Fuel injection components
   Emission control components
   Spark plug wires and spark plugs
   Ignition coils or coil packs
   Oil filter
   Engine mounts and mount brackets
   Clutch and flywheel (manual transaxle)
   Driveplate (automatic transaxle)

12 Initial start-up and break-in after overhaul

WARNING:
Have a fire extinguisher handy when starting the engine for the first time.

1 Once the engine has been installed in the vehicle, double-check the engine oil and coolant levels.

2 With the spark plugs out of the engine and the ignition system and fuel pump disabled, crank the engine until oil pressure registers on the gauge or the light goes out.

3 Install the spark plugs, hook up the plug wires and restore the ignition system and fuel pump functions.

4 Start the engine. It may take a few moments for the fuel system to build up pressure, but the engine should start without a great deal of effort.

5 After the engine starts, it should be allowed to warm up to normal operating temperature. While the engine is warming up, make a thorough check for fuel, oil and coolant leaks.

6 Shut the engine off and recheck the engine oil and coolant levels.

7 Drive the vehicle to an area with minimum traffic, accelerate from 30 to 50 mph, then allow the vehicle to slow to 30 mph with the throttle closed. Repeat the procedure 10 or 12 times. This will load the piston rings and cause them to seat properly against the cylinder walls. Check again for oil and coolant leaks.

8 Drive the vehicle gently for the first 500 miles (no sustained high speeds) and keep a constant check on the oil level. It is not unusual for an engine to use oil during the break-in period.

9 At approximately 500 to 600 miles, change the oil and filter.

10 For the next few hundred miles, drive the vehicle normally. Do not pamper it or abuse it.

11 After 2,000 miles, change the oil and filter again and consider the engine broken in

GENERAL ENGINE OVERHAUL PROCEDURES   2C-25

Specifications

General

Displacement	
Four-cylinder models	134 cubic inches (2.2 liters)
V6 models	182 cubic inches (3.0 liters)
Bore and Stroke	
Four-cylinder models	3.44 x 3.73 inches (86.0 x 94.6 mm)
V6 models	3.44 x 3.40 inches (86.0 x 85.0 mm)
Cylinder compression	Lowest cylinder must be within 75 percent of highest cylinder
Oil pressure (engine at operating temperature)	
Four-cylinder models	50 to 80 psi (344 to 551 kPa) at 1,000 rpm
V6 models	22 psi (150 kPa) at idle

Torque specifications

	Ft-lbs (unless otherwise indicated)	Nm
Subframe mounting bolts*	See Chapter 10	
Connecting rod bearing cap bolts*		
Four-cylinder models		
Step 1	18	25
Step 2	Tighten an additional 100-degrees	
V6 models		
Step 1	26	35
Step 2	Tighten an additional 45-degrees	
Step 3	Tighten an additional 15-degrees	
Lower crankcase bolts* (four-cylinder engine) (see illustration 10.19a)		
Step 1	15	20
Step 2	Tighten an additional 70-degrees	
Lower crankcase perimeter bolts		
(see illustration 10.30)	18	25
Main bearing bridge/cap bolts* (V6 engine) (see illustration 10.19b)		
Step 1 (bolts 1 through 8)	37	50
Step 2 (bolts 1 through 8)	Tighten an additional 60-degrees	
Step 3 (bolts 1 through 8)	Tighten an additional 15-degrees	
Bearing bridge adjusting sleeves	53 in-lbs	6
Outer bearing bridge bolts (V6 engines) (see illustration 10.19b)		
(bolts 9 through 12)	15	20

Bolt(s) must be replaced.

2C-26 GENERAL ENGINE OVERHAUL PROCEDURES

Notes

3

Cooling, heating and air conditioning systems

Section
1. General information
2. Antifreeze - general information
3. Thermostat - check and replacement
4. Engine cooling fans - check and replacement
5. Coolant expansion tank - removal and installation
6. Radiator - removal and installation
7. Water pump - check
8. Water pump, heater core pump (V6 models) and auxiliary pump (V6 models) - replacement
9. Engine oil cooler (V6 models) - removal and installation
10. Coolant temperature sending unit - check and replacement
11. Blower motor resistor and blower motor - replacement
12. Heater/air conditioner control assembly - removal and installation
13. Heater core - replacement
14. Air conditioning and heating system - check and maintenance
15. Air conditioning compressor - removal and installation
16. Air conditioning receiver-drier - removal and installation
17. Air conditioning condenser - removal and installation
18. Air conditioning pressure sensor - replacement

Reference to other Chapters
Coolant level check - See Chapter 1
Cooling system check - See Chapter 1
Cooling system servicing (draining, flushing and refilling) - See Chapter 1
Drivebelt check, adjustment and replacement - See Chapter 1
Underhood hose check and replacement - See Chapter 1

3-2 COOLING, HEATING AND AIR CONDITIONING SYSTEMS

1 General information

ENGINE COOLING SYSTEM

The cooling system consists of a radiator, an expansion tank, a pressure cap (located on the expansion tank), a thermostat, a cooling fan and clutch, and a belt-driven water pump (see illustrations).

The expansion tank functions somewhat differently than a conventional recovery tank. Designed to separate any trapped air in the coolant, it is pressurized by the radiator and has a pressure cap on top. The radiator on these models does not have a pressure cap.

✱✱ WARNING:

Unlike a conventional coolant recovery tank, the pressure cap on the expansion tank should never be opened after the engine has warmed up, because of the danger of severe burns caused by steam or scalding coolant.

When the engine is cold, the thermostat restricts the circulation of coolant to the engine. When the minimum operating temperature is reached, the thermostat begins to open, allowing coolant to return to the radiator.

TRANSAXLE COOLING SYSTEMS

Vehicles with an automatic transaxle are equipped with a transaxle cooler, located inside the radiator, which cools the transaxle fluid. The transaxle is connected to the cooler by a pair of hoses: one delivers hot transaxle fluid to the radiator and the other brings the cooled fluid back to the transaxle.

For more information on transaxle oil coolers, refer to Chapter 7.

1.1a Cooling, heating and air conditioning components (underhood) - four-cylinder models

1. Air conditioning refrigerant lines
2. Heater hoses-to-air conditioning/heater housing
3. Thermostat (located in thermostat housing on back of engine block)
4. Expansion tank
5. Engine cooling fan module
6. Puller fan
7. Radiator

COOLING, HEATING AND AIR CONDITIONING SYSTEMS

ENGINE OIL COOLING SYSTEM ON V6 MODELS

Besides the engine and transaxle cooling systems described above, engine heat is also dissipated through an external oil cooler that's integrated into the lubrication system. The oil cooler helps keep engine and oil temperatures within design limits under extreme load conditions.

The oil cooling system on these models consists of a radiator type housing (oil cooler) mounted between the cylinder heads, a heat exchanger inside the oil cooler and a pair of hoses that deliver oil from the engine to the oil cooler housing.

HEATING SYSTEM

The heating system consists of the heater controls, the heater core, the heater blower assembly (which houses the blower motor and the blower motor resistor), and the hoses connecting the heater core to the engine cooling system. Hot engine coolant is circulated through the heater core. When the heater mode is activated, a flap door opens to expose the heater box to the passenger compartment. A fan switch on the heater controls activates the blower motor, which forces air through the core, heating the air.

V6 models are equipped with an electric heater core pump that circulates coolant into the heater core for direct and fast heating action.

HEATER CORE PUMP AND AUXILIARY WATER PUMP (V6 MODELS)

The heater core pump is an electrically operated pump designed to circulate coolant directly to the heater core for faster heater response. The heater core pump is mounted on the back of the engine block between the heater hoses and the block outlet.

The auxiliary water pump is mounted adjacent to the puller fan in front of the radiator. This pump cycles coolant through the radiator after the engine has shut down to prevent coolant boil-over. The auxiliary pump will activate along with the puller and pusher fans to work rapidly to cool the coolant in the radiator. The auxiliary water pump only activates after the engine has been shut off.

AIR CONDITIONING SYSTEM

The air conditioning system consists of the condenser, which is mounted in front of the radiator, a receiver-drier mounted next to the condenser, the evaporator case assembly under the dash, a compressor mounted on the engine, and the plumbing connecting all of the above components.

A blower fan forces the warmer air of the passenger compartment through the evaporator core (sort of a radiator-in-reverse), transferring the heat from the air to the refrigerant. The liquid refrigerant boils off into low pressure vapor, taking the heat with it when it leaves the evaporator.

1.1b Cooling, heating and air conditioning components (underhood) - V6 models

1. Expansion tank
2. Engine cooling fan module
3. Puller fan
4. Radiator
5. Water pump (located behind the timing belt cover)

3-4 COOLING, HEATING AND AIR CONDITIONING SYSTEMS

2 Antifreeze - general information

♦ Refer to illustration 2.5

※※ WARNING:

Do not allow antifreeze to come in contact with your skin or painted surfaces of the vehicle. Rinse off spills immediately with plenty of water. Antifreeze is highly toxic if ingested. Never leave antifreeze lying around in an open container or in puddles on the floor; children and pets are attracted by its sweet smell and may drink it. Check with local authorities about disposing of used antifreeze. Many communities have collection centers which will see that antifreeze is disposed of safely. Never dump used antifreeze on the ground or pour it into drains.

➡ Note: Non-toxic antifreeze is now manufactured and available at local auto parts stores, but even this type must be disposed of properly.

The cooling system should be filled with a water/ethylene glycol based antifreeze solution, which will prevent freezing down to at least -20-degrees F (even lower in cold climates). It also provides protection against corrosion and increases the coolant boiling point. The engines in these vehicles have aluminum heads. Depending on the engine and model year, the specified coolant varies (see the Chapter 1 Specifications). The manufacturer recommends that the correct type of coolant be used and strongly urges that coolant types not be mixed.

Drain, flush and refill the cooling system at least every other year (see Chapter 1). The use of antifreeze solutions for periods of longer than two years is likely to cause damage and encourage the formation of rust and scale in the system.

Before adding antifreeze to the system, inspect all hose connections. Antifreeze can leak through very minute openings.

The exact mixture of antifreeze to water, which you should use, depends on the relative weather conditions. The mixture should contain

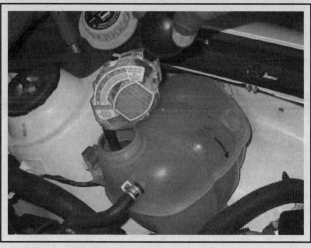

2.5 Use a hydrometer (available at auto parts stores) to test the condition of your coolant

at least 50-percent antifreeze, but should never contain more than 70-percent antifreeze. Consult the mixture ratio chart on the container before adding coolant.

Hydrometers are available at most auto parts stores to test the coolant (see illustration). Use antifreeze that meets Saturn specifications for engines with aluminum heads.

※※ WARNING:

Do not remove the coolant tank cap, drain the coolant or replace the thermostat until the engine has cooled completely.

3 Thermostat - check and replacement

CHECK

1 Before assuming the thermostat is to blame for a cooling system problem, check the coolant level, drivebelt tension (see Chapter 1) and temperature gauge operation.

2 If the engine seems to be taking a long time to warm up, based on heater output or temperature gauge operation, the thermostat is probably stuck open. Replace the thermostat with a new one.

3 If the engine runs hot, use your hand to check the temperature of the lower radiator hose. If the hose isn't hot, but the engine is, the thermostat is probably stuck closed, preventing the coolant inside the engine from escaping to the radiator. Replace the thermostat.

※※ CAUTION:

Don't drive the vehicle without a thermostat. The computer may stay in open loop and emissions and fuel economy will suffer.

4 If the lower radiator hose is hot, it means that the coolant is flowing and the thermostat is open. Consult the *Troubleshooting* section at the front of this manual for cooling system diagnosis.

REPLACEMENT

※※ WARNING:

The engine must be completely cool before beginning this procedure.

5 Disconnect the cable from the negative battery terminal (see Chapter 5, Section 1).

6 Drain the cooling system (see Chapter 1). If the coolant is relatively new and still in good condition, save it and reuse it.

Four-cylinder models

2000 through 2003 models

♦ Refer to illustrations 3.7, 3.11, 3.12 and 3.14

7 Remove the water pump drain bolt and drain any excess coolant into a container (see illustration).

8 Follow the lower radiator hose to the engine to locate the thermostat housing.

9 Loosen the hose clamp, then detach the hose from the fitting. If

COOLING, HEATING AND AIR CONDITIONING SYSTEMS

3.7 Location of the water pump drain bolt (four-cylinder models)

3.11 Remove the thermostat cover mounting bolts and separate the cover from the housing to access the thermostat (four-cylinder engine)

it's stuck, grasp it near the end with a pair of adjustable pliers and twist it to break the seal, then pull it off. If the hose is old or if it has deteriorated, cut it off and install a new one.

10 If the outer surface of the thermostat cover, which mates with the hose, is already corroded, pitted, or otherwise deteriorated, it might be damaged even more by hose removal. If it is, replace the thermostat cover.

11 Remove the fasteners and detach the thermostat cover (see illustration). If the cover is stuck, tap it with a soft-face hammer to jar it loose. Be prepared for some coolant to spill as the gasket seal is broken.

12 Note how it's installed, which end is facing up, or out and then remove the thermostat (see illustration).

13 Remove all traces of old gasket material and sealant from the housing and cover with a gasket scraper.

14 Install a new rubber gasket on the thermostat (see illustration) and install the thermostat in the housing, spring-end first.

15 Install the thermostat cover and bolts, then tighten the bolts to the torque listed in this Chapter's Specifications.

16 Reattach the radiator hose to the outlet pipe on the thermostat cover. Make sure that the hose clamp is tight. If it isn't, replace it.

17 The remaining installation is the reverse of the removal.

2004 models

18 Remove the exhaust manifold heat shield (see Chapter 2A).

19 Remove the water pump drain bolt and drain any excess coolant into a container (see illustration 3.7).

20 Raise the vehicle and support it securely on jackstands.

21 Remove the thermostat cover bolts, then separate the cover from the housing and the water pipe from the water pump, using a twisting motion.

22 Remove the inner sleeve from the thermostat housing. Note the location of the notch on the lower section of the sleeve. Note how the thermostat is installed (which end is facing up, or out), then remove the thermostat.

23 Install the thermostat cartridge in the housing, aligning the tangs on the cartridge with the bolt holes in the housing. Install the sleeve into the housing, aligning the notch on the sleeve with the groove in the housing.

24 Install a new O-ring seal onto the water pipe. Insert the water pipe into the water pump and swing the thermostat housing cover into place.

25 Install the bolts and tighten them to the torque listed in this Chapter's Specifications.

26 The remaining installation is the reverse of the removal.

3.12 Remove the thermostat from the housing, noting how it is installed (four-cylinder model shown)

3.14 Install a new rubber gasket around the perimeter of the thermostat

3-6 COOLING, HEATING AND AIR CONDITIONING SYSTEMS

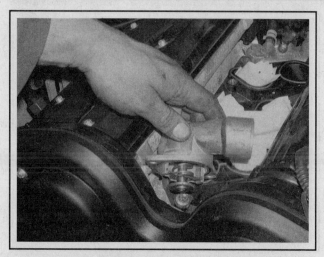

3.31 Lift the thermostat cover and the thermostat from the top of the engine block - V6 model shown

3.33 Install a new gasket

V6 models

▶ Refer to illustrations 3.31 and 3.33

27 Remove the upper intake manifold (see Chapter 2A).
28 Loosen the clamp and disconnect the radiator hose from the coolant pipe.
29 Remove the bolt that retains the engine lift bracket and the dipstick tube. Position the dipstick tube off to the side.
30 Remove the thermostat housing extension pipe. Remove the O-ring from the extension pipe.
31 Remove the fasteners and detach the thermostat cover (see illustration). If the cover is stuck, tap it with a soft-face hammer to jar it loose. Be prepared for some coolant to spill as the gasket seal is broken.
32 The thermostat can be compressed, turned and removed from the cover using the extension pipe. Remove all traces of old gasket material and sealant from the housing and cover with a gasket scraper.
33 If removed, install thermostat in the cover, making sure it is facing the proper direction and locks into place. Install a new rubber gasket in the thermostat cover (see illustration).
34 Install the thermostat cover and bolts and tighten the bolts to the torque listed in this Chapter's Specifications.
35 Reattach the thermostat housing extension pipe to the outlet pipe on the thermostat cover. Be sure to install a new O-ring.
36 The remaining installation is the reverse of the removal.

All models

37 Refill the cooling system (see Chapter 1).
38 Reconnect the battery (see Chapter 5, Section 1).
39 Start the engine and allow it to reach normal operating temperature, then check for leaks and proper thermostat operation (as described in Steps 2 through 4).

4 Engine cooling fans - check and replacement

❄ WARNING:

To avoid possible injury or damage, DO NOT operate the engine with a damaged fan. Do not attempt to repair fan blades - replace a damaged fan with a new one.

➡ Note: Always be sure to check for blown fuses before attempting to diagnose an electrical circuit problem.

CHECK

▶ Refer to illustrations 4.2 and 4.4

1 These models are equipped with a puller fan (radiator fan) mounted on the radiator located inside the engine compartment and a pusher fan (condenser fan) located in front of the condenser behind the bumper cover. A fan control module and the PCM regulate the speed and the duration of both fans according to engine temperature and air conditioning usage. The fans are protected by COOL FAN #1 and COOL FAN #2 fuses located inside the fuse/relay box in the engine compartment.
2 If the engine is overheating and the cooling fan is not coming on when the engine temperature rises to an excessive level, unplug the fan

4.2 Location of the puller and pusher fan electrical connectors

motor electrical connector(s) (see illustration) and connect the motor directly to the battery with fused jumper wires. If the fan motor doesn't come on, replace the motor.

COOLING, HEATING AND AIR CONDITIONING SYSTEMS 3-7

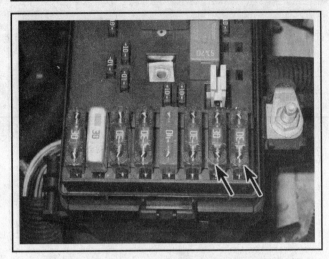

4.4 Check the condition of the COOL FAN #1 and COOL FAN #2 fuses

4.7 Lift up the cooling fan control module to release it from the bracket

3 If the radiator fan motor is okay, but it isn't coming on when the engine gets hot, the fan relay might be defective. A relay (housed inside the fan control module) is used to control a circuit by turning it on and off in response to a control decision by the PCM. These control circuits are fairly complex, and checking them should be left to a dealer service department.

4 Locate the fuses in the engine compartment fuse/relay box (see Chapter 12). Remove the fuse and check the fuses for continuity (see illustration).

5 If the fuses are okay, check all wiring and connections to the fan motor. If no obvious problems are found, the problem could be the engine coolant temperature (ECT) sensor or the fan control module. Have the cooling fan system and circuit diagnosed by a dealer service department or repair shop with the proper diagnostic equipment.

➡ **Note:** The puller fan on these models is equipped with a cooling fan motor resistor. Have the resistor checked if the fan motor does not respond to the speed variations signaled by the PCM.

REPLACEMENT

Puller fan

▶ Refer to illustration 4.7

4.11a Location of the upper left cooling fan shroud mounting bolt

6 Disconnect the cable from the negative battery terminal (see Chapter 5, Section 1).

7 Lift the fan control module off the bracket and position it off to the side without disconnecting the module connectors (see illustration).

8 Remove the battery (see Chapter 5).

9 Disconnect the fan motor electrical connector(s) (see illustration 4.2).

Four-cylinder models

▶ Refer to illustrations 4.11a and 4.11b

10 Disconnect the fan motor harness clips from the fan shroud and position the harness off to the side.

11 Remove the fan motor mounting bolts (see illustrations).

12 Remove the puller fan assembly from the radiator.

V6 models

✳✳ **WARNING:**

The engine must be completely cool before beginning this procedure.

4.11b Location of the upper right cooling fan shroud mounting bolt

3-8 COOLING, HEATING AND AIR CONDITIONING SYSTEMS

4.28 Remove the lower pusher fan mounting bolts and release the upper shroud clips by pulling down - bumper and cover removed for clarity

4.29 Remove the pusher fan from below the bumper

13 Remove the power steering fluid reservoir mounting bolts from the fan shroud and position the reservoir off to the side without disconnecting the fluid lines.

14 Disconnect the upper transaxle cooler line from the radiator (see Chapter 7B). Plug the line to prevent fluid loss.

15 Remove the cooler line retainer from the radiator to make extra room for fan removal.

16 Drain the engine coolant (see Chapter 1).

17 Remove the radiator upper hose. Loosen the hose clamp, then detach the hose from the fitting. If it's stuck, grasp it near the end with a pair of adjustable pliers and twist it to break the seal, then pull it off. If the hose is old or if it has deteriorated, cut it off and install a new one.

18 Disconnect the fan motor harness clips from the fan shroud and position the harness off to the side.

19 Remove the fan motor mounting bolts.

20 Remove the puller fan assembly by sliding the assembly up and off the radiator.

All models

21 At the time of writing, the puller fan assembly is sold only as a complete assembly. Consult with a dealer parts department or other automotive parts store.

22 Installation is the reverse of removal.

➡ **Note:** *When reinstalling the fan assembly, make sure the lower fan cutouts lock into the tabs located on the radiator.*

23 Reconnect the battery (see Chapter 5, Section 1). If you're working on a V6 model, refill the cooling system (see Chapter 1).

24 Start the engine and allow it to reach normal operating temperature, then check for leaks and proper thermostat operation (as described in Steps 2 through 4 in Section 3).

Pusher fan

♦ Refer to illustrations 4.28 and 4.29

25 Disconnect the cable from the negative battery terminal (see Chapter 5).

26 Disconnect the fan motor electrical connector(s) (see illustration 4.2).

27 Raise the vehicle and support it securely on jackstands.

28 Remove the fan motor mounting bolts (see illustration).

29 Lower the top section of the pusher fan shroud to release it from the clips and remove the pusher fan assembly from the condenser (see illustration).

30 At the time of writing, the pusher fan assembly can only be obtained as a complete assembly. Consult with a dealer parts department or other automotive parts store for parts availability.

31 Installation is the reverse of removal.

32 Reconnect the battery (see Chapter 5, Section 1).

33 Start the engine and allow it to reach normal operating temperature, then check for leaks and proper fan operation (as described in Steps 2 through 4 in Section 3).

5 Coolant expansion tank - removal and installation

♦ Refer to illustrations 5.2 and 5.3

※ WARNING:

Wait until the engine is completely cool before beginning this procedure.

1 Drain the cooling system (see Chapter 1).

2 Disconnect the expansion tank return hose and air bleed hose (see illustration). Plug the hoses to prevent leakage.

3 Remove the expansion tank mounting clip (see illustration).

4 Disconnect the Low Coolant Level sensor connector.

5 Angle the cutout away from the metal tab on the fenderwell and lift the expansion tank out of the engine compartment.

6 Clean out the tank with soapy water and a brush to remove any deposits inside. Inspect the reservoir carefully for cracks. If you find a crack, replace the reservoir.

7 Installation is the reverse of removal.

COOLING, HEATING AND AIR CONDITIONING SYSTEMS 3-9

5.2 Location of the return hose and the air bleed hose on the expansion tank

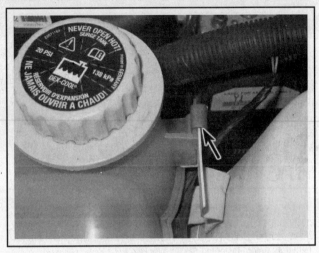

5.3 Lift up on the clip to release it from the expansion tank

6 Radiator - removal and installation

WARNING:

Wait until the engine is completely cool before beginning this procedure.

REMOVAL

▶ Refer to illustrations 6.6a, 6.6b and 6.8

1 Disconnect the cable from the negative battery terminal (see Chapter 5, Section 1).
2 Drain the cooling system (see Chapter 1). If the coolant is relatively new and in good condition, save it and reuse it.
3 Remove the puller cooling fan (see Section 4).
4 Raise the vehicle and place it securely on jackstands.
5 On automatic transaxle models, disconnect the transaxle oil cooler lines from the radiator.
6 Disconnect the upper and lower radiator hoses from the radiator (see illustrations). Loosen the hose clamp by squeezing the ends together. Hose clamp pliers work best, but regular pliers will work also. If the radiator hose is stuck, grasp it near the end with a pair of adjustable pliers and twist it to break the seal, then pull it off. If the hose is old or if it has deteriorated, cut it off and install a new one.
7 Remove the bolts from the condenser and radiator mounting brackets (see Section 17).
8 Remove the radiator support brackets from the radiator support beam and the radiator (see illustration).

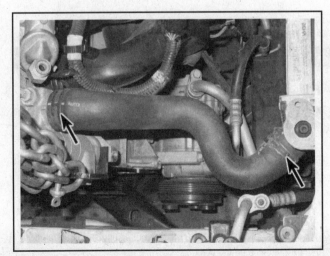

6.6a Location of the upper radiator hose clamps

6.6b Location of the lower radiator hose clamp

3-10 COOLING, HEATING AND AIR CONDITIONING SYSTEMS

9 Pull the condenser and pusher fan assembly down slightly to disengage the condenser from the radiator brackets.
10 Carefully lift out the radiator. Don't spill coolant on the vehicle or scratch the paint.
11 Make sure the rubber radiator insulators that fit on the bottom of the radiator and into the sockets in the body remain in place in the body for proper reinstallation of the radiator.
12 Remove bugs and dirt from the radiator with compressed air and a soft brush. Don't bend the cooling fins. Inspect the radiator for leaks and damage. If it requires repair, have a radiator shop do the work.

INSTALLATION

13 Inspect the rubber insulators in the lower crossmember for cracks and deterioration. Make sure that they're free of dirt and gravel. When installing the radiator, make sure that it's correctly seated on the insulators before fastening the top brackets.
14 Installation is otherwise the reverse of the removal procedure. After installation, fill the cooling system with the correct mixture of antifreeze and water (see Chapter 1).
15 Reconnect the battery (see Chapter 5, Section 1).
16 Start the engine and check for leaks. Allow the engine to reach normal operating temperature, indicated by the upper radiator hose becoming hot. Recheck the coolant level and add more if required.
17 If you're working on an automatic transaxle equipped vehicle, check and add fluid as needed.

6.8 Remove the radiator support bracket bolts and separate the brackets from the radiator and condenser assembly

7 Water pump - check

1 A failure in the water pump can cause serious engine damage due to overheating.
2 If a failure occurs in the pump seal, coolant will leak from the engine front cover (four-cylinder models) or the engine timing belt cover (V6 models).
3 Water pumps are equipped with weep or vent holes. It is possible to check the water pump weep hole using a flashlight. If a failure occurs in the pump seal, coolant will leak from the hole. Use the flashlight to find the vent hole on the water pump and check for leaks.
→ **Note:** *The vent hole on the four-cylinder engine water pump is located on top of the pump body while the vent hole on the V6 engine water pump is located under the pump body. Because the water pump is mounted behind the timing belt cover on V6 engines, use a flashlight pointed to the bottom of the cover when looking for signs of coolant leakage, or remove the cover and inspect.*
4 If the water pump shaft bearings fail, there may be a howling sound near the water pump while it's running. With the engine off, shaft wear can be felt if the water pump pulley is rocked up-and-down. Don't mistake drivebelt slippage, which causes a squealing sound, for water pump bearing failure.
5 A quick water pump performance check is to turn the heater on. If the pump is failing, it might not be able to efficiently circulate hot water all the way to the heater core as it should.

8 Water pump, heater core pump (V6 models) and auxiliary pump (V6 models) - replacement

※ WARNING:
The engine must be completely cool before beginning this procedure.

WATER PUMP

1 Disconnect the cable from the negative battery terminal (see Chapter 5).
2 Drain the cooling system (see Chapter 1).
3 Remove the drivebelt (see Chapter 1).

Four-cylinder models

▶ Refer to illustrations 8.10a, 8.10b, 8.11, 8.14a, 8.14b, 8.16a and 8.16b

4 Remove the air intake duct and the air filter housing (see Chapter 4).
5 Remove the exhaust manifold heat shield (see Chapter 2A).
6 Remove the plug from the bottom of the water pump housing (see illustration 3.7) and drain the excess coolant into a container.
7 Remove the lower radiator hose (see Section 6).
8 On 2004 models, remove the thermostat housing pipe bolt from the front of the engine block.

COOLING, HEATING AND AIR CONDITIONING SYSTEMS　3-11

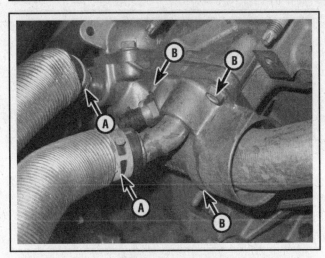

8.10a Remove the heater hoses (A) and thermostat housing mounting bolts (B) and separate the thermostat housing from the engine block and the water pump

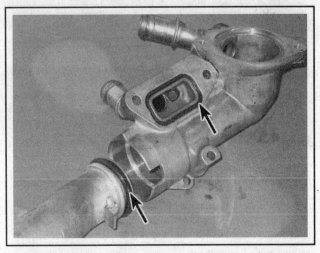

8.10b Remove the old seals and replace them with new ones

9   Disconnect the ECT harness connector (see Chapter 6).

10   Detach the heater hoses and remove the thermostat housing mounting bolts (see illustration), then separate the thermostat housing from the engine block and the water pump. Discard the old seals from the water pipe (see illustration) and replace them with new seals.

11   Remove the water pump access cover from the engine front cover (see illustration).

12   Loosen the right front wheel lug nuts. Raise the vehicle and support it securely on jackstands.

13   Remove the right front wheel, followed by the fender splash shield (see Chapter 2A).

14   Install a special holding tool onto the water pump sprocket (tool J-43651, available from specialty tool manufacturers and some dealer service departments). A tool can be fabricated if necessary (see illustrations). Be sure to lock the tool carefully, not allowing any sprocket movement. The bolts of the special tool will thread into the holes in the water pump sprocket that aren't for the sprocket bolts.

➡**Note: The water pump sprocket tool will lock the sprocket into position, allowing the balance shaft chain to remain in its timed state while the water pump is being replaced.**

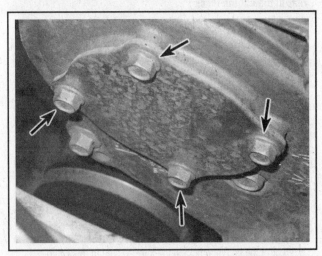

8.11 Remove the water pump access cover mounting bolts

If you're using a homemade tool like the one shown in the illustration, remove one of the water pump bolts first.

15   Remove the water pump pulley bolts.

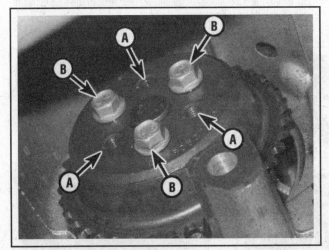

8.14a Before installing the special tool onto the water pump sprocket, note the location of the bolt holes for the sprocket tool (A) and the sprocket bolts (B)

8.14b Remove one sprocket bolt, install the special tool and lock the sprocket into position before removing the other two sprocket bolts

3-12 COOLING, HEATING AND AIR CONDITIONING SYSTEMS

8.16a Remove the two water pump bolts on the rear of the engine block . . .

8.16b . . . and the front of the engine block

16 Remove the two bolts attaching the water pump to the front and two bolts attaching to the rear of the engine block and remove the pump from the engine (see illustrations). If the water pump is stuck, gently tap it with a soft-faced hammer to break the seal.

17 Clean the bolt threads and the threaded holes in the engine and remove any corrosion or sealant. Remove all traces of old gasket material from the sealing surfaces. Remove the sealing ring from the water pump (if the same pump is to be installed).

18 Install a new sealing ring into the groove in the pump. To install the new water pump, install a guide pin (threaded stud) into the water pump pulley to align the water pump sprocket with the water pump.

19 Install the water pump mounting bolts and tighten them loosely. Install two bolts into the water pump sprocket and tighten them loosely. Remove the guide pin and install the third water pump sprocket bolt.

20 Tighten the water pump mounting bolts (two in the front and two in the rear of the engine block) to the torque listed in this Chapter's Specifications.

21 Tighten the water pump sprocket bolts to the torque listed in this Chapter's Specifications.

22 Install the water pump access cover and tighten the bolts to the torque listed in this Chapter's Specifications.

23 Install the thermostat housing and water pipe and tighten the bolts to the torque listed in this Chapter's Specifications. Lubricate the seal lightly with silicon gel before installing the water pipe into the water pump.

24 The remaining installation is the reverse of the removal. Proceed to Step 32.

V6 models

▶ Refer to illustrations 8.26a, 8.26b and 8.27

> **CAUTION:**
>
> If the water pump has been leaking coolant onto the timing belt, it will be necessary to replace the timing belt.

25 Remove the engine timing belt cover (see Chapter 2B). Inspect the condition of the timing belt.

26 Remove the water pump mounting bolts (see illustrations).

27 Clean all the gasket and O-ring surfaces on the pump (see illustration) (and the housing, if the same pump is to be reinstalled).

28 Compare the new pump to the old one to make sure that they're identical.

29 Apply a thin film of RTV sealant to hold the new gasket in place during installation. Carefully mate the pump to the block.

30 Install the water pump bolts and tighten them to the torque listed in this Chapter's Specifications.

31 The remainder of installation is the reverse of removal.

8.26a Removing the water pump mounting bolts on the V6 engine

8.26b Separate the water pump from the engine block

8.27 Remove the old gasket from the engine block

COOLING, HEATING AND AIR CONDITIONING SYSTEMS

All models

32 Refill the cooling system (see Chapter 1) when you're done.
33 Reconnect the battery (see Chapter 5, Section 1).
34 Operate the engine to check for leaks.

HEATER CORE PUMP (V6 MODELS)

→Note: The heater core pump is an electrically operated pump designed to circulate coolant directly to the heater core for faster heater response. The heater core pump is mounted on the back of the engine block.

35 Disconnect the cable from the negative battery terminal (see Chapter 5, Section 1).
36 Drain the cooling system (see Chapter 1).
37 Detach the heater hoses from the heater core pump.
38 Disconnect the heater core pump electrical connector.
39 Remove the bolt and bracket and remove the heater core pump from the engine compartment.
40 Installation is the reverse of removal.
41 Refill the cooling system (see Chapter 1) when you're done.
42 Reconnect the battery (see Chapter 5, Section 1).
43 Operate the engine to check for leaks.

AUXILIARY WATER PUMP (V6 MODELS)

→Note: The auxiliary water pump is mounted adjacent to the puller fan in front of the radiator. This pump cycles coolant through the radiator after the engine has shut down to prevent coolant boil-over. The auxiliary pump will activate along with the puller and pusher fans to work rapidly to cool the coolant in the radiator. The auxiliary water pump only activates after the engine has been shut off.

44 Disconnect the cable from the negative battery terminal (see Chapter 5, Section 1).
45 Drain the cooling system (see Chapter 1).
46 Remove the puller fan and shroud (see Section 4) from the radiator.
47 Detach the hoses from the auxiliary pump.
48 Disconnect the auxiliary pump electrical connector.
49 Remove the mounting clip and lift the pump from the engine compartment.
50 Installation is the reverse of removal.
51 Refill the cooling system (see Chapter 1) when you're done.
52 Reconnect the battery (see Chapter 5, Section 1).
53 Operate the engine to check for leaks.

9 Engine oil cooler (V6 models) - removal and installation

REMOVAL

▶ Refer to illustrations 9.4a, 9.4b, 9.5, 9.9, 9.11a, 9.11b, 9.13 and 9.14

1 Disconnect the cable from the negative battery terminal (see Chapter 5).
2 Drain the cooling system (see Chapter 1).
3 Remove the upper and lower intake manifolds and the intake manifold spacer (see Chapter 2B).
4 Remove the coolant crossover housing (see illustrations).

→Note: Be sure to remove the four upper and lower coolant crossover housing seals that may stick to the housing or the cylinder heads.

5 Remove the inlet and return lines from the oil cooler (see illustration).
6 Raise the vehicle and secure it on jackstands.

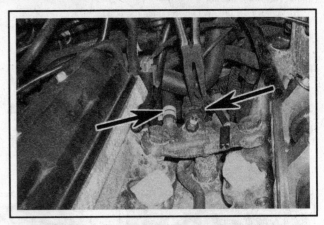

9.4a Disconnect the coolant hoses from the coolant crossover housing

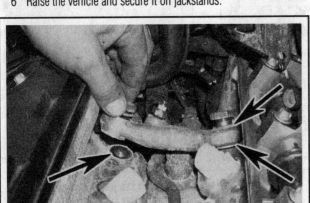

9.4b Remove the mounting bolts and lift the housing from the engine block

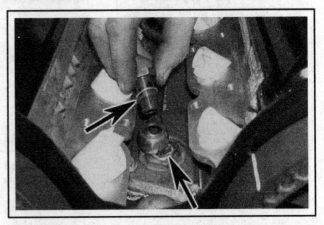

9.5 Be sure to install new sealing gaskets to the oil cooler lines

3-14 COOLING, HEATING AND AIR CONDITIONING SYSTEMS

9.9 Disconnect the oil cooler inlet and return lines from the engine block

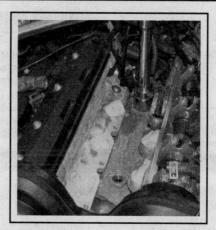

9.11a Remove the oil outlet nut . . .

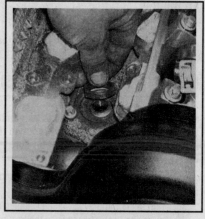

9.11b . . . and the inlet nut from the top of the oil cooler cover

7 Remove the oil filter (see Chapter 1) and the oil filter housing. The oil filter housing unscrews from the engine block.
8 Remove the crankshaft sensor (see Chapter 6).
9 Disconnect the oil cooler inlet and return lines from the engine block (see illustration). Label them for correct reassembly.
10 Lower the vehicle.
11 Remove the oil cooler inlet and outlet nuts from the top of the cooler (see illustrations).
12 Remove the oil cooler cover retaining bolts.
13 Remove the oil cooler cover (see illustration).
14 Lift the oil cooler off the engine block (see illustration).
15 Clean the area around the engine block and cylinder head. Remove any debris, deposits or material from the area.

INSTALLATION

▶ Refer to illustration 9.17

16 Install the engine oil cooler and new O-rings onto the engine block.

17 Apply a 2mm bead of RTV sealant to the perimeter of the oil cooler cover. Allow the sealant to set-up and install the cover (see illustration). Tighten the oil cooler cover bolts in sequence, starting with the inner bolts and working outward. Tighten the bolts to the torque listed in this Chapter's Specifications.
18 Install the engine oil cooler inlet and outlet nuts and torque them to the Specification listed in this Chapter.
19 Connect the inlet and outlet oil cooler pipes and tighten the fitting bolts to the Specification listed in this Chapter.

➡ Note: Be sure to use new sealing washers on either side of each pipe.

20 Raise the vehicle and support it on jackstands.
21 Install the oil cooler lines into the block above the oil filter housing (see illustration 9.9). Tighten the fittings securely.
22 The remaining installation is the reverse of the removal. Tighten the coolant crossover housing bolts to the torque listed in this Chapter's Specifications.
23 Reconnect the battery (see Chapter 5, Section 1).
24 Refill the cooling system (see Chapter 1), run the engine and check for leaks.

9.13 Lift the oil cooler cover from the engine

9.14 Lift the oil cooler off the engine block

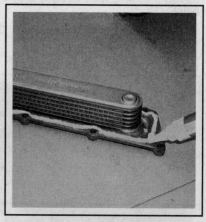

9.17 Apply a 2mm bead of RTV sealant to the perimeter of the oil cooler cover

COOLING, HEATING AND AIR CONDITIONING SYSTEMS   3-15

10 Coolant temperature sending unit - check and replacement

WARNING:
Wait until the engine is completely cool before beginning this procedure.

CHECK

1 The coolant temperature indicator system consists of a warning light or a temperature gauge on the dash and a coolant temperature sending unit mounted on the engine. On the models covered by this manual, the Engine Coolant Temperature (ECT) sensor, which is an information sensor for the Powertrain Control Module (PCM), also functions as the coolant temperature sending unit.

2 If an overheating indication occurs, check the coolant level in the system and then make sure all connectors in the wiring harness between the sending unit and the indicator light or gauge are tight.

3 When the ignition switch is turned to START and the starter motor is turning, the indicator light (if equipped) should come on. This doesn't mean the engine is overheated; it just means that the bulb is good.

4 If the light doesn't come on when the ignition key is turned to START, the bulb might be burned out, the ignition switch might be faulty or the circuit might be open.

5 As soon as the engine starts, the indicator light should go out and remain off, unless the engine overheats. If the light doesn't go out, the wire between the sending unit and the light could be grounded; the sending unit might be defective (have it checked by a dealer service department); or the ignition switch might be faulty (see Chapter 12). Check the coolant to make sure it's correctly mixed; plain water, with no antifreeze, or coolant that's mainly water, might have too low a boiling point to activate the sending unit (see Chapter 1).

REPLACEMENT

6 See Chapter 6.

11 Blower motor resistor and blower motor - replacement

WARNING:
The models covered by this manual are equipped with Supplemental Restraint systems (SRS), more commonly known as airbags. Always disarm the airbag system before working in the vicinity of any airbag system component to avoid the possibility of accidental deployment of the airbag, which could cause personal injury (see Chapter 12). Do not use a memory saving device to preserve the PCM's memory when working on or near airbag system components.

1 Disconnect the cable from the negative battery terminal (see Chapter 5, Section 1).

11.15 Location of the filter housing screws

BLOWER MOTOR RESISTOR

2 Remove the instrument panel (see Chapter 11).
3 Remove the right side heater outlet duct.
4 Disconnect the recirculation door actuator connector.
5 On vehicles equipped with automatic climate control, disconnect the air aspirator hose.
6 Remove the blower motor resistor mounting screws and then remove the resistor from the evaporator housing.
7 Disconnect the blower resistor card from the blower connector.
8 Installation is the reverse of removal.
9 Reconnect the battery (see Chapter 5, Section 1).

BLOWER MOTOR

▶ Refer to illustrations 11.15, 11.16 and 11.18

➡ **Note:** Models equipped with the automatic climate control system are also equipped with a blower motor control processor that is mounted behind the blower motor resistor.

10 Close the recirculation door by turning the ignition key on (engine not running), select the RECIRC to the OFF setting and turn the ignition key off.
11 Remove the wiper arms and the cowl cover (see Chapter 11).
12 Remove the weatherstrip from the trim along the edge of the engine compartment.
13 Remove the interior ventilation filter (see Chapter 1).
14 Remove the windshield wiper module (see Chapter 12).
15 Remove the filter housing (see illustration).
16 Remove the blower motor housing (see illustration).
17 Release the locking tabs and lift the blower motor housing.

3-16 COOLING, HEATING AND AIR CONDITIONING SYSTEMS

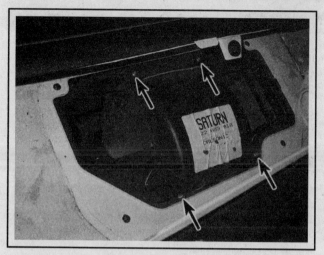

11.16 Location of the blower motor housing screws

11.18 Lift up slightly on the blower motor, disconnect the blower motor electrical connector and lift the blower motor assembly from the heater/air conditioning housing

18 Disconnect the blower motor connector and remove the blower motor from the heater/air conditioning housing (see illustration).

19 Installation is the reverse of removal. Make sure the rib on the blower motor housing slides into the slot in the heater/air conditioning housing.

20 Reconnect the battery (see Chapter 5, Section 1).

12 Heater/air conditioner control assembly - removal and installation

♦ Refer to illustrations 12.4, 12.5, 12.6 and 12.7

※※ WARNING:

The models covered by this manual are equipped with Supplemental Restraint systems (SRS), more commonly known as airbags. Always disarm the airbag system before working in the vicinity of any airbag system component to avoid the possibility of accidental deployment of the airbag, which could cause personal injury (see Chapter 12). Do not use a memory saving device to preserve the PCM's memory when working on or near airbag system components.

1 Disconnect the cable from the negative battery terminal (see Chapter 5, Section 1).
2 Remove the dashboard center trim panel (see Chapter 11).
3 Remove the radio (see Chapter 12).
4 Remove the heater/air conditioner control assembly retaining screws (see illustration).
5 Rotate the control assembly to access the backside. Disconnect the control assembly electrical connectors (see illustration).

Manual control assembly

6 Turn the temperature knob to the full hot position. Release the temperature cable by pushing the tab forward while simultaneously pushing down on the cable (see illustration).
7 Remove the temperature cable eyelet by squeezing the drive pin and lifting (see illustration).

Manual and automatic control assemblies

8 Installation is the reverse of removal.
9 Reconnect the battery (see Chapter 5, Section 1).

12.4 Remove the heater/air conditioner control mounting screws

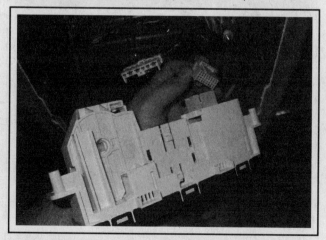

12.5 Disconnect the electrical connectors from the back of the heater/air conditioner control unit

COOLING, HEATING AND AIR CONDITIONING SYSTEMS   3-17

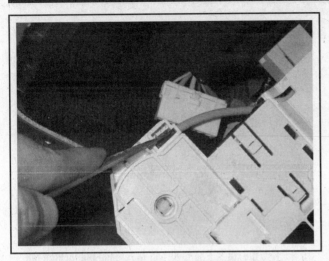

12.6 Release the temperature cable by pushing the tab forward while simultaneously pushing down on the cable

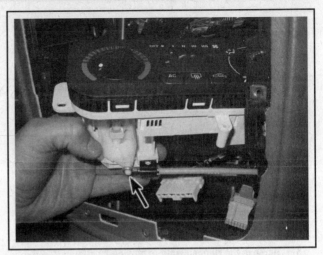

12.7 Push down on the tab on the drive pin to release the eyelet from the housing

13 Heater core - replacement

▶ Refer to illustrations 13.4a, 13.4b, 13.5, 13.6a, 13.6b, 13.7 and 13.8

✳✳ WARNING 1:

The models covered by this manual are equipped with Supplemental Restraint systems (SRS), more commonly known as airbags. Always disarm the airbag system before working in the vicinity of any airbag system component to avoid the possibility of accidental deployment of the airbag, which could cause personal injury (see Chapter 12). Do not use a memory saving device to preserve the PCM's memory when working on or near airbag system components.

✳✳ WARNING 2:

Wait until the engine is completely cool before beginning this procedure.

1 Disconnect the cable from the negative battery terminal (see Chapter 5, Section 1).
2 Drain the cooling system (see Chapter 1).
3 Disconnect the outlet heater hose at the thermostat housing (four-cylinder engines) or the heater core pump (V6 engines) and drain excess coolant from the heater core into an approved container.
4 Disconnect the heater hoses from the heater housing inlet and outlet pipes at the firewall (see illustration).
➡Note: The coolant hoses are equipped with special locking couplers at the heater housing on the firewall. Press the tab and pull the hose to disconnect it from the coupling device on the heater housing (see illustration).
5 Working in the passenger compartment, remove the right console extension (see illustration).
6 Remove the heater pipe clamps from the couplers at the heater core and remove the pipes from the heater core (see illustrations).

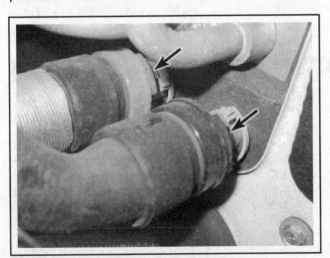

13.4a The heater hoses are connected to the heater/air conditioner housing with specialized locking couplers

13.4b Press the tab to release the coupler

3-18 COOLING, HEATING AND AIR CONDITIONING SYSTEMS

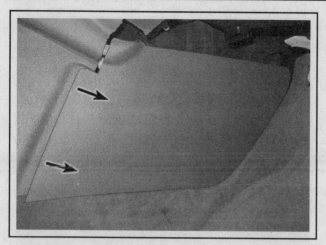

13.5 Remove the right console extension by sliding it to the rear, off the locking tabs

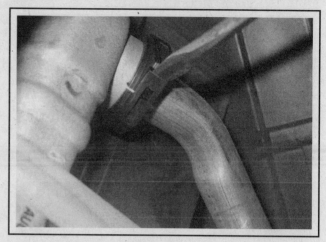

13.6a Use the tip of a screwdriver to release the locking tab on the heater pipe clamp

Place rags under the heater core pipes to catch any excess coolant that may spill.

7 Remove the heater core retaining strap mounting screw (see illustration).

8 Pull the heater core out of the housing (see illustration).

9 Installation is the reverse of removal. Don't forget to reconnect the heater core inlet and outlet hoses at the firewall.

10 Refill the cooling system when you're done (see Chapter 1).

11 Reconnect the battery (see Chapter 5, Section 1).

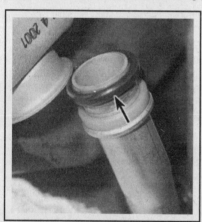

13.6b Be sure to replace the O-rings on the ends of the heater pipes

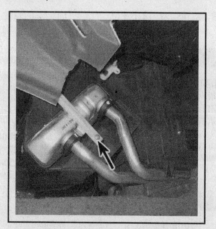

13.7 Remove the mounting screw and position the heater core strap off to the side

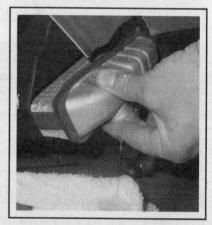

13.8 Remove the heater core from the heater/air conditioner housing

14 Air conditioning and heating system - check and maintenance

▶ Refer to illustration 14.1

WARNING:

The air conditioning system is under high pressure. Do not loosen any hose fittings or remove any components until after the system has been discharged by an air conditioning technician. Always wear eye protection when disconnecting air conditioning system fittings.

1 The following maintenance checks should be performed on a regular basis to ensure the air conditioner continues to operate at peak efficiency.

   a) *Check the compressor drivebelt. If it's worn or deteriorated, replace it* (see Chapter 1).

   b) *Check the drivebelt tension and, if necessary, adjust it* (see Chapter 1).

   c) *Check the system hoses. Look for cracks, bubbles, hard spots and deterioration. Inspect the hoses and all fittings for oil bubbles and seepage. If there's any evidence of wear, damage or leaks, replace the hose(s).*

   d) *Inspect the condenser fins for leaves, bugs and other debris. Use a "fin comb" or compressed air to clean the condenser.*

   e) *Make sure the system has the correct refrigerant charge.*

   f) *Check the evaporator housing drain tube* (see illustration) *for blockage.*

2 It's a good idea to operate the system for about 10 minutes at least once a month, particularly during the winter. Long term non-use can cause hardening, and subsequent failure, of the seals.

COOLING, HEATING AND AIR CONDITIONING SYSTEMS 3-19

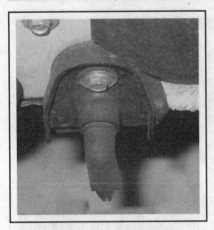

14.1 Look for the evaporator drain hose on the firewall; to remove it for cleaning or for removing the evaporator, remove the screw and separate the assembly from the firewall

14.12 Cans of R-134A refrigerant (available at auto parts stores) can be added to the low side of the air conditioning system with a simple recharging kit (four-cylinder engine shown)

14.15 Insert a thermometer in the center vent, turn on the air conditioning system and wait for it to cool down; depending on the humidity, the output air should be 30 to 40-degrees cooler than the ambient air temperature

3 Because of the complexity of the air conditioning system and the special equipment necessary to service it, in-depth troubleshooting and repairs are not included in this manual. However, simple checks and component replacement procedures are provided in this Chapter.

4 The most common cause of poor cooling is simply a low system refrigerant charge. If a noticeable drop in cool air output occurs, the following quick check will help you determine if the refrigerant level is low.

CHECKING THE REFRIGERANT CHARGE

5 Warm the engine up to normal operating temperature.

6 Place the air conditioning temperature selector at the coldest setting and the blower at the highest setting. Open the vehicle doors (to make sure the air conditioning system doesn't cycle off as soon as it cools the passenger compartment).

7 If the compressor discharge line feels warm and the compressor inlet pipe feels cool, the system is adequately charged.

8 Place a thermometer in the dashboard vent nearest the evaporator and operate the system until the indicated temperature is around 40 to 45-degrees F. If the ambient (outside) air temperature is very high, say 110-degrees F, the duct air temperature may be as high as 60-degrees F, but generally the air conditioning is 30 to 50-degrees F cooler than the ambient air.

→ Note: *Humidity of the ambient air also affects the cooling capacity of the system. Higher ambient humidity lowers the effectiveness of the air conditioning system.*

ADDING REFRIGERANT

♦ Refer to illustrations 14.12 and 14.15

9 Buy an automotive charging kit at an auto parts store. A charging kit includes a 14-ounce can of refrigerant, a tap valve and a short section of hose that can be attached between the tap valve and the system low side service valve. Because one can of refrigerant may not be sufficient to bring the system charge up to the proper level, it's a good idea to buy an additional can. Make sure that one of the cans contains red refrigerant dye. If the system is leaking, the red dye will leak out with the refrigerant and help you pinpoint the location of the leak.

☼ CAUTION:
There are two types of refrigerant used in automotive systems; R-12 - which has been widely used on earlier models - and the more environmentally-friendly R-134a used in all models covered by this manual. These two refrigerants (and their appropriate refrigerant oils) are not compatible and must never be mixed or components will be damaged. Use only R-134a refrigerant in the models covered by this manual.

10 Hook up the charging kit by following the manufacturer's instructions.

☼ WARNING:
DO NOT hook the charging kit hose to the system high side! The fittings on the charging kit are designed to fit only on the low side of the system.

11 Back off the valve handle on the charging kit and screw the kit onto the refrigerant can, making sure first that the O-ring or rubber seal inside the threaded portion of the kit is in place.

☼ WARNING:
Wear protective eyewear when dealing with pressurized refrigerant cans.

12 Remove the dust cap from the low-side charging connection and attach the quick-connect fitting on the kit hose (see illustration).

13 Warm up the engine and turn on the air conditioner. Keep the charging kit hose away from the fan and other moving parts.

14 Turn the valve handle on the kit until the stem pierces the can, then back the handle out to release the refrigerant. You should be able to hear the rush of gas. Add refrigerant to the low side of the system until the compressor discharge line feels warm and the compressor inlet pipe feels cool. Allow stabilization time between each addition.

15 If you have an accurate thermometer, place it in the center air conditioning vent (see illustration) and then note the temperature of the air coming out of the vent. A fully-charged system which is working

3-20 COOLING, HEATING AND AIR CONDITIONING SYSTEMS

correctly should cool down to about 40-degrees F. Generally, an air conditioning system will put out air that is 30 to 40-degrees F cooler than the ambient air. For example, if the ambient (outside) air temperature is very high (over 100 degrees F), the temperature of air coming out of the registers should be 60 to 70 degrees F.

16 When the can is empty, turn the valve handle to the closed position and release the connection from the low-side port. Replace the dust cap.

WARNING:

Never add more than two cans of refrigerant to the system.

17 Remove the charging kit from the can and store the kit for future use with the piercing valve in the UP position, to prevent inadvertently piercing the can on the next use.

HEATING SYSTEMS

18 If the carpet under the heater core is damp, or if antifreeze vapor or steam is coming through the vents, the heater core is leaking. Remove it (see Section 13) and install a new unit (most radiator shops will not repair a leaking heater core).

19 If the air coming out of the heater vents isn't hot, the problem could stem from any of the following causes:

a) *The thermostat is stuck open, preventing the engine coolant from warming up enough to carry heat to the heater core. Replace the thermostat (see Section 3).*

b) *There is a blockage in the system, preventing the flow of coolant through the heater core. Feel both heater hoses at the firewall. They should be hot. If one of them is cold, there is an obstruction in one of the hoses or in the heater core, or the heater control valve is shut. Detach the hoses and back flush the heater core with a water hose. If the heater core is clear but circulation is impeded, remove the two hoses and flush them out with a water hose.*

c) *If flushing fails to remove the blockage from the heater core, the core must be replaced (see Section 12).*

ELIMINATING AIR CONDITIONING ODORS

♦ Refer to illustration 14.23

20 Unpleasant odors that often develop in air conditioning systems are caused by the growth of a fungus, usually on the surface of the evaporator core. The warm, humid environment there is a perfect breed-

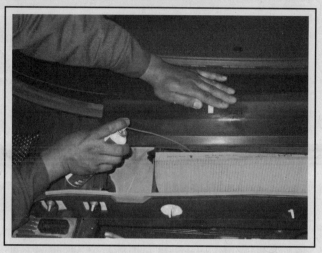

14.23 Remove the cowl cover (see Chapter 1) and insert the nozzle of the disinfectant can into the evaporator housing by shoving it past the interior ventilation filter

ing ground for mildew to develop.

21 The evaporator core on most vehicles is difficult to access, and factory dealerships have a lengthy, expensive process for eliminating the fungus by opening up the evaporator case and using a powerful disinfectant and rinse on the core until the fungus is gone. You can service your own system at home, but it takes something much stronger than basic household germ-killers or deodorizers.

22 Aerosol disinfectants for automotive air conditioning systems are available in most auto parts stores, but remember when shopping for them that the most effective treatments are also the most expensive. The basic procedure for using these sprays is to start by running the system in the RECIRC mode for ten minutes with the blower on its highest speed. Use the highest heat mode to dry out the system and keep the compressor from engaging by disconnecting the wiring connector at the compressor (see Section 15).

23 Make sure that the disinfectant can comes with a long spray hose. Point the nozzle through the interior ventilation filter chamber allowing the nozzle to protrude inside the evaporator housing (see illustration), and then spray according to the manufacturer's recommendations. Try to cover the whole surface of the evaporator core, by aiming the spray up, down and sideways. Follow the manufacturer's recommendations for the length of spray and waiting time between applications.

24 Once the evaporator has been cleaned, the best way to prevent the mildew from coming back again is to make sure your evaporator housing drain tube is clear (see illustration 14.1).

15 Air conditioning compressor - removal and installation

WARNING:

The air conditioning system is under high pressure. DO NOT loosen any fittings or remove any components until after the system has been discharged. Air conditioning refrigerant must be properly discharged into an EPA-approved container at a dealer service department or an automotive air conditioning repair facility. Always wear eye protection when disconnecting air conditioning system fittings.

➡ Note: If you are replacing the compressor, you must also replace the receiver-drier (see Section 16).

REMOVAL

♦ Refer to illustration 15.5

1 Have the air conditioning system discharged by a dealer service department or by an automotive air conditioning shop before proceeding (see **Warning** above).

COOLING, HEATING AND AIR CONDITIONING SYSTEMS   3-21

15.5 Unplug the electrical connector from the compressor clutch field coil

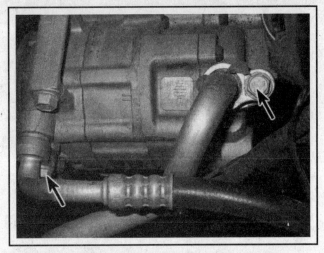

15.7 Remove the bolts and separate the inlet and outlet lines from the compressor

2   Remove the drivebelt (see Chapter 1).
3   Raise the vehicle and secure it on jackstands. Remove the right front wheel.
4   Remove the right front fender splash shield (see Chapter 11).
5   Disconnect the electrical connector from the compressor clutch field coil (see illustration).

Four-cylinder models

▶ Refer to illustrations 15.7 and 15.8

6   On 2001 through 2003 models, remove the air injection pump (see Chapter 6).
7   Disconnect the compressor inlet and outlet lines from the compressor (see illustration). Remove and discard the old O-rings.
8   Remove the compressor mounting bolts and remove the compressor (see illustration).

V6 models

9   Disconnect the compressor inlet and outlet lines from the compressor. Remove and discard the old O-rings.
10 Remove the compressor mounting bolts and remove the compressor.

INSTALLATION

11 If a new compressor is being installed, follow the directions with the compressor regarding the draining of excess oil prior to installation.

15.8 Location of the air conditioning compressor bolts (four-cylinder model shown)

12 The clutch may have to be transferred from the original to the new compressor.
13 Before reconnecting the inlet and outlet lines to the compressor, replace all manifold O-rings and lubricate them with refrigerant oil.
14 Installation is otherwise the reverse of removal.
15 Have the system evacuated, recharged and leak tested by the shop that discharged it.

16 Air conditioning receiver-drier - removal and installation

▶ Refer to illustration 16.3

WARNING:

The air conditioning system is under high pressure. DO NOT loosen any fittings or remove any components until after the system has been discharged. Air conditioning refrigerant must be properly discharged into an EPA-approved container at a dealer service department or an automotive air conditioning repair facility. Always wear eye protection when disconnecting air conditioning system fittings.

1   Have the air conditioning system discharged by a dealer service department or by an automotive air conditioning shop before proceeding (see **Warning** above).
2   Remove the condenser (see Section 17).
3   Disconnect the condenser line from the receiver-drier (see illustration).
4   Disconnect the evaporator line (see illustration 16.3). Remove and discard the old O-rings from both lines.
5   Loosen the receiver-drier mounting bracket bolt (see illustration 16.3) and then remove the receiver-drier by sliding it down through the bracket.

3-22 COOLING, HEATING AND AIR CONDITIONING SYSTEMS

6   Replace all old O-rings. Before installing the new O-rings, coat them with refrigerant oil.
7   Add 0.37 ounces of refrigerant oil to the receiver-drier.
8   Installation is otherwise the reverse of removal.
9   Take the vehicle to the shop that discharged it and have the system evacuated and recharged.

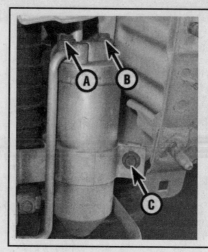

16.3 Remove the condenser line bolt (A), the evaporator line bolt (B) and loosen the bracket bolt (C) from the receiver-drier assembly

17 Air conditioning condenser - removal and installation

▶ Refer to illustrations 17.4 and 17.5

WARNING:
The air conditioning system is under high pressure. DO NOT loosen any fittings or remove any components until after the system has been discharged. Air conditioning refrigerant must be properly discharged into an EPA-approved container at a dealer service department or an automotive air conditioning repair facility. Always wear eye protection when disconnecting air conditioning system fittings.

1   Have the air conditioning system discharged by a dealer service department or by an automotive air conditioning shop before proceeding (see **Warning** above).
2   Raise the vehicle and support it securely on jackstands.
3   Remove the pusher fan from the front of the condenser (see Section 4).
4   Disconnect the refrigerant inlet and outlet lines from the condenser at the receiver-drier manifold (see illustration).
5   Remove the refrigerant line brackets from the condenser (see illustration) and position the refrigerant line away from the condenser without bending the line.
6   Remove the receiver-drier manifold bolt (see illustration 17.4).
7   Remove the left and right condenser-to-radiator bolts (see illustration 17.5).
8   Lower the condenser from the engine compartment. If you're going to reinstall the same condenser, plug the lines and store it with the line fittings facing up to prevent oil from draining out.
9   Remove the receiver-drier from the condenser (see Section 16).
10   If you're going to install a new condenser, pour 1.2 ounces of refrigerant oil of the correct type into it prior to installation.
11   Before reconnecting the refrigerant lines to the condenser, be sure to coat a pair of new O-rings with refrigerant oil, install them in the refrigerant line fittings and then tighten the condenser inlet and outlet nuts to the torque listed in this Chapter's Specifications.
12   Installation is otherwise the reverse of removal.
13   Have the system evacuated, recharged and leak tested by the shop that discharged it.

17.4 Remove the refrigerant line bolts (A) from the manifold and remove the receiver-drier manifold bolt (B) from the bracket assembly - bumper removed for clarity

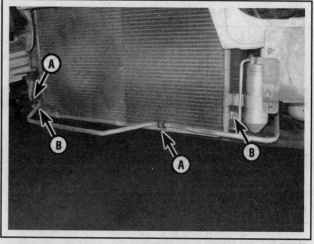

17.5 First remove the refrigerant line bracket bolts (A), then the condenser-to-radiator bolts (B) - bumper removed for clarity

COOLING, HEATING AND AIR CONDITIONING SYSTEMS

18 Air conditioning pressure sensor - replacement

→ **Note:** *The air conditioning pressure sensor is mounted in the air conditioning liquid line in the right front corner of the engine compartment. A Schrader valve in the refrigerant line prevents refrigerant loss during pressure sensor replacement.*

1. Unplug the electrical connector from the pressure sensor.
2. Unscrew the pressure sensor from the refrigerant line. Use a back-up wrench to prevent damaging the refrigerant line.
3. Lubricate the sensor O-ring with mineral oil.
4. Screw the new sensor onto the refrigerant line until hand tight, then tighten it securely.
5. Reconnect the electrical connector.

Specifications

GENERAL

Expansion tank cap pressure rating	20 psi (138 kPa)
Thermostat rating (opening to fully open temperature range)	194 to 223-degrees F (90 to 106-degrees C)
Cooling system capacity	See Chapter 1
Refrigerant type	R-134a
Refrigerant capacity	Refer to HVAC specification tag

Torque specifications

	Ft-lbs (unless otherwise indicated)	Nm
Condenser-to-radiator bolts	44 in-lbs	5
Coolant crossover housing mounting bolts (V6 engine)	22	30
Engine oil cooler inlet and outlet fitting bolts (V6 engine)	22	30
Engine oil cooler inlet and outlet nuts (V6 engine)	15	20
Engine oil cooler cover bolts (V6 engine)	22	30
Radiator bracket support bolts	53 in-lbs	8
Receiver-drier inlet and outlet line bolts	27 in-lbs	3
Receiver-drier manifold-to-radiator bracket bolt	44 in-lbs	5
Refrigerant inlet and outlet line-to-manifold bolts	15	20
Thermostat cover bolts		
Four-cylinder models	89 in-lbs	10
V6 models	15	20
Thermostat housing bolts (four-cylinder engines)	89 in-lbs	10
Water pump bolts	18	25
Water pump sprocket bolts (four-cylinder engines)	89 in-lbs	10
Water pump access cover bolts (four-cylinder engines)	89 in-lbs	10

Notes

Section

1. General information and precautions
2. Fuel pressure relief procedure
3. Fuel pump/fuel pressure - check
4. Fuel lines and fittings - general information
5. Fuel tank - removal and installation
6. Fuel tank cleaning and repair - general information
7. Fuel pump/fuel level sensor module - removal and installation
8. Fuel pump/fuel level sensor - component replacement
9. Air filter housing - removal and installation
10. Accelerator cable - removal and installation
11. Sequential Fuel Injection (SFI) system - general information
12. Sequential Fuel Injection (SFI) system - general check
13. Throttle body - inspection, removal and installation
14. Fuel pressure regulator - removal and installation
15. Fuel rail and injectors - removal and installation
16. Exhaust system servicing - general information

Reference to other Chapters

Air filter check and replacement - See Chapter 1
Catalytic converter - See Chapter 6
Exhaust system check - See Chapter 1
Fuel filter replacement - See Chapter 1
Underhood hose check and replacement - See Chapter 1

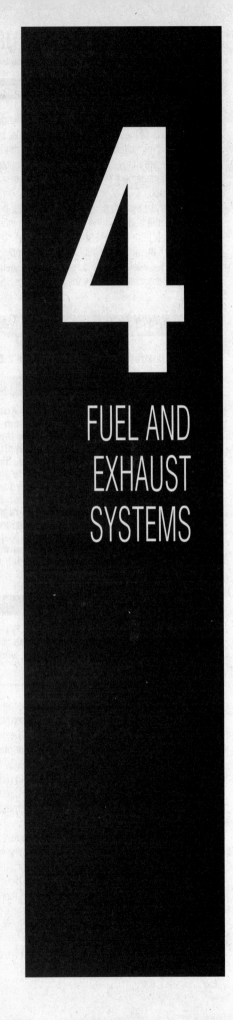

4

FUEL AND EXHAUST SYSTEMS

4-2 FUEL AND EXHAUST SYSTEMS

1 General information and precautions

This Chapter covers the removal and installation procedures for the important parts of the air intake, fuel and exhaust systems. Because emission-control systems are integral parts of the engine management system, there are many cross-references to Chapter 6. Information on the engine management system, information sensors and output actuators is in Chapter 6.

The air intake system consists of the air filter housing, the air intake duct, the throttle body and the intake manifold. Incoming air passes through the air filter element, the Mass Air Flow (MAF) sensor (V6 models only), the air intake duct, the throttle body, the intake manifold plenum and the intake manifold runners before being mixed with fuel sprayed into the intake ports by the fuel injectors.

The Sequential Fuel Injection (SFI) system consists of the fuel tank, an electric fuel pump/fuel level sending unit module mounted inside the tank, the fuel rail, the fuel injectors, the fuel pressure regulator and the metal and flexible fuel lines that connect the various components of the SFI system.

The exhaust system consists of the exhaust manifold(s), the catalytic converter(s), the resonator, the muffler and the exhaust pipes connecting these components. The system is suspended from the vehicle pan by rubber hangers. You'll find the removal and installation procedures for the exhaust manifold(s) in Chapter 2, and for the rest of the exhaust system in this Chapter. There is more information about - and the replacement procedures for - the catalytic converter(s) in Chapter 6.

2 Fuel pressure relief procedure

▶ Refer to illustrations 2.4, 2.6a, 2.6b and 2.6c

WARNING:

Gasoline is extremely flammable, so take extra precautions when you work on any part of the fuel system. Don't smoke or allow open flames or bare light bulbs near the work area, and don't work in a garage where a gas-type appliance (such as a water heater or a clothes dryer) is present. Since gasoline is carcinogenic, wear latex gloves when there's a possibility of being exposed to fuel, and, if you spill any fuel on your skin, rinse it off immediately with soap and water. Mop up any spills immediately and do not store fuel-soaked rags where they could ignite. The fuel system is under constant pressure, so, if any fuel lines are to be disconnected, the fuel pressure in the system must be relieved first. When you perform any kind of work on the fuel system, wear safety glasses and have a Class B type fire extinguisher on hand.

CAUTION:

After the fuel pressure has been relieved, it's a good idea to lay a shop towel over any fuel connection to be disassembled, to absorb the residual fuel that may leak out when servicing the fuel system.

1 The fuel system referred to in this Chapter is defined as the fuel tank and tank-mounted fuel pump/fuel gauge sender unit, the fuel filter, the fuel injectors and the metal pipes and flexible hoses of the fuel lines between these components. All these components contain fuel, which is pressurized as soon as the ignition key is turned to ON, and remains pressurized while the engine is running (and even after the ignition is switched off). And because the pressure remains for some time after the ignition has been switched off, it must be relieved before any fuel lines are disconnected.

2 On most vehicles, the fuel pressure is relieved by starting the engine and then pulling a fuel pump fuse or relay, which disables the fuel pump and stalls the engine. However, Saturn doesn't recommend this method because it might set a Diagnostic Trouble Code (DTC). Instead, it recommends that you relieve system fuel pressure by draining off the residual fuel through the Schrader valve on the fuel rail (which is also used for measuring fuel pressure, as described in the next Section).

3 Remove the fuel filler cap to relieve any pressure built-up in the fuel tank.

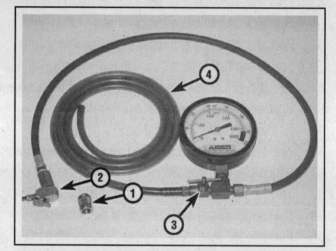

2.4 A typical fuel pressure gauge set up for relieving fuel pressure should include an screw-on adapter for the quick-release fitting (1), a quick-release fitting for connecting the gauge hose to the Schrader valve (2), a bleeder valve (3) and a bleeder hose (4) that's long enough to reach the container into which you're going to drain the fuel

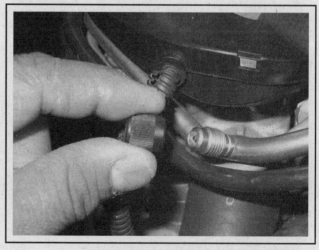

2.6a Unscrew the cap from the Schrader valve on the fuel line; this is a four-cylinder model (the test port is located near the right front corner of the cylinder head)

FUEL AND EXHAUST SYSTEMS 4-3

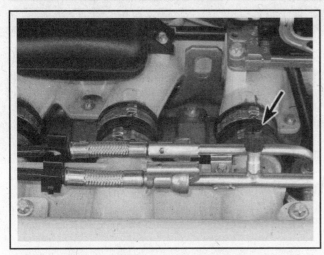

2.6b The Schrader valve test port on V6 models is located just to the rear of the intake plenum

2.6c Here's the gauge connected to the test port (this setup can be used for relieving and testing fuel pressure)

4 The recommended setup for relieving fuel pressure (see illustration) is similar to the one that you'll need to measure fuel pressure in the next Section. Your fuel pressure relief setup must include a fuel pressure gauge with a hose and fitting suitable for connecting the gauge to the Schrader valve-type test port on the fuel feed line, and a bleeder valve so that you can drain off the excess fuel. You'll also need a container approved for storing fuel and a section of plastic tubing long enough to connect the bleeder valve to the container.

5 Disconnect the cable from the negative terminal of the battery (see Chapter 5, Section 1).

6 Locate the Schrader valve on the fuel line, then unscrew the cap and connect the gauge (see illustrations). Run the hose from the bleeder valve down to an approved container, then open the bleeder valve and let the excess fuel drain into the container. When the fuel ceases dribbling out, close the bleeder valve, disconnect and remove your fuel pressure relief rig and screw the cap on the Schrader valve. The fuel pressure is now relieved. You may now open up the fuel system to service any component.

❊❊ WARNING:

This procedure merely relieves the pressure that the engine needs to run. But remember that fuel is still present in the system components, and take precautions accordingly before disconnecting any of them.

3 Fuel pump/fuel pressure - check

❊❊ WARNING:

Gasoline is extremely flammable, so take extra precautions when you work on any part of the fuel system. See the Warning in Section 2.

FUEL PUMP OPERATION CHECK

1 The fuel pump is located inside the fuel tank, which muffles its sound when the engine is running. But you can actually hear the fuel pump. Sit inside the vehicle with the windows closed, turn the ignition key to ON (not START) and listen carefully for the soft whirring sound made by the fuel pump as it's briefly turned on by the PCM to pressurize the fuel system prior to starting the engine (the sound will come from under the rear seat, because the fuel tank is located below it). You will only hear a soft whirring sound for a second or two, but that sound tells you that the pump is working. If you can't hear the pump, remove the fuel filler cap, depress the spring-loaded door inside the fuel filler neck, then have an assistant turn the ignition switch to ON while you listen for the sound of the pump operating for a couple of seconds.

2 If the pump does not come on when the ignition key is turned to ON, check the fuel pump fuse and relay (both of which are located in the engine compartment fuse and relay box). If the fuse and relay are okay, check the wiring back to the fuel pump (see Section 5 if you need help locating the fuel pump electrical connector). If the fuse, relay and wiring are okay, the fuel pump is probably defective. If the pump runs continuously with the ignition key in its ON position, the Powertrain Control Module (PCM) is probably defective. Have the PCM checked by a dealer service department or other qualified repair shop.

FUEL PRESSURE CHECK

3 To check the fuel pressure, locate the Schrader valve test port, unscrew the cap and connect a fuel pressure gauge (see illustrations 2.6a, 2.6b and 2.6c).

4 Relieve the fuel pressure (see Section 2). After you have relieved the fuel pressure, make sure that the bleeder valve on your fuel pressure gauge is CLOSED.

5 Start the engine and allow it to idle. Note the gauge reading as soon as the pressure stabilizes, and compare it with the pressure listed in this Chapter's Specifications.

  a) *If the pressure is lower than specified, suspect a restricted fuel filter* (see Chapter 1). *Also check the supply side fuel lines and hoses for kinks, blockages and leaks.*

b) *If there are no restrictions and you changed the fuel filter but the pressure is still lower than specified, check the fuel pressure regulator (see Section 14). The regulator might be stuck in the open position, which will prevent the fuel system from reaching its normal operating pressure range.*

c) *If the pressure regulator is okay, remove the fuel pump (see Section 7) and inspect the fuel inlet strainer for restrictions. If the fuel strainer is okay, the fuel pump module could be clogged or defective.*

d) *It's possible that one or more of the fuel injectors might be leaking or stuck open (but if this is happening the engine will be running very poorly and will most likely have set a trouble code). Remove the fuel rail and inspect the injectors (see Section 15).*

e) *If the fuel pressure is higher than specified, check the fuel pressure regulator (see Section 14). The vacuum signal line to the regulator might have come off or it might have a leak in it, or the regulator might be stuck in the closed position.*

f) *If the fuel pressure regulator is okay, check the return side fuel lines and hoses for kinks and obstructions.*

6 Turn off the engine. Verify that the fuel pressure loses no more than 8 psi for five minutes after the engine is turned off.

7 Relieve the fuel pressure (see Section 2), then disconnect the fuel pressure gauge. Mop up any spilled gasoline.

8 Start the engine and verify that there are no fuel leaks.

4 Fuel lines and fittings - general information

♦ Refer to illustration 4.2

✳ WARNING 1:

Gasoline is extremely flammable, so take extra precautions when you work on any part of the fuel system. See the Warning in Section 2.

✳ WARNING 2:

Before disconnecting any fuel line fittings, relieve the fuel system pressure (see Section 2) and equalize tank pressure by removing the fuel filler cap. This procedure will merely relieve the increased pressure necessary for the engine to run - remember that fuel will still be present in the system components, so you should be ready to mop up fuel spills when disconnecting fuel line fittings.

1 Always relieve the fuel pressure (see Section 2) before servicing fuel lines or fittings, then disconnect the cable from the negative battery terminal (see Chapter 5, Section 1) before proceeding.

2 The fuel supply and return lines connect the fuel pump in the fuel tank to the fuel rail on the engine. The Evaporative Emission (EVAP) system vapor lines connect the fuel tank to the EVAP canister and connect the canister to the intake manifold. The fuel and EVAP lines are secured to the underbody with small plastic brackets that are attached to the vehicle floorpan. To detach the lines from these brackets, simply turn the big screw (see illustration) counterclockwise and pull off the lower half of each bracket.

3 Whenever you're working under the vehicle, be sure to inspect all fuel and evaporative emission lines for leaks, kinks, dents and other damage. Always replace a damaged fuel or EVAP line immediately. Leaking fuel and EVAP lines will result in loss of fuel and excessive air pollution (the leaking raw fuel emits unburned hydrocarbon vapors into the atmosphere).

4 If you find signs of dirt in the lines during disassembly, disconnect all lines and blow them out with compressed air. Inspect the fuel strainer on the fuel pump pick-up unit (see Section 7) for damage and deterioration. And inspect the fuel filter (see Chapter 1).

STEEL TUBING

5 Because fuel lines used on fuel-injected vehicles are under fairly high pressure, it is critical that they be replaced with lines of equivalent specification. Never use copper or aluminum tubing to replace steel tubing. These materials cannot withstand normal vehicle vibration.

6 Some steel fuel lines have threaded fittings. When loosening these fittings to service or replace components:

a) *Hold the stationary fitting with one wrench while loosening or tightening the tubing nut with another.*

b) *If you're going to replace one of these fittings, use original equipment parts or parts that meet original equipment standards.*

PLASTIC TUBING

7 Most of the fuel (and EVAP) lines on the vehicles covered in this manual are plastic. If you ever have to replace a plastic line, use only plastic tubing meeting original equipment standards.

4.2 These big brackets that secure the fuel and EVAP lines to the underside of the vehicle are hinged so that they can be opened up to remove the lines; simply turn the big screw counterclockwise and open up the bracket

FUEL AND EXHAUST SYSTEMS 4-5

4.12 Pull the end of the retainer off the fuel line, then disengage the other end from the female side of the fitting

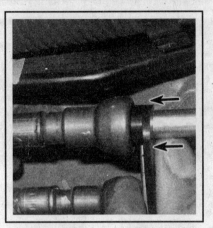

4.13a Insert the fuel line separator tool into the female side of the fitting, push it into the fitting until it releases the locking tabs inside the fitting . . .

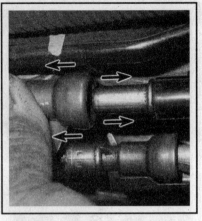

4.13b . . . then pull the two halves of the fitting apart

❊❊ CAUTION:

When removing or installing plastic fuel line tubing, be careful not to bend or twist it too much, which can damage it. And damaged fuel lines MUST be replaced! Also, be aware that the plastic fuel tubing is NOT heat resistant, so keep it away from excessive heat. Nor is it acid-proof, so don't wipe it off with a shop rag that has been used to wipe off battery electrolyte. If you accidentally spill or wipe electrolyte on plastic fuel tubing, replace the tubing.

FLEXIBLE HOSES

❊❊ WARNING:

Use only original equipment replacement hoses or their equivalent. Unapproved hoses might fail when subjected to the high operating pressures of the fuel system.

8 Don't route fuel hoses within four inches of exhaust system components or within ten inches of a catalytic converter. Make sure that no rubber hoses are installed directly against the vehicle, particularly in places where there is any vibration. If allowed to touch some vibrating part of the vehicle, a hose can easily become chafed and it might start leaking. A good rule of thumb is to maintain a minimum of 1/4-inch clearance around a hose (or metal line) to prevent contact with the vehicle underbody.

FUEL LINE AND EVAP LINE FITTINGS

9 The vehicles covered in this manual use two kinds of fuel line quick-connect fittings (metal or plastic) for most connections at the fuel pump, the fuel tank, under the vehicle and in the engine compartment. (A third type of plastic quick-connect fitting is used only at the EVAP canister and on the vent hose connection at the fuel tank for the EVAP canister vent solenoid.)

10 The procedure for releasing each type of fuel line fitting is different. But a few rules of thumb apply to all fittings:

1) Inspect the fitting for dirt. If the fitting is dirty, clean it off before disassembling it. The seals in the fitting will stick to the fuel line as they age. Twist the fitting on the line, then push and pull the fitting until it moves freely.
2) Always disconnect all fuel line fittings from a fuel system component before removing the component.
3) When disconnecting a quick-connect fitting, inspect the condition of the retainer before reconnecting the fitting. The best strategy with respect to retainers is to simply replace the retainer every time that you disconnect the fitting.
4) When you disconnect a fitting with an O-ring inside, inspect the O-ring before reconnecting the fitting. Fuel line fittings are under the same pressure as the rest of the fuel system, so to avoid leaks (and fires!) make VERY SURE that the O-ring is good condition. Even better, simply replace it.
5) In most cases, the fitting itself is a non-removable part of the fuel line, so you might have to replace an entire fuel line if a fitting is damaged or defective.

METAL COLLAR QUICK-CONNECT FITTINGS

Disconnection

▶ Refer to illustrations 4.12, 4.13a and 4.13b

➡ Note 1: You'll find these fittings at the connections between the fuel supply and return lines in the engine compartment.

➡ Note 2: You'll need a special tool set (J37088-A, or a suitable equivalent) to disconnect these fittings.

➡ Note 3: The photos accompanying the disconnection procedure depicted here shows the metal collar quick-connect fittings at a four-cylinder fuel rail, but the metal quick-connect fittings on a V6 fuel rail are identical.

11 Relieve the fuel system pressure (see Section 2).

12 Pull off the clip end of the retainer, then remove it from the fitting (see illustration).

13 Using a fuel line separator tool of the proper size (available at most auto parts stores), insert the tool into the female side of the fitting, then push it into the fitting to release the locking tabs and pull the fitting apart (see illustrations).

4-6 FUEL AND EXHAUST SYSTEMS

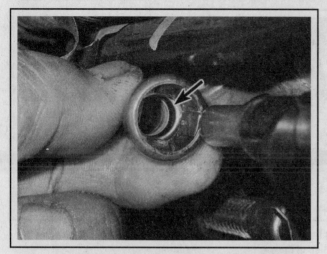

4.14 Inspect the old O-ring inside the female side of the fitting; if it's cracked, torn or deteriorated, replace it

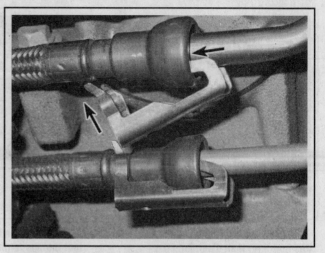

4.17 To install a retainer, insert the hooked end into the female side of the fitting, then push the clip end onto the fuel line until it snaps into place (when you're done, your retainer should look like the one already installed on the lower fitting)

Reconnection

▶ **Refer to illustrations 4.14 and 4.17**

14 Inspect the O-ring (see illustration). If it's dried out, cracked, torn or otherwise deteriorated, replace it.
15 Apply a few drops of clean engine oil to the male pipe end.
16 Push both sides of the fitting together until the retaining tabs snap into place. Pull on both sides of the fitting to verify that it's securely connected.
17 Install the retainer, making sure it clips into place (see illustration).
18 Start the engine and check for fuel leaks.

PLASTIC COLLAR QUICK-CONNECT FITTINGS

Disconnection

▶ **Refer to illustrations 4.19a and 4.19b**

19 To release this type of quick-connect fitting, depress the tabs of the retainer (see illustration). Once the retainer is released, continue pressing on the tabs while pulling the two fuel lines apart (see illustration).

4.19a To release a plastic quick-connect fitting, depress the tabs on the connector housing with a small screwdriver, then continue pressing on them . . .

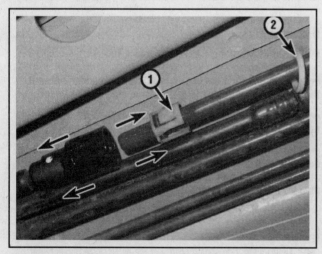

4.19b . . . until the two fuel lines are disconnected, then removed and discard the old retainer (1) and the indicator ring (2) (the indicator ring is used only during factory assembly; there is no need to reinstall it)

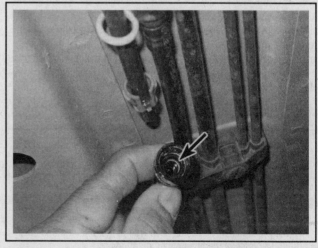

4.22 Inspect the old O-ring inside the female side of the fitting; if it's cracked, torn or deteriorated, replace it

FUEL AND EXHAUST SYSTEMS 4-7

20 Remove and discard the old retainer from the male side of the fitting.

21 Remove and discard the indicator ring from the male side of the fitting.

Reconnection

▶ Refer to illustrations 4.22 and 4.23

22 Inspect the old O-ring inside the female side of the fitting (see illustration). If it's dried out, cracked, torn or deteriorated, replace it.

23 Insert a new retainer in the female side of the fitting. Make sure that the release tabs are aligned with the "windows" of the connector (see illustration).

24 Apply a few drops of engine oil to the tip of the male fuel line.

25 Push both sides of the fitting together until the retainer release tabs snap into place.

26 Pull on both sides of the fitting to verify that it's securely connected.

27 Start the engine and check for fuel leaks.

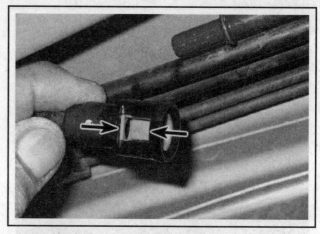

4.23 Install a new retainer in the female side of the fitting; make sure that the release tabs are aligned with the windows in the connector

5 Fuel tank - removal and installation

▶ Refer to illustrations 5.5, 5.7, 5.8a, 5.8b, 5.9, 5.10, 5.11, 5.12 and 5.14

WARNING 1:

Gasoline is extremely flammable, so take extra precautions when you work on any part of the fuel system. See the Warning in Section 2.

WARNING 2:

Before disconnecting or opening any part of the fuel system, relieve the fuel system pressure (see Section 2), and equalize the pressure inside the fuel tank by removing the fuel filler cap.

1 It's easier to remove the fuel tank when it's nearly empty. But there is no fuel tank drain plug, so if that's not possible, try to siphon out the fuel in the tank before removing the tank (see Step 5).

2 Relieve the fuel system pressure (see Section 2).

3 Disconnect the cable from the negative battery terminal (see Chapter 5, Section 1).

4 Raise the vehicle and place it securely on jackstands.

5 If there's still a lot of fuel in the tank, disconnect the quick-connect fitting for the fuel inlet line at the fuel filter (see "Fuel filter - replacement" in Chapter 1) and siphon or hand-pump the remaining fuel from the tank now (see illustration).

WARNING:

Don't start the siphoning action by mouth! Use a siphoning kit (available at most auto parts stores).

6 Remove the intermediate exhaust pipe and muffler (see Section 16).

7 Remove the rear heat shield retaining bolts (see illustration), then remove the rear heat shield.

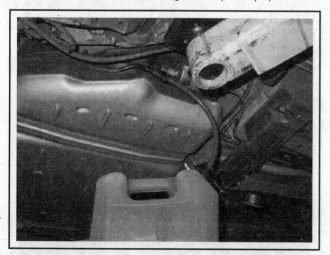

5.5 If there's still fuel in the fuel tank, disconnect the quick-connect fitting on the inlet side of the fuel filter (see Chapter 1) and siphon the excess fuel out of the tank into an approved container

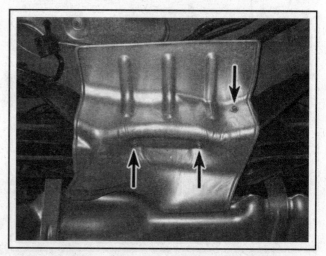

5.7 To detach the rear heat shield, remove these three retaining bolts

4-8 FUEL AND EXHAUST SYSTEMS

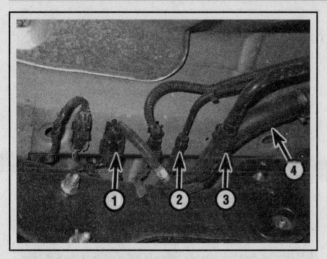

5.8a Fuel tank hoses and wiring:
1. Electrical connector for the fuel pump/fuel level sensor module (disconnect here)
2. Vent hose between fuel filler neck and fuel limit vent valve (do NOT disconnect here, see illustration 5.10)
3. Connector for the EVAP canister vent solenoid hose (disconnect here, see illustration 5.8b)
4. Fuel filler neck hose (disconnect closer to fuel tank, see illustration 5.9)

8 Disconnect the electrical connector for the fuel pump/fuel level sensor module and disconnect the connector for the EVAP canister vent solenoid hose (see illustrations).

9 Disconnect the fuel filler neck hose from the fuel tank (see illustration).

10 Disconnect the vent hose for the fuel limit vent valve at the fuel tank (see illustration).

11 Disconnect the fuel tank ground strap (see illustration).

12 Remove the rock guards (see illustration).

13 Disconnect the connector on the outlet side of the fuel filter (see illustration 5.11).

➡ **Note:** It's not absolutely necessary to disconnect the inlet side connector or to remove the fuel filter from the tank before removing the tank because the inlet side line and the filter can be removed with the fuel tank.

5.10 Disconnect the connector for the fuel limit vent valve's vent hose

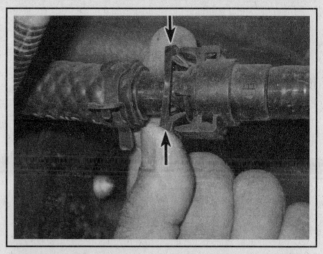

5.8b To disconnect the connector for the EVAP canister vent solenoid hose, firmly squeeze these two locking tangs together to release the connector, then pull it off

5.9 To disconnect the fuel filler neck hose, loosen this hose clamp and pull the hose off the fuel tank fitting

5.11 To detach the fuel tank ground strap, remove this nut (A); to disconnect the quick-connect fitting (B) on the outlet side of the fuel filter, see Section 4

FUEL AND EXHAUST SYSTEMS   4-9

5.12 The rock guards are retained by six bolts (right side shown, left side similar)

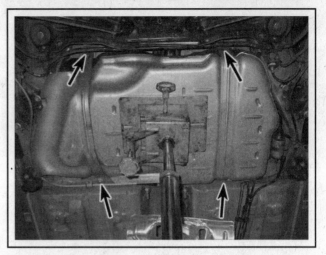

5.14 Support the fuel tank with a transmission jack; arrows indicate the locations of the fuel tank strap bolts

14 Support the fuel tank with a transmission jack, if available (see illustration), or with a floor jack. If you're going to use a floor jack, be sure to put a sturdy piece of plywood between the jack head and the fuel tank to protect the tank.

15 Remove the fuel tank strap bolts and remove the straps.

16 Lower the tank enough to have a look at the top of the tank and unclip or detach any remaining cables, hoses or lines.

17 Lower the tank the rest of the way.

18 Installation is the reverse of removal.

19 When the battery has been disconnected, the Powertrain Control Module (PCM) must relearn its former driveability and performance characteristics (see Chapter 5, Section 1 for this procedure).

6 Fuel tank cleaning and repair - general information

WARNING:

Gasoline is extremely flammable, so take extra precautions when you work on any part of the fuel system. See the Warning in Section 2.

1 The fuel tank is plastic and cannot be repaired. No reliable repair procedures are available to correct leaks or damage. Fuel tank replacement is the only approved service.

2 To remove sediment from the bottom of the tank, have the fuel tank steam-cleaned. Remove the fuel pump/level sending unit module (see Section 7) and all EVAP system components (see Chapter 6) prior to cleaning. Allow plenty of time for the tank to air dry before returning it to service.

7 Fuel pump/fuel level sensor module - removal and installation

▶ Refer to illustrations 7.4, 7.5, 7.6, 7.7 and 7.8

WARNING:

Gasoline is extremely flammable, so take extra precautions when you work on any part of the fuel system. See the Warning in Section 2.

1 Relieve the system fuel pressure (see Section 2), and equalize tank pressure by removing the fuel filler cap.

2 Disconnect the cable from the negative battery terminal (see Chapter 5, Section 1).

3 Remove the fuel tank (see Section 5).

4 Disconnect the fuel supply and return lines from the fuel pump/fuel level sending unit module (see illustration). Use a shop rag to soak up any spilled fuel.

5 Mark the orientation of the fuel pump in relation to the fuel tank (see illustration) to ensure that the fuel pump is correctly realigned when you install it again. (If you're going to install a new pump, note

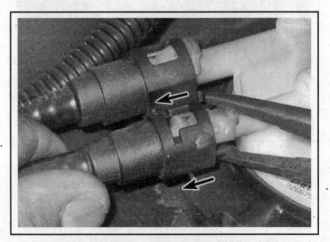

7.4 To disconnect the fuel supply and return line quick-connect fittings from the fuel pump, squeeze the legs of each retainer together and pull the fitting off the fuel pump pipe

4-10 FUEL AND EXHAUST SYSTEMS

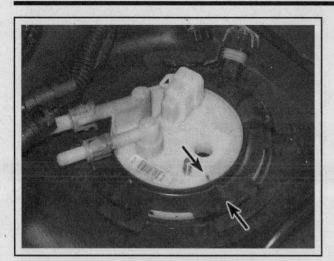

7.5 Before unscrewing the locknut that secures the fuel pump to the fuel tank, make alignment marks on the fuel pump mounting flange and on the fuel tank to ensure correct realignment of the pump when installing it again

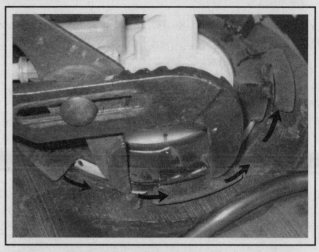

7.6 Use a large pair of water pump pliers to loosen and unscrew the fuel pump locknut; if the locknut is too tight to loosen this way, carefully tap it loose with a hammer and a brass punch

the location of your alignment mark on the old pump and make a mark at the same spot on the new unit.)

6 Using a pair of large water pump pliers, unscrew the fuel pump/fuel level sending unit module locknut by turning it counterclockwise (see illustration). If the locknut is tight, use a hammer and a brass punch to loosen it (don't use a steel punch, which could produce sparks when struck by the hammer).

7 Remove the fuel pump/fuel level sensor module (see illustration), taking care not to damage the fuel inlet strainer or the fuel level sensor float arm and float.

8 Before installing the pump, inspect the O-ring (see illustration) for cracks, tears and deterioration. If it's worn or damaged, replace it. Also inspect the fuel pump inlet strainer. Make sure that it's clean and free of debris and dirt. If it's dirty, try washing it with carburetor cleaner spray. If this filter is seriously damaged, you'll have to replace the fuel pump/fuel level sending unit module. The filter is not available separately.

9 Installation is the reverse of removal. Align the fuel pump/fuel level sending unit module with its hole in the tank and carefully insert it into the tank. Make sure that you don't damage the fuel inlet strainer, the float arm or the float during installation. If the float arm is bent, the fuel level that is indicated on the fuel level gauge on the instrument cluster will be incorrect.

10 When the battery has been disconnected, the Powertrain Control Module (PCM) must relearn its former driveability and performance characteristics (see Chapter 5, Section 1 for this procedure).

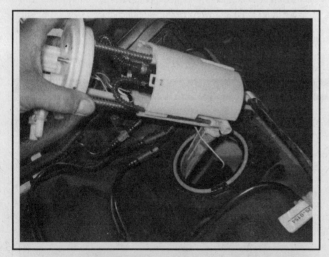

7.7 Carefully remove the fuel pump/fuel level sensor module from the fuel tank; once the module has cleared the mounting hole in the tank, angle it as shown to work the fuel pump inlet strainer and the fuel level sensor float and float arm through the hole without damaging anything

7.8 Remove and inspect the O-ring seal for the fuel pump mounting flange; if it's cracked, torn or deteriorated, replace it

FUEL AND EXHAUST SYSTEMS   4-11

8 Fuel pump/fuel level sensor - component replacement

DISASSEMBLY

► **Refer to illustrations 8.3, 8.4, 8.5 and 8.6**

1 Remove the fuel pump/fuel level sensor module (see Section 7).
2 Place the fuel pump/fuel level sensor module on a clean workbench surface.
3 Disengage the two fuel level sensor leads from the harness clips on the side of the module (see illustration).
4 Disconnect the positive and ground wires from the top of the fuel pump (see illustration).
5 Disengage the locking tab with a small screwdriver (see illustration) and detach the connector housing from the bottom of the fuel pump cover.
6 To remove the fuel level sensor unit, depress the locking tab and slide the fuel level sensor unit toward the upper end of the fuel pump assembly (see illustration).

REASSEMBLY

7 Slide the fuel level sensor unit into place until you hear a click (pull up lightly to verify that the sender unit is locked into place).
8 To verify that the float arm on the fuel level sensor unit is correctly aligned, stand the fuel pump module on a flat horizontal surface and verify that the float lays flat on the surface.
9 Installation is otherwise the reverse of removal.

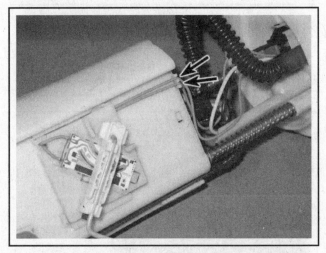

8.3 Detach the two wires for the fuel level sensor from these two clips on the module

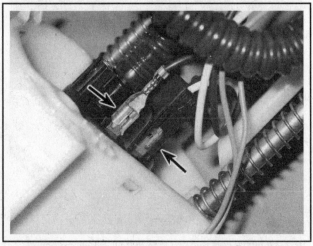

8.4 Disconnect the positive and ground wires from the top of the pump

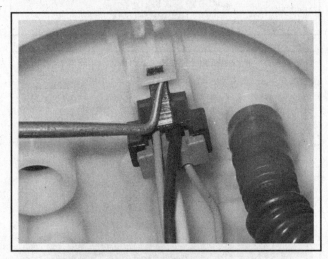

8.5 To detach the connector housing from the bottom of the fuel pump cover, disengage the locking tab with an awl (shown) or with a small screwdriver

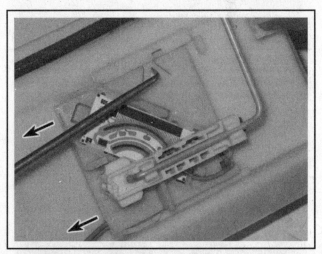

8.6 To detach the fuel level sensor unit from the fuel pump assembly, depress the locking tab with a pointed tool and slide the sensor unit toward the upper end of the fuel pump until it's free of its retaining rails

4-12 FUEL AND EXHAUST SYSTEMS

9 Air filter housing - removal and installation

FOUR-CYLINDER MODELS

Air intake duct

▶ Refer to illustration 9.2

1 Make sure that the ignition key is turned to OFF.
2 Disconnect the electrical connector from the Intake Air Temperature (IAT) sensor (see illustration).
3 Disconnect the PCV hose from the air intake duct.
4 Loosen the hose clamp screws at both ends of the air intake duct and remove the duct.
5 Installation is the reverse of removal.

Air filter housing

▶ Refer to illustrations 9.7 and 9.8

6 Remove the air intake duct (see Steps 1 through 4).
7 Remove the air filter housing mounting bolt from the backside of the filter housing (see illustration).
8 Detach the wiring harness from the clip, work the air filter housing locator pin out of its mounting bracket grommet (see illustration) and remove the filter housing.
9 Inspect the rubber grommet before installing the air filter housing. If it's cracked, torn or otherwise damaged, replace it.
10 Installation is the reverse of removal. Make sure that the air filter housing locator pin is correctly seated in its grommet.

V6 MODELS

Air intake duct

▶ Refer to illustration 9.12

11 Make sure that the ignition key is turned to OFF.
12 Disconnect the electrical connector from the Mass Air Flow/Intake Air Temperature (MAF/IAT) sensor (see illustration).

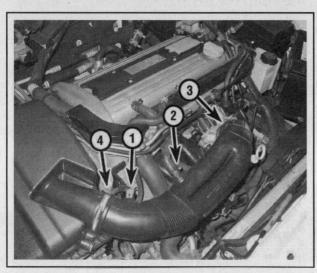

9.2 Air intake duct removal details (four-cylinder models)
1 Intake Air Temperature (IAT) sensor electrical connector
2 PCV hose
3 Hose clamp at throttle body
4 Hose clamp at air filter housing

9.7 To detach the air filter housing from the vehicle, remove this bolt

9.8 To remove the air filter housing, detach the wiring harness from this clip (1) and work the locator pin out from this mounting bracket grommet (2) by sliding the air filter housing toward the firewall

9.12 To remove the air intake duct on V6 models, disconnect the electrical connector from the Mass Air Flow/Intake Air Temperature (MAF/IAT) sensor, then loosen the hose clamp screws at each end of the duct

FUEL AND EXHAUST SYSTEMS   4-13

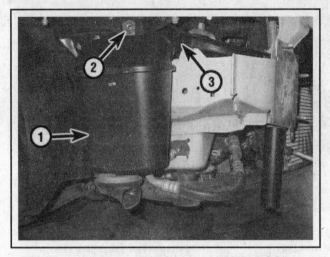

9.21 To remove the resonator (1) from the fresh air inlet duct, remove this bolt (2) and detach the resonator from the fresh air inlet duct (3) by simply pulling it off (four-cylinder model shown, V6 models similar)

9.22 To remove the fresh air inlet duct (1), pull off the smaller resonator (2), pull the duct locator pin (3) out of its mounting grommet in the radiator crossmember, then carefully work the inlet duct out of the vehicle (four-cylinder model shown, V6 models similar)

13 Loosen the hose clamp screws and remove the air intake duct and MAF/IAT sensor as a single assembly.
14 If you're replacing the MAF/IAT sensor, refer to Chapter 6.
15 Installation is the reverse of removal.

Air filter housing

16 Remove the air intake duct (see Steps 11 through 13).
17 The remainder of the air filter housing removal procedure is similar to the procedure for removing the air filter housing on four-cylinder models (see Steps 6 through 9).
18 Installation is the reverse of removal.

ALL MODELS

Fresh air inlet duct and resonator assembly

▶ Refer to illustrations 9.21 and 9.22

19 Remove the bumper cover and the bumper beam (see Chapter 11).
20 Remove the right headlight assembly (see Chapter 12).
21 Remove the resonator mounting bolt (see illustration) and remove the resonator by pulling it off the fresh air inlet duct.
22 Remove the fresh air inlet duct (see illustration).
23 Installation is the reverse of removal.

10 Accelerator cable - removal and installation

▶ Refer to illustrations 10.2, 10.3, 10.4, 10.5 and 10.6

➡ **Note:** *The following procedure applies only to four-cylinder models. V6 models do not use an accelerator cable.*

1 Make sure that the ignition key is turned to OFF.
2 Disconnect the accelerator cable from the throttle cam (see illustration).
3 Disengage the accelerator cable from the cable bracket (see illustration).
4 Trace the accelerator cable back to the firewall and disengage it from any clips or cable guides (see illustration).

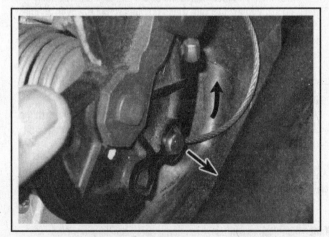

10.2 To disengage the accelerator cable from the throttle cam, rotate the cam counterclockwise, align the accelerator cable with the slot in the cam and pull the cable end plug out of its socket in the cam

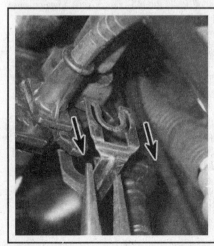

10.3 To disengage the accelerator cable from the cable bracket, pull off this retainer, then pull the cable out of its slot in the cable bracket

4-14 FUEL AND EXHAUST SYSTEMS

10.4 Before proceeding, note the routing of the accelerator cable and locate the cable clamps; to disengage the accelerator cable from each of the hinged cable clamps, pry open the clamp, open it up and remove the cable

10.5 To disengage the lower end of the accelerator cable from the accelerator pedal arm, push the upper end of the arm forward slightly and slide the cable out the slot

10.6 To detach the rubber bushing on the accelerator cable from the grommet at the firewall, simply pull it out

5   Working inside the passenger compartment, disengage the lower end of the accelerator cable from the accelerator pedal (see illustration).
6   Disengage the cable grommet from the firewall (see illustration).
7   Pull the cable through the firewall and remove it.
8   Installation is the reverse of removal.
9   Have an assistant depress and release the accelerator pedal while you verify that the throttle valve moves smoothly and easily from the fully closed to the fully open position and back again.

11 Sequential Fuel Injection (SFI) system - general information

These models are equipped with a Sequential Fuel Injection (SFI) system. The SFI system consists of three basic sub-systems: the air induction system, the fuel system and the electronic control system.

➡ **Note:** *Refer to Chapter 6 for further information on the components of the electronic control system.*

AIR INDUCTION SYSTEM

The air induction system consists of the fresh air inlet duct and resonator(s), the air filter housing, the Mass Air Flow (MAF) sensor (V6 models), the Intake Air Temperature (IAT) sensor (on the air intake duct on four-cylinder models, an integral part of the MAF sensor on V6 models), the air intake duct, the throttle body, the air intake plenum and the intake manifold. The MAF sensor is an information sensor for the Powertrain Control Module (PCM). The MAF sensor uses a heated wire system to send the PCM an analog (constantly variable) voltage signal corresponding to the volume of air passing into the engine. The IAT sensor measures the temperature of the intake air. The PCM uses these signals to calculate the mass (density) of air entering the engine. For more information about the IAT and MAF sensors, refer to Chapter 6.

The driver controls the throttle plate inside the throttle body, through the accelerator pedal. As the throttle plate opens, the amount of air that can pass through the system increases. As the throttle plate opens, the Throttle Position (TP) sensor, which is located on the end of the throttle plate shaft, opens with it. And as more air enters the engine, the MAF sensor signal changes too. In response to these two signals from TP and MAF sensors, the PCM opens each injector for a longer duration to increase the amount of fuel delivered to the inlet ports. For more information about the TP sensor, refer to Chapter 6.

On four-cylinder models, the accelerator pedal is connected to the throttle plate by an accelerator cable. On V6 models, there is no accelerator cable. The accelerator pedal is equipped with an Accelerator Pedal Position (APP) sensor, a potentiometer that monitors the angle (or position) of the accelerator pedal. This information is monitored by the PCM, which in turn commands an electric motor inside the throttle body to open or close the throttle plate in accordance with the position of the accelerator pedal.

FUEL SYSTEM

An electric fuel pump located inside the fuel tank supplies fuel under pressure to the fuel rail, which distributes fuel evenly to all injectors. A filter between the fuel pump and the fuel rail protects the components of the system. From the fuel rail, fuel is injected into the intake ports, just above the intake valves, by a fuel injector.

The amount of fuel supplied by the injectors is precisely controlled by injector "drivers" inside the PCM. The injector drivers, which are turned on and off by the PCM, control the ground side of each injector circuit: When the ground path is closed, the injectors are on; when the ground path is open, the injectors are off. The PCM uses signals from the Crankshaft Position (CKP) sensor and (on V6 models) the Camshaft Position (CMP) sensor to determine when to trigger each injector in cylinder firing order (hence the term "sequential injection"). This precise control of injector timing produces more power, better fuel economy and lower exhaust emissions.

To prevent fuel starvation, there is always more fuel in the fuel rail than the injectors can use, even under heavy acceleration. When the pressure exceeds a certain threshold, it forces open a spring loaded diaphragm inside the fuel pressure regulator and returns the excess fuel to the fuel tank. The fuel pressure regulator is also equipped with a

FUEL AND EXHAUST SYSTEMS   4-15

vacuum line connected to intake manifold vacuum. When the throttle plate is closed under deceleration or at idle, intake manifold vacuum is high, which opens the diaphragm inside the pressure regulator and returns the excess fuel to the fuel tank.

ELECTRONIC CONTROL SYSTEM

The PCM controls the SFI system and the engine management system. It receives signals from an array of information sensors that monitor such variables as intake air mass and temperature, coolant temperature, engine speed, crankshaft position, acceleration/deceleration, and exhaust gas oxygen content. These signals help the PCM determine the injection duration necessary for the optimal air/fuel ratio. These sensors and various PCM-controlled output actuators (relays, solenoids, etc.) are located throughout the engine compartment. For further information regarding the PCM, the engine management system, the information sensors and the output actuators, see Chapter 6.

12 Sequential Fuel Injection (SFI) system - general check

♦ Refer to illustrations 12.7 and 12.8

WARNING:

Gasoline is extremely flammable, so take extra precautions when you work on any part of the fuel system. See the Warning in Section 2.

1 Inspect the SFI system electrical connectors. Verify that all ground wire connections are tight. Loose connectors and poor grounds can cause many problems that resemble more serious malfunctions.

2 Verify that the battery is fully charged (see Chapters 1 and 5 for help with the battery). The PCM, information sensors and output actuators depend on a steady and adequate voltage to function correctly.

3 Inspect the air filter element (see Chapter 1). A dirty or partially blocked filter will severely impede performance and economy.

4 Inspect any fuses for the circuit you're checking. If you find a blown fuse, replace it and note whether it blows again. If it does, look for a short in the circuit.

5 Inspect the air intake duct, the throttle body and the intake manifold for leaks, which will cause an excessively lean mixture. Also inspect all vacuum hoses connected to the intake manifold and to the throttle body.

6 Remove the air intake duct (see Section 9) and inspect the throttle body for dirt, carbon, varnish or other residue inside the bore of the throttle body, particularly around the throttle plate. If it's dirty, clean it with carburetor cleaner spray and a shop towel.

7 With the engine running, place an automotive stethoscope against each injector, one at a time, and listen for a clicking sound that indicates operation (see illustration). If you don't have a stethoscope, you can place the tip of a long screwdriver against the injector and listen through the handle. If you hear the injectors operating but there is a misfire condition present, the electrical circuits are functioning, but the injectors might be dirty or fouled from carbon deposits. Commercial cleaning products might help. If not, then you might have to replace the injectors (see Section 15).

8 If you can't hear an injector operating, disconnect the injector electrical connector and measure the resistance across the terminals of each injector connector with an ohmmeter (see illustration). Compare your measurement with the resistance value listed in this Chapter's Specifications. If the indicated resistance of any injector is outside the specified range of resistance, replace the injector.

9 If an injector is not operating, but its resistance is within the specified range, then the circuit between the PCM and the injector might be faulty, or the driver for the injector, which is located inside the PCM, might be defective. If an injector driver is bad, replace the PCM. Have the PCM and the engine management system tested first by a dealer service department or other qualified repair shop before replacing the PCM.

12.7 Place the tip of an automotive stethoscope against each injector housing and listen for a clicking sound, which indicates that it's operating correctly

12.8 To measure the resistance of an injector solenoid coil winding, touch the tips of your ohmmeter probes to the two terminals on the *injector side* of the injector electrical connector

4-16 FUEL AND EXHAUST SYSTEMS

13 Throttle body - inspection, removal and installation

INSPECTION

◆ Refer to illustration 13.2

1 Verify that the throttle linkage operates smoothly.
2 Remove the air intake duct from the throttle body, open the throttle plate and inspect the throttle body bore for carbon and residue build-up. If it's dirty, clean it with solvent or carburetor cleaner (see illustration). Make sure that the solvent or carb cleaner is safe for oxygen sensor systems and catalytic converters.

✳✳ CAUTION:

Do not clean the Throttle Position (TP) sensor or the Idle Air Control (IAC) motor with the solvent. Also, do NOT use a metal brush to clean the bore of the throttle body, which is protected by a special coating. Scrubbing the bore with a stiff brush could ruin the coating. Instead, wipe out the bore with a clean shop rag and a little solvent.

REMOVAL AND INSTALLATION

Four-cylinder models

◆ Refer to illustrations 13.5, 13.8 and 13.9

✳✳ WARNING:

Wait until the engine is completely cool before beginning this procedure.

3 Disconnect the cable from the negative battery terminal (see Chapter 5, Section 1).
4 Remove the air intake duct (see Section 9).
5 Disconnect the electrical connectors from the Idle Air Control (IAC) valve, the Manifold Absolute Pressure (MAP) sensor and the Throttle Position (TP) sensor (see illustration).
6 Disconnect the EVAP canister purge hose, the power brake booster vacuum hose, the Air Injection Reaction (AIR) check valve hose and the fuel pressure regulator vacuum hose from the throttle body (see illustration 13.5).
7 Disconnect the accelerator cable from the throttle cam (see Section 10). If the vehicle is equipped with cruise control, disconnect the cruise control cable from the throttle cam too. (Disconnect the cruise control cable from the throttle cam the same way you disconnected the accelerator cable.)
8 Remove the throttle body mounting bolts and nuts (see illustration) and remove the throttle body.
9 Remove the throttle body gasket (see illustration) and inspect it. If the gasket isn't cracked, torn or otherwise deteriorated, it's okay to reuse it. But if it's damaged or worn, replace it. (If the vehicle is fairly old, it's a good idea to replace this gasket regardless of its apparent condition.)

13.2 To clean the bore of the throttle body with carburetor cleaner, remove the air intake duct, open the throttle plate and spray off any carbon, sludge or varnish residue that has built up on the walls of the throttle bore; it's okay to wipe down the bore with a clean shop rag but don't use a stiff wire brush to do so

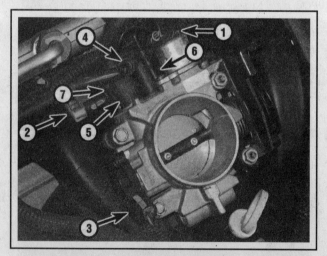

13.5 Throttle body removal details (four-cylinder models)
1 Idle Air Control (IAC) valve electrical connector
2 Manifold Absolute Pressure (MAP) sensor electrical connector
3 Throttle Position (TP) sensor electrical connector
4 EVAP canister purge hose
5 Power brake booster vacuum hose
6 Air Injection Reaction (AIR) check valve hose
7 Fuel pressure regulator vacuum hose

13.8 To detach the throttle body from the intake manifold, remove the mounting bolts and nuts (four-cylinder models)

FUEL AND EXHAUST SYSTEMS   4-17

13.9 Remove and inspect the throttle body's O-ring type gasket; if it's in good condition it's okay to reuse it, but if it's cracked, torn or otherwise deteriorated, replace it

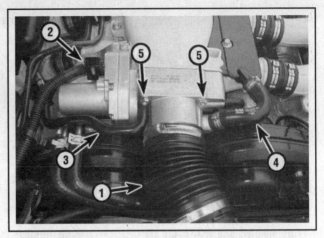

13.16 To remove the throttle body on V6 models, remove or disconnect the:

1. Air intake duct (see Section 9)
2. Throttle body electrical connector
3. Vacuum hose
4. PCV hose
5. Upper throttle body mounting nuts (lower nuts not visible in this photo)

10 If necessary, clean the throttle body as outlined in Step 2.

11 Installation is the reverse of removal. Be sure to tighten the throttle body mounting bolts and nuts to the torque listed in this Chapter's Specifications.

12 When the battery has been disconnected, the PCM must relearn its former driveability and performance characteristics (see Chapter 5, Section 1 for this procedure).

13 Start the engine and verify that the throttle body operates correctly and that there are no air leaks.

V6 models

▶ Refer to illustration 13.16

14 Disconnect the cable from the negative battery terminal (see Chapter 5, Section 1).

15 Remove the air intake duct (see Section 9).

16 Disconnect the throttle body electrical connector, the vacuum hose and the PCV hose (see illustration).

17 Loosen the screws on the hose clamps that secure the intake manifold plenum to the intake runners and remove the intake plenum bolts (see "Intake manifold - removal and installation" in Chapter 2B).

18 Disconnect the EGR pipe passage bolt.

19 Rotate the intake manifold plenum up slightly, then disconnect the coolant hoses, EVAP hose and vacuum hose from the throttle body.

20 Remove the throttle body mounting nuts (see illustration 13.16) and remove the throttle body.

21 Remove and discard the old throttle body gasket.

22 If necessary, clean the throttle body as outlined in Step 2.

23 Install a new throttle body gasket.

24 Installation is the reverse of removal. Be sure to tighten the throttle body mounting nuts to the torque listed in this Chapter's Specifications.

25 Reconnect the battery (see Chapter 5, Section 1).

14 Fuel pressure regulator - removal and installation

FOUR-CYLINDER MODELS

▶ Refer to illustrations 14.3, 14.5a and 14.5b

✱✱ WARNING:

Gasoline is extremely flammable, so take extra precautions when you work on any part of the fuel system. See the Warning in Section 2.

1 Relieve the fuel system pressure (see Section 2).

2 Disconnect the cable from the negative battery terminal (see Chapter 5, Section 1).

3 Disconnect the vacuum hose from the fuel pressure regulator (see illustration).

4 Remove the two bolts that attach the fuel pressure regulator to the fuel rail (see illustration 14.3) and remove the pressure regulator.

5 Remove the old O-ring and plastic retainer from the fuel pressure regulator (see illustration), remove the old O-ring from the pressure

14.3 To remove the fuel pressure regulator from the fuel rail on four-cylinder models, disconnect the vacuum hose and remove the two mounting bolts

4-18 FUEL AND EXHAUST SYSTEMS

14.5a Remove and discard the old O-ring (1) and the plastic retainer (2) from the fuel pressure regulator and discard them; always use a new O-ring and retainer when installing the pressure regulator (four-cylinder models)

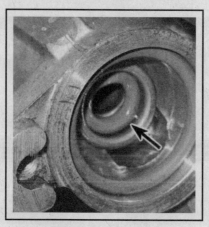

14.5b Remove the old O-ring from the pressure regulator bore in the end of the fuel rail, discard it and replace it with a new one (four-cylinder models)

14.11 To remove the fuel pressure regulator from the fuel rail on V6 models, disconnect the vacuum hose, remove the clamp bolt and nut, then pull the regulator assembly straight down from its housing; be prepared to mop up spilled fuel with a shop rag

regulator bore (see illustration), discard all three and replace them with new ones.

6 Installation is the reverse of removal. Be sure to tighten the fuel pressure regulator mounting bolts to the torque listed in this Chapter's Specifications.

7 When the battery has been disconnected, the PCM must relearn its former driveability and performance characteristics (see Chapter 5, Section 1 for this procedure).

8 Start the engine and verify that there are no fuel leaks.

V6 MODELS

▶ **Refer to illustration 14.11**

9 Relieve the fuel system pressure (see Section 2).

10 Disconnect the cable from the negative battery terminal (see Chapter 5, Section 1).

11 Disconnect the vacuum hose from the fuel pressure regulator (see illustration).

12 Remove the clamp bolt and nut (see illustration 14.11).

13 Remove the fuel pressure regulator from its housing (the housing is a permanent part of the fuel rail). Be prepared to mop up any spilled fuel with a shop rag.

14 Remove and discard the two old fuel pressure regulator O-rings.

15 Installation is the reverse of removal. Be sure to install new O-rings and tighten the clamp bolt/nut securely.

16 When the battery has been disconnected, the PCM must relearn its former driveability and performance characteristics (see Chapter 5, Section 1 for this procedure).

17 Start the engine and verify that there are no fuel leaks.

15 Fuel rail and injectors - removal and installation

1 Relieve the fuel system pressure (see Section 2) and equalize tank pressure by removing the fuel filler cap.

2 Disconnect the cable from the negative battery terminal (see Chapter 5, Section 1).

FOUR-CYLINDER MODELS

▶ **Refer to illustrations 15.5, 15.6, 15.7a, 15.7b, 15.8, 15.9, 15.10a, 15.10b, 15.11 and 15.12**

3 Remove the air intake duct (see Section 9).

4 Disconnect the vacuum line from the fuel pressure regulator (see illustration 14.3).

5 Disconnect the EVAP canister purge hose and the coolant hose from the loom near the right front corner of the valve cover, then disconnect the PCV hose (see illustration). After disconnecting the hoses, push them out of the way.

6 Remove the engine wiring harness cover (see illustration) from the valve cover, then set the wiring harness aside.

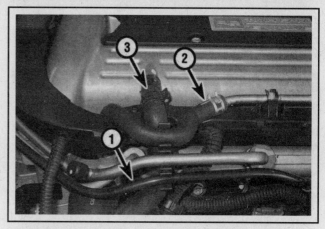

15.5 To access the fuel rail, detach the EVAP canister purge hose (1) and the coolant hose (2) from the hose guide, remove the hose guide and disconnect the PCV hose (3) from the valve cover, then push the hoses out of the way

FUEL AND EXHAUST SYSTEMS   4-19

15.6 To detach the engine wiring harness from the valve cover, remove these two nuts from the fuel rail mounting studs, then set the harness aside to give yourself enough room to remove the fuel rail

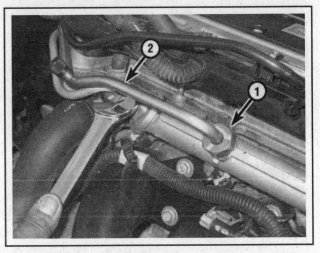

15.7a To disconnect the fuel supply (1) and return (2) lines from the fuel rail, unscrew these fittings

15.7b After unscrewing the fuel supply and return line fittings, remove and discard these O-rings; be sure to install new O-rings before reconnecting the fuel lines to the fuel rail

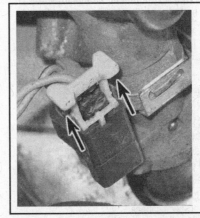

15.8 To disconnect the electrical connector from each fuel injector, pull up on this lock, then unplug the connector

7 Disconnect the fuel supply line and return lines from the fuel rail (see illustration). Remove and discard the old O-rings (see illustration).
8 Disconnect the electrical connectors from the fuel injectors (see illustration).
9 Remove the fuel rail mounting nuts (see illustration).
10 Remove the fuel rail and injectors as a single assembly (see illustration). After removing the fuel rail and injectors, look inside each

15.9 To detach the fuel rail from the cylinder head, remove these two nuts from the fuel rail mounting studs

15.10a Remove the fuel rail and injectors as a single assembly; you might have to wiggle each injector a little to work it free from its mounting hole

4-20 FUEL AND EXHAUST SYSTEMS

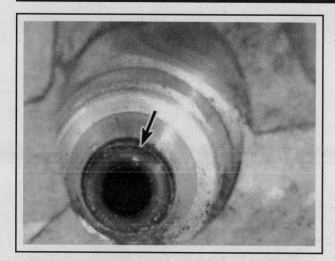

15.10b After removing the fuel rail and injectors, check each hole and make sure that there are no O-rings in the holes (sometimes the lower O-ring on an injector comes off when the injector is removed and stays in the hole)

15.11 To release an injector retainer, free it from the small lugs on each side of the injector and pull it off

injector hole and make sure that no O-rings remain in the holes (see illustration).

11 Remove the retainer that secures each fuel injector to the fuel rail and pull out the injector (see illustration).

12 Remove the old O-rings from each injector (see illustration) and discard them. Always install new O-rings on the injectors before reassembling the injectors and the fuel rail.

13 Installation is otherwise the reverse of removal. To ensure that the new injector O-rings are not damaged when the injectors are installed into the fuel rail and into the intake manifold, lubricate them with clean engine oil. And be sure to tighten the fuel rail mounting bolts to the torque listed in this Chapter's Specifications.

14 When the battery has been disconnected, the PCM must relearn its former driveability and performance characteristics (see Chapter 5, Section 1 for this procedure).

15 Start the engine and verify that there are no fuel leaks.

V6 MODELS

16 Disconnect the fuel lines from the fuel rail (see Section 4).

17 Remove the intake manifold plenum (see *"Intake manifold - removal and installation"* in Chapter 2B).

18 Disconnect the main electrical connector to the fuel injector harness, disconnect the electrical connectors from the injectors and set the harness aside.

19 Remove the fuel rail mounting bolts.

20 Disconnect the vacuum line from the fuel pressure regulator (see illustration 14.11).

21 Carefully disengage the injectors from the lower intake manifold and lift the fuel rail and all six injectors from the engine as a single assembly.

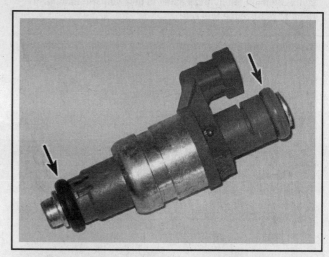

15.12 Remove and discard the old O-rings from each injector and install new ones

22 Remove the injectors from the fuel rail.

23 Remove and discard the old injector O-rings. Always install new O-rings on the injectors before reassembling the injectors and the fuel rail.

24 Installation is the reverse of removal. To ensure that the new injector O-rings are not damaged when the injectors are installed into the fuel rail and into the intake manifold, lubricate them with clean engine oil. Be sure to tighten the fuel rail mounting bolts to the torque listed in this Chapter's Specifications.

25 When the battery has been disconnected, the PCM must relearn its former driveability and performance characteristics (see Chapter 5, Section 1 for this procedure).

26 Start the engine and verify that there are no fuel leaks.

FUEL AND EXHAUST SYSTEMS

16 Exhaust system servicing - general information

INSPECTION

> **WARNING:**
> Inspect and repair exhaust system components only after allowing the exhaust components to cool completely. This applies particularly to the catalytic converter, which operates at very high temperatures. Also, when working under the vehicle, make sure it is securely supported on jackstands.

1 The exhaust system consists of the exhaust manifold(s), the catalytic converter(s), the exhaust pipes, the resonator, the muffler and all brackets, hangers and clamps. Inspect the exhaust system regularly to ensure that it remains safe and quiet. Look for any damaged or bent parts, open seams, holes, loose connections, excessive corrosion or other defects which could allow exhaust fumes to enter the vehicle. Also check the catalytic converter(s) when you inspect the exhaust system. Inspect the catalytic converter heat shield(s) for cracks, dents and loose or missing fasteners. If a heat shield is damaged, the converter might also be damaged. Damaged or deteriorated exhaust system components should not be repaired; they should be replaced with new parts.

2 Before trying to disassemble any exhaust components, spray the fasteners with a penetrating oil to help ease removal. If the exhaust system components are extremely corroded or rusted together, welding equipment will probably be required to remove them. The convenient way to accomplish this is to have a muffler repair shop remove the corroded sections with a cutting torch. If, however, you want to save money by doing it yourself (and you don't have a welding outfit with a cutting torch), simply cut off the old components with a hacksaw. If you have compressed air, special pneumatic cutting chisels can also be used. If you decide to tackle the job at home, be sure to wear safety goggles to protect your eyes from metal chips and work gloves to protect your hands.

3 Here are some simple guidelines to follow when repairing the exhaust system:
   a) *Work from the back to the front when removing exhaust system components.*
   b) *Apply penetrating oil to the exhaust system component fasteners to make them easier to remove.*
   c) *Use new gaskets, hangers and clamps when installing exhaust systems components.*
   d) *Apply anti-seize compound to the threads of all exhaust system fasteners at reassembly.*
   e) *Be sure to allow sufficient clearance between newly installed parts and all points on the underbody to avoid overheating the floor pan and possibly damaging the interior carpet and insulation. Pay particularly close attention to the catalytic converter and heat shield.*

COMPONENT REPLACEMENT

Rubber exhaust hangers
▶ Refer to illustration 16.4

4 The exhaust system is attached to the body with mounting brackets and rubber hangers (see illustration). Anytime you must raise the vehicle to perform any under-vehicle service, make sure that you inspect the exhaust hangers. Look for cracks, tears and deterioration. If a hanger is worn or damaged, replace it.

Catalytic converters

5 The procedures for replacing the catalytic converter(s) are in Chapter 6.

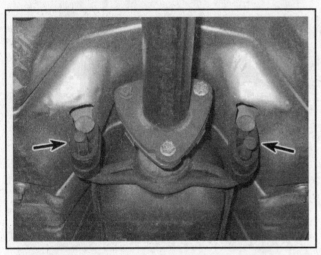

16.4 Typical rubber exhaust hangers used on Saturn vehicles

Specifications

Fuel pressure
    Four-cylinder engine   50 to 60 psi (345 to 414 kPa)
    V6 engine   39 to 49 psi (269 to 338 kPa)
Fuel injector resistance
    Four-cylinder engine
        2000 through 2003   12 to 13 ohms
        2004   11 to 14 ohms
    V6 engine   10.3 to 17.3 ohms
        2000 and 2001   14 to 23 ohms
        2002 and 2003   15.6 to 16.4 ohms
        2004   13 to 15.5 ohms

Torque specifications

	Ft-lbs (unless otherwise indicated)	Nm
Fuel pressure regulator retaining bolts		
Four-cylinder engine	44 in-lbs	5
V6 engine	N/A	
Fuel rail mounting bolts		
Four-cylinder engine	89 in-lbs	10
V6 engine	71 in-lbs	8
Throttle body mounting bolts/nuts		
Four-cylinder engine	89 in-lbs	10
V6 engine	71 in-lbs	8

5
ENGINE ELECTRICAL SYSTEMS

Section

1 General information, precautions and battery disconnection
2 Battery - emergency jump starting
3 Battery - check, removal and installation
4 Battery cables - replacement
5 Ignition system - general information
6 Ignition system - check
7 Ignition control module (four-cylinder models) - replacement
8 Ignition coil pack - removal and installation
9 Charging system - general information and precautions
10 Charging system - check
11 Alternator - removal and installation
12 Starting system - general information and precautions
13 Starter motor and circuit - check
14 Starter motor - removal and installation

5-2 ENGINE ELECTRICAL SYSTEMS

1 General information, precautions and battery disconnection

The engine electrical systems include all ignition, charging and starting components. Because of their engine-related functions, these components are discussed separately from chassis electrical devices such as the lights, the instruments, etc. (which are included in Chapter 12).

PRECAUTIONS

Always observe the following precautions when working on the electrical system:

a) Be extremely careful when servicing engine electrical components. They are easily damaged if checked, connected or handled improperly.
b) Never leave the ignition switched on for long periods of time when the engine is not running.
c) Never disconnect the battery cables while the engine is running.
d) Maintain correct polarity when connecting battery cables from another vehicle during jump starting - see the "Booster battery (jump) starting" Section at the front of this manual.
e) Always disconnect the negative cable from the battery before working on the electrical system, but read the following battery disconnection procedure first.

It's also a good idea to review the safety-related information regarding the engine electrical systems located in the "Safety first!" Section at the front of this manual, before beginning any operation included in this Chapter.

BATTERY DISCONNECTION

The battery is located in the engine compartment on all vehicles covered by this manual. To disconnect the battery for service procedures that require battery disconnection, simply disconnect the cable from the negative battery terminal. Make sure that you isolate the cable to prevent it from coming into contact with the battery negative terminal.

Some vehicle systems (radio, alarm system, power door locks, etc.) require battery power all the time, either to enable their operation or to maintain control unit memory (Powertrain Control Module, automatic transaxle control module, etc.), which would be lost if the battery were to be disconnected. So before you disconnect the battery, note the following points:

a) Before connecting or disconnecting the cable from the negative battery terminal, make sure that you turn the ignition key and the lighting switch to their OFF positions. Failure to do so could damage semiconductor components.
b) On a vehicle with power door locks, it is a wise precaution to remove the key from the ignition and to keep it with you, so that it does not get locked inside if the power door locks should engage accidentally when the battery is reconnected!
c) After the battery has been disconnected, then reconnected (or a new battery has been installed) on vehicles with an automatic transaxle, the Transaxle Control Module (TCM) will need some time to relearn its adaptive strategy. As a result, shifting might feel firmer than usual. This is a normal condition and will not adversely affect the operation or service life of the transaxle. Eventually, the TCM will complete its adaptive learning process and the shift feel of the transaxle will return to normal.
d) The engine management system's PCM has some learning capabilities that allow it to adapt or make corrections in response to minor variations in the fuel system in order to optimize driveability and idle characteristics. However, the PCM might lose some or all of this information when the battery is disconnected. The PCM must go through a relearning process before it can regain its former driveability and performance characteristics. Until it relearns this lost data, you might notice a difference in driveability, idle and/or (if you have an automatic) shift "feel." To facilitate this relearning process, refer to "Enabling the PCM to relearn" below.

MEMORY SAVERS

Devices known as "memory savers" (typically, small 9-volt batteries) can be used to avoid some of the above problems. A memory saver is usually plugged into the cigarette lighter, and then you can disconnect the vehicle battery from the electrical system. The memory saver will deliver sufficient current to maintain security alarm codes and - maybe, but don't count on it! - PCM memory. It will also run "unswitched" (always on) circuits such as the clock and radio memory, while isolating the car battery in the event that a short circuit occurs while the vehicle is being serviced.

※ WARNING:
If you're going to work around any airbag system components, disconnect the battery and do not use a memory saver. If you do, the airbag could accidentally deploy and cause personal injury.

※ CAUTION:
Because memory savers deliver current to operate unswitched circuits when the battery is disconnected, make sure that the circuit that you're going to service is actually open before working on it!

ENABLING THE PCM TO RELEARN

After the battery has been disconnected or the PCM has been replaced, perform the following procedure in order to facilitate PCM relearning:

1 Start the engine and allow it to warm up to its normal operating temperature.
2 Drive the vehicle at part-throttle, under moderate acceleration and idle conditions, until normal performance returns.
3 Park the vehicle and apply the parking brake with the engine running.
4 On vehicles equipped with a manual transaxle, put the shift lever in NEUTRAL. On vehicles equipped with an automatic transaxle, put the shift lever in DRIVE.
5 Allow the engine to idle for about two minutes, or until the idle stabilizes. Make sure that the engine is at its normal operating temperature.

ENGINE ELECTRICAL SYSTEMS 5-3

2 Battery - emergency jump starting

Refer to the *Booster battery (jump) starting* procedure at the front of this manual.

3 Battery - check, removal and installation

> **⚠ WARNING:**
>
> Hydrogen gas is produced by the battery, so keep open flames and lighted cigarettes away from it at all times. Always wear eye protection when working around a battery. Rinse off spilled electrolyte immediately with large amounts of water.

CHECK

♦ Refer to illustrations 3.1a, 3.1b and 3.1c

1 A battery cannot be accurately tested until it is at or near a fully charged state. Disconnect the negative battery cable from the battery and perform the following tests:

a) **Battery state of charge test** - Visually inspect the indicator eye (if equipped) on the top of the battery. If the indicator eye is dark in color, charge the battery as described in Chapter 1. If the battery is equipped with removable caps, check the battery electrolyte. The electrolyte level should be above the upper edge of the plates. If the level is low, add distilled water. DO NOT OVERFILL. The excess electrolyte may spill over during periods of heavy charging. Test the specific gravity of the electrolyte using a hydrometer (see illustration). Remove the caps and extract a sample of the electrolyte and observe the float inside the barrel of the hydrometer. Follow the instructions from the tool manufacturer and determine the specific gravity of the electrolyte for each cell. A fully charged battery will indicate approximately 1.270 (green zone) at 68-degrees F (20-degrees C). If the specific gravity of the electrolyte is low (red zone), charge the battery as described in Chapter 1.

3.1a Use a battery hydrometer to draw electrolyte from the battery cell; this hydrometer is equipped with a thermometer to make temperature corrections

b) **Open circuit voltage test** - Using a digital voltmeter, perform an open circuit voltage test (see illustration). Connect the negative probe of the voltmeter to the negative battery post and the positive probe to the positive battery post. The battery voltage should be greater than 12.5 volts. If the battery is less than the specified voltage, charge the battery before proceeding to the next test. Do not proceed with the battery load test until the battery is fully charged.

c) **Battery load test** - An accurate check of the battery condition can only be performed with a load tester (available at most auto

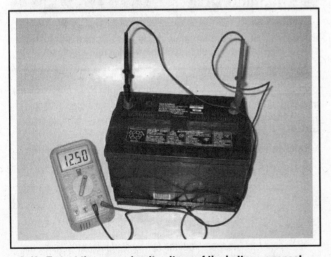

3.1b To test the open-circuit voltage of the battery, connect the black probe of a voltmeter to the negative terminal and the red probe to the positive terminal of the battery; if the battery is fully charged, the voltmeter should indicate about 12.5 volts (depending on the outside air temperature)

3.1c Some battery load testers are equipped with an ammeter, which enables you to impose a precise load on the battery (less expensive testers, like this one, have only a load switch and a voltmeter)

5-4 ENGINE ELECTRICAL SYSTEMS

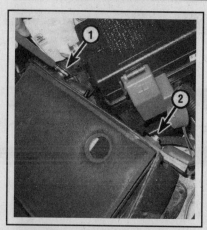

3.2 Whenever you must disconnect the cables from the battery terminals, ALWAYS disconnect the cable from the negative terminal (1) FIRST, then disconnect the cable from the positive terminal (2)

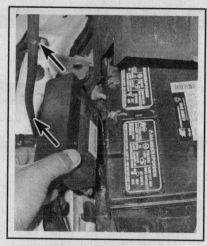

3.4 To disengage the fan control module from its mounting bracket, simply slide it up and off the bracket

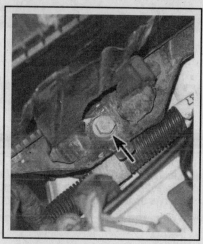

3.5 Remove the bolt and hold-down bracket, then lift the battery out

parts stores). This test evaluates the ability of the battery to operate the starter and other accessories during periods of heavy amperage draw (load). Install a special battery load-testing tool onto the battery terminals (see illustration). Load test the battery according to the tool manufacturer's instructions. This tool utilizes a carbon pile to increase the load demand (amperage draw) on the battery. Maintain the load on the battery for 15 seconds and observe that the battery voltage does not drop below 9.6 volts. If the battery condition is weak or defective, the tool will indicate this condition immediately.

➡ Note: Cold temperatures will cause the minimum voltage requirements to drop slightly. Follow the chart given in the tool manufacturer's instructions to compensate for cold climates. Minimum load voltage for freezing temperatures (32-degrees F/0-degrees C) should be approximately 9.1 volts.

d) *Battery drain test* - This test will indicate whether there's a constant drain on the vehicle's electrical system that can cause the battery to discharge. Make sure all accessories are turned Off. If the vehicle has an underhood light, verify that it's working properly, then disconnect it. Connect one lead of a digital ammeter to the disconnected negative battery cable clamp and the other lead to the negative battery post. A drain of approximately 100 milliamps or less is considered normal (due to the engine control computers, clocks, digital radios and other components that normally cause a key-off battery drain). An excessive drain (approximately 500 milliamps or more) will cause the battery to discharge. The problem circuit or component can be located by removing the fuses, one at a time, until the excessive drain stops and normal drain is indicated on the meter.

REPLACEMENT

Battery

▸ Refer to illustrations 3.2, 3.4 and 3.5

2 Remove the protective cover (if equipped) from the positive battery terminal, then disconnect both cables from the battery terminals (see illustration).

✷✷ WARNING:

Always disconnect the negative cable first and connect it last, or you might accidentally short the battery with the tool you're using to loosen the cable clamps.

3 Remove the battery insulating cover.
4 Remove the fan control module (see illustration).
5 Remove the hold-down nuts and the hold-down bracket (see illustration).
6 Lift out the battery. Use a battery-lifting strap that attaches to the battery posts to lift the battery safely and easily.

➡ Note: If you need to access components underneath the battery tray, refer to Steps 10 through 15.

7 Installation is the reverse of removal.

✷✷ WARNING:

When connecting the battery cables, always connect the positive cable first and the negative cable last to avoid a short circuit caused by the tool used to tighten the cable clamps.

8 When the battery has been disconnected, the PCM must relearn its former driveability and performance characteristics (see Section 1).

Battery tray

▸ Refer to illustrations 3.10 and 3.15

9 Remove the battery (see Steps 1 through 6).
10 Detach the coolant tank hose from the engine compartment fuse box cover (see illustration).
11 Unlock the clip that secures the fuse box harness to the body (see illustration 3.10) and disengage the harness from the clip.
12 Disconnect the cable from the remote starter terminal (see illustration 3.10).
13 Disconnect the ground cable from the body (see illustration 3.10).
14 Release the tabs on the ends of the fuse box (see illustra-

ENGINE ELECTRICAL SYSTEMS   5-5

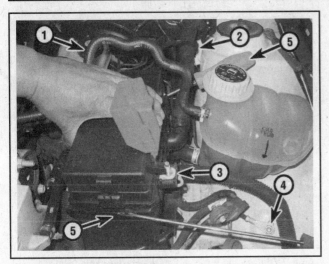

3.10 Battery tray removal details
1. *Detach the coolant hose from the clips on the back of the fuse box*
2. *Unlock and open the clip that secures the fuse box wiring harness to the body*
3. *Flip up the red plastic cover and disconnect the cable from the remote positive terminal*
4. *Disconnect the ground cable*
5. *Pry open the two locking tabs on the front and the rear of the fuse box*

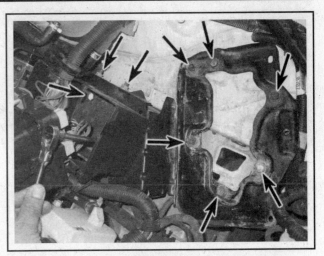

3.15 Unbolt the fuse box from the battery tray, then remove the battery tray mounting fasteners

tion 3.10) and tilt the fuse box toward the engine (the fuse box is hinged on the engine compartment side of the box).

15 Remove the two bolts that attach the fuse box to the battery tray mounting bracket and remove the two nuts and six bolts that attach the battery tray to the body (see illustration).

16 Installation is the reverse of removal.

4 Battery cables - replacement

▶ **Refer to illustrations 4.2, 4.4a, 4.4b and 4.4c**

1 Periodically inspect the entire length of each battery cable for damage, cracked or burned insulation and corrosion. Poor battery cable connections can cause starting problems and decreased engine performance.

2 Check the cable-to-terminal connections at the ends of the cables for cracks, loose wire strands and corrosion (see illustration). The presence of white, fluffy deposits under the insulation at the cable terminal connection is a sign that the cable is corroded and should be replaced. Check the terminals for distortion, missing mounting bolts and corrosion (see Chapter 1 for further information regarding battery cable maintenance).

3 When removing the cables always disconnect the negative cable from the negative battery post first and hook it up last or the tool used to loosen the cable clamps may short the battery. Even if only the positive cable is being replaced, be sure to disconnect the negative cable from the negative battery post first.

4 Before disconnecting any cables, note the routing of both cables to ensure correct installation. Disconnect the old cables from the battery terminals (see Section 3), then disconnect them at the other end(s). Trace each cable from the battery down to its lower end and disconnect it. The smaller ground cable is bolted to the body (see illustration 3.10) and the larger ground cable is attached to the

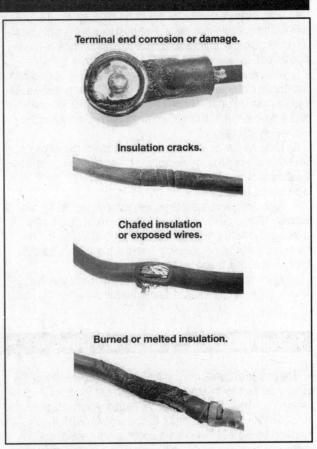

4.2 Typical battery cable problems

5-6 ENGINE ELECTRICAL SYSTEMS

4.4a To detach the larger ground cable from the transaxle, remove this nut

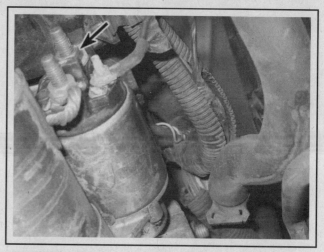

4.4b To detach the lower end of the starter cable from the starter solenoid, remove this nut (four-cylinder model shown, V6 models similar)

transaxle (see illustration). The starter cable is connected to two terminals on the starter motor solenoid (see illustration). Both cables are bundled together with a large engine compartment wiring harness. Although you could remove or cut off the electrical tape that bundles this harness together, remove the old battery cables and retape the harness, we don't recommend doing so (and neither does Saturn). That's because you will create some serious driveability problems if you accidentally cut or damage one of the small wires in the harness. Instead, remove the battery to give yourself some room to work (see Section 3). Then locate the point at which the two cables disappear into the harness and the point at which they emerge from the harness down below, cut off the old cables at those points (see illustration) and tape up the exposed ends to prevent corrosion.

5   Positive cables are almost always red and larger in cross-section; ground cables are usually black and smaller in cross-section. But if you're replacing either or both of the cables, take them with you when buying new cables. It is vitally important that you replace the cables with the identical parts.

6   Clean the threads of the starter solenoid and/or ground connection with a wire brush to remove rust and corrosion. Apply a light coat of battery terminal corrosion inhibitor or petroleum jelly to the threads to prevent future corrosion.

7   Attach the lower ends of the cables first, then connect the positive cable to the positive battery post (don't reconnect the ground cable to the negative battery post until you're completely finished). Before connecting a new cable to the battery, make sure that it reaches the battery post without having to be stretched.

8   Where the cables parallel the wiring harness containing the remains of the old cables, attach them to the harness with several cable

4.4c After you've disconnected the upper and lower ends of both old cables, cut the cables where they disappear into the big wiring harness (shown) and where they come out the other end of the harness; be sure to tape up the cut ends to protect the remaining sections of old cable from corrosion

ties. If either cable is supposed to be secured by any brackets or clips, make sure that you reattach them.

9   After both cables are completely installed, reconnect the ground cable to the negative battery post.

10 When the battery has been disconnected, the PCM must relearn its former driveability and performance characteristics (see Section 1).

5 Ignition system - general information

The ignition system consists of the ignition control module(s), the ignition coil pack(s), the spark plugs, the Camshaft Position (CMP) sensor (V6 models only), the Crankshaft Position (CKP) sensor, the knock sensor(s) and the Powertrain Control Module (PCM). Four-cylinder models are equipped with a single "coil-over-plug" coil pack/ignition control module that fits directly onto the four spark plugs. V6 models have two "coil-over-plug" coil packs/ignition control modules (one for each cylinder head). Neither engine's ignition system uses a distributor or spark plug wires. On four-cylinder models, the ignition control module is mounted on the ignition coil pack, and it can be replaced separately. V6 models do not have separate ignition control modules; they're an integral part of each coil pack.

The coil pack assembly on four-cylinder models consists of two ignition coils. Each coil is connected to the spark plugs for two cylinders by a short boot, which contains a coil spring that carries voltage to the plugs. And each coil fires the two plugs to which it's connected

ENGINE ELECTRICAL SYSTEMS 5-7

simultaneously. One coil fires the plugs for cylinders 1 and 4; the other coil fires the plugs for cylinders 2 and 3. When the piston in cylinder No. 1 is approaching Top Dead Center (TDC) on the compression stroke, the piston in cylinder No. 4 is on the exhaust stroke. When the first ignition coil fires No. 1 cylinder on the compression stroke, most of the spark voltage goes to that cylinder because the pressure - and therefore the resistance - is high in that cylinder (the higher the resistance, the higher the voltage needed to jump the gap from the spark plug's center electrode to ground). Conversely, the piston in cylinder No. 4 produces no pressure, and therefore no resistance, so little voltage is needed to jump the gap from the spark plug's center electrode to ground. The ignition coil for cylinders 2 and 3 works the same way. This design is known as a "waste spark" ignition. The ignition control module on four-cylinder models houses the "driver modules" that turn the ignition coils on and off by closing and opening their ground paths. The timing of these drivers is controlled by the PCM. The ignition control module on four-cylinder coil packs is serviceable separately from the coil pack.

The two coil pack assemblies (one coil pack per cylinder head) on V6 models consist of three ignition coils per coil pack (one for each spark plug). In this design, each coil fires one spark plug. The coils are connected to the spark plugs by short boots with coil springs inside to carry the voltage from the coil to the plug. Each coil pack includes an ignition control module that houses three control circuits, one for each coil. The PCM turns on each ignition coil by grounding the control circuit for that coil. This control circuit is "pulse-width modulated," i.e. its "on" interval is controlled by the PCM. The ignition control modules on V6 coil packs are not separately serviceable. If a coil pack or its ignition module fails, you must replace the entire unit.

The CKP, CMP and knock sensors are information sensors used by the PCM to control ignition timing and other engine operating parameters. The PCM also uses a number of other information sensors to make decisions regarding the correct ignition timing. These other sensors include the Throttle Position (TP) sensor, the Engine Coolant Temperature (ECT) sensor, the Mass Air Flow (MAF) sensor, the Intake Air Temperature (IAT) sensor, the Vehicle Speed Sensor (VSS) and the transmission gear position sensor or Transmission Range (TR) switch. For more information on the CKP, CMP, knock and these other sensors, refer to Chapter 6.

6 Ignition system - check

▶ Refer to illustration 6.4

※ WARNING:

Because of the very high voltage generated by the ignition system (as much as 40,000 volts), use extreme care when you're servicing ignition components such as the ignition coil pack and spark plugs.

➡ **Note 1:** The ignition system components on these models are difficult to diagnose. In the event of ignition system failure, if the checks do not clearly indicate the source of the ignition system problem, have the vehicle tested by a dealer service department or other qualified auto repair facility.

➡ **Note 2:** For the following test, you'll need a calibrated spark tester (available at auto parts stores) and four (four-cylinder models) or three (V6 models) spark plug wires to connect the coil high-tension terminals to the spark tester and to the other three spark plugs.

1 If a malfunction occurs and the vehicle won't start, do not immediately assume that the ignition system is causing the problem. First, check the following items:

   a) Make sure the battery cable clamps, where they connect to the battery, are clean and tight.
   b) est the condition of the battery (see Section 3). If it does not pass all the tests, replace it with a new battery.
   c) Check the wiring and connections for the ignition control module and for the ignition coil pack.
   d) Check the related fuses inside the fuse box (see Chapter 12). If they're burned, determine the cause and repair the circuit.

2 If the engine turns over but won't start, verify that here is sufficient secondary ignition voltage to fire the spark plug as follows.

3 Remove the ignition coil pack (see Section 8), then remove the spark plug boots from the coil pack (to remove the boots, simply pull them off).

4 Connect the spark plug wires to the coil high-tension terminals. On four-cylinder models, connect the calibrated spark tester to the spark plug wire for the No. 1 cylinder. On V6 models, connect the spark tester to the spark plug wire for the No. 1 cylinder (coil pack for rear cylinder head) or the No. 2 cylinder (coil pack for front cylinder head). Then connect the tester to the spark plug (see illustration). On four-cylinder models, connect the other three spark plug wires to the other three spark plugs. On V6 models, connect the other two plug wires to the other two plugs (V6 models).

5 Crank the engine while watching the tester. If the tester flashes, sufficient voltage is reaching the spark plug to fire it.

6.4 Here's the setup used for checking to see if the ignition coil is sending power to the spark plug. If the coil is delivering power to the plug, the tester will flash

※ CAUTION:

Do NOT crank the engine or allow it to run for more than five seconds; running the engine for more than five seconds may set a Diagnostic Trouble Code (DTC) for a cylinder misfire.

5-8 ENGINE ELECTRICAL SYSTEMS

6 Repeat this test on the remaining cylinders.

7 Proceed on this basis until you have verified that there's a good spark from each coil terminal. If there is, then you have verified that the two coils (four-cylinder models) or three coils (V6 models) in the coil pack are functioning correctly.

8 If there is no spark from a coil terminal, then either the coil is bad, the plug wire is bad or a connection at one end of the plug wire is loose. Assuming that you're using new plug wires or known good wires, then the coil is probably defective. Also inspect the coil pack electrical connector. Make sure that it's clean, tight and in good condition.

9 If all the coils are firing correctly, but the engine misfires when the coil pack is connected to the spark plugs via the boots, then one or more of the plugs might be fouled. Remove and check the spark plugs or install new ones (see Chapter 1). Also inspect the boots carefully for corrosion (high resistance) or deterioration of the insulation (low resistance). If any of the boots look damaged or deteriorated, replace them as a set.

10 No further testing of the ignition system is possible without special tools. If the problem persists, have the ignition system tested by a dealer service department.

7 Ignition control module (four-cylinder models) - replacement

▶ Refer to illustrations 7.2 and 7.4

1 Make sure that the ignition key is turned to OFF.
2 Disconnect the electrical connector from the ignition control module (see illustration).
3 Remove the ignition control module mounting screws.
4 Remove the ignition control module and the "interconnect" (see illustration).
5 If you're replacing the ignition control module, disconnect the interconnect from the module and plug it into the new module. The interconnect is an adapter plug that connects the terminals on the ignition control module to the terminals on the coil pack assembly. You'll have to swap it to the new module if you're replacing the old module. Either end of the interconnect can be plugged into the ignition control module or the ignition coil pack. But pay attention to how the plug is oriented in relation to the terminals because it only goes in one way. One side of the interconnect - and one side of the terminals on the ignition control module and on the ignition coil pack - has rounded corners and the other side has square corners. The interconnect is equipped with a weather-resistant grommet. Make sure that this grommet is in good shape. If it's cracked, torn or deteriorated, replace it.

6 Installation is the reverse of removal. Tighten the ignition control module mounting screws to the torque listed in this Chapter's Specifications.

7.2 To remove the ignition control module, disconnect the electrical connector and remove the mounting screws

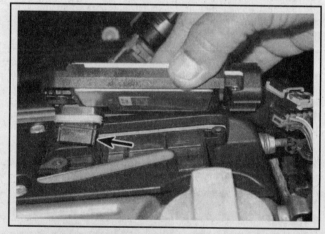

7.4 To remove the ignition control module, pull it straight up; to remove the "interconnect," simply disconnect it from the module and plug it into the new module

8 Ignition coil pack - removal and installation

FOUR-CYLINDER ENGINES

▶ Refer to illustrations 8.3, 8.4 and 8.5

1 Make sure that the ignition key is turned to OFF.
2 Disconnect the electrical connector from the ignition control module (see illustration 7.2). It's not necessary to remove the ignition control module from the ignition coil pack in order to remove the coil pack assembly, which is something you must do in order to remove the valve cover or to service the cylinder head components. However, if you're going to replace the ignition coil pack, you'll have to remove the ignition control module in order to remove the cover from the ignition coil pack (see illustration 7.2).
3 Remove the four ignition coil pack mounting bolts (see illustration).
4 Pull the ignition coil pack straight up, detaching the spark plug

ENGINE ELECTRICAL SYSTEMS

8.3 To detach the ignition coil pack assembly from the valve cover on a four-cylinder model, remove these four bolts (the ignition control module is already removed in this photo, but it's not necessary to do so unless you're planning to replace the coil pack)

8.4 To remove the ignition coil pack assembly from the valve cover on a four-cylinder model, grasp it firmly and pull straight up; the boots should come off with the coil pack (if any of them stay with the spark plugs, simply pull them off the plugs)

boots (see illustration).

5 If you're replacing the ignition coil pack, remove the cover (see illustration) and install it on the new coil pack.

6 If you're replacing the ignition coil pack, remove the four boots from the coil pack and inspect them for cracks, tears and deterioration. If any of the boots are damaged, replace them.

7 Before installing the boots on the ignition coil pack, coat the interior of each boot with silicone dielectric compound.

8 Installation is otherwise the reverse of removal. Tighten the ignition coil pack mounting bolts to the torque listed in this Chapter's Specifications.

V6 MODELS

▶ Refer to illustration 8.10

9 Disconnect the cable from the negative battery terminal (see Section 1).

10 Disconnect the electrical connector from the ignition coil pack

(see illustration).

11 Disconnect the rubber intake boots and remove the intake runners for the front or rear cylinder head (see "Intake manifold - removal and installation" in Chapter 2B).

12 Remove the ignition coil pack mounting bolts.

13 Remove the ignition coil pack assembly.

14 If you're replacing the ignition coil pack, remove the cover and install it on the new coil pack.

15 If you're replacing the ignition coil pack, remove the three boots from the coil pack and inspect them for cracks, tears and deterioration. If any of the boots are damaged, replace them.

16 Before installing the boots on the ignition coil pack coat the interior of each boot with silicone dielectric compound.

17 Installation is otherwise the reverse of removal. Tighten the ignition coil pack mounting bolts to the torque listed in this Chapter's Specifications.

18 When the battery has been disconnected, the PCM must relearn its former driveability and performance characteristics (see Section 1).

8.5 To separate the cover from ignition coil pack, remove these three screws (four-cylinder models)

8.10 To get to a coil pack (arrow) on V6 models, the intake runners will have to be removed

5-10 ENGINE ELECTRICAL SYSTEMS

9 Charging system - general information and precautions

The charging system supplies electrical power for the ignition system, the lights, the radio, the electronic control systems and all other electrical components on the car. The charging system consists of the battery, the alternator (with an integral voltage regulator), the Powertrain Control Module (PCM), the charge indicator lamp on the instrument panel cluster, a fusible link (located inline between the starter solenoid terminal and the alternator) and the wiring between all the components.

The alternator generates alternating current (AC), which is rectified to direct current (DC) to charge the battery and supply power to other electrical systems. The alternator is driven by a drivebelt at the front of the engine (right side of the vehicle). The alternator is located on the front side of both four-cylinder and V6 engines. The voltage regulator limits the alternator charging voltage by regulating the current supplied to the alternator field circuit. The regulator is a solid-state electronic assembly mounted inside the alternator. The regulator is not separately replaceable on these vehicles. If it's defective, you must replace the alternator.

The alternator drivebelt, the battery, and all charging system wires and connections should be inspected at the intervals listed in Chapter 1.

Be very careful when making any circuit connections and note the following:

a) Never start the engine with a battery charger connected.
b) Never disconnect a battery cable with the engine running.
c) Always disconnect both battery cables before using a battery charger: negative cable first, positive cable last. After you're done, the Powertrain Control Module (PCM) must relearn before it can optimize driveability and performance (see Section 1 for this procedure).

10 Charging system - check

▶ **Refer to illustration 10.2**

1 If the charging system malfunctions, don't immediately assume that the alternator is causing the problem. First check the following items:

a) Ensure that the battery cable connections at the battery are clean and tight.
b) If the battery is not a maintenance-free type, check the electrolyte level and specific gravity. If the electrolyte level is low, add clean, mineral-free tap water. If the specific gravity is low, charge the battery.
c) Check the alternator wiring and connections.
d) Check the drivebelt condition and tension (see Chapter 1).
e) Check the alternator mounting bolts for looseness.
f) Run the engine and check the alternator for abnormal noise.

2 Use a voltmeter to check the battery voltage with the engine off. It should be at least 12 volts (see illustration).

3 Start the engine and check the battery voltage again. It should now be approximately 13.5 to 14.5 volts.

4 If the charging voltage reading is zero, inspect the condition of the fusible link that's located in the wire between the alternator and the starter solenoid terminals. If a fusible link is badly blown, it will be obvious that a meltdown has occurred. If a visual inspection is inconclusive, use a continuity tester or ohmmeter to determine whether there is continuity through the fusible link. If the fusible link is blown, replace the fusible link and the wire in which it's located as a single assembly (available at dealer parts departments). The correct size of the fusible link should be printed on the outside of the link. Make sure that you obtain the correct fusible link and wire for the application. After replacing the fusible link, check the charging voltage again.

5 If the voltage reading is more or less than the specified charging voltage, the voltage regulator is defective. Replace the alternator (the voltage regulator cannot be replaced separately).

6 The charging system (battery) light on the instrument cluster lights up when the ignition key is turned to ON, but it should go out

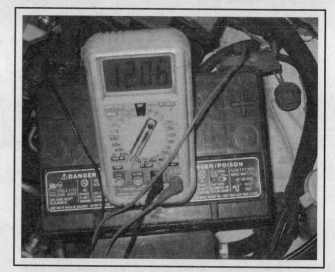

10.2 To measure standing voltage, put your multimeter in the volts mode, connect the positive probe of the meter to the positive battery terminal and the negative probe to the negative terminal and note the reading, which should be around 12 volts; to measure charging voltage, turn on the engine and note the reading again, which should then be about 13.5 volts to 14.5 volts

when the engine starts.

7 If the charging system light stays on after the engine has been started, there is a problem with the charging system. Before replacing the alternator, check the battery condition, alternator belt tension and electrical cable connections.

8 If replacing the alternator doesn't restore voltage to the specified range, have the charging system tested by a dealer service department or other qualified repair shop.

ENGINE ELECTRICAL SYSTEMS 5-11

11 Alternator - removal and installation

FOUR-CYLINDER MODELS

◆ **Refer to illustration 11.4**

1 Disconnect the cable from the negative battery terminal (see Section 1).
2 Remove the air intake duct (see "Air filter housing - removal and installation" in Chapter 4).
3 Remove the accessory drivebelt (see Chapter 1).
4 Disconnect the electrical connectors from the alternator (see illustration).
5 Remove the alternator mounting bolts (see illustration 11.4).
6 Installation is the reverse of removal. Tighten the alternator mounting bolts to the torque listed in this Chapter's Specifications.
7 When the battery has been disconnected, the PCM must relearn its former driveability and performance characteristics (see Section 1).

V6 MODELS

8 Disconnect the cable from the negative battery terminal (see Section 1).
9 Remove the accessory drivebelt and the drivebelt tensioner (see Chapter 1).
10 Remove the upper alternator mounting bolts.
11 Raise the front of the vehicle and place it securely on jackstands.
12 Turn the wheels to the right to provide access to the lower alternator mounting bolt.
13 Remove the lower alternator mounting bolt.

11.4 To remove the alternator from a four-cylinder engine, disconnect the two electrical connectors and remove the four mounting bolts

14 Pull the alternator *away* from the engine block far enough to disconnect the electrical connectors from the alternator.
15 Remove the alternator.
16 Installation is the reverse of removal. Tighten the alternator mounting bolts to the torque listed in this Chapter's Specifications.
17 When the battery has been disconnected, the PCM must relearn its former driveability and performance characteristics (see Section 1).

12 Starting system - general information and precautions

The starting system consists of the battery, the ignition switch, the clutch start switch (manual transaxles), the Transmission Range (TR) switch (automatic transaxles), the starter motor solenoid, the starter motor and the wires that connect these components. The solenoid is located on top of and is an integral part of the starter motor. The starter is located on the front left side of the engine block on four-cylinder models and on the left rear side of the block on V6 models.

The starter motor on a vehicle with a manual transaxle can be operated only when the clutch pedal is depressed. The starter on a vehicle with an automatic transaxle can be operated only when the shift lever is in PARK or NEUTRAL. When the ignition key is turned to the START position, it closes the starter control circuit, which sends battery voltage to the clutch start switch or TR switch. If the clutch pedal is depressed (manual transaxle) or the shift lever is in PARK or NEUTRAL (automatic transaxle), battery voltage is sent to the starter solenoid terminal, which energizes the solenoid, which moves a lever that engages the starter pinion gear with the flywheel ring gear to crank the engine.

Always observe the following precautions when working on the starting system:

a) *Excessive cranking of the starter motor can overheat it and cause serious damage. Never operate the starter motor for more than 15 seconds at a time without pausing for at least two minutes to allow it to cool.*
b) *The starter is connected directly to the battery and could arc or cause a fire if mishandled, overloaded or short-circuited.*
c) *Always detach the cable from the negative battery terminal before working on the starting system.*

13 Starter motor and circuit - check

◆ **Refer to illustrations 13.3 and 13.4**

1 If a malfunction occurs in the starting circuit, do not immediately assume that the starter is causing the problem. First, check the following items:
  a) *Make sure that the battery cable clamps are clean and tight where they connect to the battery.*
  b) *Check the condition of the battery cables (see Section 4). Replace any defective battery cables with new parts.*
  c) *Test the condition of the battery (see Section 3). If it does not pass all the tests, replace it with a new battery.*
  d) *Check the starter solenoid wiring and connections. Refer to the wiring diagrams at the end of Chapter 12.*
  e) *Check the starter mounting bolts for tightness.*

5-12 ENGINE ELECTRICAL SYSTEMS

13.3 Use an inductive-type ammeter to measure the current draw

f) *Make sure that the shift lever is in PARK or NEUTRAL (automatic transaxle) or the clutch pedal is pressed (manual transaxle).*
g) *On vehicles with an automatic transaxle, check the adjustment of the Transmission Range (TR) switch (see Chapter 6). On vehicles with a manual transaxle, make sure that the clutch start switch is correctly installed* (see Chapter 8).

2 If the starter motor does not operate when the ignition switch is turned to the START position, check for battery voltage to the solenoid. Connect a test light or voltmeter to the starter solenoid switched terminal (the small wire) while an assistant turns the ignition switch to the START position. If voltage is not available, check the starting system circuit (see the wiring diagrams at the end of Chapter 12). If voltage is available but the starter motor does not operate, remove the starter from the engine compartment (see Section 14) and bench test the starter (see Step 4).

3 If the starter turns over slowly, check the starter cranking voltage and the current draw from the battery. This test must be performed with the starter assembly on the engine. Crank the engine over (for 10 seconds or less) and observe the battery voltage. It should not drop below 8.0 volts on manual transaxle models or 8.5 volts on automatic transaxle models. Also, observe the current draw using an amp meter (see illustration). It should not exceed 400 amps or drop below 250 amps.

✱ CAUTION:
The battery cables might overheat because of the large amount of current being drawn from the battery. Discontinue the testing until the starting system has cooled down.

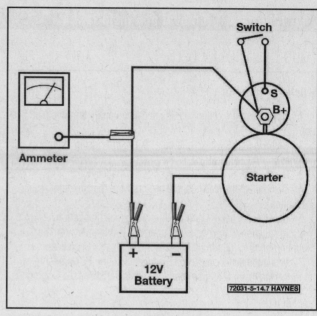

13.4 Starter motor bench testing details

If the starter motor cranking amp values are not within the correct range, replace it with a new unit. There are several conditions that may affect the starter cranking potential. The battery must be in good condition and the battery cold-cranking rating must not be under-rated for the particular application. Be sure to check the battery specifications carefully. The battery terminals and cables must be clean and not corroded. Also, in cases of extreme cold temperatures, make sure the battery and/or engine block is warmed before performing the tests.

4 If the starter is receiving voltage but does not activate, remove and check the starter/solenoid assembly on the bench (see illustration). Most likely the solenoid is defective. In some rare cases, the engine may be seized so be sure to try and rotate the crankshaft pulley (see Chapter 2) before proceeding. With the starter/solenoid assembly mounted in a vise on the bench, install one jumper cable from the negative battery terminal to the body of the starter. Install the other jumper cable from the positive battery terminal to the B+ terminal on the starter. Install a starter switch and apply battery voltage to the solenoid S terminal (for 10 seconds or less) and see if the solenoid plunger, shift lever and overrunning clutch extends and rotates the pinion drive. If the pinion drive extends but does not rotate, the solenoid is operating but the starter motor is defective. If there is no movement but the solenoid clicks, the solenoid and/or the starter motor is defective. If the solenoid plunger extends and rotates the pinion drive, the starter/solenoid assembly is working properly.

14 Starter motor - removal and installation

FOUR-CYLINDER MODELS

▶ Refer to illustrations 14.3, 14.4a and 14.4b

1 Turn the ignition key to OFF, then disconnect the cable from the negative battery terminal (see Section 1).

2 Raise the front of the vehicle and place it securely on jackstands.
3 Disconnect the battery cable (the larger cable) and the starter control cable (the smaller cable) from the starter motor solenoid terminals (see illustration).
4 Remove the starter motor mounting bolts (see illustrations) and remove the starter motor.

ENGINE ELECTRICAL SYSTEMS

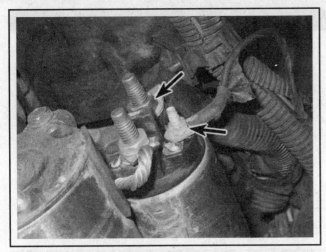

14.3 Disconnect the battery cable and the starter solenoid wire from the starter motor solenoid terminals (four-cylinder model)

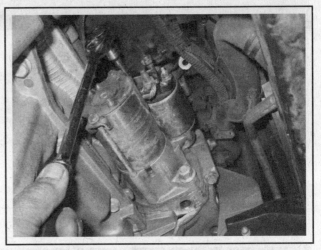

14.4a To detach the starter motor from a four-cylinder engine, use a socket and extension to remove the upper mounting bolt . . .

14.4b . . . then remove the lower mounting bolt

5 Installation is the reverse of removal. Tighten the starter motor mounting bolts to the torque listed in this Chapter's Specifications.

6 When the battery has been disconnected, the PCM must relearn its former driveability and performance characteristics (see Section 1).

V6 MODELS

7 Turn the ignition key to OFF, then disconnect the cable from the negative battery terminal (see Section 1).

8 Loosen the right front wheel lug nuts, raise the front of the vehicle, place it securely on jackstands, then remove the right front wheel.

9 Disconnect the battery cable (the larger cable) and the starter control cable (the smaller cable) from the starter motor solenoid terminals.

10 Detach the wiring harness clip from the engine block and push the harness aside to provide access to the starter motor.

11 Remove the lower starter motor mounting bolt.

12 Lower the vehicle.

13 Remove the upper starter motor mounting bolt, then remove the starter from the engine.

14 Installation is the reverse of removal. Tighten the starter motor mounting bolts to the torque listed in this Chapter's Specifications.

15 When the battery has been disconnected, the PCM must relearn its former driveability and performance characteristics (see Section 1).

ENGINE ELECTRICAL SYSTEMS

Specifications

General

Firing order
- Four-cylinder engine   1-3-4-2
- V6 engine   1-2-3-4-5-6

Cylinder numbering (from drivebelt end to transaxle end)
- Four-cylinder engine   1-2-3-4
- V6 engine
  - Rear bank   1-3-5
  - Front bank   2-4-6

Ignition timing   Not adjustable

Torque specifications

	Ft-lbs (unless otherwise indicated)	Nm
Alternator mounting bolts		
Four-cylinder models	15	20
V6 models	30	40
Ignition coil pack mounting bolts		
Four-cylinder models	89 in-lbs	10
V6 models	71 in-lbs	8
Ignition control module mounting screws	13 in-lbs	1.5
Starter motor mounting bolts (all models)	30	40

6

EMISSIONS AND ENGINE CONTROL SYSTEMS

Section
1. General information
2. On-Board Diagnosis (OBD) system and Diagnostic Trouble Codes (DTCs)
3. Accelerator Pedal Position (APP) sensor (V6 models) - replacement
4. Camshaft Position (CMP) sensor (V6 models) - replacement
5. Crankshaft Position (CKP) sensor - replacement
6. Engine Coolant Temperature (ECT) sensor - replacement
7. Inlet Air Temperature (IAT) sensor - replacement
8. Knock sensor - replacement
9. Manifold Absolute Pressure (MAP) sensor - replacement
10. Mass Air Flow (MAF) sensor (V6 models) - replacement
11. Output Shaft Speed (OSS) sensor - replacement
12. Oxygen sensors - general information and replacement
13. Throttle Position (TP) sensor - replacement
14. Transmission Range (TR) switch - replacement
15. Powertrain Control Module (PCM) - removal and installation
16. Idle Air Control Valve (IAC) valve (four-cylinder models) - replacement
17. Intake Manifold Runner Control (IMRC) system (V6 models) - general information and component replacement
18. Air Injection Reaction (AIR) system (2001 through 2003 four-cylinder models) - general information and component replacement
19. Catalytic converters - general information, check and replacement
20. Evaporative Emissions Control (EVAP) system - general information and component replacement
21. Exhaust Gas Recirculation (EGR) system (V6 models) - general information and component replacement
22. Positive Crankcase Ventilation (PCV) system - general information, inspection and component replacement

6-2 EMISSIONS AND ENGINE CONTROL SYSTEMS

1 General information

♦ Refer to illustration 1.7

The emission control systems and components are an integral part of the engine management system, which is called the Sequential Fuel Injection (SFI) system (see Chapter 4 for more information on the SFI system). The SFI system also includes all the government-mandated diagnostic features of the second generation of on-board diagnostics, which is known as On-Board Diagnostics II (OBD-II).

At the center of the SFI and OBD-II systems is the on-board computer, which is known as the Powertrain Control Module (PCM). Using a variety of information sensors, the PCM monitors all of the important engine operating parameters (temperature, speed, load, etc.). It also uses an array of output actuators (such as the ignition coils, fuel injectors and various solenoids and relays) to respond to and alter these parameters as necessary to maintain optimal performance, economy and emissions. The principal emission control systems used on the vehicles covered in this manual include the:

Air injection Reaction (AIR) system (2001 through 2003 four-cylinder models)
Catalytic converters
Evaporative Emission Control (EVAP) system
Exhaust Gas Recirculation (EGR) system (V6 models)
Intake Manifold Runner Control (IMRC) system (V6 models)
Positive Crankcase Ventilation (PCV) system

The Sections in this Chapter include general descriptions and component replacement procedures for most of the information sensors and output actuators, as well as the important components that are part of the systems listed above. Refer to Chapter 4 for more information on the air intake, fuel and exhaust systems, and to Chapter 5 for information on the ignition system. Refer to Chapter 1 for any scheduled maintenance for emission-related systems and components.

The procedures in this Chapter are intended to be practical, affordable and within the capabilities of the home mechanic. The diagnosis of most engine and emission control functions and driveability problems requires specialized tools, equipment and training. When servicing emission devices or systems becomes too difficult or requires special test equipment, consult a dealer service department.

Although engine and emission control systems are very sophisticated on late-model vehicles, you can do most of the regular maintenance and some servicing at home with common tune-up and hand tools and relatively inexpensive meters. Because of the Federally mandated warranty that covers the emission control system, check with a dealer about warranty coverage before working on any emission-related

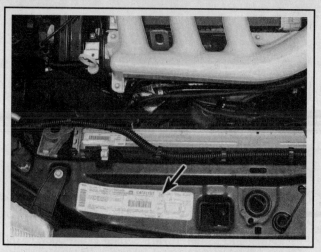

1.7 The Vehicle Emissions Control Information (VECI) label, which is located on the radiator crossmember, specifies critical tune-up and/or adjustment procedures, if applicable (there aren't any on this vehicle), and provides a vacuum hose diagram that shows how vacuum-controlled emission components are connected

systems. After the warranty has expired, you may wish to perform some of the component replacement procedures in this Chapter to save money. Remember that the most frequent cause of emission and driveability problems is a loose electrical connector or a broken wire or vacuum hose, so always check the electrical connections, the electrical wiring and the vacuum hoses first.

Pay close attention to any special precautions given in this Chapter. Remember that illustrations of various systems might not exactly match the system installed on the vehicle on which you're working because of changes made by the manufacturer during production or from year to year.

A Vehicle Emission Control Information (VECI) label (see illustration) is located in the engine compartment. This label contains emission-control and engine tune-up specifications and adjustment information. It also includes a vacuum hose routing diagram for emission-control components. When servicing the engine or emission systems, always check the VECI label in your vehicle. If any information in this manual contradicts what you read on the VECI label on your vehicle, always defer to the information on the VECI label.

2 On Board Diagnostic (OBD) system and Diagnostic Trouble Codes (DTCs)

SCAN TOOL INFORMATION

♦ Refer to illustrations 2.1 and 2.2

1 Hand-held scanners (see illustration) are the most powerful and versatile tools for analyzing engine management systems used on later model vehicles. Early model scanners handle codes and some diagnostics for many systems. Each brand scan tool must be examined carefully to match the year, make and model of the vehicle you are working on. Often, interchangeable cartridges are available to access the particular manufacturer (Chrysler, Ford, GM, Honda, Toyota etc.). Some manufacturers will specify by continent (Asia, Europe, USA, etc.).

➡Note: **An aftermarket generic scanner should work with any model covered by this manual. Before purchasing a generic scan tool, contact the manufacturer of the scanner you're planning to buy and verify that it will work properly with the OBD-II system you want to scan. If necessary, of course, you can always have the codes extracted by a dealer service department or an independent repair shop with a professional scan tool.**

EMISSIONS AND ENGINE CONTROL SYSTEMS   6-3

2.1 Scanners like the Actron Scantool and the AutoXray XP240 are powerful diagnostic aids - programmed with comprehensive diagnostic information, they can tell you just about anything you want to know about your engine management system

2.2 Trouble code readers like the Actron OBD-II diagnostic tester simplify the task of extracting the trouble codes

2 With the advent of the Federally mandated emission control system, which is known as On-Board Diagnostics-II (OBD-II), specially designed scanners were developed. Several tool manufacturers have released OBD-II scan tools for the home mechanic (see illustration).

OBD-II SYSTEM GENERAL DESCRIPTION

3 All vehicles covered by this manual are equipped with the OBD-II system. This system consists of the on-board computer, known as the Powertrain Control Module (PCM), and information sensors that monitor various functions of the engine and send a constant stream of data to the PCM during engine operation. Unlike earlier on-board diagnostics systems, the OBD-II system doesn't just monitor everything, store Diagnostic Trouble Codes (DTCs) and illuminate a Malfunction Indicator Light (MIL) when there's a problem. *It even predicts the probable failure of systems and components when their data starts to become suspicious!*

4 The PCM is the "brain" of the electronically controlled OBD-II system. It receives data from a number of information sensors and switches. Based on the data that it receives from the sensors, the PCM constantly alters engine operating conditions to optimize driveability, performance, emissions and fuel economy. It does so by turning on and off and by controlling various output actuators such as relays, solenoids, valves and other devices. On 2000 and 2001 four-cylinder models, the PCM is mounted in the right end of the dash, behind the glove box. On 2002 and later four-cylinder models, it's located at the right rear corner of the engine compartment. On V6 models, the PCM is located on top of the intake manifold. The PCM data can only be accessed with an OBD-II scan tool plugged into the 16-pin Data Link Connector (DLC), which is located under the dashboard, near the steering column.

5 If your vehicle is still under warranty, virtually every fuel, ignition and emission control component in the OBD-II system is covered by a Federally mandated emissions warranty that is longer than the warranty covering the rest of the vehicle. Vehicles sold in California and in some other states have even longer emissions warranties than other states. Read your owner's manual for the terms of the warranty protecting the emission-control systems on your vehicle. It isn't a good idea to "do-it-yourself" at home while the vehicle emission systems are still under warranty because owner-induced damage to the PCM, the sensors and/or the control devices might VOID this warranty. So as long as the emission systems are still warranted, take the vehicle to a dealer service department if there's a problem.

INFORMATION SENSORS

6 **Accelerator Pedal Position (APP) sensor (V6 models)** - The APP sensor, which is located at the top of the accelerator pedal, is part of the electronic accelerator control system on V6 models, which do not use a conventional accelerator cable. The APP sensor constantly monitors the angle of the accelerator pedal and sends this data to the PCM, which controls an output actuator known as the throttle motor (located in the throttle body) to open or close the throttle plate inside the throttle body to the correct position. The APP sensor consists of a couple of identical "potentiometers," variable resistors that receive a reference voltage from the PCM and return a signal voltage to the PCM that's proportional to the angle of the accelerator pedal. One of the potentiometers is redundant, and serves as a back-up in the event that the primary potentiometer fails. The PCM compares the signal outputs from both potentiometers to assess the accuracy of the primary potentiometer's signal.

7 **Camshaft Position (CMP) sensor** - The CMP sensor, which is used on V6 models, is a Hall effect switching device that produces a signal that the PCM uses to monitor the position of the exhaust camshaft on the front cylinder head. This data enables the CMP sensor to identify the number 1 cylinder so that it can time the firing order of the spark plugs and the firing sequence of the fuel injectors. The CMP sensor is located on the right end of the front cylinder head, near the end of the exhaust camshaft.

8 **Crankshaft Position (CKP) sensor** - The CKP sensor is a permanent magnet generator (also known as a variable reluctance sensor) that produces a variable AC voltage signal that the PCM uses to determine the position of the crankshaft. The PCM uses data from the CKP sensor (and from the CMP sensor on V6 models) to synchronize ignition timing with fuel injector timing, to control spark knock and to detect misfires. On four-cylinder models, the CKP sensor is located on the front of the block, right above the starter motor. On V6 models, the CKP sensor is located on the front of the block, near the flywheel.

6-4 EMISSIONS AND ENGINE CONTROL SYSTEMS

9 **Engine Coolant Temperature (ECT) sensor** - The ECT sensor is a thermistor (temperature-sensitive variable resistor) that sends a voltage signal to the PCM, which uses this data to determine the temperature of the engine coolant. The ECT sensor tells the PCM when the engine is sufficiently warmed up to go into closed loop, helps the PCM control the air/fuel mixture ratio and ignition timing, and also helps the PCM determine when to turn the Exhaust Gas Recirculation (EGR) system on and off. On four-cylinder models, the ECT sensor is located on the thermostat housing, which is located on the left rear corner of the cylinder head. On V6 models, the ECT sensor is located on the coolant crossover, which is located at the left end of the engine, between the cylinder heads.

10 **Fuel tank pressure sensor** - The fuel tank pressure sensor, which is located on top of the fuel tank (on the mounting flange of the fuel pump/fuel level sending unit module), measures the fuel tank pressure when the PCM tests the EVAP system. It's also used to control fuel tank pressure by signaling the EVAP system to purge the tank when the pressure becomes excessive.

11 **Inlet Air Temperature (IAT) sensor** - The IAT sensor, which is located at the left rear corner of the intake manifold, monitors the temperature of the air entering the engine and sends a signal to the PCM. On four-cylinder models, the IAT sensor is located on the air intake duct that connects the air filter housing to the throttle body. On V6 models, the IAT sensor is an integral part of the Mass Air Flow (MAF) sensor, which is located between the air filter housing and the air intake duct.

12 **Knock sensor** - The knock sensor is a "piezoelectric" crystal that oscillates in proportion to engine vibration. (The term *piezoelectric* refers to the property of certain crystals that produce a voltage when subjected to a mechanical stress.) The oscillation of the piezoelectric crystal produces a voltage output that is monitored by the PCM, which retards the ignition timing when the oscillation exceeds a certain threshold. When the engine is operating normally, the knock sensor oscillates consistently and its voltage signal is steady. When detonation occurs, engine vibration increases, and the oscillation of the knock sensor exceeds a design threshold. (Detonation is an uncontrolled explosion, after the spark occurs at the spark plug, which spontaneously combusts the remaining air/fuel mixture, resulting in a "pinging" or "knocking" sound.) If allowed to continue, the engine can be damaged. On four-cylinder models, the knock sensor is located on the block, near the starter motor. On V6 models, there are two knock sensors. Knock sensor No. 1 is located on the left end of the rear side of the block; knock sensor No. 2 is located on the right end of the front side of the block.

13 **Manifold Absolute Pressure (MAP) sensor** - The MAP sensor, which is located on the intake manifold, monitors the pressure or vacuum downstream from the throttle plate, inside the intake manifold. The MAP sensor measures intake manifold pressure and vacuum on the absolute scale, i.e. from zero instead of from sea-level atmospheric pressure (14.7 psi). The MAP sensor converts the absolute pressure into a variable voltage signal that changes with the pressure. The PCM uses this data to determine engine load so that it can alter the ignition advance and fuel enrichment.

14 **Mass Air Flow (MAF) sensor** - Used only on V6 models, the MAF sensor is the means by which the PCM measures the amount of intake air drawn into the engine. It uses a hot-wire sensing element to measure the amount of air entering the engine. The wire is constantly maintained at a specified temperature above the ambient temperature of the incoming air by electrical current. As intake air passes through the MAF sensor and over the hot wire, it cools the wire, and the control system immediately corrects the temperature back to its constant value. The current required to maintain the constant value is used by the PCM to determine the amount of air flowing through the MAF sensor. The MAF sensor also includes an integral Intake Air Temperature (IAT) sensor. The two components cannot be serviced separately; if either sensor is defective, replace the MAF sensor. The MAF sensor is located between the air filter housing and the air intake duct.

15 **Output Shaft Speed (OSS) sensor** - The OSS sensor is a magnetic pick-up coil located on the right side of the automatic transaxle, near the inner CV joint. The OSS sensor provides the PCM with information about the rotational speed of the output shaft in the transmission. The PCM uses this information to control the torque converter and to calculate speed scheduling and the correct operating pressure for the transaxle.

16 **Oxygen sensors** - An oxygen sensor is a galvanic battery that generates a small variable voltage signal in proportion to the difference between the oxygen content in the exhaust stream and the oxygen content in the ambient air. The PCM uses the voltage signal from the upstream oxygen sensor to maintain a "stoichiometric" air/fuel ratio of 14.7:1 by constantly adjusting the "on-time" of the fuel injectors. On four-cylinder models, there are *two* oxygen sensors: the *upstream* sensor is located on the exhaust manifold and a *downstream* oxygen sensor is located behind the catalyst. On V6 models, there are four oxygen sensors: one *upstream* sensor in each exhaust manifold, and a *downstream* sensor below each catalyst.

17 **Throttle Position (TP) sensor** - The TP sensor is a potentiometer that receives a constant voltage input from the PCM and sends back a voltage signal that varies in relation to the opening angle of the throttle plate inside the throttle body. This voltage signal tells the PCM when the throttle is closed, half-open, wide open or anywhere in between. The PCM uses this data, along with information from other sensors, to calculate injector "pulse width" (the interval of time during which an injector solenoid is energized by the PCM). The TP sensor is located on the throttle body, on the end of the throttle plate shaft. On V6 models, there are actually two TP sensors. One of them is redundant, and serves as a back-up in the event that the primary TP sensor fails. The PCM also compares the voltage signals from the two TP sensors in order to gauge the accuracy of the primary sensor and to predict its demise. On these models, the TP sensors are an integral part of the throttle body and cannot be serviced separately.

18 **Transmission Range (TR) switch** - The TR switch is located on top of the transaxle. The TR switch functions like a conventional Park/Neutral Position (PNP) switch: it prevents the engine from starting in any gear other than Park or Neutral, and it closes the circuit for the back-up lights when the shift lever is moved to Reverse. The PCM also sends a voltage signal to the transmission range switch, which uses a series of step-down resistors that act as a voltage divider. The PCM monitors the voltage output signal from the switch, which corresponds to the position of the manual lever. Thus the PCM is able to determine the gear selected and is able to determine the correct pressure for the electronic pressure control system of the transaxle.

OUTPUT ACTUATORS

19 **AIR pump solenoid** - The AIR pump solenoid is a component in the Air Injection Reaction (AIR) system used on 2001 through 2003 California four-cylinder models. The PCM-controlled AIR pump solenoid, which is located on the left side of the engine compartment (near

EMISSIONS AND ENGINE CONTROL SYSTEMS

the ECT sensor), controls intake manifold vacuum to the AIR shutoff valve. When the AIR pump solenoid is energized by the PCM, it allows intake vacuum to open the shutoff valve, which allows air from the AIR pump to enter the exhaust manifold.

20 **EVAP canister purge solenoid** - The EVAP canister purge solenoid (or "purge valve") is located between the right end of the engine and the right strut tower. The EVAP purge solenoid is normally closed. But when ordered to do so by the PCM, it allows the fuel vapors that are stored in the EVAP canister to be drawn into the intake manifold, where they're mixed with intake air, then burned along with the normal air/fuel mixture, under certain operating conditions. The PCM-controlled EVAP canister purge control solenoid valve also controls this vapor flow.

21 **EVAP canister vent solenoid** - The EVAP canister vent solenoid, which is located near the top of the fuel filler neck, is part of the EVAP system's leak diagnostics. The vent solenoid is normally open, to allow outside air to flow through the vent, through the EVAP canister and into the fuel tank, which maintains atmospheric pressure inside the fuel tank. But when energized by the PCM, the vent solenoid closes and seals off the EVAP system for inspection and maintenance tests and for OBD-II leak and pressure tests.

22 **Exhaust Gas Recirculation (EGR) valve** - When the engine is put under a load (hard acceleration, passing, going up a steep hill, pulling a trailer, etc.), combustion chamber temperature increases. When combustion chamber temperature exceeds 2500-degrees F (1380-degrees C), excessive amounts of oxides of nitrogen (NOx) are produced. NOx is a precursor of photochemical smog. When combined with hydrocarbons (HC), other "reactive organic compounds" (ROCs) and sunlight, it forms ozone, nitrogen dioxide and nitrogen nitrate and other nasty stuff. The PCM-controlled EGR valve allows exhaust gases to be recirculated back to the intake manifold where they dilute the incoming air/fuel mixture, which lowers the combustion chamber temperature and decreases the amount of NOx produced during high-load conditions. The EGR valve, which is used only on V6 models, is located on the right rear corner of the engine, behind the throttle body and above the timing belt cover.

23 **Fuel injectors** - The fuel injectors, which spray a fine mist of fuel into the intake ports, where it is mixed with incoming air, are inductive coils under PCM control. The injectors are installed in the intake ports that connect the intake manifold runners to the combustion chambers. For more information about the injectors, see Chapter 4.

24 **Idle Air Control (IAC) valve** - The IAC valve, which is used on four-cylinder models, controls the amount of air allowed to bypass the throttle plate when the throttle plate is at its (nearly closed) idle position. The IAC valve is controlled by the PCM. When the engine is placed under an additional load at idle (high power steering pressure or running the air conditioning compressor during low-speed maneuvers, for example), the engine can run roughly, stumble and even stall. To prevent this from happening, the PCM opens the IAC valve to increase the idle speed enough to overcome the extra load imposed on the engine. The IAC valve is mounted on the throttle body.

25 **Ignition coils** - The ignition coils are under the control of the Powertrain Control Module (PCM). Four-cylinder models use a coil pack assembly consisting of two ignition coils, each of which fires two cylinders. V6 models use a pair of coil packs (one for each cylinder head), each of which consists of three ignition coils. Both units are known as a "coil-over-plug" design because they're mounted directly over the spark plugs (there are no spark plug wires). For more information about the ignition coils, see Chapter 5.

26 **Intake Manifold Runner Control (IMRC) solenoid** - The IMRC solenoid is a device used by the PCM to control the Intake Manifold Runner Control (IMRC) system on V6 models. The IMRC solenoid, which is normally closed, controls intake vacuum to a vacuum diaphragm that controls a flap inside the intake manifold. When the IMRC solenoid and the flap are closed, the flap directs incoming air through a set of shorter intake runners. But when engine speed reaches 3950 rpm (or a certain load threshold), the PCM energizes the IMRC solenoid, which allows intake vacuum to reach the diaphragm, which opens the flap inside the intake manifold and sends the incoming air through a set of longer runners. The IMRC solenoid is located on the air intake manifold plenum.

OBTAINING AND CLEARING DIAGNOSTIC TROUBLE CODES (DTCS)

27 All models covered by this manual are equipped with on-board diagnostics. When the PCM recognizes a malfunction in a monitored emission control system, component or circuit, it turns on the Malfunction Indicator Light (MIL) on the dash. The PCM will continue to display the MIL until the problem is fixed and the Diagnostic Trouble Code (DTC) is cleared from the PCM's memory. You'll need a scan tool to access any DTCs stored in the PCM.

28 Before outputting any DTCs stored in the PCM, thoroughly inspect ALL electrical connectors and hoses. Make sure that all electrical connections are tight, clean and free of corrosion. And make sure that all hoses are correctly connected, fit tightly and are in good condition (no cracks or tears). Also, make sure that the engine is tuned up. A poorly running engine is probably one of the biggest causes of emission-related malfunctions. Often, simply giving the engine a good tune-up will correct the problem.

Accessing the DTCs

▶ Refer to illustration 2.29

29 On these models, all of which are equipped with On-Board Diagnostic II (OBD-II) systems, the Diagnostic Trouble Codes (DTCs) can only be accessed with a scan tool (see illustration 2.1). Simply plug the connector of the scan tool into the Data Link Connector (DLC) or diagnostic connector (see illustration), which is located under the lower edge of the dash, just to the right of the steering column. Then follow the instructions included with the scan tool to extract the DTCs.

2.29 The 16-pin Data Link Connector (DLC), also referred to as the diagnostic connector, is located under the left part of the dash

6-6 EMISSIONS AND ENGINE CONTROL SYSTEMS

30 Once you have outputted all of the stored DTCs look them up on the accompanying DTC chart.

31 After troubleshooting the source of each DTC make any necessary repairs or replace the defective component(s).

Clearing the DTCs

32 Clear the DTCs with the scan tool in accordance with the instructions provided by the scan tool's manufacturer.

DIAGNOSTIC TROUBLE CODES

33 The accompanying tables are a list of the Diagnostic Trouble Codes (DTCs) that can be accessed by a do-it-yourselfer working at home (there are many, many more DTCs available to dealerships with proprietary scan tools and software, but those codes cannot be accessed by a generic scan tool). If, after you have checked and repaired the connectors, wire harness and vacuum hoses (if applicable) for an emission-related system, component or circuit, the problem persists, have the vehicle checked by a dealer service department.

OBD-II DIAGNOSTIC TROUBLE CODES (DTCS)

➡ Note: Not all trouble codes apply to all models.

Code	Probable cause
P0030 or P0050	Oxygen sensor heater control circuit
P0031 or P0051	Oxygen sensor heater control circuit, low voltage
P0032 or P0052	Oxygen sensor heater control circuit, high voltage
P0036 or P0056	Oxygen sensor heater control circuit
P0037 or P0057	Oxygen sensor heater control circuit, low voltage
P0038 or P0058	Oxygen sensor heater control circuit, high voltage
P0101	Mass Air Flow (MAF) sensor performance
P0102	Mass Air Flow (MAF) sensor circuit, low voltage
P0103	Mass Air Flow (MAF) sensor circuit, high voltage
P0106	Manifold Absolute Pressure (MAP) sensor circuit
P0107	Manifold Absolute Pressure (MAP) sensor circuit, low voltage
P0108	Manifold Absolute Pressure (MAP) sensor circuit, high voltage
P0112	Intake Air Temperature (IAT) sensor circuit, low voltage
P0113	Intake Air Temperature (IAT) sensor circuit, high voltage
P0116	Engine Coolant Temperature (ECT) sensor performance
P0117	Engine Coolant Temperature (ECT) sensor circuit, low voltage
P0118	Engine Coolant Temperature (ECT) sensor circuit, high voltage
P0121	Throttle Position (TP) sensor 1 performance
P0122	Throttle Position (TP) sensor 1 circuit, low voltage (V6 models)
P0123	Throttle Position (TP) sensor 1 circuit, high voltage (V6 models)
P0125	Engine coolant temperature insufficient for closed loop fuel control
P0128	Engine coolant temperature below thermostat-regulated temperature
P0130	Upstream oxygen sensor circuit (four-cylinder models)
P0130 or P0150	Oxygen sensor circuit (V6 models)
P0131	Upstream oxygen sensor circuit, low voltage (four-cylinder models)

EMISSIONS AND ENGINE CONTROL SYSTEMS   6-7

Code	Probable cause
P0131 or P0151	Oxygen sensor circuit, low voltage (V6 models)
P0132	Upstream oxygen sensor circuit, high voltage (four-cylinder models)
P0132 or P0152	Oxygen sensor circuit, high voltage (V6 models)
P0133	Upstream oxygen sensor circuit, slow response (four-cylinder models)
P0133 or P0153	Oxygen sensor circuit, slow response (V6 models)
P0134	Upstream oxygen sensor circuit, insufficient activity (four-cylinders)
P0134 or P0154	Oxygen sensor circuit, insufficient activity (V6 models)
P0135 or P0141	Oxygen sensor heater performance (four-cylinder models)
P0135 or P0155	Oxygen sensor heater performance (V6 models)
P0136	Downstream oxygen sensor circuit (four-cylinder models)
P0136 or P0156	Oxygen sensor circuit (V6 models)
P0137	Downstream oxygen sensor circuit, low voltage (four-cylinder models)
P0137 or P0157	Oxygen sensor circuit, low voltage (V6 models)
P0138	Downstream oxygen sensor circuit, high voltage (four-cylinder models)
P0138 or P0158	Oxygen sensor circuit, high voltage (V6 models)
P0139 or P0159	Oxygen sensor circuit, slow response (V6 models)
P0140	Downstream oxygen sensor circuit, insufficient activity (four-cylinders)
P0140 or P0160	Oxygen sensor circuit, insufficient activity (V6 models)
P0141 or P0161	Oxygen sensor heater performance (V6 models)
P0171	Fuel trim system lean (four-cylinder models)
P0171 or P0174	Fuel trim system lean (V6 models)
P0172	Fuel trim system rich (four-cylinder models)
P0172 or P0175	Fuel trim system rich (V6 models)
P0201	Injector no. 1 circuit malfunction
P0202	Injector no. 2 circuit malfunction
P0203	Injector no. 3 circuit malfunction
P0204	Injector no. 4 circuit malfunction
P0205	Injector no. 5 circuit malfunction
P0206	Injector no. 6 circuit malfunction
P0261, P0264, P0267, P0270, P0273 or P0276	Injector control circuit, low voltage
P0262, P0265, P0268, P0271, P0274 or P0277	Injector control circuit, high voltage
P0301 through P0304	Engine misfire detected (four-cylinder models)
P0301 through P0306	Engine misfire detected (V6 models)
P0313	Misfire detected with low fuel level
P0315	Crankshaft position system variation not learned

6-8 EMISSIONS AND ENGINE CONTROL SYSTEMS

OBD-II DIAGNOSTIC TROUBLE CODES (DTCS)

➡ Note: Not all trouble codes apply to all models.

Code	Probable cause
P0324	Knock sensor (KS) module performance
P0326	Knock sensor (KS) performance (four-cylinder models)
P0327	Knock sensor (KS) circuit, low voltage (four-cylinder models)
P0327 or P0332	Knock sensor (KS) circuit, low voltage (V6 models)
P0328 or P0333	Knock sensor (KS) circuit, high voltage (V6 models)
P0335	Crankshaft Position (CKP) sensor circuit
P0336	Crankshaft Position (CKP) sensor performance
P0340	Camshaft Position (CMP) sensor circuit
P0341	Camshaft Position (CMP) sensor performance
P0342	Camshaft Position (CMP) sensor circuit, low voltage
P0343	Camshaft Position (CMP) sensor circuit, high voltage
P0420	Catalyst system, low efficiency
P0421 or P0431	Warm-up catalyst, low efficiency
P0442	Evaporative Emission (EVAP) system, small leak detected
P0443	Evaporative Emission (EVAP) system, purge solenoid control circuit
P0446	Evaporative Emission (EVAP) vent system performance
P0449	Evaporative Emission (EVAP) vent solenoid control circuit
P0451	Fuel Tank Pressure (FTP) sensor performance
P0452	Fuel Tank Pressure (FTP) sensor circuit, low voltage
P0453	Fuel Tank Pressure (FTP) sensor circuit, high voltage
P0455	Evaporative Emission (EVAP) system, leak detected
P0458	EVAP system purge solenoid control circuit, low voltage
P0459	EVAP system purge solenoid control circuit, high voltage
P0462	Fuel level sensor circuit, low voltage
P0463	Fuel level sensor circuit, high voltage
P0480 or P0481	Cooling fan relay control circuit
P0496	EVAP system flow during non-purge
P0498	EVAP system vent solenoid control circuit, low voltage
P0499	EVAP system vent solenoid control circuit, high voltage
P0502	Vehicle Speed Sensor (VSS) circuit, low voltage
P0503	Vehicle Speed Sensor (VSS) circuit, intermittent malfunction
P0506	Idle speed low

EMISSIONS AND ENGINE CONTROL SYSTEMS

Code	Probable cause
P0507	Idle speed high
P0530, P0532 or P0533	Air conditioning refrigerant pressure sensor circuit
P0560	System voltage
P0562	System voltage low
P0563	System voltage high
P0571	Cruise control brake switch circuit
P0601	Control module Read Only Memory (ROM)
P0601 through P0607	Powertrain Control Module (PCM) internal performance problem
P0602	Powertrain Control Module (PCM) not programmed
P0603	Powertrain Control Module (PCM) long-term memory reset
P0604	Transmission Control Module (TCM) Random Access Memory (RAM)
P0621	Alternator L-terminal circuit
P0627	Fuel pump relay control circuit
P0628	Fuel pump relay control circuit, low voltage
P0629	Fuel pump relay control circuit, high voltage
P0638	Throttle Actuator Control (TAC) command performance
P0645, P0646 or P0647	Air conditioning clutch relay control circuit
P0700	Transmission Control Module (TCM) requested MIL illumination
P0705	Transmission Range (TR) switch circuit
P0711	Transmission Fluid Temperature (TFT) sensor performance
P0712	Transmission Fluid Temperature (TFT) sensor circuit, low voltage
P0713	Transmission Fluid Temperature (TFT) sensor circuit, high voltage
P0716	Input speed sensor performance
P0717	Input speed sensor circuit, low voltage
P0719	Brake switch circuit, low voltage
P0724	Brake switch circuit, high voltage
P0727	Engine speed, no signal
P0730	Incorrect gear ratio
P0731	Incorrect 1st gear ratio
P0732	Incorrect 2nd gear ratio
P0733	Incorrect 3rd gear ratio
P0734	Incorrect 4th gear ratio
P0741	Torque Converter Clutch (TCC) system, stuck off
P0742	Torque Converter Clutch (TCC) system, stuck on
P0748	Pressure Control (PC) solenoid control circuit

6-10 EMISSIONS AND ENGINE CONTROL SYSTEMS

OBD-II DIAGNOSTIC TROUBLE CODES (DTCS)

➡ Note: Not all trouble codes apply to all models.

Code	Probable cause
P0751	1-2 shift solenoid valve performance, no 1st or 4th gear
P0752	1-2 shift solenoid valve performance, no 2nd or 3rd gear
P0753	1-2 shift solenoid valve control circuit
P0756	2-3 shift solenoid valve performance, no 1st or 2nd gear
P0757	2-3 shift solenoid valve performance, no 3rd or 4th gear
P0758	2-3 shift solenoid control circuit

3 Accelerator Pedal Position (APP) sensor (V6 models) - replacement

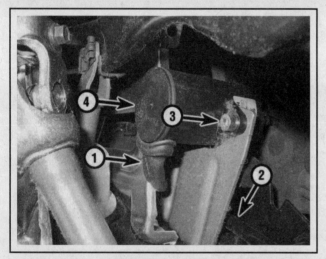

3.4 The Accelerator Pedal Position (APP) sensor (V6 models) is located on a mounting bracket near the top of the accelerator pedal arm. To remove it:

1. Trace the electrical lead up to the connector and disconnect it (connector not visible in this photo)
2. Follow the upper end of the accelerator pedal arm all the way to the top, where it's connected to the APP sensor linkage arm by a link rod. Remove the retainer and disconnect the link rod's Heim joint from the APP sensor linkage arm by prying it loose with a screwdriver
3. Remove the two APP sensor mounting bolts (other bolt, on front side of sensor, not visible in this photo)
4. Remove the APP sensor from its mounting bracket

▶ Refer to illustration 3.4

1 Turn the ignition key to OFF.
2 Remove the left lower closeout panel (see Chapter 11).
3 Remove the left lower heater duct assembly (see Chapter 3).
4 Using a flashlight, locate the APP sensor on its mounting bracket at the top of the accelerator pedal arm (see illustration).
5 Disconnect the electrical connector from the APP sensor.
6 Follow the accelerator pedal arm all the way to the top, where it's connected to the APP sensor linkage arm by a link rod and Heim joint (spherical rod bearing). Remove the retainer and disconnect the link rod Heim joint from the APP sensor linkage arm.

✴ CAUTION:

Do NOT remove the APP sensor's linkage arm (the actuator arm to which the link rod is attached). It's difficult to realign the linkage arm correctly once it's been removed, and an incorrectly attached linkage arm will affect accelerator pedal travel.

7 Remove the two APP sensor mounting bolts and detach the APP sensor from the sensor mounting bracket.
8 Installation is the reverse of removal. Tighten the APP sensor mounting bolts to the torque listed in this Chapter's Specifications.

EMISSIONS AND ENGINE CONTROL SYSTEMS 6-11

4 Camshaft Position (CMP) sensor (V6 models) - replacement

▶ Refer to illustration 4.3

1 Turn the ignition key to OFF.
2 Locate the CMP sensor on the right end of the cylinder head, near the exhaust camshaft.
3 Disconnect the CMP sensor electrical connector (see illustration).
4 Disconnect the electrical connector for knock sensor No. 2 (see illustration 4.3), then detach the knock sensor No. 2 electrical lead from the clip on the CMP sensor and set the knock sensor lead aside.
5 Remove the CMP sensor retaining bolt (see illustration 4.3).
6 Remove the CMP sensor.
7 Installation is the reverse of removal. Be sure to tighten the CMP sensor bolt securely.

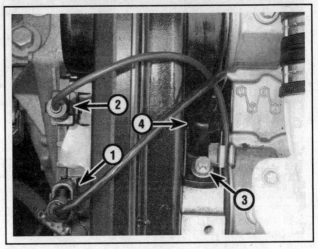

4.3 Camshaft Position (CMP) sensor removal details (V6 models):

1 *Disconnect the CMP sensor electrical connector*
2 *Disconnect the knock sensor No. 2 electrical connector, detach the knock sensor electrical lead from the clip on the CMP sensor and set the sensor lead aside*
3 *Remove the CMP sensor retaining bolt*
4 *Remove the CMP sensor*

5 Crankshaft Position (CKP) sensor - replacement

FOUR-CYLINDER MODELS

▶ Refer to illustrations 5.4 and 5.6

1 Turn the ignition key to OFF.
2 Raise the vehicle and place it securely on jackstands.
3 Remove the starter motor (see Chapter 5).
4 Disconnect the electrical connector from the CKP sensor (see illustration).
5 Unscrew the CKP sensor mounting bolt (see illustration 5.4) and remove the CKP sensor.
6 Even if you're planning to reuse the old CKP sensor, be sure to remove the old O-ring (see illustration) and discard it. Always install a new O-ring when installing the CKP sensor.
7 Installation is the reverse of removal. Be sure to tighten the CKP sensor mounting bolt securely.

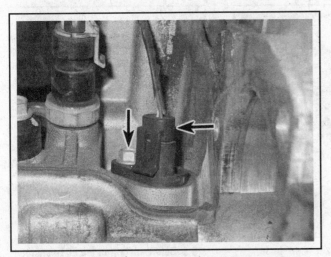

5.4 To remove the CKP sensor from a four-cylinder engine, disconnect the electrical connector and remove the sensor mounting bolt

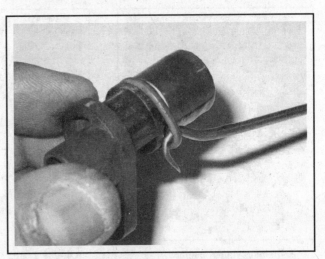

5.6 Be sure to remove the old O-ring from the CKP sensor; always install the sensor with a new O-ring (even if you're installing the old sensor)

6-12 EMISSIONS AND ENGINE CONTROL SYSTEMS

5.11 Removing the CKP sensor on a V6 model (oil filter removed for clarity)

V6 MODELS

▶ Refer to illustration 5.11

8 Turn the ignition key to OFF.
9 Disconnect the electrical connector from the CKP sensor.
10 Raise the vehicle and place it securely on jackstands.
11 Locate the CKP sensor on the left front end of the block, near the oil filter. Remove the CKP sensor mounting bolt and remove the CKP sensor (see illustration).
12 Even if you're planning to reuse the old CKP sensor, be sure to remove the old O-ring (see illustration 5.6) and discard it. Always install a new O-ring when installing the CKP sensor.
13 Installation is the reverse of removal. Be sure to tighten the CKP sensor mounting bolt securely.

6 Engine Coolant Temperature (ECT) sensor - replacement

※※ WARNING:

Wait until the engine is completely cool before beginning this procedure.

FOUR-CYLINDER MODELS

▶ Refer to illustrations 6.3, 6.4 and 6.5

1 Turn the ignition key to OFF.
2 Drain the cooling system (see Chapter 1). (It's not necessary to fully drain the coolant, but it must be drained to a level that's below the level of the ECT sensor.)
3 Disconnect the electrical connector from the ECT sensor (see illustration).
4 Unscrew the ECT sensor (see illustration) and remove it.
5 Wrap the threads of the ECT sensor with Teflon tape (see illustration).
6 Installation is the reverse of removal. Be sure to tighten the ECT sensor to the torque listed in this Chapter's Specifications.
7 Refill the cooling system when you're done (see Chapter 1).

V6 MODELS

▶ Refer to illustration 6.10

8 Turn the ignition key to OFF.

6.3 On four-cylinder engines, disconnect the electrical connector from the ECT sensor, which is located at the lower left corner of the cylinder head, on the thermostat housing

6.4 Use a deep socket, a six-inch extension and a ratchet to unscrew the ECT sensor from four-cylinder engines

6.5 Before installing the ECT sensor, wrap the threads of the sensor with Teflon tape to prevent leaks

EMISSIONS AND ENGINE CONTROL SYSTEMS 6-13

9 Drain the cooling system (see Chapter 1). (It's not necessary to fully drain the coolant, but it must be drained to a level that's below the level of the ECT sensor.)

10 Locate the ECT sensor (see illustration) in the coolant crossover, which is located at the left end of the engine, between the cylinder heads.

11 Disconnect the electrical connector from the ECT sensor.

12 Unscrew the ECT sensor and remove it.

13 Wrap the threads of the new ECT sensor with Teflon tape (see illustration 6.5).

14 Installation is the reverse of removal. Be sure to tighten the ECT sensor to the torque listed in this Chapter's Specifications.

15 Refill the cooling system when you're done (see Chapter 1).

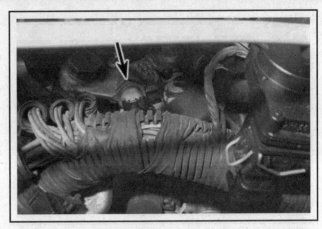

6.10 On V6 engines, the ECT sensor is located in the coolant crossover

7 Inlet Air Temperature (IAT) sensor - replacement

FOUR-CYLINDER MODELS

▸ Refer to illustration 7.2

1 Make sure that the ignition key is turned to OFF.

2 Disconnect the electrical connector from the IAT sensor (see illustration).

3 Remove the IAT sensor from the air intake duct.

4 Inspect the rubber insulator grommet for cracks, tears and deterioration. If it's damaged, replace it.

5 Installation is the reverse of removal.

V6 MODELS

6 On V6 models, the IAT sensor is an integral part of the Mass Air Flow (MAF) sensor (see Section 10).

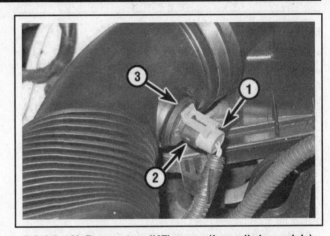

7.2 Inlet Air Temperature (IAT) sensor (four-cylinder models)

1 Electrical connector
2 Inlet Air Temperature (IAT) sensor
3 Rubber insulator grommet

8 Knock sensor - replacement

FOUR-CYLINDER MODELS

▸ Refer to illustration 8.4

1 Make sure that the ignition key is turned to OFF.

2 Raise the front of the vehicle and place it securely on jackstands.

3 Remove the starter motor (see Chapter 5).

4 Disconnect the knock sensor electrical connector (see illustration).

5 Remove the knock sensor retaining bolt (see illustration 8.4).

6 Remove the knock sensor.

7 Installation is the reverse of removal. Be sure to tighten the knock sensor retaining bolt to the torque listed in this Chapter's Specifications.

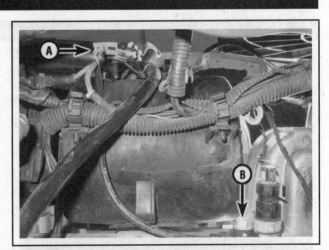

8.4 To remove the knock sensor on four-cylinder models, disconnect the electrical connector (A) and remove the sensor retaining bolt (B)

6-14 EMISSIONS AND ENGINE CONTROL SYSTEMS

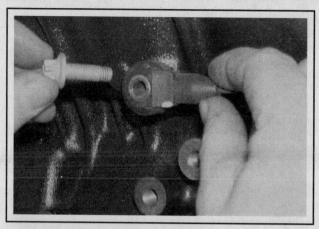

8.17 To detach either of the knock sensors on a V6, simply remove the retaining bolt (knock sensor No. 1 shown, knock sensor No. 2 identical)

V6 MODELS

Knock sensor No. 1

♦ Refer to illustration 8.17

➡Note: Knock sensor No. 1 is located on the rear side of the block.

8 Turn the ignition key to OFF, then disconnect the cable from the negative battery terminal (see Chapter 5, Section 1).
9 Disconnect the knock sensor No. 1 electrical connector.
10 Remove the accessory drivebelt and the belt tensioner pulley (see Chapter 1).
11 Remove the upper alternator mounting bolt (see Chapter 5).
12 Raise the front of the vehicle and place it securely on jackstands.
13 Remove the right front splash shield screws and pushpins and remove the splash shield (see Chapter 11).
14 Remove the accessory drivebelt tensioner (see Chapter 1).
15 Disconnect the electrical connectors from the alternator and remove the lower alternator mounting bolt (see Chapter 5). Move the alternator aside to provide access to the knock sensor.
16 Note the routing of the knock sensor No. 1 wire harness, then detach the harness from any routing clips.
17 Remove the knock sensor No. 1 retaining bolt (see illustration).
18 Remove the knock sensor.
19 Installation is the reverse of removal. Be sure to tighten the knock sensor retaining bolt to the torque listed in this Chapter's Specifications.
20 When you're done, reconnect the cable to the negative battery terminal (see Chapter 5, Section 1).

Knock sensor No. 2

➡Note: Knock sensor No. 2 is located on the front side of the block.

21 Turn the steering wheel all the way to the right, turn the ignition key to OFF, then disconnect the cable from the negative battery terminal (see Chapter 5, Section 1).
22 Disconnect the knock sensor No. 2 electrical connector (see illustration 4.3).
23 Detach the knock sensor No. 2 electrical lead from the CMP sensor clip.
24 Disengage the sensor lead from the attachment clip that's located about six inches down the wire from the CMP sensor clip.
25 Tie a four-foot-long piece of string or wire to the knock sensor No. 2 electrical connector and the other end to a fixed component near the CMP sensor.
26 Loosen - but don't completely back out - the power steering pump pulley bolts (see Chapter 10).
27 Remove the accessory drivebelt (see Chapter 1).
28 Remove the power steering pump pulley bolts and the power steering pump pulley (see Chapter 10).
29 Raise the vehicle and place it securely on jackstands.
30 Remove the knock sensor No. 2 retaining bolt (see illustration 8.17).
31 Guide the knock sensor out from underneath with the string or wire.
32 Disconnect the knock sensor from the string or wire, but do NOT disconnect the string or wire from the engine yet.
33 Using the same string or wire that you used to remove the old knock sensor, tie the new knock sensor to the string or wire and guide it back into place.
34 Installation is otherwise the reverse of removal. Be sure to tighten the knock sensor No. 2 retaining bolt to the torque listed in this Chapter's Specifications.

9 Manifold Absolute Pressure (MAP) sensor - replacement

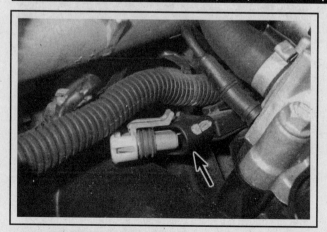

9.2 The MAP sensor electrical connector is located near the throttle body

FOUR-CYLINDER MODELS

♦ Refer to illustrations 9.2, 9.3, 9.4 and 9.5

1 Turn the ignition key to OFF.
2 Disconnect the MAP sensor electrical connector (see illustration).

EMISSIONS AND ENGINE CONTROL SYSTEMS   6-15

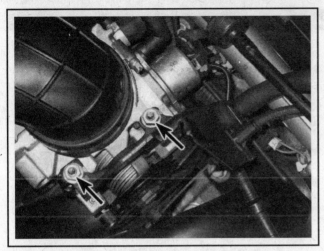

9.3 To detach the accelerator/cruise control cable bracket from the throttle body, remove these two nuts

9.4 To remove the MAP sensor, pull it up and out of its mounting hole in the intake manifold; upper arrows indicate the MAP sensor retainer, which is part of the accelerator/cruise control cable bracket; lower arrow indicates the pipe for the EVAP canister purge solenoid hose, which has been removed for clarity (disconnecting this hose will enable you to see better and give you a little more room to work)

9.5 Remove and discard the old MAP sensor sealing grommet and install a new grommet before installing the MAP sensor

3 Remove the accelerator/cruise control cable bracket nuts (see illustration). You don't have to remove the accelerator/cruise control cable bracket, but you have to raise it up slightly and push it aside, because the MAP sensor retainer is an integral part of the bracket (see illustration 9.4).
4 Remove the MAP sensor (see illustration).
5 Remove the old sealing grommet from the MAP sensor (see illustration) and discard it. Always use a new grommet when installing the MAP sensor, regardless of whether you're installing the old sensor or a new one.
6 Installation is the reverse of removal.

V6 MODELS

♦ Refer to illustration 9.8

7 Turn the ignition key to OFF.

9.8 To detach the MAP sensor from the intake manifold on a V6 model, disconnect the electrical connector and remove the sensor retaining bolt

8 Disconnect the MAP sensor electrical connector (see illustration).
9 Remove the MAP sensor retaining bolt (see illustration 9.8) and remove the MAP sensor.
10 Remove the old sealing grommet from the MAP sensor and discard it. Always use a new grommet when installing the MAP sensor, regardless of whether you're installing the old sensor or a new one.
11 Installation is the reverse of removal. Be sure to tighten the MAP sensor retaining bolt securely.

6-16 EMISSIONS AND ENGINE CONTROL SYSTEMS

10 Mass Air Flow (MAF) sensor (V6 models) - replacement

10.2 Mass Air Flow/Intake Air Temperature (MAF/IAT) sensor removal details
1   MAF/IAT sensor electrical connector
2   Air intake duct hose clamp screws
3   MAF/IAT sensor hose clamp screws

▸ Refer to illustration 10.2

1 Turn the ignition key to OFF.
2 Disconnect the electrical connector from the MAF sensor (see illustration).
3 Loosen the hose clamp screws at each end of the air intake duct and remove the intake duct and the MAF sensor as a single assembly.
4 Loosen the hose clamp screws at each end of the MAF sensor and remove the MAF sensor.
5 Installation is the reverse of removal. Be sure to tighten the MAF sensor and air intake duct hose clamp screws securely.

11 Output Shaft Speed (OSS) sensor - replacement

▸ Refer to illustrations 11.3 and 11.4

➡ Note: This sensor is used only on models with an automatic transaxle.

1 Turn the ignition key to OFF.
2 Raise the front of the vehicle and place it securely on jackstands.
3 Disconnect the electrical connector from the OSS sensor (see illustration).
4 Disengage the electrical harness from the bracket that's attached to the OSS sensor (see illustration).
5 Remove the OSS sensor retaining bolt (see illustration 11.4) and remove the OSS sensor.
6 Remove the old O-ring from the OSS sensor and discard it. Be sure to use a new O-ring when installing the OSS sensor (even if you're planning to reuse the old OSS sensor).
7 Installation is the reverse of removal. Be sure to tighten the OSS sensor retaining bolt securely.

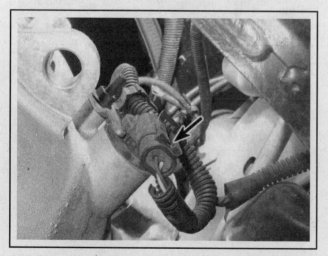

11.3 Disconnect the electrical connector from the OSS sensor

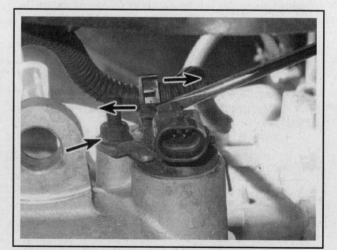

11.4 To disengage the clip that secures the electrical harness to the OSS sensor, push the lower half of the clip to the left with a small screwdriver until it releases from the upper half of the clip; to remove the OSS sensor, remove the retaining bolt

EMISSIONS AND ENGINE CONTROL SYSTEMS

12 Oxygen sensors - general information and replacement

GENERAL INFORMATION

1 An oxygen sensor is a galvanic battery that produces a very small voltage output in response to the amount of oxygen in the exhaust gases. This voltage signal is the "input" side of the feedback loop between the oxygen sensor and the Powertrain Control Module (PCM). Without it, the PCM would be unable to correct the injector on-time (which determines the air/fuel ratio) to maintain the "perfect" (known as *stoichiometric*) air/fuel ratio of 14.7:1 that the catalyst needs for optimal operation.

ON-BOARD DIAGNOSTICS II (OBD-II)

2 All vehicles covered by this manual have On-Board Diagnostics II (OBD-II) engine management systems, which means they have the ability to verify the accuracy of the basic feedback loop between the oxygen sensor and the PCM. They accomplish this by using an oxygen sensor *ahead of* the catalytic converter and another oxygen sensor *behind* the catalytic converter. By comparing the amount of oxygen in the post-catalyst exhaust gas to the oxygen content of the exhaust gas before it enters the catalyst, the PCM can determine the efficiency of the converter.

Four-cylinder models

3 All four-cylinder vehicles covered by this manual have two heated oxygen sensors: one *upstream* sensor (ahead of the catalytic converter) and a *downstream* oxygen sensor (after the catalyst).

V6 models

4 There are *four* heated oxygen sensors, one upstream and one downstream sensor for each cylinder head, on V6 models. On these models, there are three catalysts: a smaller "fast-light-off" cat right below each exhaust manifold and another larger downstream catalyst. The oxygen sensors are located upstream and downstream in relation to the two smaller catalysts.

All models

5 The upstream and downstream oxygen sensors on all models are heated to speed up the warm-up time during which the sensors are unable to produce an accurate voltage signal. The circuit for each oxygen sensor heater is controlled by the PCM, which opens the ground side of the circuit to shut off the heater as soon as the sensor reaches its normal operating temperature.

6 Special care must be taken whenever a sensor is serviced.
   a) *Oxygen sensors have a permanently attached pigtail and an electrical connector that cannot be removed. Damaging or removing the pigtail or electrical connector will render the sensor useless.*
   b) *Keep grease, dirt and other contaminants away from the electrical connector and the louvered end of the sensor.*
   c) *Do not use cleaning solvents of any kind on an oxygen sensor.*
   d) *Oxygen sensors are extremely delicate. Do not drop a sensor or throw it around or handle it roughly.*
   e) *Make sure that the silicone boot on the sensor is installed in the correct position. Otherwise, the boot might melt and it might prevent the sensor from operating correctly.*

REPLACEMENT

→**Note: Because it is installed in the exhaust manifold or pipe, both of which contract when cool, an oxygen sensor can be very difficult to loosen when the engine is cold. Rather than risk damage to the sensor or its mounting threads, start and run the engine for a minute or two, then shut it off. Be careful not to burn yourself during the following procedure.**

Four-cylinder models

Upstream oxygen sensor

▶ Refer to illustrations 12.8 and 12.9

7 Turn the ignition key to OFF.
8 Disconnect the upstream oxygen sensor electrical connector (see illustration), which is located near the left rear corner of the engine.
9 Locate the oxygen sensor on the left side of the exhaust manifold. Using an oxygen sensor socket (available at most automotive retailers), unscrew the upstream oxygen sensor (see illustration). If the sensor is difficult to loosen, spray some penetrant onto the sensor threads and allow it to soak in for awhile.

12.8 The electrical connector for the upstream oxygen sensor on four-cylinder models is located at the left rear corner of the engine

12.9 Use an oxygen sensor socket to unscrew the upstream oxygen sensor from the exhaust manifold on four-cylinder models

6-18 EMISSIONS AND ENGINE CONTROL SYSTEMS

12.14 On four-cylinder models, the electrical connector for the downstream oxygen sensor is located on top of the right rear corner of the engine subframe

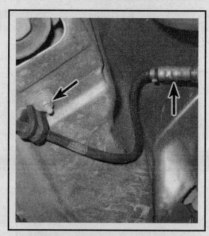

12.15 Remove this nut to detach the clip for the downstream oxygen sensor lead, then unscrew the downstream oxygen sensor (because there's plenty of room to work down here, you can use a big wrench instead of an oxygen sensor socket to remove the sensor)

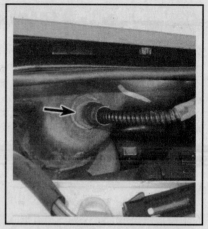

12.19 On V6 models, the upstream oxygen sensors are located in the exhaust manifold (this is the upstream oxygen sensor for the front cylinder bank, but the upstream oxygen sensor for the rear bank is identical in appearance and location, except that it's on the rear exhaust manifold)

10 If you're going to install the old sensor, apply anti-seize compound to the threads of the sensor to facilitate future removal. If you're going to install a new oxygen sensor, it's not necessary to apply anti-seize compound to the threads. The threads on new sensors already have anti-seize compound on them.

11 Installation is otherwise the reverse of removal. Be sure to tighten the sensor to the torque listed in this Chapter's Specifications.

Downstream oxygen sensor

▶ Refer to illustrations 12.14 and 12.15

12 Turn the ignition key to OFF.
13 Raise the vehicle and place it securely on jackstands.
14 Disconnect the oxygen sensor electrical connector (see illustration), which is located on top of the extreme right rear corner of the engine subframe.
15 Detach the harness clip for the downstream oxygen sensor lead (see illustration), then unscrew the downstream oxygen sensor, which is located right behind the catalytic converter. If the sensor is difficult to loosen, spray some penetrant onto the sensor threads and allow it to soak in for awhile.
16 If you're going to install the old sensor, apply anti-seize compound to the threads of the sensor to facilitate future removal. If you're going to install a new oxygen sensor, it's not necessary to apply anti-seize compound to the threads. The threads on new sensors already have anti-seize compound on them.
17 Installation is otherwise the reverse of removal. Be sure to tighten the sensor to the torque listed in this Chapter's Specifications.

V6 models

Upstream oxygen sensor (front cylinder bank)

▶ Refer to illustration 12.19

18 Turn the ignition key to OFF.

19 Disconnect the upstream oxygen sensor electrical connector, which is located to the left of the engine, right next to the connector for the downstream oxygen sensor. If you have difficulty finding the electrical connector, trace the electrical lead from the upstream sensor (see illustration) back to the connector.
20 Using an oxygen sensor socket (available at most automotive retailers), unscrew the upstream oxygen sensor from the exhaust manifold. If the sensor is difficult to loosen, spray some penetrant onto the sensor threads and allow it to soak in for awhile.
21 If you're going to install the old sensor, apply anti-seize compound to the threads of the sensor to facilitate future removal. If you're going to install a new oxygen sensor, it's not necessary to apply anti-seize compound to the threads. The threads on new sensors already have anti-seize compound on them.
22 Installation is otherwise the reverse of removal. Be sure to tighten the sensor to the torque listed in this Chapter's Specifications.

Downstream oxygen sensor (front cylinder bank)

23 Turn the ignition key to OFF.
24 Disconnect the downstream oxygen sensor electrical connector, which is located to the left of the engine, right next to the connector for the upstream oxygen sensor. (If you have difficulty finding the electrical connector, trace the electrical lead from the downstream sensor back to the connector.)
25 Raise the front of the vehicle and place it securely on jackstands.
26 Locate the downstream oxygen sensor for the front cylinder bank right below the "pup" (fast-light) catalyst, then unscrew the downstream oxygen sensor. If the sensor is difficult to loosen, spray some penetrant onto the sensor threads and allow it to soak in for awhile.
27 If you're going to install the old sensor, apply anti-seize compound to the threads of the sensor to facilitate future removal. If you're going to install a new oxygen sensor, it's not necessary to apply anti-seize compound to the threads. The threads on new sensors already have anti-seize compound on them.

EMISSIONS AND ENGINE CONTROL SYSTEMS   6-19

28 Installation is otherwise the reverse of removal. Be sure to tighten the sensor to the torque listed in this Chapter's Specifications.

Upstream oxygen sensor (rear cylinder bank)

29 Turn the ignition key to OFF.
30 Locate the upstream oxygen sensor in the exhaust manifold (see illustration 12.19), then trace the sensor electrical lead back to the electrical connector and disconnect the connector.
31 Using an oxygen sensor socket, unscrew the oxygen sensor. If the sensor is difficult to loosen, spray some penetrant onto the sensor threads and allow it to soak in awhile.
32 If you're going to install the old sensor, apply anti-seize compound to the threads of the sensor to facilitate future removal. If you're going to install a new oxygen sensor, it's not necessary to apply anti-seize compound to the threads. The threads on new sensors already have anti-seize compound on them.
33 Installation is otherwise the reverse of removal. Be sure to tighten the sensor to the torque listed in this Chapter's Specifications.

Downstream oxygen sensor (rear cylinder bank)

34 Turn the ignition key to OFF.
35 Locate the downstream oxygen sensor below the "pup" (fast-light) catalyst, which is located right below the exhaust manifold, then trace the electrical lead from the sensor back to the electrical connector and disconnect the connector.
36 Raise the front of the vehicle and place it securely on jackstands.
37 Using a wrench, unscrew and remove the downstream oxygen sensor. If the sensor is difficult to loosen, spray some penetrant onto the sensor threads and allow it to soak in for awhile.
38 If you're going to install the old sensor, apply anti-seize compound to the threads of the sensor to facilitate future removal. If you're going to install a new oxygen sensor, it's not necessary to apply anti-seize compound to the threads. The threads on new sensors already have anti-seize compound on them.
39 Installation is otherwise the reverse of removal. Be sure to tighten the downstream oxygen sensor to the torque listed in this Chapter's Specifications.

13 Throttle Position (TP) sensor - replacement

FOUR-CYLINDER MODELS

▶ **Refer to illustrations 13.2 and 13.3**

1 Turn the ignition key to OFF.
2 Disconnect the TP sensor electrical connector (see illustration).
3 Remove the TP sensor mounting screws and remove the TP sensor (see illustration).
4 Remove the old TP sensor O-ring and discard it. Install a new O-ring before you install the TP sensor, whether you're installing the old sensor or a new unit.
5 When installing the TP sensor, make sure that the flats on the throttle plate shaft are aligned with the flats inside the TP sensor (see illustration 13.3).
6 Before installing the TP sensor mounting screws, coat the threads of the screws with a thread locking compound.
7 Installation is otherwise the reverse of removal. Be sure to tighten the TP sensor mounting screws securely.

V6 MODELS

8 On V6 models, the TP sensor is an integral part of the throttle body and cannot be removed separately. If the TP sensor must be replaced, replace the throttle body assembly (see Chapter 4).

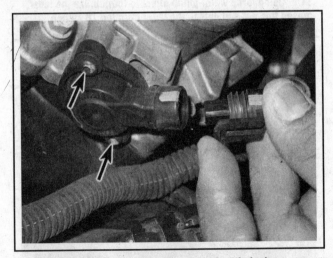

13.2 To remove the TP sensor from the throttle body on a four-cylinder model, disconnect the electrical connector and remove the sensor mounting screws

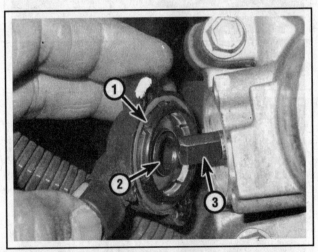

13.3 After removing the TP sensor, remove and discard the old sensor O-ring (1); when installing the TP sensor, make sure that the flat sides of the hub (2) are aligned with the flat sides of the throttle plate shaft (3)

6-20 EMISSIONS AND ENGINE CONTROL SYSTEMS

14 Transmission Range (TR) switch - replacement

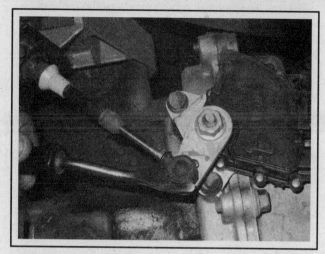

14.3 Use a screwdriver or a trim panel removal tool (shown) to pop the Heim joint on the end of the shift control cable loose from the TR switch lever

REMOVAL

▶ Refer to illustrations 14.3, 14.4, 14.5 and 14.6

➡ Note: You'll need a special alignment tool (J41545, or a suitable equivalent) to adjust the TR switch.

1 Set the parking brake, then place the shift lever in the NEUTRAL position.
2 Locate the TR switch on top of the transaxle.
3 Disconnect the shift control cable from the TR switch lever (see illustration).
4 Disconnect the electrical connector from the TR switch (see illustration).
5 Remove the TR switch lever nut (see illustration) and remove the lever.
6 Remove the TR switch mounting bolts (see illustration) and remove the switch.

14.5 Using large water pump pliers to immobilize the lever, loosen this nut and detach the lever from the TR switch

14.4 Release this lock, then disconnect the electrical connector from the TR switch

14.6 To detach the TR switch from the transaxle, remove these two mounting bolts

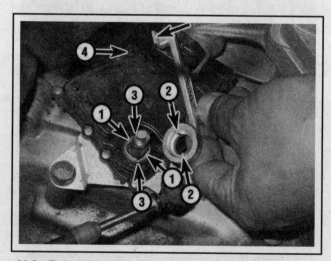

14.8a To install the TR switch, align the flats (1) on the transaxle shift shaft with the flats on the switch, slide the switch onto the shaft and loosely install the switch mounting bolts; to install the alignment tool, align the lugs (2) on the tool with the notches (3) in the switch and align the lug on the other end of the tool with the raised ridge (4) on the switch . . .

EMISSIONS AND ENGINE CONTROL SYSTEMS

INSTALLATION

▶ Refer to illustrations 14.8a and 14.8b

➡ Note: The following procedure applies to a new or old TR switch that's being installed on the transaxle, as well as a TR switch that's already installed, but out of adjustment.

7 Make sure that the shift lever is still in NEUTRAL.
8 To install the TR switch, align the flats on the transaxle shift shaft with the flats on the TR switch (see illustration), then loosely install the TR switch mounting bolts. Install the special TR switch alignment tool (available through specialty tool dealers and some dealership parts departments) (see illustration) and rotate the TR switch until the tool falls into place. When the TR switch is correctly aligned, tighten the switch mounting bolts to the torque listed in this Chapter's Specifications.
9 Installation is otherwise the reverse of removal. Be sure to tighten the TR switch lever nut to the torque listed in this Chapter's Specifications (and hold it with your water pump pliers while doing so).

14.8b . . . then install the tool and rotate the switch slightly until the alignment tool drops into place (and looks like this), then tighten the TR switch mounting bolts to the torque listed in this Chapter's Specifications

15 Powertrain Control Module (PCM) - removal and installation

✲✲ CAUTION:

To avoid electrostatic discharge damage to the PCM, handle the PCM only by its case. Do not touch the electrical terminals during removal and installation. If available, ground yourself to the vehicle with an anti-static ground strap, available at computer supply stores.

➡ Note 1: The procedures in this section apply only to removing and installing the PCM that is already installed in your vehicle. If you need a new PCM, it must be programmed with new software and calibrations. This procedure requires the use of GM's TECH-2 scan tool and GM's latest PCM-programming software, so you WILL NOT BE ABLE TO REPLACE THE PCM AT HOME.

➡ Note 2: The PCM is a highly reliable component and rarely requires replacement. Because the PCM is the most expensive part of the engine management system, you should be absolutely positive that it has failed before replacing it. If in doubt, have the system tested by an experienced driveability technician at a dealer service department or other qualified repair shop.

FOUR-CYLINDER MODELS

2000 and 2001 models

▶ Refer to illustrations 15.3 and 15.7

1 Disconnect the cable from the negative terminal of the battery (see Chapter 5, Section 1).
2 Remove the rear hood seal and remove the right side of the cowl trim panel (see Chapter 11).
3 To access the PCM electrical connectors, remove the small access panel (see illustration) located in the floor of the right side cowl area.
4 Reach through the access panel hole, remove the electrical connector boot nuts and pull back the boots far enough to access the PCM electrical connectors. Then disconnect the electrical connectors from the PCM and pull the harnesses up through the hole and into the engine compartment.
5 Remove the glovebox and the right lower closeout panel (see Chapter 11).
6 Remove the right lower heater duct (see Chapter 3).

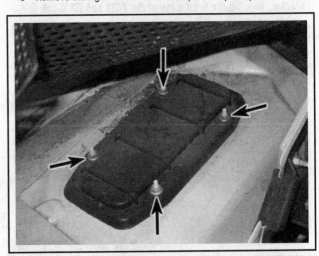

15.3 To access the PCM electrical connectors on 2000 and 2001 four-cylinder models, remove this access panel

6-22 EMISSIONS AND ENGINE CONTROL SYSTEMS

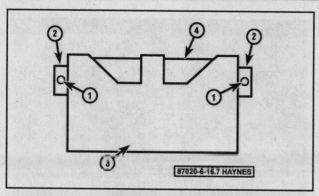

15.7 Powertrain Control Module (PCM) and mounting bracket details (2000 and 2001 four-cylinder models)

1. PCM mounting bracket bolts
2. Release locking tabs by depressing them with a 90-degree pick tool
3. PCM mounting bracket
4. PCM

15.13 To remove the PCM from its mounting bracket, release the two retaining clips on the front edge of the PCM mounting bracket, then lift the front edge of the PCM and lift it out

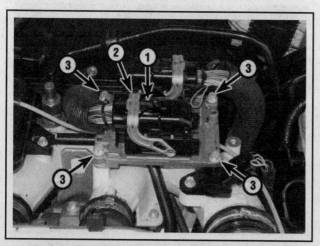

15.18 To disconnect the electrical connectors from the PCM on V6 models, depress the locking tab on top of each connector (1), push the locking lever (2) all the way to its released position (opposite side of the connector from the wiring harness), then pull the connector straight up. To detach the PCM from its mounting bracket, remove these four mounting bolts (3)

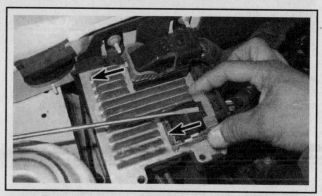

15.12 To release the PCM electrical connectors on 2002 and later four-cylinder models, depress the locking tab on top of each connector with a screwdriver, push the locking lever all the way to the right (toward the fender) to its released position, then pull the connector straight up. In this photo, the locking lever on the connector closer to the firewall has already been released and the other locking lever is being unlocked

7 Remove the PCM mounting bracket bolts (see illustration), release the locking tabs and remove the bracket and PCM as a single assembly.

8 Separate the PCM from the PCM mounting bracket.

9 Installation is the reverse of removal.

10 When the battery has been disconnected, the PCM must relearn its former driveability and performance characteristics (see Chapter 5, Section 1).

2002 and later models

▶ **Refer to illustrations 15.12 and 15.13**

11 Disconnect the cable from the negative battery terminal (see Chapter 5, Section 1).

12 Disconnect the electrical connectors from the PCM (see illustration).

13 Release the two retaining clips on the front edge of the PCM mounting bracket (see illustration).

14 To remove the PCM from its mounting bracket, lift the front edge of the PCM and slide it toward the front of the vehicle.

15 Installation is the reverse of removal.

16 When the battery has been disconnected, the PCM must relearn its former driveability and performance characteristics (see Chapter 5, Section 1).

V6 MODELS

▶ **Refer to illustration 15.18**

17 Disconnect the cable from the negative battery terminal (see Chapter 5, Section 1).

18 To disconnect the electrical connectors from the PCM (see illustration), depress the locking tab on top of each connector, push the locking lever all the way to the opposite side of the connector from the wiring harness to its released position, then pull the connector straight up.

19 Remove the PCM mounting bolts and remove the PCM.

20 Installation is the reverse of removal.

21 When the battery has been disconnected, the PCM must relearn its former driveability and performance characteristics (see Chapter 5, Section 1).

EMISSIONS AND ENGINE CONTROL SYSTEMS

16 Idle Air Control (IAC) valve (four-cylinder models) - replacement

▶ Refer to illustrations 16.2, 16.4 and 16.7

1 Turn the ignition key to OFF.
2 Disconnect the electrical connector from the IAC valve (see illustration).
3 Remove the IAC valve mounting screws (see illustration 16.2) and remove the IAC valve.
4 Remove the old IAC valve O-ring (see illustration) and inspect it. The O-ring should be free of cracks, tears and deterioration, and it must not be distorted. If the O-ring is damaged in any way, be sure to replace it. (It's a good idea to just install a new O-ring whether the old one looks good or not. A damaged or worn O-ring can cause an air leak.) Also, inspect the bore of the IAC valve mounting hole and the air bypass passages inside the throttle body. If the bore and/or the passages are clogged with carbon deposits, remove the throttle body (see Chapter 4) and clean it.
5 If you're going to clean the IAC valve while it's removed, soak a clean shop towel in solvent, squeeze it out and wipe off any dirt or oil from the valve. But do NOT soak the IAC valve in any type of liquid solvent or you will ruin it! Also, do NOT push or pull on the IAC valve's pintle shaft (the part that protrudes from the IAC valve) or you will damage the threads of the worm drive.
6 If you're going to replace the IAC valve, make SURE that you obtain the correct IAC valve. GM IAC valves are very similar in appearance, but the diameter and length of the pintle shaft, and the size and shape of the IAC valve body, are designed for specific applications. It's a good idea to take the old IAC valve with you when buying a new unit.
7 If you're going to install a new IAC valve, measure the distance between the tip of the IAC valve and the mounting flange (see illustration). If the measurement exceeds 1.10 inches (28 mm), use your fingers to carefully and slowly retract the pintle until the distance is 0.79 inch (20 mm) or less. (The force that it takes to retract the pintle on a new IAC valve will not harm the worm drive.)
8 Installation is the reverse of removal. Be sure to tighten the IAC valve mounting screws securely.

16.2 To remove the IAC valve from the throttle body, disconnect the electrical connector, then remove the two IAC valve mounting screws

16.4 Be sure to remove and discard the old IAC valve O-ring from the IAC valve mounting hole in the throttle body; always install a new O-ring before installing the IAC valve

16.7 Before installing a new IAC valve, measure the distance between the IAC valve mounting flange and the tip of the pintle shaft. If it exceeds 1.10 inches (28 mm), carefully retract the pintle with your fingers until the distance is no more than 0.79 inch (20 mm)

17 Intake Manifold Runner Control (IMRC) system (V6 models) - general information and component replacement

GENERAL INFORMATION

1 When intake air is drawn into the cylinders at idle or at low engine speeds, less air is needed because the cylinders don't need to be filled so often or so quickly. So at idle and at low engine speeds, the air drawn into an engine with smaller intake runners will have a higher velocity than one with larger intake runners. However, at higher engine speeds, smaller intake runners would prevent the cylinders from filling quickly enough and would therefore limit power. Most intake manifold designs are a compromise between the conflicting demands of low and high engine speeds.
2 The Intake Manifold Runner Control (IMRC) system helps to maintain a uniformly higher intake air velocity throughout the engine's operating range. Higher intake air velocity promotes better vaporization of the fuel sprayed into the stream of incoming air by the fuel injectors, which means more complete combustion, more power, better fuel economy and less emissions.
3 The IMRC system consists of a special intake manifold with a vacuum-controlled tuning valve inside the manifold, a vacuum diaphragm to control the tuning valve, a PCM-controlled solenoid that controls vacuum to the diaphragm, and the PCM itself. When the engine is idling or operating below 3950 rpm the IMRC solenoid is closed. When the IMRC solenoid is closed, the tuning valve inside the manifold directs incoming air through a shorter path. Directing incom-

6-24 EMISSIONS AND ENGINE CONTROL SYSTEMS

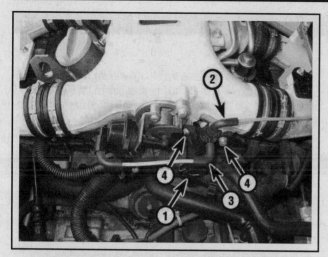

17.4 Intake Manifold Runner Control (IMRC) solenoid removal details

1. Electrical connector
2. Vacuum line from intake manifold
3. Vacuum line to actuator
4. Mounting screws

speed reaches 3950 rpm (or a certain load threshold), the PCM energizes the IMRC solenoid, which allows intake vacuum to reach the actuator. When the diaphragm inside the actuator moves, it moves a link rod connected to it, which rotates the tuning valve inside the intake manifold. When the tuning valve rotates, it sends the incoming air through a longer intake path designed to handle a larger volume of air. At that point, the volume of air drawn into the cylinders is sufficient to promote good velocity even through the longer intake path. And the longer intake path enhances performance during heavy acceleration or high cruising speeds.

COMPONENT REPLACEMENT

Intake Manifold Runner Control (IMRC) solenoid

◆ Refer to illustration 17.4

4 Disconnect the IMRC solenoid electrical connector (see illustration).
5 Disconnect the vacuum lines from the IMRC solenoid.
6 Remove the IMRC solenoid mounting screws and remove the solenoid.
7 Installation is the reverse of removal.

Intake manifold

8 Refer to Chapter 2B.

ing air through a shorter intake path at low engine speeds promotes higher intake air velocities because the incoming air can move more quickly through the intake manifold to fill the cylinders. When engine

18 Air Injection Reaction (AIR) system (2001 through 2003 four-cylinder models) - general information and component replacement

GENERAL INFORMATION

1 The Air Injection Reaction (AIR) system reduces hydrocarbons (HC) and carbon monoxide (CO) during cold starts, when the catalytic converter has not yet reached operating temperature, and helps warm up the catalyst more quickly by pumping air into the hot exhaust stream, which raises the temperature of the exhaust gases. When the hot exhaust gases are combined with more air, oxygen in the air helps to burn up any HC and CO not already consumed during combustion.

2 The AIR system consists of the air pump, air pump solenoid, shutoff valve, a couple of check valves, pipes and hoses connecting the pump, shutoff valve and check valve, a relay (located in the engine compartment fuse and relay box) and the Powertrain Control Module (PCM).

3 The air pump is an electrically-driven device, the circuit for which is controlled by the air pump relay, which in turn is controlled by the PCM. When the Engine Coolant Temperature (ECT) sensor indicates coolant temperature between 36 and 104-degrees Fahrenheit (between 2 and 40-degrees Centigrade), the Intake Air Temperature (IAT) sensor indicates an air temperature over 39-degrees F (4-degrees C) and engine speed is less than 4450 rpm, the PCM energizes the AIR relay, which closes the circuit to the air pump and to the air pump solenoid simultaneously.

4 The air pump solenoid controls vacuum to the shutoff valve. When the PCM turns on the relay, a valve inside the air pump solenoid opens and intake manifold vacuum pulls up the shutoff valve diaphragm. When the shutoff valve diaphragm moves up, low-pressure air from the air pump is pumped through an inline check valve into the exhaust manifold. (The one-way check valve prevents exhaust gas from entering the air pump.) Another check valve, between the manifold vacuum supply pipe (on the throttle body) and the air pump solenoid, maintains vacuum to the solenoid for a specified interval of time when manifold vacuum decreases.

5 Because it never operates for more than a minute at a time, the AIR system should prove to be reliable and trouble free. Nonetheless, if an AIR component ever fails, one of the following procedures will tell you how to replace it.

18.6 The AIR vacuum check valve is located in the vacuum line between the throttle body and the AIR pump solenoid; to remove the vacuum check valve, simply disconnect the vacuum lines from both ends of the valve

EMISSIONS AND ENGINE CONTROL SYSTEMS

18.8 When installing the AIR vacuum check valve, make sure that the side that's labeled "VAC" is facing toward the vacuum source (the throttle body)

18.9 To remove the AIR pump solenoid, disconnect the electrical connector and the vacuum hoses, then remove the retaining nut

18.14 To remove the AIR shutoff valve, disconnect the AIR pump solenoid vacuum line (1), disconnect the AIR pump hose (2) and disconnect the hose (3) that connects the shutoff valve to the check valve

COMPONENT REPLACEMENT

AIR vacuum check valve

♦ Refer to illustrations 18.6 and 18.8

6 Locate the AIR vacuum check valve in the vacuum line between the throttle body and the AIR pump solenoid valve (see illustration).

7 To remove the AIR vacuum check valve, simply disconnect the vacuum lines from both ends of the valve.

8 Installation is the reverse of removal. When installing the AIR vacuum check valve, make sure that the side that's labeled "VAC" (see illustration) is facing toward the vacuum source, i.e. toward the throttle body.

AIR pump solenoid

♦ Refer to illustration 18.9

9 Locate the AIR pump solenoid at the left end of the engine (see illustration).
10 Disconnect the electrical connector from the AIR pump solenoid.
11 Disconnect the vacuum hoses from the AIR pump solenoid.
12 Remove the AIR pump solenoid retaining nut.
13 Installation is the reverse of removal.

AIR shut-off valve

♦ Refer to illustration 18.14

14 Locate the AIR shutoff valve behind the valve cover (see illustration).
15 Disconnect the AIR pump solenoid vacuum line from the AIR shutoff valve.
16 Disconnect the AIR pump hose from the shutoff valve.
17 Disconnect the hose that connects the shutoff valve to the AIR check valve.
18 Installation is the reverse of removal.

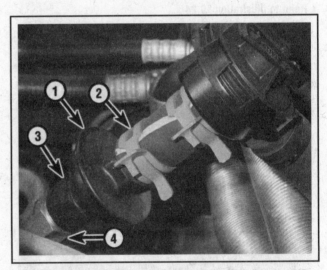

18.19 To remove the AIR check valve (1), disconnect the AIR shutoff valve hose (2), then unscrew the check valve from the AIR pipe with a wrench on the hex nut (3) of the check valve and a back-up wrench on the hex nut (4) of the AIR pipe

AIR check valve

♦ Refer to illustration 18.19

19 Locate the AIR check valve right below the AIR shutoff valve (see illustration).
20 Disconnect the AIR shutoff valve hose from the check valve.
21 Using a back-up wrench on the AIR pipe hex nut, unscrew the AIR check valve from the AIR pipe.
22 When installing the check valve, it's a good idea to apply anti-seize compound to the threads of the check valve and the AIR pipe to facilitate future removal.
23 Installation is otherwise the reverse of removal.

6-26 EMISSIONS AND ENGINE CONTROL SYSTEMS

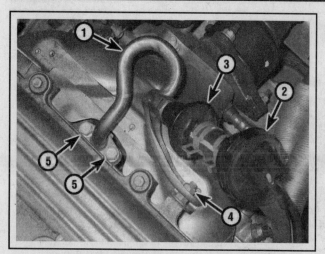

18.24 To remove the AIR pipe (1), remove the AIR shutoff valve (2), the AIR check valve (3), the support bracket nut (4) and the two bolts (5) that attach the pipe to the exhaust manifold

18.30 To detach the AIR pump mounting bracket from the engine, remove these three bolts

AIR pipe

▶ Refer to illustration 18.24

24 Locate the AIR pipe between the AIR check valve and the exhaust manifold (see illustration).

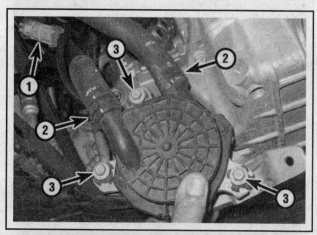

18.32 After pulling down the AIR pump and mounting bracket assembly, disconnect the electrical connector (1), disconnect the inlet and outlet hoses (2); to separate the pump from its mounting bracket, remove the three mounting bolts (3)

25 Remove the AIR shutoff valve (see Steps 14 through 17) and the AIR check valve (see Steps 19 through 21).
26 Remove the AIR pipe support bracket nut and the two AIR pipe mounting bolts that attach the pipe to the exhaust manifold and remove the AIR pipe.
27 Remove and discard the old gasket from the AIR pipe mounting flange.
28 Installation is the reverse of removal. Be sure to use a new gasket and to tighten the AIR pipe mounting bolts to the torque listed in this Chapter's Specifications.

AIR pump

▶ Refer to illustrations 18.30 and 18.32

29 Raise the front of the vehicle and place it securely on jackstands.
30 Locate the AIR pump at the right front corner of the engine oil pan (see illustration).
31 Remove the three AIR pump mounting bracket bolts.
32 Pull down the AIR pump and bracket assembly and disconnect the electrical connector and the inlet and outlet hoses (see illustration).
33 To separate the AIR pump from the mounting bracket, remove the three pump mounting bolts.

19 Catalytic converters - general information, check and replacement

➡ **Note:** *Because of a Federally-mandated extended warranty which covers emission-related components such as the catalytic converter, check with a dealer service department before replacing the converter at your own expense.*

GENERAL DESCRIPTION

1 A catalytic converter (or catalyst) is an emission control device in the exhaust system that reduces certain pollutants in the exhaust gas stream. There are two types of converters. An oxidation catalyst reduces hydrocarbons (HC) and carbon monoxide (CO). A reduction catalyst reduces oxides of nitrogen (NOx). Catalysts that can reduce *all three pollutants* are known as "three-way catalysts." The models covered by this manual are equipped with three-way catalysts.

CHECK

2 The test equipment for a catalytic converter (a "loaded-mode" dynamometer and a 5-gas analyzer) is expensive. If you suspect that the converter on your vehicle is malfunctioning, take it to a dealer or

EMISSIONS AND ENGINE CONTROL SYSTEMS   6-27

19.5 To disconnect the upper end of the catalytic converter/exhaust pipe assembly from the exhaust manifold flange, remove these three nuts

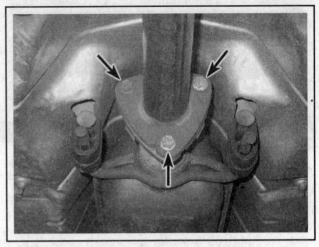

19.7 To disconnect the rear end of the catalytic converter/exhaust pipe assembly from the resonator pipe flange, remove these three bolts

authorized emission inspection facility for diagnosis and repair.

3 Whenever you raise the vehicle to service underbody components, inspect the converter for leaks, corrosion, dents and other damage. Carefully inspect the welds and/or flange bolts and nuts that attach the front and rear ends of the converter to the exhaust system. If you note any damage, replace the converter.

4 Although catalytic converters don't break too often, they can become clogged or even plugged up. The easiest way to check for a restricted converter is to use a vacuum gauge to diagnose the effect of a blocked exhaust on intake vacuum.

   a) *Connect a vacuum gauge to an intake manifold vacuum source (see Chapter 2).*
   b) *Warm the engine to operating temperature, place the transaxle in Park (automatic models) or Neutral (manual models) and apply the parking brake.*
   c) *Note the vacuum reading at idle and jot it down.*
   d) *Quickly open the throttle to near its wide-open position and then quickly get off the throttle and allow it to close. Note the vacuum reading and jot it down.*
   e) *Do this test three more times, recording your measurement after each test.*
   f) *If your fourth reading is more than one in-Hg lower than the reading that you noted at idle, the exhaust system might be restricted (the catalytic converter could be plugged, OR an exhaust pipe or muffler could be restricted).*

REPLACEMENT

Four-cylinder models

▶ Refer to illustrations 19.5 and 19.7

※※ WARNING:

Make sure that the exhaust system is completely cooled down before proceeding. If the vehicle has been driven recently, the catalytic converter can be hot enough to cause serious burns.

5 Open the hood and remove the three upper exhaust pipe-to-exhaust manifold flange nuts (see illustration).

6 Raise the vehicle and place it securely on jackstands.
7 Remove the three bolts that attach the exhaust pipe behind the catalytic converter to the resonator pipe flange (see illustration).
8 Remove the catalytic converter.
9 Remove and discard the old flange gaskets.
10 Installation is the reverse of removal. Be sure to use new gaskets at both mounting flanges. Use new nuts at the front flange and new bolts at the rear flange. Coat the threads of the nuts and bolts with anti-seize compound to facilitate future removal. Tighten the fasteners to the torque listed in this Chapter's Specifications.

V6 models

Front catalyst

11 Disconnect the downstream oxygen sensor electrical connector.
12 Raise the front of the vehicle and place it securely on jackstands.
13 Remove the downstream oxygen sensor (see Section 12).
14 Remove the three front catalyst-to-front exhaust manifold flange nuts.
15 Remove the three front catalyst-to-rear catalyst flange bolts.
16 Remove the front catalyst/exhaust pipe assembly.
17 Remove and discard the old flange gaskets from both mounting flanges.
18 Be sure to use new gaskets at both mounting flanges.
19 Use new nuts at the front flange and new bolts at the rear flange. Coat the threads of the nuts and bolts with anti-seize compound to facilitate future removal. Tighten the fasteners to the torque listed in this Chapter's Specifications.
20 Installation is otherwise the reverse of removal. Be sure to tighten the downstream oxygen sensor to the torque listed in this Chapter's Specifications.

Rear catalyst

21 Remove the four catalyst heat shield bolts and remove the heat shield.
22 Disconnect the downstream oxygen sensor electrical connector.
23 Raise the vehicle and place it securely on jackstands.
24 Remove the downstream oxygen sensor (see Section 12).
25 Remove the three rear catalyst-to-rear exhaust manifold flange nuts.
26 Remove the three front catalyst-to-rear catalyst flange bolts.

6-28 EMISSIONS AND ENGINE CONTROL SYSTEMS

27 Remove the three rear catalyst-to-resonator flange bolts.
28 Remove the rear catalyst/exhaust pipe assembly.
29 Remove and discard the old flange gaskets from all three mounting flanges.
30 Be sure to use new gaskets at all three mounting flanges.
31 Use new nuts at the front flange and new bolts at both lower flanges. Coat the threads of the nuts and bolts with anti-seize compound to facilitate future removal. Tighten the fasteners to the torque listed in this Chapter's Specifications.
32 Installation is otherwise the reverse of removal. Be sure to tighten the downstream oxygen sensor to the torque listed in this Chapter's Specifications.

20 Evaporative emissions control (EVAP) system - general information and component replacement

GENERAL DESCRIPTION

1 The **Evaporative Emissions Control (EVAP)** system prevents fuel system vapors (which contain unburned hydrocarbons) from escaping into the atmosphere. On warm days, vapors trapped inside the fuel tank expand until the pressure reaches a certain threshold, at which point the fuel vapors are routed from the fuel tank through the fuel vapor vent valve and the fuel vapor control valve to the EVAP canister, where they're stored temporarily, until they can be consumed by the engine during normal operation. When the conditions are right (engine warmed up, vehicle up to speed, moderate or heavy load on the engine, etc.) the Powertrain Control Module (PCM) opens the canister purge solenoid, which allows the fuel vapors to be drawn from the canister into the intake manifold, where they mix with the air/fuel mixture before being consumed in the combustion chambers. This system is complex and virtually impossible to troubleshoot without the right tools and training. However, the following description should give you a good idea of how it works:

2 The **EVAP canister** is located under the vehicle, on top of the fuel tank. The EVAP canister, which contains activated carbon, is the repository for storing the fuel vapors. You'll have to raise the vehicle and lower the fuel tank to inspect or replace the canister (or the fuel tank pressure sensor) but the canister is designed to be maintenance-free and should last the life of the vehicle.

3 The **fuel tank pressure sensor**, which is located on top of the mounting flange for the in-tank fuel pump/fuel level sending unit module, monitors the pressure inside the tank, and transmits its measurement to the PCM during an OBD-II leak test.

4 The **EVAP canister vent solenoid**, which is mounted near the upper end of the fuel filler neck, is normally open. But it seals off the EVAP system for inspection and maintenance (I/M 240) testing and for OBD-II leak and pressure tests.

5 The **EVAP canister purge solenoid**, which is under the control of the Powertrain Control Module (PCM), regulates the flow of vapors being purged from the EVAP canister into the intake manifold. The canister purge solenoid is normally closed. It opens only when directed to do so by the PCM, which uses the availability of intake manifold vacuum and data from various information sensor inputs to determine when and how long to open the valve. The interval of time during which the purge valve is opened by the PCM is known as its "duty cycle." The purge valve is located on the right side of the engine compartment, between the timing cover and the right strut tower.

General system checks

6 The most common symptom of a faulty EVAP system is a strong fuel odor (particularly during hot weather). If you smell fuel while driving or (more likely) right after you park the vehicle and turn off the engine, check the fuel filler cap first. Make sure that it's screwed onto the fuel filler neck all the way. If the odor persists, inspect all EVAP hose connections, both in the engine compartment and under the vehicle. You'll have to raise the vehicle and place it securely on jackstands to inspect most of the EVAP system, since it's located under the vehicle. Be sure to inspect each hose attached to the canister for damage and leakage along its entire length. Repair or replace as necessary. Inspect the canister for damage and look for fuel leaking from the bottom. If fuel is leaking or the canister is otherwise damaged, replace it.

7 Poor idle, stalling, and poor driveability can be caused by a defective fuel vapor vent valve or canister purge solenoid, a damaged canister, cracked hoses, or hoses connected to the wrong tubes. Fuel loss or fuel odor can be caused by fuel leaking from fuel lines or hoses, a cracked or damaged canister, or a defective vapor valve.

8 To check for excessive fuel vapor pressure in the fuel tank, remove the gas cap and listen for the sound of pressure release. If the fuel tank emits a "whooshing" sound when you open the filler cap, fuel tank vapor pressure is excessive. Inspect the canister vapor hoses and the canister inlet port for blockage or collapsed hoses. Also inspect the vapor vent valve. A complete test can only be done with a proprietary OBD-II scan tool (see Section 2), which will run a series of checks to detect excessive pressure. You'll have to take the vehicle to a dealer service department to have the EVAP system professionally diagnosed.

COMPONENT REPLACEMENT

EVAP canister purge solenoid

▶ Refer to illustration 20.10

9 Turn the ignition key to OFF.
10 Locate the EVAP canister purge solenoid on the right side of the engine compartment, near the right strut tower (see illustration).

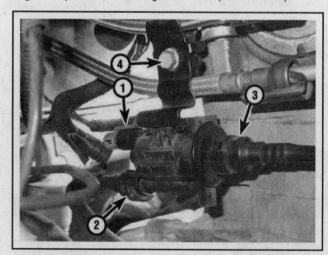

20.10 To remove the EVAP purge solenoid, disconnect the electrical connector (1), the inlet purge line fitting (2), the outlet purge line fitting (3) and remove the mounting bracket bolt (4) (to disconnect the purge line fittings, push the locking tab, then pull off the connector)

EMISSIONS AND ENGINE CONTROL SYSTEMS   6-29

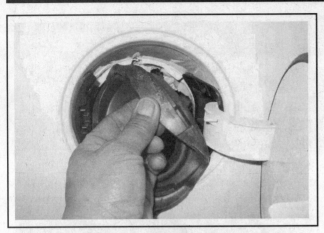

20.18 Remove the sealing boot from the fuel filler neck

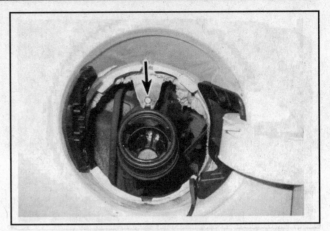

20.19 Loosen the screw above the fuel filler neck

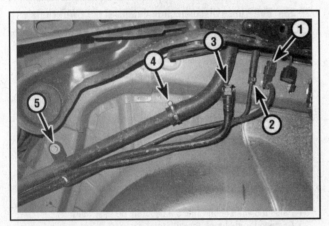

20.20 To remove the fuel filler neck, disconnect or remove the following items:

1 EVAP canister vent solenoid electrical connector
2 Fuel filler neck EVAP vapor line
3 EVAP canister vent solenoid line quick-connect fitting (see Chapter 4)
4 Fuel filler neck hose clamp
5 Fuel filler neck tube retaining bracket bolt

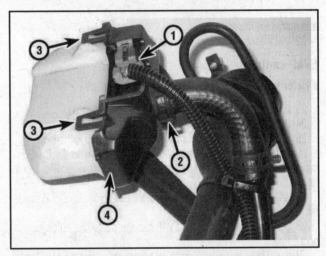

20.21a To remove the EVAP canister vent solenoid from the fuel filler neck assembly, disconnect the electrical connector (1), loosen the vent hose clamp (2) and pull off the hose, disengage the two locking tabs (3), remove the solenoid housing cover (4) . . .

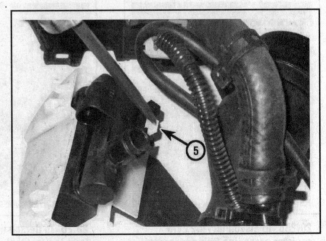

20.21b . . . then disengage the locking tang (5) . . .

11 Disconnect the canister purge solenoid electrical connector.
12 Disconnect the inlet and outlet purge line fittings from the canister purge solenoid. Cap the lines to prevent dirt, dust and moisture from entering the EVAP system while the lines are open.
13 Remove the canister purge solenoid mounting bracket bolt and remove the purge solenoid and bracket.
14 Separate the canister purge solenoid from its mounting bracket.
15 Installation is the reverse of removal.

EVAP canister vent solenoid

▶ Refer to illustrations 20.18, 20.19, 20.20, 20.21a, 20.21b and 20.21c

16 Loosen the right rear wheel lug nuts, raise the rear of the vehicle, place it securely on jackstands and remove the right rear wheel.
17 Remove the liner from the right rear wheelhousing.
18 Remove the sealing boot from the fuel filler neck (see illustration).
19 Loosen the screw that's located right above the fuel filler neck

(see illustration).
20 Remove the fuel filler neck assembly (see illustration).
21 Remove the EVAP canister vent solenoid from the fuel filler neck

6-30 EMISSIONS AND ENGINE CONTROL SYSTEMS

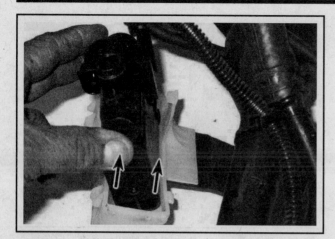

20.21c ... and lift the vent solenoid out of its housing

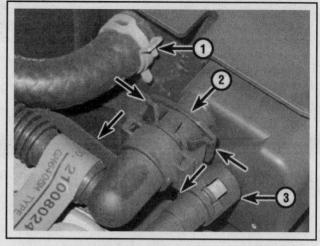

20.25a EVAP hose disconnection details at the EVAP canister

1. This is the vent hose that goes to the EVAP canister vent solenoid; to disconnect it, loosen the hose clamp and pull off the hose
2. This is the vapor hose that goes to the fill limit vent valve (located right next to the EVAP canister); to disconnect it, squeeze the two locking tangs together and pull off the connector
3. This is the EVAP canister purge line that goes to the canister purge solenoid in the engine compartment; to disconnect it, refer to the next illustration

assembly (see illustrations).

22 Installation is the reverse of removal.

EVAP canister

▶ Refer to illustrations 20.25a, 20.25b and 20.26

23 Raise the vehicle and place it securely on jackstands.
24 Remove the fuel tank (see Chapter 4).
25 Disconnect the hoses from the EVAP canister (see illustrations).
26 Cut the retaining strap (see illustration) and remove the EVAP canister. Discard the old retaining strap and obtain a new one for installation.
27 Use a new retaining strap to secure the new (or old) EVAP canister.
28 Use a new retainer for the EVAP canister purge line quick-connect fitting.
29 If you have any questions about how to reconnect the EVAP canister purge line quick-connect fitting, refer to "Fuel lines and fittings - general information" in Chapter 4.
30 Installation is otherwise the reverse of removal.

Fuel tank pressure sensor

▶ Refer to illustration 20.33

31 Raise the vehicle and place it securely on jackstands.
32 Remove the fuel tank (see Chapter 4).
33 Disconnect the electrical connector from the fuel tank pressure sensor (see illustration).
34 Remove the fuel tank pressure sensor mounting nut and remove the sensor.
35 Installation is the reverse of removal.

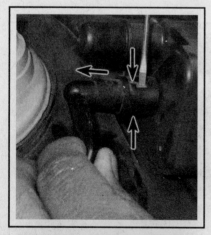

20.25b To disconnect the EVAP canister purge line fitting, depress the two locks by pushing them in with a small screwdriver, then pull off the fitting. Remove and discard the retainer (always use a new retainer when reconnecting this fitting)

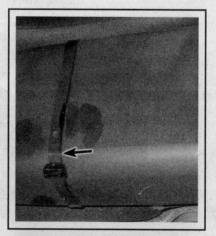

20.26 To detach the EVAP canister from the fuel tank, cut this retaining strap and remove the canister; discard the old retaining strap and obtain a new strap

20.33 To detach the fuel tank pressure sensor from the fuel pump/fuel level sending unit mounting flange, disconnect the electrical connector and remove the sensor mounting nut

EMISSIONS AND ENGINE CONTROL SYSTEMS

21 Exhaust Gas Recirculation (EGR) system (V6 models) - general information and component replacement

GENERAL DESCRIPTION

1 Oxides of nitrogen, nitrogen oxide, or simply NOx, is a compound that is formed in the combustion chambers when the oxygen and nitrogen in the incoming air mix together. NOx is a natural byproduct of high combustion chamber temperatures (2500-degrees F and higher). When NOx is emitted from the tailpipe, it mixes with reactive organic compounds (ROCs), hydrocarbons (HC) and sunlight to form ozone and photochemical smog.

2 The EGR system reduces NOx by recirculating exhaust gases from the exhaust manifold, through the EGR valve and intake manifold, then back to the combustion chambers, where it mixes with the incoming air/fuel mixture before being consumed. These recirculated exhaust gases "dilute" the incoming air/fuel mixture, which cools the combustion chambers, thereby reducing NOx emissions.

3 The EGR system consists of the Powertrain Control Module (PCM), the EGR valve and various information sensors (ECT, TP, MAP, IAT, RPM and VSS sensors) that the PCM uses to determine when to open the EGR valve. When the PCM closes the power/control circuit for the EGR valve, a solenoid inside the EGR valve is energized. This creates an electromagnetic field, which causes an armature to pull up, lifting the pintle off its seat. The exhaust gas then flows from the exhaust manifold port to the intake manifold.

4 Once activated by the PCM, the EGR valve uses a position feedback circuit to control the position of the pintle valve. The feedback circuit, which functions like a potentiometer, puts out a variable output voltage signal with an operating range between 0.5 and 5.4 volts. This variable output enables the PCM to control the position of the pintle with a high degree of precision. A pintle position sensor monitors the position of the pintle, and the PCM adjusts the current to match the actual pintle position to the optimal pintle position.

5 If there is too much EGR flow at idle, cruise or during cold running conditions, the engine will stop after a cold start, stop at idle after deceleration, surge during cruising speeds or idle roughly. If there is too little EGR flow, combustion chamber temperature can become too high during acceleration or under a heavy load, which can cause spark knock (detonation) and/or engine overheating.

COMPONENT REPLACEMENT

EGR valve

▶ Refer to illustration 21.6

6 Disconnect the electrical connector (see illustration) from the EGR valve solenoid.

7 Remove the EGR valve mounting bolts.

8 Remove the EGR valve and the old gasket. Discard the old gasket.

9 Clean the gasket mating surfaces on the intake manifold and on the EGR valve mounting flange and install a new gasket.

10 Installation is otherwise the reverse of removal. Tighten the EGR valve mounting bolts securely.

EGR pipe

▶ Refer to illustrations 21.11 and 21.12

11 Unscrew the threaded fitting (see illustration) that connects the EGR pipe to the intake manifold, near the EGR valve.

12 Unscrew the threaded fitting (see illustration) that connects the EGR pipe to the exhaust manifold.

13 Remove the EGR pipe.

14 Clean out the EGR pipe with a wire bottlebrush, then blow it out with compressed air.

15 Coat the threads of the EGR pipe fittings with anti-seize compound. Installation is otherwise the reverse of removal. Be sure to tighten the EGR pipe's threaded fittings securely.

21.6 To remove the EGR valve from a V6 model, disconnect the electrical connector and remove the two mounting bolts

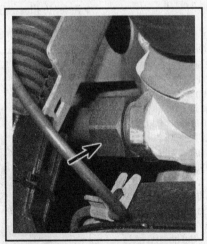

21.11 The threaded fitting that connects the upper end of the EGR pipe to the intake manifold is located directly below the EGR valve

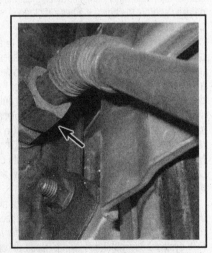

21.12 The threaded fitting that connects the EGR pipe to the exhaust manifold is located on the exhaust manifold for the rear cylinder head

22 Positive Crankcase Ventilation (PCV) system - general information, inspection and component replacement

GENERAL INFORMATION

1 The Positive Crankcase Ventilation (PCV) system reduces hydrocarbon emissions by scavenging crankcase vapors, which are rich in unburned hydrocarbons.

2 On four-cylinder models, the PCV system consists of a single hose between the valve cover and the air intake duct, and a crankcase ventilation housing (or simply, the vent housing) permanently affixed to the underside of the valve cover. Crankcase blow-by vapors are directed through internal passages in the engine block and cylinder head up to a vent housing, from which they're drawn through the hose into the air intake duct, then through the throttle body and manifold and into the combustion chambers where they're consumed along with the air/fuel mixture.

3 On V6 models, the PCV system consists of an engine vent adapter (located on the intake manifold), a crankcase ventilation housing or vent housing (located at the left end of the rear cylinder head) and the vent hoses connecting the vent adapter to the vent housing. There are two hoses in this system between the engine vent adapter and the vent housing: one hose carries fresh air from the intake manifold to the vent housing and the other carries a mixture of fresh air and crankcase vapors back to the intake manifold. The vent housing meters the flow of crankcase blow-by gases from the crankcase to the intake manifold in proportion to intake manifold vacuum. At idle, when intake vacuum is high, it restricts the flow, but as intake vacuum decreases, it allows a greater flow. If crankcase pressures become too high, the vent housing allows excess vapors to backflow to the intake manifold.

CHECK

4 An engine that is operated without a properly functioning crankcase ventilation system can be damaged. So anytime you're servicing the engine, be sure to inspect the PCV system hose(s) for cracks, tears and other damage. Disconnect the hose(s) and check for damage and obstructions. If a hose is clogged, clean it out. If you're unable to clean it satisfactorily, replace it.

5 A plugged PCV hose might cause any or all of the following conditions: A rough idle, stalling or a slow idle speed, oil leaks or sludge in the engine. So if the engine is running roughly, stalling and *idling at a lower than normal speed*, or is losing oil, or has oil in the throttle body or air intake manifold plenum, or has a build-up of sludge, a PCV system hose might be clogged. Repair or replace the hose(s) as necessary. And clean or replace the vent housing. On four-cylinder engines, remove the valve cover (see Chapter 2A) and inspect the vent housing. Make sure that it's clean by blowing it out with compressed air. On V6 engines, remove the vent housing (see below) and inspect the gasket. If the gasket is damaged, replace it.

6 A leaking PCV hose might cause any or all of the following conditions: a rough idle, stalling or a high idle speed. So if the engine is running roughly, stalling and *idling at a higher than normal speed*, a PCV system hose might be leaking. Repair or replace the hose(s) as necessary.

7 Here's an easy functional check of the PCV system:
  1 Disconnect the PCV hose.
  2 Start the engine and let it warm up to its normal idle.

3 Verify that there is vacuum at the PCV hose. If there is no vacuum, look for a plugged hose or manifold port. Also look for a hose that collapses when it's blocked (i.e. when vacuum is applied). Replace clogged or deteriorated hoses.
4 Remove the engine oil dipstick and install a vacuum gauge on the upper end of the dipstick tube.
5 Block off the PCV system fresh air passage.
6 Run the engine at 1500 rpm for 30 seconds, then read the vacuum gauge while the engine is running at 1500 rpm.
7 If there's vacuum present, the crankcase ventilation system is operating correctly.
8 If there's NO vacuum present, the engine might be drawing in outside air. The PCV system won't function correctly unless the engine is a sealed system. Inspect the valve cover(s), oil pan gasket or other sealing areas for leaks.
9 If the vacuum gauge indicates positive pressure, look for a plugged hose or engine blow-by.

8 If the PCV system is functioning correctly, but there's evidence of engine oil in the throttle body or air filter housing, it could be caused by excessive crankcase pressure. Have the crankcase pressure tested by a dealer service department or other repair shop.

9 In this type of PCV system, excessive blow-by (caused by worn rings, pistons and/or cylinders, or by constant heavy loads) is discharged into the intake manifold and consumed. If you discover heavy sludge deposits or a dilution of the engine oil, even though the PCV system is functioning correctly, look for other causes (see *Troubleshooting* and refer to Chapter 2C) and correct them as soon as possible.

COMPONENT REPLACEMENT

Four-cylinder models

PCV hose
▶ Refer to illustration 22.10

10 Disconnect the hose from the air intake duct (see illustration).

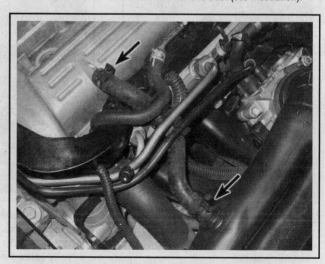

22.10 To remove the PCV hose from a four-cylinder engine, simply loosen the two hose clamps and pull off the hose

EMISSIONS AND ENGINE CONTROL SYSTEMS

11 Disconnect the hose from the valve cover.
12 Installation is the reverse of removal.

Crankcase vent housing

13 Remove the valve cover (see Chapter 2A). The vent housing is an integral component of the valve cover (it's riveted to the underside of the valve cover). It cannot be replaced separately.
14 Installation is the reverse of removal.

V6 models

Crankcase vent housing

15 Disconnect the cable from the negative battery terminal.
16 Remove the air intake duct (see "Air filter housing - removal and installation" in Chapter 4).
17 Remove the rear intake runner (see Chapter 2B).
18 Disconnect the electrical connector from the throttle body (see Chapter 4).
19 Disconnect the electrical connector and the vacuum lines from the Intake Manifold Runner Control (IMRC) solenoid (see Section 17).
20 Disconnect the vacuum hose from the fuel pressure regulator (see Chapter 4).
21 Loosen the front intake runner-to-intake plenum hose clamp and remove the plenum mounting bolts (see Chapter 2B).
22 Remove the EGR manifold bolts (see Section 21).
23 Lift the plenum slightly and disconnect the throttle body coolant hoses (see Chapter 4).
24 Stuff clean shop rags into the exposed intake ports to prevent dirt, dust and moisture from entering the engine.
25 Remove the intake plenum.
26 Disconnect the electrical connector for the rear ignition coil pack (see Chapter 5).
27 Disconnect the oil separator and vacuum hoses from the vent housing.
28 Detach the coolant line bracket from the vent housing.
29 Remove the vent housing mounting bolts and remove the vent housing.
30 Remove and discard the old vent housing gasket.
31 The vent housing contains a rubber one-way reed valve, which is the device that maintains the positive crankcase pressure. Inspect the reed valve for cracks and damage. If it's damaged or seriously worn, replace it.
32 Be sure to thoroughly clean the gasket surfaces of the cylinder head and the vent housing and to use a new gasket when installing the vent housing.
33 Before installing the vent housing mounting bolts, coat the threads with LOCTITE, 518, or a suitable equivalent. Be sure to tighten the vent housing mounting bolts securely.
34 Installation is otherwise the reverse of removal.

Specifications

Torque specifications

	Ft-lbs (unless otherwise indicated)	Nm
AIR pipe mounting bolts	156 in-lbs	18
Catalytic converter		
Four-cylinder models		
Front flange nuts	22	30
Rear flange bolts	180 in-lbs	20
V6 models		
Front catalyst		
Lower flange bolts	180 in-lbs	20
Upper flange nuts	25	30
Rear catalyst		
Lower flange bolts	180 in-lbs	20
Upper flange nuts	25	30
Knock sensor retaining bolt		
Four-cylinder engine	19	25
V6 engine (both knock sensors)	180 in-lbs	20
Oxygen sensors		
Four-cylinder models (upstream and downstream sensors)	33	45
V6 models		
Upstream sensors	37	50
Downstream sensors	33	45
Transmission Range (TR) switch		
TR switch mounting bolts	180 in-lbs	20
TR switch lever retaining nut	26	35

7A

MANUAL TRANSAXLE

Section
1   General information
2   Back-up light switch - replacement
3   Transaxle shift linkage assembly - removal and installation
4   Manual transaxle - removal and installation
5   Manual transaxle overhaul - general information

7A-2 MANUAL TRANSAXLE

1 General information

The vehicles covered by this manual are equipped with either a 5-speed manual or a 4-speed automatic transaxle. This Part of Chapter 7 contains information on the manual transaxle. Service procedures for the automatic transaxle are contained in Part B.

Because of the complexity of the manual transaxle and the specialized equipment necessary to perform most service operations, this Chapter contains only those procedures related to general diagnosis, adjustment and removal and installation procedures.

If the transaxle requires major repair work, it should be left to a dealer service department or an automotive or transmission repair shop. Once properly diagnosed, however, you can remove and install the transaxle yourself and save the expense, and have the repair work done by a transmission shop.

TRANSAXLE OVERHAUL

Because of the complexity of the assembly, possible unavailability of replacement parts and special tools necessary, internal repair procedures for the transaxle are not recommended for the home mechanic. The bulk of the information in this Chapter is devoted to removal and installation procedures.

2 Back-up light switch - replacement

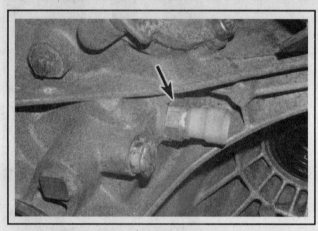

2.1 The back-up light switch is located above the left-side driveaxle

▶ Refer to illustration 2.1

1 The back-up light switch is located above the left-side driveaxle (see illustration).
2 Disconnect the electrical connector from the back-up light switch.
3 Unscrew the switch from the case and remove the gasket.
4 Install a new gasket, screw in the switch and tighten it securely.
5 Reconnect the electrical connector.
6 Check the operation of the of the back-up lights.

3 Transaxle shift linkage assembly - removal and installation

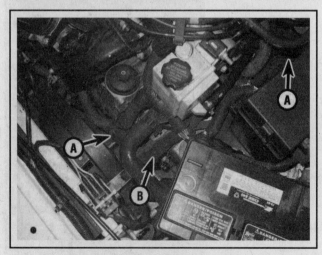

3.3 Separate the A.I.R. connecting hoses (A) from the A.I.R. tube (B)

▶ Refer to illustrations 3.3, 3.5, 3.6a, 3.6b and 3.7

1 Working inside the passenger compartment, lift up the shifter boot.
2 Move the gearshift lever to the far right, between fifth and reverse gears. Hold the lever in place by inserting a 3/8-inch punch or drill bit into the locating hole at the right side of the shifter base.
3 Working inside the engine compartment, separate the air injection reaction system tube from its connecting hoses and set the tube aside (see illustration).
4 Lock the transmission selector mechanism in position by pressing in the spring-loaded locking pin on the selector mechanism cover, which is located on the top of the transaxle.
5 Mark the relationship of the shift linkage rod to the shift linkage assembly then loosen the pinch-bolt (see illustration).
6 Depress the detent mechanism on the retaining pin securing the transaxle selector lever to the shift linkage assembly (see illustrations).

MANUAL TRANSAXLE 7A-3

➡ **Note:** A spring loaded detent mechanism secures the retaining pin. Damage to the pin will occur if the detent mechanism is not depressed before removing the pin. If the pin is not damaged during removal, it may be reused.

7 Remove the clips securing the shift linkage assembly to the transaxle (see illustration).

8 Remove the linkage assembly from the vehicle. Examine the linkage assembly for wear, replace worn or damaged parts.

9 Installation is the reverse of removal noting the following points:

a) When installing the shift linkage assembly, guide the transaxle selector lever and shift linkage rod in place before securing the linkage assembly with the mounting clips.

b) Tighten the shifter linkage rod-to-shift linkage assembly pinch-bolt to the torque listed in this Chapter's Specifications.

c) Remove the punch from the gearshift lever. The transaxle locking pin will automatically release when the lever is moved into the reverse position. At the top of the transaxle, ensure the transaxle locking pin has released.

3.5 Loosen the shifter linkage rod-to-shift linkage pinch-bolt

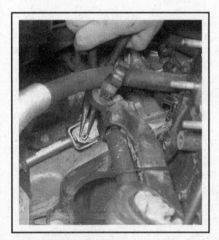

3.6a Remove the pin connecting the linkage . . .

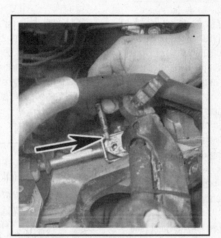

3.6b . . . noting that it's necessary to press the detent mechanism

3.7 Carefully slide off the retaining clips securing the transaxle shift linkage assembly

4 Manual transaxle - removal and installation

➡ **Note:** The vehicle must be raised and supported so that it is high enough to allow the transaxle to come out the bottom of the engine compartment. Moreover, the vehicle must be supported by the uni-body and not the subframe, as subframe removal is necessary for transaxle removal.

REMOVAL

▸ **Refer to illustrations 4.6a and 4.6b**

1 Remove the battery and the battery tray (see Chapter 5).

✳ **CAUTION:**

Always disconnect the negative cable first and hook it up last or the battery may be shorted by the tool being used to loosen the cable terminals.

2 Clearly label, then unplug, all transaxle electrical connectors. Remove any related brackets as necessary.

3 Remove the A.I.R. pump electrical connectors, related hoses and the air pump and bracket (see Chapter 6).

4 Disconnect the electrical connector for the back-up light switch, then disconnect the wire harness from the transaxle (see Section 2).

7A-4 MANUAL TRANSAXLE

4.6a Use a small screwdriver to pry out the retaining clip . . .

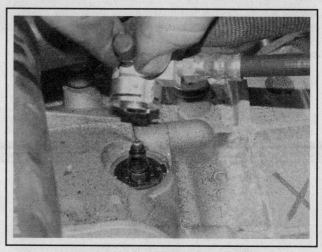

4.6b . . . and disconnect the clutch hydraulic fitting

5 Remove the transaxle shift linkage assembly (see Section 3).
6 Remove the clutch hydraulic fitting at the transaxle (see illustrations). Plug the fittings to prevent fluid loss and contamination.
7 Remove the upper rear transaxle support, then mark the position of the left transaxle mount and remove it from the vehicle (see Chapter 2).
8 Support the radiator by fastening it to the upper radiator support.
9 At the top of the transaxle remove the lockpin or hole plug depending on which is installed.
10 Loosen the front wheel lug nuts and the driveaxle/hub nuts (see Chapter 8).
11 Raise the vehicle and support it securely on jackstands placed under the unibody frame rails (see Chapter 10, Section 21 for jackstand placement).
12 Support the engine (from above only) using an engine hoist (see Chapter 2).
13 Drain the transaxle fluid (see Chapter 1).
14 Remove both driveaxles (see Chapter 8).
15 Support the transaxle with a transmission jack. Secure the transaxle to the jack with straps or chains so it doesn't fall off during removal.
16 Remove the subframe (see Chapter 10).
17 Remove the front and rear transaxle mounts.
18 Disconnect the air conditioning line from the right side of the vehicle chassis.
19 Remove the upper and lower bolts that attach the transaxle to the engine.

20 Lower the engine slightly using the engine hoist, enough to let the transaxle clear the engine compartment rail.
21 Make a final check that all wires, hoses, brackets and mounts have been disconnected from the transaxle, then slide the transaxle jack toward the side of the vehicle until the transaxle input shaft is clear of the engine. Make sure you keep the transaxle level as you do this.

INSTALLATION

22 Installation of the transaxle is a reversal of the removal procedure, but note the following points:
   a) Apply a film of high-temperature grease to the splines of the transaxle input shaft.
   b) Tighten the transaxle mounting bolts to the torque listed in this Chapter's Specifications.
   c) Tighten the driveaxle/hub nuts to the torque value listed in the Chapter 8 Specifications.
   d) Tighten the subframe mounting bolts to the torque values listed in the Chapter 10 Specifications.
   e) Tighten the wheel lug nuts to the torque listed in the Chapter 1 Specifications.
   f) If equipped, use a new spring loaded locking pin for the top of the transaxle.
   g) Fill the transaxle with the correct type and amount of manual transmission fluid as described in Chapter 1.

5 Manual transaxle overhaul - general information

1 Overhauling a manual transaxle is a difficult job for the do-it-yourselfer. It involves the disassembly and reassembly of many small parts. Numerous clearances must be precisely measured and, if necessary, changed with select-fit spacers and snap-rings. As a result, if transaxle problems arise, it can be removed and installed by a competent do-it-yourselfer, but overhaul should be left to a transmission repair shop. Rebuilt transaxles may be available - check with your dealer parts department and auto parts stores. At any rate, the time and money involved in an overhaul is almost sure to exceed the cost of a rebuilt unit.

2 Nevertheless, it's not impossible for an inexperienced mechanic to rebuild a transaxle if the special tools are available and the job is done in a deliberate step-by-step manner so nothing is overlooked.

3 The tools necessary for an overhaul include internal and external snap-ring pliers, a bearing puller, a slide hammer, a set of pin punches, a dial indicator and possibly a hydraulic press. In addition, a large, sturdy workbench and a vise or transaxle stand will be required.

4 During disassembly of the transaxle, make careful notes of how each piece comes off, where it fits in relation to other pieces and what holds it in place.

5 Before taking the transaxle apart for repair, it will help if you have some idea what area of the transaxle is malfunctioning. Certain problems can be closely tied to specific areas in the transaxle, which can make component examination and replacement easier. Refer to the *Troubleshooting* Section at the front of this manual for information regarding possible sources of trouble.

7A-6 MANUAL TRANSAXLE

Specifications

General

Transaxle oil type	See Chapter 1
Transaxle oil capacity	See Chapter 1

Torque specifications

	Ft-lbs (unless otherwise indicated)	Nm
Transaxle to engine mounting bolts	18	65
Shifter linkage rod-to-shift linkage pinch-bolt		
Step 1	108 in-lbs	12
Step 2	tighten an additional 180-degrees	

Section
1. General information
2. Diagnosis - general
3. Driveaxle oil seals - replacement
4. Shift cable - removal, installation and adjustment
5. Shift lever and shift interlock solenoid - replacement
6. Transaxle Control Module (TCM) - removal and installation
7. Automatic transaxle - removal and installation
8. Automatic transaxle overhaul - general information

7B
AUTOMATIC TRANSAXLE

7B-2 AUTOMATIC TRANSAXLE

1 General information

All information on the automatic transaxle is included in this Part of Chapter 7. Information for the manual transaxle can be found in Part A of this Chapter.

Because of the complexity of the automatic transaxles and the specialized equipment necessary to perform most service operations, this Chapter contains only those procedures related to general diagnosis, adjustment and removal and installation procedures.

If the transaxle requires major repair work, it should be left to a dealer service department or an automotive or transmission repair shop. Once properly diagnosed, however, you can remove and install the transaxle yourself and save the expense, and have the repair work done by a transmission shop.

2 Diagnosis - general

1 Automatic transaxle malfunctions may be caused by five general conditions:
  a) Poor engine performance
  b) Improper adjustments
  c) Hydraulic malfunctions
  d) Mechanical malfunctions
  e) Malfunctions in the computer or its signal network

2 Diagnosis of these problems should always begin with a check of the easily repaired items: fluid level and condition (see Chapter 1), shift cable adjustment and shift lever installation. Next, perform a road test to determine if the problem has been corrected or if more diagnosis is necessary. If the problem persists after the preliminary tests and corrections are complete, additional diagnosis should be performed by a dealer service department or other qualified transmission repair shop. Refer to the *Troubleshooting* Section at the front of this manual for information on symptoms of transaxle problems.

PRELIMINARY CHECKS

3 Drive the vehicle to warm the transaxle to normal operating temperature.

4 Check the fluid level as described in Chapter 1:
  a) *If the fluid level is unusually low, add enough fluid to bring the level within the designated area of the dipstick, then check for external leaks* (see following).
  b) *If the fluid level is abnormally high, drain off the excess, and then check the drained fluid for contamination by coolant. The presence of engine coolant in the automatic transmission fluid indicates that a failure has occurred in the internal radiator oil cooler walls that separate the coolant from the transmission fluid* (see Chapter 3).
  c) *If the fluid is foaming, drain it and refill the transaxle, then check for coolant in the fluid, or a high fluid level.*

5 Check the engine idle speed.

➡**Note: If the engine is malfunctioning, do not proceed with the preliminary checks until repairs have been made and the engine runs normally.**

6 Check and adjust the shift cable, if necessary (see Section 4).

7 If hard shifting is experienced, inspect the shift cable under the center console and at the shift lever on the transaxle (see Section 4).

FLUID LEAK DIAGNOSIS

8 Most fluid leaks are easy to locate visually. Repair usually consists of replacing a seal or gasket. If a leak is difficult to find, the following procedure may help.

9 Identify the fluid. Make sure it's transmission fluid and not engine oil or brake fluid (automatic transmission fluid is a deep red color).

10 Try to pinpoint the source of the leak. Drive the vehicle several miles, then park it over a large sheet of cardboard. After a minute or two, you should be able to locate the leak by determining the source of the fluid dripping onto the cardboard.

11 Make a careful visual inspection of the suspected component and the area immediately around it. Pay particular attention to gasket mating surfaces. A mirror is often helpful for finding leaks in areas that are hard to see.

12 If the leak still cannot be found, clean the suspected area thoroughly with a degreaser or solvent, then dry it thoroughly.

13 Drive the vehicle for several miles at normal operating temperature and varying speeds. After driving the vehicle, visually inspect the suspected component again.

14 Once the leak has been located, the cause must be determined before it can be properly repaired. If a gasket is replaced but the sealing flange is bent, the new gasket will not stop the leak. The bent flange must be straightened.

15 Before attempting to repair a leak, check to make sure that the following conditions are corrected or they may cause another leak.

➡**Note: Some of the following conditions cannot be fixed without highly specialized tools and expertise. Such problems must be referred to a qualified transmission shop or a dealer service department.**

Gasket leaks

16 Check the pan periodically. Make sure the bolts are tight, no bolts are missing, the gasket is in good condition and the pan is flat (dents in the pan may indicate damage to the valve body inside).

17 If the pan gasket is leaking, the fluid level or pressure may be too high, the vent may be plugged, the pan bolts may be too tight, the pan sealing flange may be warped, the sealing surface of the transaxle housing may be damaged, the gasket may be damaged or the transaxle casting may be cracked or porous. If sealant instead of gasket material has been used to form a seal between the pan and the transaxle housing, it may be the wrong type of sealant.

Seal leaks

18 If a transaxle seal is leaking, the fluid level may be too high, the vent may be plugged, the seal bore may be damaged, the seal itself may be damaged or improperly installed, the surface of the shaft protruding through the seal may be damaged or a loose bearing may be causing excessive shaft movement.

19 Make sure the dipstick tube seal is in good condition and the tube is properly seated. Periodically check the area around the sensors for leakage. If transmission fluid is evident, check the seals for damage.

AUTOMATIC TRANSAXLE 7B-3

Case leaks

20 If the case itself appears to be leaking, the casting is porous and will have to be repaired or replaced.

21 Make sure the oil cooler hose fittings are tight and in good condition.

Fluid comes out vent pipe or fill tube

22 If this condition occurs the possible causes are: the transaxle is overfilled; there is coolant in the fluid; the dipstick is incorrect; the vent is plugged or the drain-back holes are plugged.

3 Driveaxle oil seals - replacement

▶ Refer to illustrations 3.3, 3.4, 3.7

1 The driveaxle oil seals are located on the sides of the transaxle, where the inner ends of the driveaxles enter the transaxle. If you suspect that a driveaxle oil seal is leaking, raise the vehicle and support it securely on jackstands. If the seal is leaking, you'll see lubricant on the side of the transaxle, below the seal.

2 Remove the driveaxle (see Chapter 8).

3 Note the distance from the top edge of the oil seal's metal casing to the surface of the transaxle (see illustration). This measurement is important for seal installation in Step 7.

4 Use a screwdriver or prybar to carefully pry the seal out of the transaxle bore. If the seal cannot be removed with these tools, a special seal removal tool (available at most auto parts stores) will be required (see illustration).

5 Compare the old seal to the new one to be sure it's the correct one.

6 Coat the lips and the outside of the new seal with multi-purpose grease.

7 Using a seal installer or equivalent tool, install the new oil seal. Drive it squarely into the bore until it matches the measurement obtained in step three (see illustration).

8 Install the driveaxle (see Chapter 8).

9 Lower the vehicle and check the fluid level in the transaxle, adding as necessary (see Chapter 1).

3.3 Measure the distance of the seal's edge from the bore

3.4 Using a large screwdriver or prybar, carefully pry the oil seal out of the transaxle (if you can't remove the oil seal with a screwdriver or prybar, you may need to obtain a special seal removal tool - available at most auto parts stores - to do the job

3.7 Using a seal installer or a large deep socket as a drift, drive the new seal squarely into the bore and make sure that it's completely seated; lubricate the lip of the new seal with multi-purpose grease

4 Shift cable - removal, installation and adjustment

▶ Refer to illustrations 4.2, 4.4, 4.6, 4.10, 4.11, 4.12 and 4.13

✱ WARNING:

The models covered by this manual are equipped with a Supplemental Restraint System (SRS), more commonly known as airbags. Always disarm the airbag system before working in the vicinity of any airbag system component to avoid the possibility of accidental deployment of the airbag, which could cause personal injury (see Chapter 12).

REMOVAL

1 In the event of hard shifting, disconnect the cable at the transaxle (Steps 9 through 11) and operate the shifter from the driver's seat. If the shift lever moves smoothly through all positions with the cable disconnected, then the cable should be adjusted as described at the end of this Section. To remove and install the shift cable, perform the following:

7B-4 AUTOMATIC TRANSAXLE

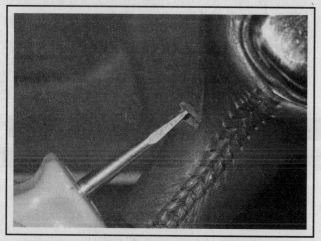

4.2 Remove the shift knob retainer with a screwdriver

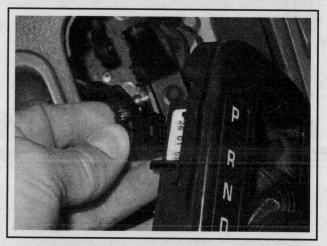

4.4 Disconnect the electrical connector to the shift indicator in the center trim bezel

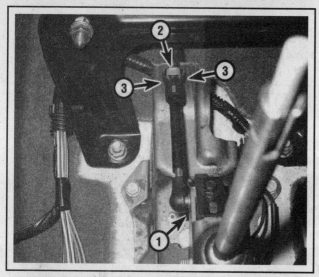

4.6 Shift cable and bracket details (passenger compartment)
1. Pry the end of the shift cable from the lever with a screwdriver
2. Pry the shift cable retaining clip up and away with a screwdriver
3. Depress the tabs while pushing forward to release the cable

4.10 The shift cable can be found under various components and hoses

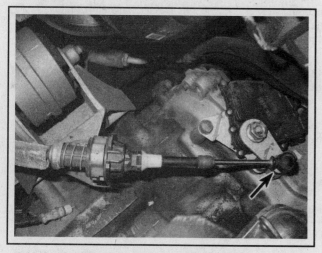

4.11 Separate the shift cable from the shift lever on the transaxle

2 Working inside the vehicle, remove the shift knob retainer, then remove the knob (see illustration).

3 Pull up on the shift lever boot until the shift trim bezel disengages from the center console.

4 Disconnect the shift indicator electrical connector and remove the shift trim and boot together (see illustration).

5 Remove the center console (see Chapter 11).

6 Use a flat-blade screwdriver and pry the shift cable end from the shift lever pin (see illustration).

7 Pry the shift cable retaining clip up and away from the shift cable (see illustration 4.6).

8 Depress the tabs on the shift cable while pushing it forward to release it from its bracket on the shift lever assembly, then note its path to the firewall (see illustration 4.6).

9 Raise the hood and place a blanket over the left (driver's) fender to protect it.

10 Locate the shift lever in the engine compartment near the top of the transaxle (see illustration).

11 Disconnect the shift cable end by prying it up and away from the shift lever (see illustration).

12 Depress and hold the two tabs on the shift cable and push it through its bracket (see illustration).

AUTOMATIC TRANSAXLE 7B-5

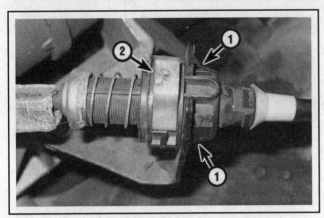

4.12 To remove the shift cable from its bracket at the transaxle, depress the tabs (1) and push the cable through the bracket (some models also have a retaining clip). (2) is the lock tab for the adjuster

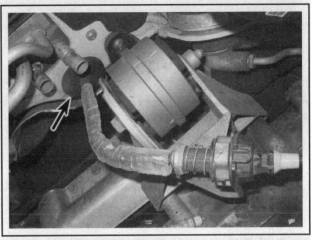

4.13 Trace the shift cable to the firewall, then separate the grommet from the firewall

→Note: *Some models also have a retaining clip which must be removed.*

13 Trace the shift cable to the engine-compartment firewall, then pry the large rubber cable grommet from the firewall (see illustration).

14 Carefully remove the shift cable by pulling it through the hole in the firewall from the engine compartment side.

INSTALLATION

15 Installation is the reverse of removal, with attention paid to the following points:
  a) Install the shift cable to its original path within the engine compartment and passenger cabin.
  b) The cable end fittings must be firmly snapped back into place onto their respective levers. Correctly installed, they will remain attached even with moderate force to pull them off again.
  c) Adjust the shift cable as described in the following steps.

ADJUSTMENT

16 Confirm that the shift control lever (in the vehicle) is in the PARK position.

17 Detach the shift cable from the transaxle shift lever in the engine compartment (see illustration 4.11).

18 Make sure that the transaxle shift lever is in PARK by confirming it's as far counterclockwise as possible (rotated away from the firewall). A firm grasp will move the shift lever when the cable is disconnected.

19 Pry the lock tab up on the shift cable adjustment assembly with a small screwdriver (see illustration 4.12).

20 Reconnect the end of the shift cable to the transaxle shift lever (see Step 15b). Push the lock tab back into place.

21 Shift the transaxle into all gear positions to make sure the cable is functioning properly. Readjust if necessary. Also make sure the engine only starts in PARK and NEUTRAL.

5 Shift lever and shift interlock solenoid - replacement

♦ Refer to illustration 5.3

※ WARNING:

The models covered by this manual are equipped with a Supplemental Restraint System (SRS), more commonly known as airbags. Always disarm the airbag system before working in the vicinity of any airbag system component to avoid the possibility of accidental deployment of the airbag, which could cause personal injury (see Chapter 12).

SHIFT LEVER ASSEMBLY

1 Remove the center console (see Chapter 11).

2 Remove the shift cable end from the pin on the shift lever assembly (see Section 4, Steps 6 and 7).

3 Disconnect any electrical connectors to the shift lever assembly (see illustration).

4 Remove the four shift lever assembly mounting bolts and lift the assembly out (see illustration 5.3).

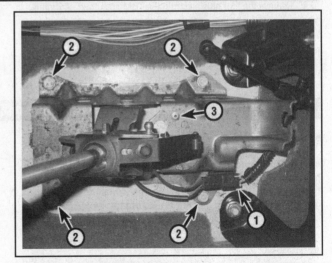

5.3 The shift lever assembly
  1 Electrical connector
  2 Shift lever assembly mounting bolts
  3 Rivet holding the shift interlock solenoid

7B-6 AUTOMATIC TRANSAXLE

5 Transfer any parts as necessary if the assembly is being replaced.
6 Installation is reverse of removal.

SHIFT INTERLOCK SOLENOID

➡ **Note:** Removal and installation of the solenoid can be made easier by removing the shift lever assembly.

7 Disconnect the cable from the negative terminal of the battery (see Chapter 5, Section 1).
8 Remove center console (see Chapter 11).
9 Disconnect the electrical connector on the shift lever assembly (see illustration 5.3).
10 Use a fine pick to carefully remove the solenoid wire terminal from the connector.
11 Use a 5/32-inch drill bit to drill out the rivet attaching the solenoid to the shift lever assembly (see illustration 5.3).
12 When the solenoid comes loose, separate the linkage and move it underneath the shift lever assembly towards the dash to remove it.
13 The installation is in the reverse of removal with attention paid to the following points:
  a) Be certain to thread the solenoid wire and terminal back through the shift lever assembly to its original path.
  b) A rivet gun will be required to reattach the solenoid to the shift lever assembly.
  c) Reconnect the battery (see Chapter 5, Section 1).

6 Transaxle Control Module (TCM) - removal and installation

◆ **Refer to illustration 6.2**

※※ **CAUTION:**
The TCM is Electro-Static Discharge (ESD) sensitive and could easily become damaged by static electricity discharged from your body. Be sure to properly ground yourself and the TCM before handling it. Avoid touching the electrical terminals of the TCM.

➡ **Note 1:** Do not interchange TCM's from a different year vehicle. If you are replacing a TCM with a comparable unit, take the vehicle to your local dealer service department or other qualified transmission shop to have it calibrated for your vehicle.

➡ **Note 2:** The TCM on four-cylinder engines is integrated with the vehicle's PCM (Powertrain Control Module) and serviced as a single unit (see Chapter 6 for PCM replacement).

1 Disconnect the cable from the negative terminal of the battery (see Chapter 5, Section 1).
2 Lift up the right end of the hood seal and cowl cover (see illustration).
3 Remove the four nuts securing the TCM cover and lift it off the studs.
4 Depress the locking tab on the top of the TCM connector to rotate the connector lever down towards the TCM, then separate the connector from the unit.
5 Depress the locking tab that holds the TCM in its carrier and lift it straight up and out.
6 Installation is the reverse of removal. Reconnect the battery (see Chapter 5, Section 1).

6.2 Lift up the cowl and seal above the TCM (V6 models only)

7 Automatic transaxle - removal and installation

◆ **Refer to illustrations 7.12, 7.17, 7.19, 7.20 and 7.21**

➡ **Note:** The vehicle must be raised and supported so that it is high enough to allow the transaxle to come out the bottom of the engine compartment. Moreover, the vehicle must be supported by the uni-body and not the subframe, as subframe removal is necessary for transaxle removal.

REMOVAL

1 Place the vehicle in PARK with the parking brake engaged.
2 Remove the battery and the battery tray (see Chapter 5).

※※ **CAUTION:**
Always disconnect the negative cable first and hook it up last or the battery may be shorted by the tool being used to loosen the cable terminals.

3 Clearly label, and then unplug, all transaxle electrical connectors. Remove any related brackets as necessary.

➡ **Note:** On V6 models, a wire harness bracket is fastened behind an engine to transaxle mounting bolt. It will be removed later.

AUTOMATIC TRANSAXLE 7B-7

4 Disconnect the shift cable from the shift lever at the transaxle (see Section 4).

5 Remove the shift cable bracket from the rear powertrain mount and tie it (with the cable) to the firewall.

6 Remove the Output Shaft Speed (OSS) sensor (See Chapter 6).

7 On V6 models, remove the drivebelt, the drivebelt tensioner and alternator (see Chapter 5).

8 Loosen the front wheel lug nuts and the driveaxle/hub nuts (see Chapter 8).

9 Place the vehicle securely on jackstands (See Chapter 10, Section 21 for jackstand placement).

10 Drain the transaxle fluid (see Chapter 1).

11 Remove both driveaxles (see Chapter 8).

12 Remove the transaxle oil cooler line assembly (from the side of the transaxle) by first removing the retaining nut, then pulling the assembly straight out to avoid damaging any seals (see illustration).

13 On 2001 through 2003 four-cylinder models, remove the A.I.R. pump electrical connectors, related hoses and the air pump and bracket (see Chapter 6).

14 Support the engine (from above only) using an engine hoist or engine support tool (see Chapter 2).

15 Remove the subframe (see Chapter 10).

16 Support the transaxle with a transmission jack. Secure the transaxle to the jack with straps or chains so it doesn't fall off during removal.

17 Remove the brace (on the passenger's side of the transaxle) connecting the engine to the transaxle (see illustration).

18 Remove the starter (see Chapter 5).

19 Remove the torque converter-to-driveplate bolts. Turn the crankshaft 120-degrees at a time for access to each bolt. After all three bolts are removed, carefully push the torque converter into the bellhousing so it doesn't stay with the engine when the transaxle is removed.

➡ **Note: On V6 models there are six bolts. Only the three bolts that are recessed in the driveplate need to be removed (see illustration).**

20 Mark the relationship of the torque converter to the driveplate so they can be installed in the same position (see illustration).

21 Remove the two lower transaxle-to-engine bolts.

22 Mark the position of the driver's side powertrain mount bolts and then remove them (see Chapter 2A).

7.12 The transaxle cooler lines are retained to the transaxle by a single nut

23 Remove the remaining bolts that attach the transaxle to the engine.

24 Make a final check that all wires, hoses, brackets and mounts have been disconnected from the transaxle, then slide the transaxle jack toward the side of the vehicle until the transaxle is clear of the engine locating dowels. Make sure you keep the transaxle level as you do this.

INSTALLATION

25 Installation of the transaxle is a reversal of the removal procedure, but note the following points:

a) As the torque converter is reinstalled, ensure that the drive tangs at the center of the torque converter hub engage with the recesses in the automatic transaxle fluid pump inner gear. This can be confirmed by turning the torque converter while pushing it towards the transaxle. If it isn't fully engaged, it will "clunk" into place.

b) When installing the transaxle, make sure the match-marks you made on the torque converter and driveplate line up.

c) Install all of the driveplate-to-torque converter bolts before tightening any of them.

7.17 The engine-to-transaxle brace

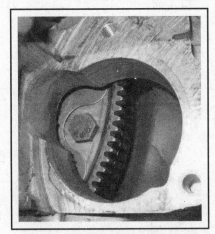

7.19 Remove the torque converter-to-driveplate bolts

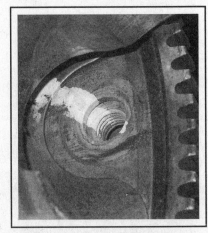

7.20 Mark the relationship of the torque converter to the driveplate

7B-8 AUTOMATIC TRANSAXLE

d) Tighten the driveplate-to-torque converter bolts to the torque listed in this Chapter's Specifications.
e) Tighten the transaxle mounting bolts to the torque listed in this Chapter's Specifications.
f) Tighten the driveaxle/hub nuts to the torque value listed in the Chapter 8 Specifications.
g) Tighten the wheel lug nuts to the torque listed in the Chapter 1 Specifications.
h) Fill the transaxle with the correct type and amount of automatic transmission fluid as described in Chapter 1.
i) Shift transaxle into all gear positions to confirm shift cable adjustment. Readjust if necessary (see Section 4).

8 Automatic transaxle overhaul - general information

In the event of a problem occurring, it will be necessary to establish whether the fault is electrical, mechanical or hydraulic in nature, before any repair work can be considered. Diagnosis requires detailed knowledge of the transaxle's operation and construction as well as access to specialized test equipment. Because of these factors, diagnosis and repair of the transaxle itself are beyond the scope of this manual. It is therefore essential that problems with the automatic transaxle are referred to a dealer service department or other qualified repair facility for assessment.

Note that a faulty transaxle should not be removed before the vehicle has been diagnosed by a knowledgeable technician equipped with the proper tools, as troubleshooting must be performed with the transaxle installed in the vehicle.

Specifications

General

Lubricant type and capacity .... See Chapter 1

Torque specifications

	Ft-lbs (unless otherwise indicated)	Nm
Torque converter-to-driveplate bolts		
Four-cylinder models	33	45
V6 models	48	65
Transaxle-to-engine bolts	48	65
Transaxle oil cooler line assembly nut	71 in-lbs	8
Powertrain mount-to-mounting bracket bolts	41	55

Section

1. General information
2. Clutch - description and check
3. Clutch master cylinder - removal and installation
4. Clutch release cylinder - replacement
5. Clutch hydraulic system - bleeding
6. Clutch components - removal, inspection and installation
7. Clutch start switch - check and replacement
8. Driveaxles - removal and installation
9. Driveaxle boot replacement and CV joint inspection

8
CLUTCH AND DRIVEAXLES

8-2 CLUTCH AND DRIVEAXLES

1 General information

The information in this Chapter deals with the components that transmit power to the front wheels, except for the transaxle, which is dealt with in Chapter 7A and 7B. For the purposes of this Chapter, these components are grouped into two categories: clutch and driveaxles. Separate Sections within this Chapter offer general descriptions and checking procedures for both groups.

Since nearly all the procedures covered in this Chapter involve working under the vehicle, make sure it's securely supported on sturdy jackstands or on a hoist where the vehicle can be easily raised and lowered.

2 Clutch - description and check

1   All vehicles with a manual transaxle have a single dry plate, diaphragm spring-type clutch. The clutch disc has a splined hub which allows it to slide along the splines of the transaxle input shaft. The clutch and pressure plate are held in contact by spring pressure exerted by the diaphragm in the pressure plate.

2   The clutch release system is operated by hydraulic pressure. The hydraulic release system consists of the clutch pedal, a master cylinder and a shared common reservoir with the brake master cylinder, a release (or slave) cylinder and the hydraulic line connecting the two components. The release cylinder and release bearing are integral parts of a single assembly which is installed concentric to the input shaft and is bolted to the transaxle.

3   When the clutch pedal is depressed, a pushrod pushes against brake fluid inside the master cylinder, applying hydraulic pressure to the release cylinder, which pushes the release bearing against the diaphragm fingers of the clutch pressure plate.

4   Terminology can be a problem when discussing the clutch components because common names are in some cases different from those used by the manufacturer. For example, the driven plate is also called the clutch plate or disc, the clutch release bearing is sometimes called a throwout bearing, the release cylinder is sometimes called the slave cylinder.

5   Unless you're replacing components with obvious damage, do these preliminary checks to diagnose clutch problems:

   a) *The first check should be of the fluid level in the master cylinder. If the fluid level is low, add fluid as necessary and inspect the hydraulic system for leaks. If the master cylinder reservoir is dry, bleed the system as described in Section 5 and recheck the clutch operation.*
   b) *To check "clutch spin-down time," run the engine at normal idle speed with the transaxle in Neutral (clutch pedal up - engaged). Disengage the clutch (pedal down), wait several seconds and shift the transaxle into Reverse. No grinding noise should be heard. A grinding noise would most likely indicate a bad pressure plate or clutch disc.*
   c) *To check for complete clutch release, run the engine (with the parking brake applied to prevent vehicle movement) and hold the clutch pedal approximately 1/2-inch from the floor. Shift the transaxle between 1st gear and Reverse several times. If the shift is rough, component failure is indicated.*
   d) *Visually inspect the pivot bushing at the top of the clutch pedal to make sure there's no binding or excessive play.*

3 Clutch master cylinder - removal and installation

REMOVAL

♦ **Refer to illustration 3.6**

1   Locate the pin that secures the clutch master cylinder pushrod to the clutch pedal. It's accessible from inside the vehicle, under the dash on the driver's side. Remove the driver's side lower trim panel (see Chapter 11), then remove the clip and pin that secure the clutch master cylinder pushrod to the clutch pedal.

2   Working inside the engine compartment, remove as much fluid as you can from the brake master cylinder reservoir with a syringe, such as an old turkey baster (the clutch master cylinder is supplied with fluid from the brake fluid reservoir).

※ **WARNING 1:**

If a baster is used, never again use it for the preparation of food.

※ **WARNING 2:**

Don't depress the brake pedal until the reservoir has been refilled, otherwise air may be introduced into the brake hydraulic system.

3   Place rags under the fluid fittings and prepare caps or plastic bags to cover the ends of lines or fittings once they are disconnected.

※ **CAUTION:**

Brake fluid will damage paint. Cover all body parts and be careful not to spill fluid during this procedure.

4   Disconnect the clutch master cylinder supply hose from the brake master cylinder.

5   Release the clutch master cylinder-to-release cylinder hydraulic line from the body retaining clips.

CLUTCH AND DRIVEAXLES 8-3

6 Remove the line clip at the master cylinder hydraulic line fitting, then separate the hydraulic line from the cylinder (see illustration).

7 Remove the nuts securing the clutch master cylinder to the firewall and remove the master cylinder from the vehicle.

INSTALLATION

8 Place the master cylinder pushrod through the firewall and install the mounting fasteners finger tight.

9 Working inside the vehicle, connect the master cylinder pushrod to the clutch pedal and install the pin.

10 Tighten the clutch master cylinder mounting nuts to the torque listed in this Chapter's Specifications.

11 Attach the fluid feed hose to the brake reservoir and tighten the hose clamp.

12 Connect the hydraulic line fitting to the clutch master cylinder and install the retaining clip at the fitting.

13 Fill the reservoir with brake fluid conforming to DOT 3 specifications and bleed the clutch system as outlined in Section 5.

14 Also check the feel of the brake pedal. If it feels spongy when depressed, bleed the brake hydraulic system (see Chapter 9).

3.6 Remove the clip securing the hydraulic line to the clutch master cylinder

4 Clutch release cylinder - replacement

▶ **Refer to illustration 4.2**

1 Remove the transaxle (see Chapter 7, Part A).

2 Using a flare-nut wrench, disconnect the hydraulic line fitting from the release cylinder (see illustration).

3 Remove the fasteners securing the release cylinder to the transaxle and slide the release cylinder assembly off the transaxle input shaft.

4 Installation is the reverse of removal, noting the following points:
  a) Lubricate the release cylinder oil seal with a smear of transmission fluid.
  b) Apply a thread locking compound to the release cylinder mounting fasteners and tighten them to the torque listed in this Chapter's Specifications.
  c) Tighten the hydraulic fitting securely, using a flare-nut wrench.
  d) Install the transaxle (see Chapter 7).
  e) Check the fluid level in the brake fluid reservoir, adding brake fluid conforming to DOT 3 specifications until the level is correct.
  f) Bleed the system as described in Section 5.

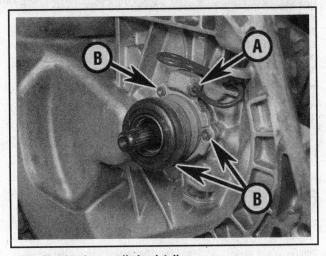

4.2 Clutch release cylinder details
A Clutch hydraulic line fitting
B Clutch release cylinder mounting fasteners

5 Clutch hydraulic system - bleeding

▶ **Refer to illustration 5.3**

1 Bleed the hydraulic system whenever any part of the system has been removed or the fluid level has fallen so low that air has been drawn into the master cylinder. The bleeding procedure is very similar to bleeding a brake system.

2 Fill the brake master cylinder reservoir with new brake fluid conforming to DOT 3 specifications.

✳✳ CAUTION:

Do not re-use any of the fluid coming from the system during the bleeding operation or use fluid which has been inside an open container for an extended period of time.

3 Have an assistant depress the clutch pedal and hold it. Open the

8-4 CLUTCH AND DRIVEAXLES

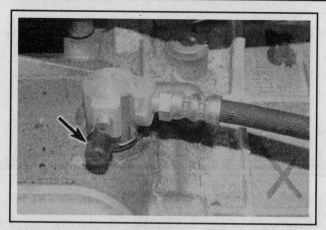

5.3 The clutch release cylinder bleeder valve is located at a fitting on top of the transaxle

bleeder valve at the release cylinder fitting (see illustration), allowing fluid and any air escape. Close the bleeder valve when the flow of fluid (and bubbles) ceases. Once closed, have your assistant release the pedal.

4 Continue this process until all air is evacuated from the system, indicated by a solid stream of fluid being ejected from the bleeder valve each time with no air bubbles. Keep a close watch on the fluid level inside the brake master cylinder reservoir - if the level drops too far, air will get into the system and you'll have to start all over again.

➡ **Note: Wash the area with water to remove any spilled brake fluid.**

5 Check the brake fluid level again, and add some, if necessary, to bring it to the appropriate level. Check carefully for proper operation before placing the vehicle into normal service.

6 Clutch components - removal, inspection and installation

❄ WARNING:

Dust produced by clutch wear is hazardous to your health. DO NOT blow it out with compressed air and DO NOT inhale it. DO NOT use gasoline or petroleum-based solvents to remove the dust. Brake system cleaner should be used to flush the dust into a drain pan. After the clutch components are wiped clean with a rag, dispose of the contaminated rags and cleaner in a covered, marked container.

REMOVAL

▸ **Refer to illustration 6.4**

1 Access to the clutch components is normally accomplished by removing the transaxle, leaving the engine in the vehicle. If the engine is being removed for major overhaul, check the clutch for wear and replace worn components as necessary. However, the relatively low cost of the clutch components compared to the time and trouble spent gaining access to them warrants their replacement anytime the engine or transaxle is removed, unless they are new or in near-perfect condition. The following procedures are based on the assumption the engine will stay in place.

2 Remove the transaxle from the vehicle (see Chapter 7, Part A). Support the engine while the transaxle is out. Preferably, an engine support fixture or a hoist should be used to support it from above.

3 To support the clutch disc during removal, install a clutch alignment tool through the clutch disc hub.

4 Carefully inspect the flywheel and pressure plate for indexing marks. The marks are usually an X, an O or a black mark. If they cannot be found, scribe or paint marks yourself so the pressure plate and the flywheel will be in the same alignment during installation (see illustration).

5 Turning each bolt a little at a time, loosen the pressure plate-to-flywheel bolts. Work in a criss-cross pattern until all spring pressure is

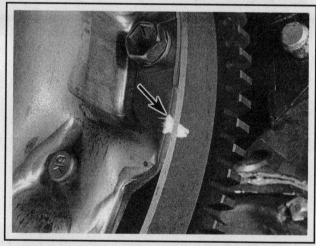

6.4 Mark the relationship of the pressure plate to the flywheel (if you're planning to re-use the old pressure plate)

relieved. Then hold the pressure plate securely and completely remove the bolts, followed by the pressure plate and clutch disc.

INSPECTION

▸ **Refer to illustrations 6.8, 6.10a and 6.10b**

6 Ordinarily, when a problem occurs in the clutch, it can be attributed to wear of the clutch driven plate assembly (clutch disc). However, all components should be inspected at this time.

7 Inspect the flywheel for cracks, heat checking, grooves and other obvious defects. If the imperfections are slight, a machine shop can machine the surface flat and smooth, which is highly recommended regardless of the surface appearance. Refer to Chapter 2 for the flywheel removal and installation procedure.

CLUTCH AND DRIVEAXLES 8-5

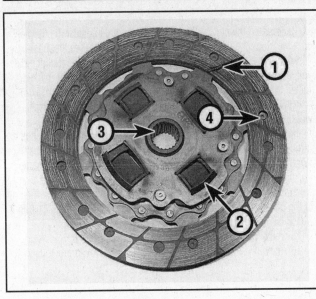

6.8 **The clutch disc**
1 **Lining** - *this will wear down in use*
2 **Springs or dampers** - *check for cracking and deformation*
3 **Splined hub** - *the splines must not be worn and should slide smoothly on the transmission input shaft splines*
4 **Rivets** - *these secure the lining and will damage the flywheel or pressure plate if allowed to contact the surfaces*

NORMAL FINGER WEAR

EXCESSIVE FINGER WEAR

BROKEN OR BENT FINGERS

6.10a **Replace the pressure plate if excessive wear or damage is noted**

8 Inspect the lining on the clutch disc. There should be at least 1/16-inch of lining above the rivet heads. Check for loose rivets, distortion, cracks, broken springs/dampers and other obvious damage (see illustration). As mentioned above, ordinarily the clutch disc is routinely replaced, so if in doubt about the condition, replace it with a new one.

9 The clutch release cylinder/release bearing should also be replaced along with the clutch disc (see Section 4).

10 Check the machined surfaces and the diaphragm spring fingers of the pressure plate (see illustrations). If the surface is grooved or otherwise damaged, replace the pressure plate. Also check for obvious damage, distortion, cracking, etc. Light glazing can be removed with emery cloth or sandpaper. If a new pressure plate is required, new and re-manufactured units are available.

INSTALLATION

▸ *Refer to illustration 6.12*

11 Before installation, clean the flywheel and pressure plate machined surfaces with brake cleaner, lacquer thinner or acetone. It's important that no oil or grease is on these surfaces or the lining of the clutch disc. Handle the parts only with clean hands.

6.10b **Inspect the pressure plate surface for excessive score marks, cracks and signs of overheating**

8-6 CLUTCH AND DRIVEAXLES

6.12 Center the clutch disc in the pressure plate with a clutch alignment tool

12 Position the clutch disc and pressure plate against the flywheel with the clutch held in place with an alignment tool (see illustration). Make sure the disc is installed properly (most replacement clutch discs will be marked "flywheel side" or something similar - if not marked, install the clutch disc with the damper springs toward the transaxle).

13 Tighten the pressure plate-to-flywheel bolts only finger tight, working around the pressure plate.

14 Center the clutch disc by ensuring the alignment tool extends through the splined hub and into the pilot bearing in the crankshaft. Wiggle the tool up, down or side-to-side as needed to center the disc. Install and tighten the pressure plate-to-flywheel bolts a little at a time, working in a criss-cross pattern to prevent distorting the cover. After all of the bolts are snug, tighten them to the torque listed in this Chapter's Specifications. Remove the alignment tool.

15 Install the clutch release cylinder (see Section 4).

16 Install the transaxle and all components removed previously.

7 Clutch start switch - check and replacement

1 In the passenger compartment, remove the driver's side lower trim panel (see Chapter 11).

2 Remove the push pin fastener securing the heater duct, then remove the duct.

CHECK

3 Verify that the engine will not start when the clutch pedal is released. Now, depress the clutch pedal - the engine should start.

4 Locate the switch on the clutch master cylinder pushrod and unplug the electrical connector.

5 Using an ohmmeter, verify that there is continuity between the proper terminals of the clutch start switch when the pedal is depressed. There should be no continuity when the pedal is released.

6 If the switch does not work as described, replace it.

REPLACEMENT

7 Unplug the electrical connector from the switch.

8 Using a feeler gauge, release the switch locking tab between the mounting bracket and switch. Pull the switch downward and remove it from the bracket.

9 Installation is the reverse of removal. The switch is self-adjusting, so there's no need for adjustment.

10 Verify that the engine doesn't start when the clutch pedal is released, and does start when the pedal is depressed.

8 Driveaxles - removal and installation

8.1 Remove the cotter pin from the driveaxle/hub nut

✱✱ WARNING:
The manufacturer recommends replacing the driveaxle/hub nut and cotter pin with new ones whenever they are removed.

REMOVAL

▶ Refer to illustrations 8.1, 8.4a, 8.4b, 8.7 and 8.10

1 Set the parking brake. Remove the wheel cover or hubcap. Remove the cotter pin from the driveaxle/hub nut (see illustration).

2 Loosen the driveaxle/hub nut with a large socket and breaker bar, but don't remove it yet.

3 Loosen the front wheel lug nuts, raise the vehicle and support it securely on jackstands. Remove the wheel.

CLUTCH AND DRIVEAXLES   8-7

8.4a Remove the driveaxle/hub nut . . .

8.4b . . . and washer

8.7 Use a soft-faced hammer to loosen the driveaxle from the hub splines

4 Remove the driveaxle/hub nut and washer (see illustrations).
5 Disconnect the tie-rod end from the steering knuckle (see Chapter 10).
6 Separate the control arm from the steering knuckle (see Chapter 10).
7 To loosen the driveaxle from the hub splines, tap the end of the driveaxle with a soft-faced hammer (see illustration). If the driveaxle is stuck in the hub splines and won't move, it may be necessary to push it from the hub with a puller.
8 Pull out on the steering knuckle and detach the driveaxle from the hub. Suspend the outer end of the driveaxle on a bungee cord or piece of wire.
9 Before you remove the driveaxle, look for lubricant leakage in the area around the differential seal. If there's evidence of a leak, you'll want to replace the seal after removing the driveaxle (see Chapter 7B).
10 Position a prybar against the inner joint and carefully pry the joint off the transaxle side gear (see illustration). Do not use the driveaxle to pull the on the inner joint. Doing so might damage the inner joint components. Remove the driveaxle assembly, being careful not to over-extend the inner joint or damage the axleshaft boots.

➡ Note: If you're removing a driveaxle on the right side, the transaxle stub shaft may come out with the driveaxle. Inspect the driveaxle oil seal for damage if this occurs. Replace as necessary.

11 Should it become necessary to move the vehicle while the driveaxle is out, place a large bolt with two large washers (one on each side of the hub) through the hub and tighten the nut securely.

INSTALLATION

♦ Refer to illustration 8.12

12 Installation is the reverse of removal, but with the following additional points:
   a) Remove the old set-ring from the transaxle stub shaft or driveaxle (depending on which side you're working on) and install a new one (see illustration).
   b) Apply a film of multi-purpose grease around the splines of the joints.
   c) When installing the driveaxle, hold the driveaxle straight out, then push it in sharply to seat the driveaxle set-ring. To make sure the

8.10 Using a large prybar, carefully pry the inner end of the driveaxle from the transaxle

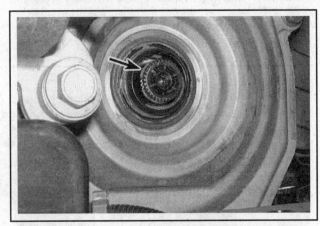
8.12 Remove the old driveaxle set-ring and replace it with a new one

set-ring is properly seated, attempt to pull the inner CV joint housing out of the transaxle by hand. If the set-ring is properly seated the inner joint will not move out.
   d) Clean all foreign matter from the driveaxle outer CV joint threads and coat the splines with multi-purpose grease. Guide the driveaxle

8-8 CLUTCH AND DRIVEAXLES

into the hub splines and install the new driveaxle/hub nut. Tighten the nut securely but not to the specified torque at this time.
e) Reconnect the control arm and tie-rod end, then tighten the suspension fasteners to the torque listed in the Chapter 10 Specifications.
f) Install the wheel and lug nuts, then lower the vehicle.
g) Tighten the driveaxle/hub nut to the torque listed in this Chapter's Specifications. Install a new cotter pin.
h) Tighten the wheel lug nuts to the torque listed in the Chapter 1 Specifications.
i) Add transaxle lubricant if it was drained or if any fluid spilled out (see Chapter 1).

9   Driveaxle boot replacement

➡ **Note:** If the CV joints or boots must be replaced, explore all options before beginning the job. Complete, rebuilt driveaxles are available on an exchange basis, eliminating much time and work. Whichever route you choose to take, check on the cost and availability of parts before disassembling the vehicle.

INNER CV JOINT

1 Remove the driveaxle (see Section 8).
2 Mount the driveaxle in a vise with wood-lined jaws, to prevent damage to the axleshaft. Check the CV joints for excessive play in the radial direction, which indicates worn parts. Check for smooth operation throughout the full range of motion for each CV joint. If a boot is torn, the recommended procedure is to disassemble the joint, clean the components and inspect for damage due to loss of lubrication and possible contamination by foreign matter. If the CV joint is in good condition, lubricate it with CV joint grease and install a new boot.

Disassembly

♦ **Refer to illustrations 9.5, 9.6 and 9.7**

3 Cut the boot clamps with side-cutters, then remove and discard them.
4 Using a screwdriver, pry up on the edge of the boot, pull it off the CV joint housing and slide it down the axleshaft, exposing the tri-pot spider assembly. Pull the CV joint housing straight off.
➡ **Note:** When removing the housing, hold the rollers in place on the spider trunnion to prevent the rollers and the needle bearings from falling free.
5 Remove the spider assembly snap-ring with a pair of snap-ring pliers (see illustration).
6 Mark the tri-pot to the axleshaft to ensure that they are reassembled properly (see illustration).
7 Use a hammer and a brass drift to drive the spider assembly from the axleshaft (see illustration).
8 Slide the boot off the shaft.

Inspection

9 Thoroughly clean all components with solvent until the old CV joint grease is completely removed. Inspect the bearing surfaces of the inner tri-pots and housings for cracks, pitting, scoring and other signs of wear. If any part of the inner CV joint is worn, you must replace the entire driveaxle assembly (inner tri-pot joint, axleshaft and outer CV joint). The only components that can be purchased separately are the boots themselves and the boot clamps.

Reassembly

♦ **Refer to illustrations 9.10a, 9.10b, 9.10c, 9.10.d, 9.11 and 9.12**

10 Wrap the splines on the inner end of the axleshaft with electrical or duct tape to protect the boots from the sharp edges of the splines, then slide the clamps and boot onto the axleshaft (see illustration). Remove the tape and place the tri-pot spider on the axleshaft with the chamfer toward the shaft (see illustration). Tap the spider onto the shaft (aligning the marks made in Step 6) with a brass drift until it's seated, then install the snap-ring. Apply grease to the tri-pot assembly and inside the housing (see illustration). Insert the tri-pot into the housing and pack the remainder of the grease around the tri-pot (see illustration).

9.5 Remove the snap-ring with a pair of snap-ring pliers

9.6 Mark the relationship of the tri-pot bearing assembly to the axleshaft

9.7 Drive the tri-pot joint off the axleshaft with a brass punch and hammer; be careful not to damage the bearing surfaces or the splines on the shaft

9.10a Wrap the axleshaft splines with electrical tape to prevent damaging the boot as it's slid onto the shaft

9.10b Install the tri-pot spider on the axleshaft (make sure your match mark is facing out and aligned with the mark on the axleshaft)

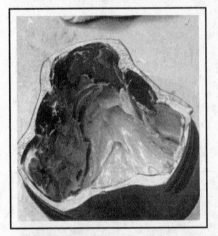

9.10c Place grease at the bottom of the CV joint housing

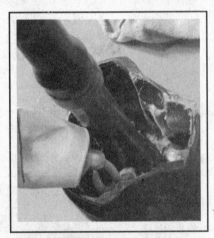

9.10d Install the boot and clamps onto the axleshaft, then insert the tri-pot into the housing, followed by the rest of the grease

9.11 Make sure that the thinnest groove on the axleshaft is the only one showing

11 Slide the boot into place, making sure the raised bead on the inside of the seal boot is positioned in the groove on the interconnecting shaft. If the driveaxle has multiple locating grooves on the shaft, position the boot so only one of the grooves (the thinnest) is exposed (see illustration). Position the sealing boot into the groove on the tri-pot housing retaining groove.

12 Position the CV joint mid-way through its travel, then equalize the pressure in the boot (see illustration).

13 Make sure each end of the boot is seated properly, and the boot is not distorted.

14 Two types of clamps are used on the inner CV joint. If a crimp-type clamp is used, clamp the new boot clamps onto the boot with a special crimping tool (available at most automotive parts stores). Place the crimping tool over the bridge of each new boot clamp, then tighten the nut on the crimping tool until the jaws are closed. If a low-profile latching type clamp is used, place the prongs of the clamping tool (available at most automotive parts stores) in the holes of the clamp and squeeze the tool together until the top band latches behind the tabs of the lower band.

15 The driveaxle is now ready for installation (see Section 8).

9.12 Equalize the pressure inside the boot by inserting a screwdriver between the boot and the CV joint housing

8-10 CLUTCH AND DRIVEAXLES

9.17 Strike the edge of the CV joint housing sharply with a soft-faced hammer to dislodge the CV joint from the shaft

9.22a Apply the CV joint grease through the splined hole . . .

OUTER CV JOINT

♦ Refer to illustrations 9.17, 9.22a and 9.22b

Removal

16 Cut the boot clamps with side-cutters, then remove and discard them.
17 Strike the edge of the CV joint housing sharply with a soft-face hammer to dislodge the outer CV joint from the axleshaft (see illustration). Remove and discard the bearing retainer clip from the axleshaft.
18 Slide the outer CV joint boot off the axleshaft.

Inspection

19 Thoroughly clean all components with solvent until the old CV grease is completely removed. Inspect the bearing surfaces of the inner tri-pods and housings for cracks, pitting, scoring, and other signs of wear. If any part of the outer CV joint is worn, you must replace the entire driveaxle assembly (inner CV joint, axleshaft and outer CV joint).

Reassembly

20 Slide a new sealing boot clamp and sealing boot onto the axleshaft. It's a good idea to wrap the axleshaft splines with tape to prevent damaging the boot (see illustration 9.10a).
21 Place a new bearing retainer clip onto the axleshaft.
22 Pack the outer CV joint with CV joint grease, then put the remaining grease into the sealing boot (see illustrations).
23 Align the splines on the axleshaft with the splines on the outer CV joint assembly and using a soft-faced hammer, gently drive the CV joint onto the axleshaft until the CV joint is seated to the axleshaft.
24 Seat the ends of the boot on the axleshaft and the CV joint housing, then equalize the pressure in the boot (see illustration 9.12).

25 Two types of clamps are used on the outer CV joint. If a crimp-type clamp is used, clamp the new boot clamps onto the boot with a special crimping tool (available at most automotive parts stores). Place the crimping tool over the bridge of each new boot clamp, then tighten the nut on the crimping tool until the jaws are closed. If a low profile latching type clamp is used, place the prongs of the clamping tool (available at most automotive parts stores) in the holes of the clamp and squeeze the tool together until the top band latches behind the tabs of the lower band.
26 Install the driveaxle as outlined in Section 8.

9.22b . . .then insert a wooden dowel (slightly smaller in diameter than the hole) into the hole and push down - the dowel will force the grease into the joint. Repeat this until the joint is packed

CLUTCH AND DRIVEAXLES

Specifications

Clutch fluid type — See Chapter 1

Torque specifications

	Ft-lbs (unless otherwise indicated)	Nm
Clutch master cylinder mounting nuts	15	20
Clutch pressure plate-to-flywheel bolts*	132 in-lbs	15
Clutch release cylinder mounting fasteners	89 in-lbs	10
Driveaxle hub/nut*		
Step 1	74 to 118	100 to 160
Step 2	Loosen nut until it's able to turn by hand	
Step 3	15	20
Step 4	Turn additional 90-degrees	
Wheel lug nuts	See Chapter 1	

Bolt(s) must be replaced

Notes

Section
1   General information
2   Anti-lock Brake System (ABS) - general information
3   Disc brake pads - replacement
4   Disc brake caliper - removal and installation
5   Brake disc - inspection, removal and installation
6   Drum brake shoes/parking brake shoes - replacement
7   Wheel cylinder - removal and installation
8   Master cylinder - removal and installation
9   Brake hoses and lines - inspection and replacement
10 Brake hydraulic system - bleeding
11 Power brake booster - removal and installation
12 Parking brake - adjustment
13 Brake light switch - replacement

9
BRAKES

1 General information

The vehicles covered by this manual are equipped with hydraulically operated front and rear brake systems. The front brakes are disc type and the rear brakes are disc or drum type. Both the front and rear brakes are self adjusting. The disc brakes automatically compensate for pad wear, while the drum brakes incorporate an adjustment mechanism which is activated as the parking brake is applied.

HYDRAULIC SYSTEM

The hydraulic system consists of two separate circuits. The master cylinder has separate reservoir chambers for the two circuits, and, in the event of a leak or failure in one hydraulic circuit, the other circuit will remain operative. A dual proportioning valve on the firewall provides brake balance between the front and rear brakes.

POWER BRAKE BOOSTER

The power brake booster, utilizing engine manifold vacuum and atmospheric pressure to provide assistance to the hydraulically operated brakes, is mounted on the firewall in the engine compartment.

PARKING BRAKE

The parking brake operates the rear brakes only, through cable actuation. It's activated by a lever mounted in the center console.

SERVICE

After completing any operation involving disassembly of any part of the brake system, always test drive the vehicle to check for proper braking performance before resuming normal driving. When testing the brakes, perform the tests on a clean, dry, flat surface. Conditions other than these can lead to inaccurate test results.

Test the brakes at various speeds with both light and heavy pedal pressure. The vehicle should stop evenly without pulling to one side or the other. Avoid locking the brakes, because this slides the tires and diminishes braking efficiency and control of the vehicle.

Tires, vehicle load and wheel alignment are factors which also affect braking performance.

PRECAUTIONS

There are some general cautions and warnings involving the brake system on this vehicle:

a) *Use only brake fluid conforming to DOT 3 specifications.*
b) *The brake pads and linings contain fibers which are hazardous to your health if inhaled. Whenever you work on brake system components, clean all parts with brake system cleaner. Do not allow the fine dust to become airborne. Also, wear an approved filtering mask.*
c) *Safety should be paramount whenever any servicing of the brake components is performed. Do not use parts or fasteners which are not in perfect condition, and be sure that all clearances and torque specifications are adhered to. If you are at all unsure about a certain procedure, seek professional advice. Upon completion of any brake system work, test the brakes carefully in a controlled area before putting the vehicle into normal service. If a problem is suspected in the brake system, don't drive the vehicle until it's fixed.*

2 Anti-lock Brake System (ABS) - general information

GENERAL INFORMATION

◆ **Refer to illustration 2.2**

1 The anti-lock brake system is designed to maintain vehicle steerability, directional stability and optimum deceleration under severe braking conditions on most road surfaces. It does so by monitoring the rotational speed of each wheel and controlling the brake line pressure to each wheel during braking. This prevents the wheels from locking up.

2 The ABS system has three main components - the wheel speed sensors, the electronic control unit (ECU) and the hydraulic unit (see illustration). Four wheel speed sensors - one at each wheel - send a variable voltage signal to the control unit, which monitors these signals, compares them to its program and determines whether a wheel is about to lock up. When a wheel is about to lock up, the control unit signals the hydraulic unit to reduce hydraulic pressure (or not increase it further) at that wheel's brake caliper. Pressure modulation is handled by electrically-operated solenoid valves.

3 If a problem develops within the system, an "ABS" warning light will glow on the dashboard. Sometimes, a visual inspection of the ABS system can help you locate the problem. Carefully inspect the ABS wiring harness. Pay particularly close attention to the harness and connections near each wheel. Look for signs of chafing and other damage caused by incorrectly routed wires. If a wheel sensor harness is damaged, the sensor must be replaced.

2.2 The ABS hydraulic unit

BRAKES 9-3

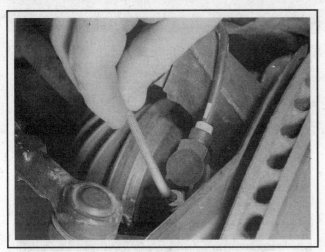

2.9a Remove the sensor mounting bolt . . .

2.9b . . . then remove the sensor

※ WARNING:

Do NOT try to repair an ABS wiring harness. The ABS system is sensitive to even the smallest changes in resistance. Repairing the harness could alter resistance values and cause the system to malfunction. If the ABS wiring harness is damaged in any way, it must be replaced.

※ CAUTION:

Make sure the ignition is turned off before unplugging or reattaching any electrical connections.

DIAGNOSIS AND REPAIR

4 If a dashboard warning light comes on and stays on while the vehicle is in operation, the ABS system requires attention. Although special electronic ABS diagnostic testing tools are necessary to properly diagnose the system, you can perform a few preliminary checks before taking the vehicle to a dealer service department.

   a) Check the brake fluid level in the reservoir.
   b) Verify that the computer electrical connectors are securely connected.
   c) Check the electrical connectors at the hydraulic control unit.
   d) Check the fuses.
   e) Follow the wiring harness to each wheel and verify that all connections are secure and that the wiring is undamaged.

5 If the above preliminary checks do not rectify the problem, the vehicle should be diagnosed by a dealer service department or other qualified repair shop. Due to the complex nature of this system, all actual repair work must be done by a qualified automotive technician.

WHEEL SPEED SENSOR - REMOVAL AND INSTALLATION

▶ Refer to illustrations 2.9a and 2.9b

6 Loosen the wheel lug nuts, raise the vehicle and support it securely on jackstands. Remove the wheel.
7 Make sure the ignition key is turned to the Off position.
8 Trace the wiring back from the sensor, detaching all brackets and clips while noting its correct routing, then disconnect the electrical connector.
9 Remove the mounting bolt and carefully pull the sensor out from the knuckle or brake backing plate (see illustrations).
10 Installation is the reverse of the removal procedure. Tighten the mounting bolt securely.
11 Install the wheel and lug nuts, tightening them securely. Lower the vehicle and tighten the lug nuts to the torque listed in the Chapter 1 Specifications.

3 Disc brake pads - replacement

▶ Refer to illustration 3.5

※ WARNING:

Disc brake pads must be replaced on both front or rear wheels at the same time - never replace the pads on only one wheel. Also, the dust created by the brake system is harmful to your health. Never blow it out with compressed air and don't inhale any of it. An approved filtering mask should be worn when working on the brakes. Do not, under any circumstances, use petroleum-based solvents to clean brake parts. Use brake system cleaner only!

1 Remove the cap from the brake fluid reservoir.
2 Loosen the wheel lug nuts, raise the end of the vehicle you're working on and support it securely on jackstands. Block the wheels at the opposite end.
3 Remove the wheels. Work on one brake assembly at a time, using the assembled brake for reference if necessary.
4 Inspect the brake disc carefully as outlined in Section 5. If machining is necessary, follow the information in that Section to remove the disc, at which time the pads can be removed as well.
5 Push the piston back into its bore to provide room for the new brake pads. A C-clamp can be used to accomplish this (see illustra-

9-4 BRAKES

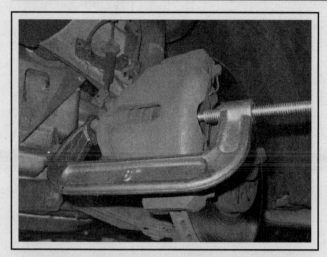

3.5 Before removing the caliper, be sure to depress the piston into the bottom of its bore in the caliper with a large C-clamp to make room for the new pads

3.6a Always wash the brakes with brake cleaner before disassembling anything

3.6b Remove the brake caliper anti-rattle clip

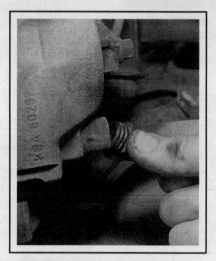

3.6c Remove these covers to locate the caliper guide pins . . .

3.6d . . . then unscrew the pins

3.6e Remove the caliper . . .

tion). As the piston is depressed to the bottom of the caliper bore, the fluid in the master cylinder will rise. Make sure that it doesn't overflow. If necessary, siphon off some of the fluid.

FRONT

▶ Refer to illustrations 3.6a through 3.6l

6 Follow the accompanying photos (illustrations 3.6a through 3.6l), for the actual pad replacement procedure. Be sure to stay in order and read the caption under each illustration.

7 After the job has been completed, firmly depress the brake pedal a few times to bring the pads into contact with the disc. Check the level of the brake fluid, adding some if necessary. Check the operation of the brakes carefully before placing the vehicle into normal service.

Brakes 9-5

3.6f ... and use a piece of wire to tie it to the coil spring - never let the caliper hang by the brake hose

3.6g Unclip the inner pad from the piston ...

3.6h ... then remove the outer pad from the caliper mounting bracket

3.6i Clip the new inner pad into the piston

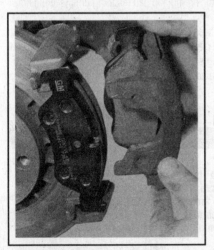

3.6j Install the outer pad, then position the caliper on the mounting bracket

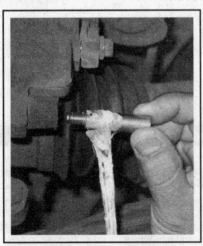

3.6k Before installing the guide pins, clean them off, then apply a light coat of high-temperature grease to the pins

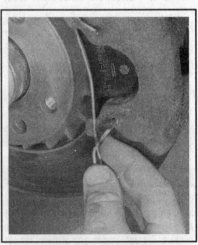

3.6l Tighten the guide pins to the torque listed in this Chapter's Specifications, then install the anti-rattle clip

9-6 BRAKES

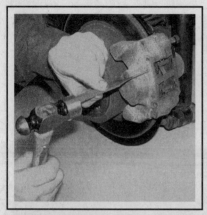

3.8a Using a punch, drive out the upper pad retaining pin and partially drive out the lower one (just enough to unseat it)

3.8b Remove the anti-rattle spring, followed by the lower retaining pin

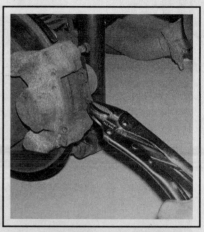

3.8c Using a pair of pliers, remove the inboard . . .

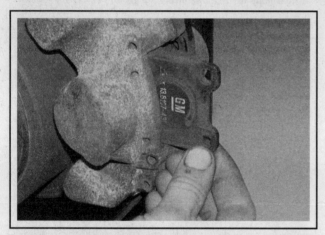

3.8d . . . and outboard pads from the caliper

3.10 Drive the brake pad retaining pins in from the inboard side of the caliper until they're seated

REAR

♦ Refer to illustrations 3.8a, 3.8b, 3.8c, 3.8d and 3.10

8 Wash the brake with brake system cleaner (see illustration 3.6a), then follow the accompanying photos (3.8a through 3.8d), for the actual pad removal procedure. Be sure to stay in order and read the caption under each illustration.

9 Insert the new pads in the caliper by reversing the removal procedure.
10 Install the anti-rattle spring and pad retaining pins (see illustration).
11 After the job has been completed, firmly depress the brake pedal a few times to bring the pads into contact with the disc. Check the level of the brake fluid, adding some if necessary. Check the operation of the brakes carefully before placing the vehicle into normal service.

4 Disc brake caliper - removal and installation

✳✳ WARNING:
Dust created by the brake system is harmful to your health. Never blow it out with compressed air and don't inhale any of it. An approved filtering mask should be worn when working on the brakes. Do not, under any circumstances, use petroleum-based solvents to clean brake parts. Use brake system cleaner only.

➡ Note: If replacement is indicated (usually because of fluid leakage), it is recommended that the calipers be replaced, not overhauled. New and factory rebuilt units are available on an exchange basis, which makes this job quite easy. Always replace the calipers in pairs - never replace just one of them.

Brakes

FRONT

Removal

▶ Refer to illustration 4.2

1 Loosen - but don't remove - the lug nuts on the front wheels. Raise the front of the vehicle and place it securely on jackstands. Remove the front wheels.

2 Disconnect the brake line from the caliper and plug it to keep contaminants out of the brake system and to prevent losing any more brake fluid than is necessary (see illustration).

➡ Note: If you're simply removing the caliper for access to other components, don't disconnect the hose.

3 Remove the caliper guide pins.
4 Detach the caliper from its mounting bracket.

Installation

5 Install the caliper by reversing the removal procedure. Remember to replace the sealing washers on either side of the brake line fitting with new ones. Tighten the caliper guide pins and the brake line banjo fitting bolt to the torque listed in this Chapter's Specifications.

6 Bleed the brake system (see Section 10).

7 Install the wheels and lug nuts and lower the vehicle. Tighten the wheel lug nuts to the torque listed in the Chapter 1 Specifications.

REAR

Removal

▶ Refer to illustrations 4.9 and 4.10

8 Loosen - but don't remove - the lug nuts on the rear wheels. Raise the rear of the vehicle and place it securely on jackstands. Remove the rear wheels.

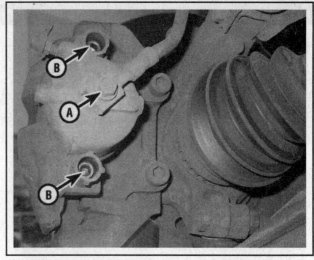

4.2 Brake caliper details:

A Brake line fitting B Caliper guide pins

9 Unscrew the brake line threaded fitting with a flare-nut wrench and detach the brake line from the caliper (see illustration).

10 Remove the caliper mounting bolts (see illustration).

11 Detach the caliper from its mounting bracket.

Installation

12 Install the caliper by reversing the removal procedure. Tighten the caliper mounting bolts to the torque listed in this Chapter's Specifications. Tighten the brake line fitting securely.

13 Bleed the brake system (see Section 10).

14 Install the wheels and lug nuts. Lower the vehicle and tighten the lug nuts to the torque listed in the Chapter 1 Specifications.

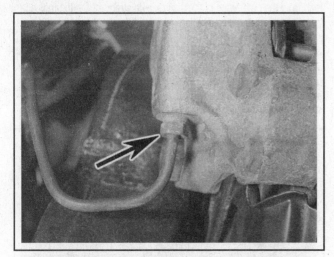

4.9 Unscrew the brake line threaded fitting, then detach it from the caliper

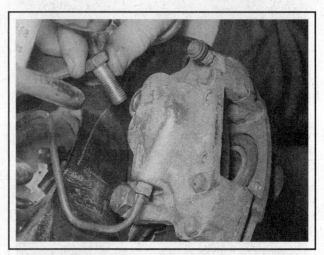

4.10 Remove the caliper mounting bolts

9-8 BRAKES

5 Brake disc - inspection, removal and installation

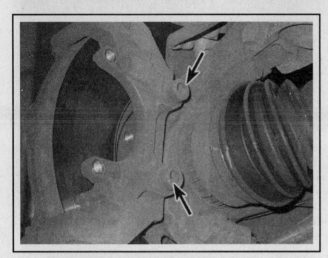

5.2 Front caliper mounting bracket bolts

✱ WARNING:

The dust created by the brake system is harmful to your health. Never blow it out with compressed air and don't inhale any of it. An approved filtering mask should be worn when working on the brakes. Do not, under any circumstances, use petroleum-based solvents to clean brake parts. Use brake system cleaner only!

INSPECTION

▸ **Refer to illustrations 5.2, 5.3, 5.4a, 5.4b and 5.5**

1   Loosen the wheel lug nuts, raise the vehicle and support it securely on jackstands. Remove the wheel and install the lug nuts to hold the disc in place against the hub flange.

➡**Note: If the lug nuts don't contact the disc when screwed on all the way, install washers under them. If you're checking the rear disc, release the parking brake.**

2   Remove the brake caliper as outlined in Section 4. It isn't necessary to disconnect the brake hose. After removing the caliper bolts, suspend the caliper out of the way with a piece of wire. If you're working on a front disc, remove the two caliper mounting bracket-to-steering knuckle bolts (see illustration) and remove the mounting bracket.

3   Visually inspect the disc surface for score marks and other damage. Light scratches and shallow grooves are normal after use and may not always be detrimental to brake operation, but deep scoring requires disc removal and refinishing by an automotive machine shop. Be sure to check both sides of the disc (see illustration). If pulsating has been noticed during application of the brakes, suspect disc runout.

4   To check disc runout, place a dial indicator at a point about 1/2-inch from the outer edge of the disc (see illustration). Set the indicator to zero and turn the disc. The indicator reading should not exceed the specified allowable runout limit. If it does, the disc should be refinished by an automotive machine shop.

➡**Note: The discs should be resurfaced regardless of the dial indicator reading, as this will impart a smooth finish and ensure a perfectly flat surface, eliminating any brake pedal pulsation or other undesirable symptoms related to questionable**

5.3 The brake pads on this vehicle were obviously neglected, as they wore down completely and cut deep grooves into the disc - wear this severe means the disc must be replaced

5.4a To check disc runout, mount a dial indicator as shown and rotate the disc

5.4b Using a swirling motion, remove the glaze from the disc surface with sandpaper or emery cloth

BRAKES 9-9

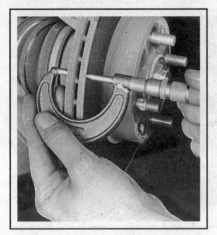

5.5 Use a micrometer to measure disc thickness

5.6a If the disc retaining screws are stuck, use an impact screwdriver to loosen them

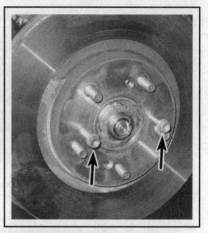

5.6b If the disc is stuck, thread two bolts into the disc and tighten to force the disc off the hub

discs. At the very least, if you elect not to have the discs resurfaced, remove the glaze from the surface with emery cloth or sandpaper, using a swirling motion (see illustration).

5 It's absolutely critical that the disc not be machined to a thickness under the specified minimum thickness. The minimum (or discard) thickness is cast or stamped into the disc. The disc thickness can be checked with a micrometer (see illustration).

REMOVAL

♦ Refer to illustrations 5.6a and 5.6b

6 Remove the lug nuts which were installed to hold the disc in place, or remove the disc retaining screws (see illustration) and remove the disc from the hub. If the disc is stuck to the hub and won't come off, thread two bolts into the holes provided (see illustration) and tighten them. Alternate between the bolts, turning them a couple of turns at a time, until the disc is free. Remove the disc from the hub.

INSTALLATION

7 Place the disc in position over the threaded studs. Install the disc retaining screws and tighten them securely.

8 Install the caliper mounting bracket (front brakes only) and caliper, tightening the bolts to the torque values listed in this Chapter's Specifications.

➡ **Note: Before installing the caliper mounting bracket-to-steering knuckle bolts, coat the threads of the bolts with a thread locking compound.**

9 Install the wheel and lug nuts, then lower the vehicle to the ground. Tighten the lug nuts to the torque listed in the Chapter 1 Specifications. Depress the brake pedal a few times to bring the brake pads into contact with the disc. Bleeding won't be necessary unless the brake hose was disconnected from the caliper. Check the operation of the brakes carefully before driving the vehicle.

6 Drum brake shoes/parking brake shoes - replacement

DRUM BRAKE SHOES

※ WARNING:

Brake shoes must be replaced on both rear wheels at the same time - never replace the shoes on only one side. Also, the dust created by the brake system is harmful to your health. Never blow it out with compressed air and don't inhale any of it. An approved filtering mask should be worn when working on the brakes. Do not, under any circumstances, use petroleum-based solvents to clean brake parts. Use brake system cleaner only!

※ CAUTION:

Whenever the brake shoes are replaced, the return and hold-down springs should also be replaced. Due to the continuous heating/cooling cycle the springs are subjected to, they can lose tension over a period of time and may allow the shoes to drag on the drum and wear at a much faster rate than normal.

Removal

♦ Refer to illustrations 6.5, 6.6, 6.7, 6.9a, 6.9b, 6.11, 6.13 and 6.15

1 Loosen the wheel lug nuts, raise the rear of the vehicle and support it securely on jackstands. Block the front wheels to keep the vehicle from rolling.
2 Release the parking brake.
3 Remove the wheel.

➡ **Note: All four rear brake shoes must be replaced at the same time, but to avoid mixing up parts, work on only one brake assembly at a time.**

4 Remove the brake drum.

➡ **Note: If the brake drum cannot be easily pulled off, make sure the parking brake is completely released. If the drum still cannot be pulled off, the brake shoes will have to be retracted. This is done by first removing the plug from the backing plate.**

9-10 BRAKES

6.5 Details of the rear drum brake assembly

1. Upper return spring
2. Adjuster assembly
3. Self adjuster lever
4. Return spring retaining bracket
5. Hold-down spring
6. Self adjuster spring
7. Wheel cylinder

With the plug removed, push the adjuster lever towards the drum to release the shoes from the adjuster. The drum should now come off.

5 Clean the brake shoe assembly with brake system cleaner before beginning work. Note the location and orientation of all components before beginning work, as an aid to reassembly (see illustration).

6 Using a pair of pliers, carefully unhook the upper return spring and remove it from the brake shoes (see illustration).

7 Remove the return spring retaining bracket, followed by the self adjuster lever and spring (see illustration).

8 Pull the upper ends of the brake shoes apart, and remove the adjuster from between the shoes.

9 Remove the leading and trailing shoe hold-down springs (see illustrations).

10 Detach the lower return spring from the leading shoe, then remove the shoe.

11 Remove the trailing shoe, then detach the cable from the parking

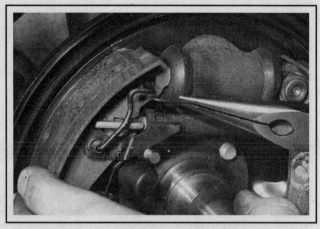

6.6 Unhook the upper return spring and remove it from the brake shoes

6.7 Remove the return spring retaining bracket (A), followed by the self adjuster lever (B) and spring

brake lever (see illustration).

12 Do not depress the brake pedal with the shoes removed. As a precaution against the wheel cylinder pistons dropping out, wrap a strong elastic band around them.

6.9a Push down and give the spring cup a quarter-turn . . .

6.9b . . . then remove the spring and withdraw the retainer pin

6.11 Disengage the cable from the parking brake lever

Brakes 9-11

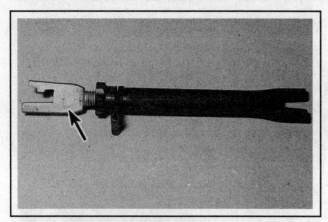

6.13 The adjuster assembly is marked L for left or R for right, depending on which side of the vehicle it is installed on - don't mix them up!

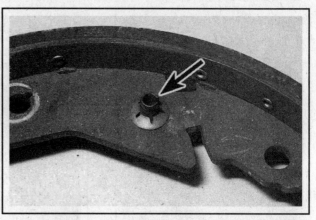

6.15 It may be necessary to transfer the adjuster lever from the original leading shoe to the new one. If so, be sure to use a new retaining clip

6.17 Lubricate the brake shoe contact areas on the backing plate with high-temperature grease

6.19a Connect the lower return spring to the bottom of each shoe, then position the leading shoe on the backing plate . . .

13 Dismantle the brake adjuster (see illustration). Clean the adjuster screw then lubricate the threads and ends with high-temperature grease.

14 Examine the return springs. If they are distorted, or if they have seen extensive service, replacement is advisable. Weak springs may cause the brakes to drag and wear out prematurely.

15 If a new parking brake lever was not supplied with the new shoes, transfer the lever from the old shoe. The lever may be secured with a pin and retaining clip, or by a rivet, which will have to be drilled out. It may also be necessary to transfer the adjuster lever pivot pin and clip from the original leading shoe to the new shoe (see illustration).

16 Peel back the rubber boots and check the wheel cylinder for fluid leaks or other damage. Ensure that both cylinder pistons are free to move easily. Refer to Section 7, if necessary, for information on wheel cylinder replacement.

Installation

▶ **Refer to illustrations 6.17, 6.19a, 6.19b, 6.21, 6.22 and 6.23**

17 Prior to installation, clean the backing plate thoroughly. Lubricate the brake shoe contact areas on the backing plate with high-temperature grease or anti-seize compound (see illustration).

18 Ensure that the parking brake cable is correctly engaged on the

6.19b . . . and secure with the hold-down spring

trailing shoe's parking brake lever then position the shoe on the backing plate and secure it with the hold-down spring.

19 Hook the lower return spring onto the trailing shoe, then engage the leading shoe with the return spring. Position the leading shoe on the backing plate and secure it with its hold-down spring (see illustrations).

9-12 BRAKES

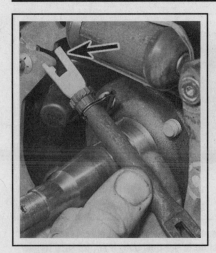

6.21 Install the brake adjuster, noting that the longer, straight part of the fork must be behind the leading shoe

6.22 Install the adjuster lever and spring, making sure that the spring is correctly engaged in the hole of the leading shoe

6.23 Make sure both shoes are correctly aligned with the wheel cylinder, then install the upper return spring

20 Screw the adjuster wheel fully onto the forked end of the adjuster, so that the adjuster assembly is set to its shortest possible length. Back the wheel off a half a turn, and check that it is free to rotate easily.

21 Place the brake adjuster into position between the brake shoes. Make sure that both ends of the adjuster are correctly engaged with the shoes, noting that the forked end of the adjuster must be positioned so that its longer, straight fork is behind the leading shoe (see illustration).

22 Install the adjuster lever spring on the leading shoe and adjuster lever then position the lever on its pivot pin (see illustration). Check that the lever and spring are correctly installed, then secure the lever in position with the return spring retaining bracket, making sure the spring ends are securely located in the retaining pin and shoe.

23 Remove the rubber band from the wheel cylinder. Make sure that both shoes are correctly positioned on the wheel cylinder pistons, then install the upper return spring (see illustration).

24 Before reinstalling the drum, it should be checked for cracks, score marks, deep scratches and hard spots, which will appear as small discolored areas. If the hard spots cannot be removed with sandpaper or emery cloth, or if any of the other conditions listed above exist, the drum must be taken to an automotive machine shop to have it resurfaced.

➡Note: *Professionals recommend resurfacing the drums each time a brake job is done. Resurfacing will eliminate the possibility of out-of-round drums. If the drums are worn so much that they can't be resurfaced without exceeding the maximum allowable diameter, then new ones will be required. At the very least, if you elect not to have the drums resurfaced, remove the glaze from the surface with emery cloth or sandpaper using a swirling motion.*

25 Install the brake drum on the hub flange.
26 Mount the wheel and install the lug nuts.
27 Repeat the operation on the remaining brake.
28 Once both sets of rear shoes have been replaced, lower the vehicle and tighten the lug nuts to the torque listed in the Chapter 1 Specifications.
29 With the parking brake fully released, adjust the lining-to-drum clearance by repeatedly depressing the brake pedal at least 20 to 25 times.

30 While depressing the pedal, have an assistant listen to the rear drums to check that the adjuster assembly is functioning correctly; if so, a clicking sound will be emitted by the adjuster lever as the pedal is depressed.

31 Make a number of forward and reverse stops and operate the parking brake to adjust the brakes until satisfactory pedal action is obtained.

32 Check the operation of the brakes carefully before driving the vehicle in traffic.

PARKING BRAKE SHOES (MODELS WITH REAR DISC BRAKES)

❋❋ WARNING:

The dust created by the brake system is harmful to your health. Never blow it out with compressed air and don't inhale any of it. An approved filtering mask should be worn when working on the brakes. Do not, under any circumstances, use petroleum-based solvents to clean brake parts. Use brake system cleaner only!

Replacement

▶ Refer to illustrations 6.37a through 6.37p and 6.39

33 Loosen the rear wheel lug nuts, raise the rear of the vehicle and support it securely on jackstands. Block the front wheels and remove the rear wheels. Release the parking brake.

➡Note: *All four parking brake shoes must be replaced at the same time, but to avoid mixing up parts, work on only one brake assembly at a time.*

34 Remove the rear calipers (see Section 4). Support the caliper assemblies with a coat hanger or heavy wire and don't disconnect the brake line from the caliper.

35 Remove the rear discs (see Section 5).

36 Clean the parking brake assembly with brake system cleaner.

37 Follow the accompanying illustrations for the parking brake shoe replacement procedure (see illustrations 6.37a through 6.37p). Be sure to stay in order and read the caption under each illustration.

Brakes 9-13

6.37a Using a pair of pliers . . .

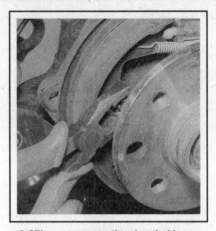

6.37b . . . remove the shoe hold-down springs. Push down and give the spring cup a quarter-turn, then release and lift off the spring and withdraw the retainer pin

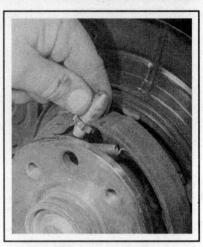

6.37c Note the positions of all the brake components, then remove the adjuster

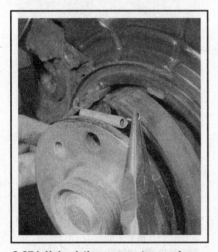

6.37d Unhook the upper return spring

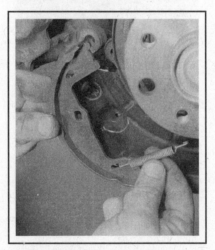

6.37e Unhook the lower return spring and remove the rear . . .

6.37f . . . and front shoes from the backing plate

6.37g Prior to installation of the new shoes, better access during installation is possible by removing the rear hub and bearing assembly as described in Chapter 10. With the hub removed, temporarily hold the backing plate in position using two bolts

6.37h Clean the backing plate thoroughly, then lubricate the brake shoe contact areas on the backing plate with high-temperature grease or anti-seize compound

9-14 BRAKES

6.37i Position the front shoe on the backing plate . . .

6.37j . . . and secure it with the hold-down spring

6.37k Hook the lower return spring to each shoe

6.37l Engage the bottom of the rear shoe with the lever . . .

6.37m . . . then position the shoe on the backing plate and install the hold-down spring

6.37n Install the adjuster between the upper end of the shoes (the star wheel part of the adjuster must be closer to the rear shoe, as shown)

6.37o Install the upper return spring on the shoes

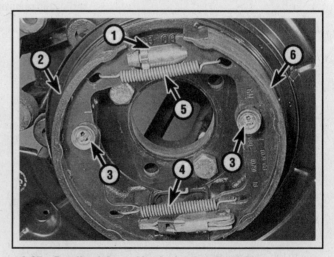

6.37p Details of the parking brake shoe assembly

1 Adjuster screw
2 Rear shoe
3 Hold-down spring
4 Lower return spring
5 Upper return spring
6 Front shoe

38 Install the disc and caliper. Tighten the caliper mounting bolts to the torque listed in this Chapter's Specifications.

39 Using a screwdriver inserted through the adjusting hole in the disc, turn the adjuster star wheel until the brake shoes slightly drag as the disc is rotated, then back-off the adjuster until the shoes don't drag (see illustration).

40 Mount the wheel and install the lug nuts. Lower the vehicle and tighten the lug nuts to the torque listed in the Chapter 1 Specifications.

41 Pull up on the parking brake lever four times with approximately 100 lbs. (45 kg) of force to set the parking brake shoes.

42 Adjust the parking brake cable (see Section 12).

43 Check the operation of the brakes carefully before driving the vehicle in traffic.

6.39 Using a screwdriver through the hole in the disc, turn the adjuster star wheel (disc removed for clarity)

7 Wheel cylinder - removal and installation

→Note: **If replacement is indicated (usually because of fluid leakage or sticky operation), it is recommended that the wheel cylinders be replaced, not overhauled. Always replace the wheel cylinders in pairs - never replace just one of them.**

REMOVAL

1 Loosen the wheel lug nuts, raise the rear of the vehicle and support it securely on jackstands. Block the front wheels to keep the vehicle from rolling.

2 Release the parking brake and remove the wheel.

3 Remove the brake drum as described in Section 6.

4 Carefully remove the brake shoe upper return spring, and remove it from both shoes (see illustration 6.6). Pull the upper ends of the shoes away from the wheel cylinder to disengage them from the pistons.

5 Remove all dirt and foreign material from around the wheel cylinder.

6 At the rear of the backing plate, unscrew the brake line fitting, using a flare-nut wrench, if available. Don't pull the brake line away from the wheel cylinder (it could become kinked).

7 Remove the wheel cylinder mounting bolt.

8 Detach the wheel cylinder from the brake backing plate and immediately plug the brake line to prevent fluid loss and contamination.

INSTALLATION

9 Place the wheel cylinder in position and install the bolt finger tight. Connect the brake line to the cylinder, being careful not to cross thread the fitting. Tighten the wheel cylinder mounting bolt to the torque listed in this Chapter's Specifications. Now tighten the brake line fitting securely.

10 Position the brake shoes on the wheel cylinder pistons, then carefully install the upper return spring on the shoes (see Section 6).

11 Install the brake drum on the hub flange, then install the wheel and lug nuts.

12 Bleed the brake system as outlined in Section 10. Check the operation of the brakes carefully before driving the vehicle.

13 Lower the vehicle to the ground, then tighten the lug nuts to the torque listed in the Chapter 1 Specifications.

8 Master cylinder - removal and installation

REMOVAL

♦ **Refer to illustration 8.5**

1 The master cylinder is located in the engine compartment, mounted to the power brake booster.

2 Remove as much fluid as you can from the reservoir with a syringe, such as an old turkey baster.

※※ **WARNING:**

If a baster is used, never again use it for the preparation of food.

3 Place rags under the fluid fittings and prepare caps or plastic bags to cover the ends of the lines once they are disconnected.

9-16 BRAKES

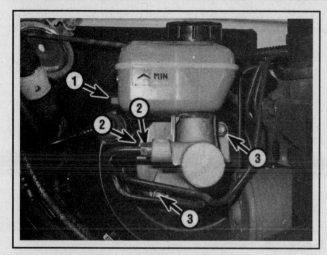

8.5 Brake master cylinder mounting details:

1. Brake fluid switch electrical connector
2. Brake lines
3. Mounting nuts

✴✴ CAUTION:

Brake fluid will damage paint. Cover all pained surfaces around the work area and be careful not to spill fluid during this procedure.

4 On manual transaxle models, disconnect the clutch master cylinder supply hose from the brake fluid reservoir.

5 Loosen the fittings at the ends of the brake lines where they enter the master cylinder (see illustration). To prevent rounding off the corners on these nuts, the use of a flare-nut wrench, which wraps around the nut, is preferred. Pull the brake lines slightly away from the master cylinder and plug the ends to prevent contamination.

6 Disconnect the electrical connector at the brake fluid level switch on the master cylinder reservoir, then remove the nuts attaching the master cylinder to the power booster. Pull the master cylinder off the studs and out of the engine compartment. Again, be careful not to spill the fluid as this is done.

7 If a new master cylinder is being installed, unclip the reservoir mounting tabs then pull up on the reservoir to remove it from the master cylinder. Transfer the reservoir to the new master cylinder.

➡Note: Be sure to install new seals when transferring the reservoir.

INSTALLATION

▸ Refer to illustration 8.9

8 Bench bleed the new master cylinder before installing it. Mount the master cylinder in a vise, with the jaws of the vise clamping on the mounting flange.

9 Attach a pair of master cylinder bleeder tubes to the outlet ports of the master cylinder (see illustration).

10 Fill the reservoir with brake fluid of the recommended type (see Chapter 1).

11 Slowly push the pistons into the master cylinder (a large Phillips screwdriver can be used for this) - air will be expelled from the pressure chambers and into the reservoir. Because the tubes are submerged in fluid, air can't be drawn back into the master cylinder when you release the pistons.

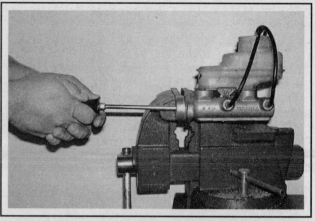

8.9 The best way to bleed air from the master cylinder before installing it on the vehicle is with a pair of bleeder tubes that direct brake fluid into the reservoir during bleeding

12 Repeat the procedure until no more air bubbles are present.

13 Remove the bleed tubes, one at a time, and install plugs in the open ports to prevent fluid leakage and air from entering. Install the reservoir cap.

14 Install the master cylinder over the studs on the power brake booster and tighten the attaching nuts only finger tight at this time.

➡Note: Be sure to install a new O-ring onto the sleeve of the master cylinder.

15 Thread the brake line fittings into the master cylinder. Since the master cylinder is still a bit loose, it can be moved slightly in order for the fittings to thread in easily. Do not strip the threads as the fittings are tightened.

16 Fully tighten the mounting nuts, then the brake line fittings. Tighten the nuts to the torque listed in this Chapter's Specifications.

17 Connect the brake fluid switch electrical connector. On models with a manual transaxle, connect the clutch master cylinder supply hose.

18 Fill the master cylinder reservoir with fluid, then bleed the master cylinder and the brake system as described in Section 10. To bleed the cylinder on the vehicle, have an assistant depress the brake pedal and hold the pedal to the floor. Loosen the fitting to allow air and fluid to escape. Repeat this procedure on both fittings until the fluid is clear of air bubbles.

✴✴ CAUTION:

Have plenty of rags on hand to catch the fluid - brake fluid will ruin painted surfaces. After the bleeding procedure is completed, rinse the area under the master cylinder with clean water.

19 Test the operation of the brake system carefully before placing the vehicle into normal service.

✴✴ WARNING:

Do not operate the vehicle if you are in doubt about the effectiveness of the brake system. It is possible for air to become trapped in the anti-lock brake system hydraulic control unit, so, if the pedal continues to feel spongy after repeated bleedings or the BRAKE or ANTI-LOCK light stays on, have the vehicle towed to a dealer service department or other qualified shop to be bled with the aid of a scan tool.

BRAKES 9-17

9 Brake hoses and lines - inspection and replacement

1 About every six months, with the vehicle raised and placed securely on jackstands, the flexible hoses which connect the steel brake lines with the front and rear brake assemblies should be inspected for cracks, chafing of the outer cover, leaks, blisters and other damage. These are important and vulnerable parts of the brake system and inspection should be complete. A light and mirror will be needed for a thorough check. If a hose exhibits any of the above defects, replace it with a new one.

FLEXIBLE HOSES

▸ Refer to illustration 9.3

2 Clean all dirt away from the ends of the hose.
3 To disconnect a brake hose from the brake line, unscrew the metal tube nut with a flare nut wrench, then remove the U-clip from the female fitting at the bracket and remove the hose from the bracket (see illustration).
4 Disconnect the hose from the caliper, discarding the sealing washers on either side of the fitting.
5 Using new sealing washers, attach the new brake hose to the caliper or wheel cylinder.
6 To reattach a brake hose to the metal line, insert the end of the hose through the frame bracket, make sure the hose isn't twisted, then attach the metal line by tightening the tube nut fitting securely. Install the U-clip at the frame bracket.
7 Carefully check to make sure the suspension or steering components don't make contact with the hose. Have an assistant push down on the vehicle and also turn the steering wheel lock-to-lock during inspection.
8 Bleed the brake system (see Section 10).

METAL BRAKE LINES

9 When replacing brake lines, be sure to use the correct parts. Don't use copper tubing for any brake system components. Purchase steel brake lines from a dealer parts department or auto parts store.
10 Prefabricated brake line, with the tube ends already flared and fittings installed, is available at auto parts stores and dealer parts departments. These lines can be bent to the proper shapes using a tubing bender.
11 When installing the new line make sure it's well supported in the brackets and has plenty of clearance between moving or hot components.
12 After installation, check the master cylinder fluid level and add fluid as necessary. Bleed the brake system as outlined in Section 10 and test the brakes carefully before placing the vehicle into normal operation.

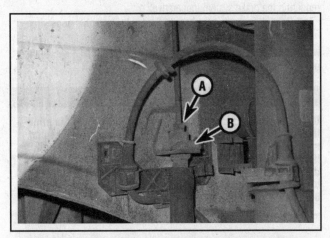

9.3 Using a flare nut wrench, unscrew the threaded fitting on the brake line (A), then pry the U-clip (B) off the end of the hose and separate the hose from the bracket

10 Brake hydraulic system - bleeding

▸ Refer to illustration 10.8

※ WARNING 1:

If air has found its way into the hydraulic control unit, the system must be bled with the use of a scan tool. If the brake pedal feels "spongy" even after bleeding the brakes, or the ABS light on the instrument panel does not go off, or if you have any doubts whatsoever about the effectiveness of the brake system, have the vehicle towed to a dealer service department or other repair shop equipped with the necessary tools for bleeding the system.

※ WARNING 2:

Wear eye protection when bleeding the brake system. If the fluid comes in contact with your eyes, immediately rinse them with water and seek medical attention.

➥ Note: Bleeding the brake system is necessary to remove any air that's trapped in the system when it's opened during removal and installation of a hose, line, caliper, wheel cylinder or master cylinder.

1 It will probably be necessary to bleed the system at all four brakes if air has entered the system due to a low fluid level, or if the brake lines have been disconnected at the master cylinder.
2 If a brake line was disconnected only at a wheel, then only that caliper or wheel cylinder must be bled.
3 If a brake line is disconnected at a fitting located between the master cylinder and any of the brakes, that part of the system served by the disconnected line must be bled.
4 Remove any residual vacuum (or hydraulic pressure) from the brake power booster by applying the brake several times with the engine off.
5 Remove the master cylinder reservoir cap and fill the reservoir with brake fluid. Reinstall the cap.

9-18 BRAKES

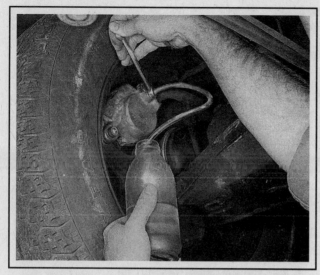

10.8 When bleeding the brakes, a hose is connected to the bleed screw at the caliper or wheel cylinder and submerged in brake fluid - air will be seen as bubbles in the tube and container (all air must be expelled before moving to the next wheel)

→Note: Check the fluid level often during the bleeding operation and add fluid as necessary to prevent the fluid level from falling low enough to allow air bubbles into the master cylinder.

6 Have an assistant on hand, as well as a supply of new brake fluid, an empty clear plastic container, a length of plastic, rubber or vinyl tubing to fit over the bleeder valve and a wrench to open and close the bleeder valve.

7 Beginning at the right rear wheel, loosen the bleeder screw slightly, then tighten it to a point where it's snug but can still be loosened quickly and easily.
8 Place one end of the tubing over the bleeder screw fitting and submerge the other end in brake fluid in the container (see illustration).
9 Have the assistant slowly depress the brake pedal and hold it in the depressed position.
10 While the pedal is held depressed, open the bleeder screw just enough to allow a flow of fluid to leave the valve. Watch for air bubbles to exit the submerged end of the tube. When the fluid flow slows after a couple of seconds, tighten the screw and have your assistant release the pedal.
11 Repeat Steps 9 and 10 until no more air is seen leaving the tube, then tighten the bleeder screw and proceed to the left front wheel, the left rear wheel and the right front wheel, in that order, and perform the same procedure. Be sure to check the fluid in the master cylinder reservoir frequently.
12 Never use old brake fluid. It contains moisture which can boil, rendering the brake system inoperative.
13 Refill the master cylinder with fluid at the end of the operation.
14 Check the operation of the brakes. The pedal should feel solid when depressed, with no sponginess. If necessary, repeat the entire process.

WARNING:

Do not operate the vehicle if you are in doubt about the effectiveness of the brake system. It is possible for air to become trapped in the anti-lock brake system hydraulic control unit, so, if the pedal continues to feel spongy after repeated bleedings or the BRAKE or ANTI-LOCK light stays on, have the vehicle towed to a dealer service department or other qualified shop to be bled with the aid of a scan tool.

11 Power brake booster - removal and installation

OPERATING CHECK

1 Depress the brake pedal several times with the engine off and make sure that there is no change in the pedal reserve distance.
2 Depress the pedal and start the engine. If the pedal goes down slightly, operation is normal.

AIRTIGHTNESS CHECK

3 Start the engine and turn it off after one or two minutes. Depress the brake pedal several times slowly. If the pedal goes down farther the first time but gradually rises after the second or third depression, the booster is airtight.
4 Depress the brake pedal while the engine is running, then stop the engine with the pedal depressed. If there is no change in the pedal reserve travel after holding the pedal for 30 seconds, the booster is airtight.

REMOVAL AND INSTALLATION

▶ Refer to illustrations 11.9 and 11.10

5 Disassembly of the power unit requires special tools and is not ordinarily performed by the home mechanic. If a problem develops, it's recommended that a new or factory rebuilt unit be installed.
6 Remove the master cylinder (see Section 8). On vehicles equipped with ABS, unbolt the hydraulic control unit from the inner fender panel.
7 Disconnect the vacuum hose check valve where it attaches to the power brake booster.
8 In the passenger compartment, remove the driver's side lower trim panel (see Chapter 11).
9 Remove the clip and retaining pin securing the booster pushrod to the brake pedal arm (see illustration).
10 Working inside the engine compartment, remove the bolts securing the brake booster to the brake booster mounting bracket (see illustration).

→Note: The top mounting bolt can be accessed by opening the interior ventilation filter access panel and removing the rubber plug at the bottom of the cowl.

11 Carefully guide the booster unit away from the bracket and out of the engine compartment.
12 To install the booster, place it into position and tighten the retaining bolts to the torque listed in this Chapter's Specifications. Connect the pushrod to the brake pedal and install the clip.

Brakes 9-19

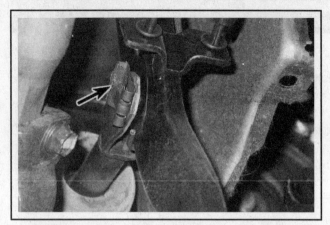

11.9 Remove the clip and retaining pin from the brake pedal arm

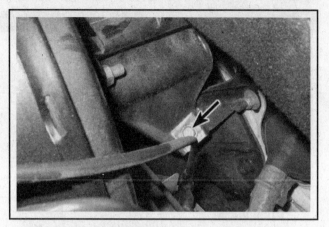

11.10 Remove the brake booster-to-brake booster bracket mounting bolts (two bolts not visible in photo)

13 Install the master cylinder. Reconnect the vacuum hose. On models with ABS, reposition the hydraulic control unit and install the mounting nuts, tightening them securely.

14 Bleed the brakes (see Section 10).
15 Carefully test the operation of the brakes before placing the vehicle in normal service.

12 Parking brake - adjustment

▶ **Refer to illustration 12.4**

1 The parking brake lever, when properly adjusted, should travel three to five clicks. If it travels less than specified, there's a chance the parking brake might not be releasing completely and might be dragging on the drum or disc. If the lever can be pulled more than specified, the parking brake may not hold adequately on an incline, allowing the car to roll.

2 Block the front wheels, raise the rear of the vehicle and support it securely on jackstands. Pull on the lever until you hear the third click.

3 To gain access to the parking brake cable adjuster, remove the parking brake lever boot (see Chapter 11).

4 Turn the adjusting nut on the lever (see illustration) clockwise while rotating the rear wheels. Stop turning the nut when the brakes start to drag heavily on the rear wheels.

5 Release the parking brake and check to see the brakes don't drag when the rear wheels are turned. The travel on the parking brake lever travel should be as listed in Step 1 when properly adjusted.

12.4 Turn this adjusting nut to adjust the parking brake

13 Brake light switch - replacement

▶ **Refer to illustration 13.2**

1 The brake light switch is located at the top of the brake pedal bracket inside the driver's side footwell. Remove the driver's side lower trim panel (see Chapter 11).

2 Disconnect the electrical connector at the switch then rotate the switch clockwise 90-degrees and remove the switch (see illustration).

3 To install the new switch, rotate the switch counterclockwise 90-degrees while inserting it into the mounting bracket, then connect the electrical connector.

4 Depress and hold the brake pedal, pull the plunger outward then release the brake pedal.

5 Apply the brake pedal, release the pedal and verify that the brake lights go off when the pedal is released.

13.2 The brake light switch is located next to the cruise control switch

BRAKES

Specifications

General
Brake fluid type	See Chapter 1

Disc brakes
Brake pad minimum thickness	See Chapter 1
Disc lateral runout limit	0.004 inch
Disc minimum thickness	Cast into disc
Parallelism (thickness variation) limit	0.0006 inch

Drum brakes
Maximum drum diameter	Cast into drum
Shoe lining minimum thickness	See Chapter 1

Torque specifications
	Ft-lbs (unless otherwise indicated)	Nm
Brake hose banjo fitting bolt	30	40
Caliper guide pins (mounting bolts)		
Front	22	30
Rear	59	80
Caliper mounting bracket bolts (front)	70	95
Master cylinder mounting nuts	18	25
Power brake booster-to- mounting bolts	15	20
Wheel cylinder mounting bolt	80 in-lbs	9
Wheel lug nuts	See Chapter 1	

10 SUSPENSION AND STEERING SYSTEMS

Section
1. General information
2. Strut assembly (front) - removal, inspection and installation
3. Strut/coil spring - replacement
4. Stabilizer bar, bushings and links (front) - removal and installation
5. Control arm - removal, inspection and installation
6. Balljoints - check and replacement
7. Steering knuckle and hub removal (front) - removal and installation
8. Hub and bearing assembly (front)
9. Shock absorber/coil spring assembly (rear) - removal and installation
10. Shock absorber/coil spring assembly (rear) - replacement
11. Stabilizer bar, bushings and links (rear) - removal and installation
12. Suspension arms (rear) - removal and installation
13. Hub and bearing assembly (rear) - removal and installation
14. Steering wheel - removal and installation
15. Steering column - removal and installation
16. Tie-rod ends - removal and installation
17. Steering gear boots - removal and installation
18. Steering gear - removal and installation
19. Power steering pump - removal and installation
20. Power steering system - bleeding
21. Subframe - removal and installation
22. Wheels and tires - general information
23. Wheel alignment - general information

10-2 SUSPENSION AND STEERING SYSTEMS

1 General information

Refer to illustrations 1.1 and 1.2

The front suspension is a MacPherson strut design. The upper end of each strut is attached to the vehicle's body strut support. The lower end of the strut is connected to the upper end of the steering knuckle. The steering knuckle is attached to a balljoint mounted on the outer end of the suspension control arm. A stabilizer bar connected to each strut and mounted to the suspension crossmember reduces body roll during cornering (see illustration).

The rear suspension employs a trailing arm, two lateral suspension arms, and shock absorbers with coil springs (see illustration). A stabilizer bar is clamped to a suspension support and connected to the trailing arms by two links.

The power-assisted rack-and-pinion steering gear is attached to the front suspension subframe. The steering gear actuates the tie-rods, which are attached to the steering knuckles. The steering column is designed to collapse in the event of an accident.

Frequently, when working on the suspension or steering system components, you may come across fasteners which seem impossible to loosen. These fasteners on the underside of the vehicle are continually subjected to water, road grime, mud, etc., and can become rusted or "frozen" in place, making them extremely difficult to remove. In order to unscrew these stubborn fasteners without damaging them (or other components), be sure to use lots of penetrating oil and allow it to soak in for a while. Using a wire brush to clean exposed threads will also ease removal of the nut or bolt and prevent damage to the threads. Sometimes a sharp blow with a hammer and punch will break the bond between a nut and bolt threads, but care must be taken to prevent the punch from slipping off the fastener and ruining the threads. Heating the stuck fastener and surrounding area with a torch sometimes helps too, but isn't recommended because of the obvious dangers associated with fire. Long breaker bars and extension, or "cheater," pipes will increase leverage, but never use an extension pipe on a ratchet - the ratcheting mechanism could be damaged. Sometimes tightening the nut or bolt first will help to break it loose. Fasteners that require drastic measures to remove should always be replaced with new ones.

Since most of the procedures dealt with in this Chapter involve jacking up the vehicle and working underneath it, a good pair of jackstands will be needed. A hydraulic floor jack is the preferred type of jack to lift the vehicle, and it can also be used to support certain components during various operations.

1.1a Front suspension and steering components

1 Subframe
2 Control arm
3 Tie-rod end
4 Steering knuckle
5 Balljoint
6 Strut

SUSPENSION AND STEERING SYSTEMS   10-3

> **☼☼ WARNING:**
> Never, under any circumstances, rely on a jack to support the vehicle while working on it. Whenever any of the suspension or steering fasteners are loosened or removed they must be inspected and, if necessary, replaced with new ones of the same part number or of original equipment quality and design. Torque specifications must be followed for proper reassembly and component retention. Never attempt to heat or straighten any suspension or steering components. Instead, replace any bent or damaged part with a new one.

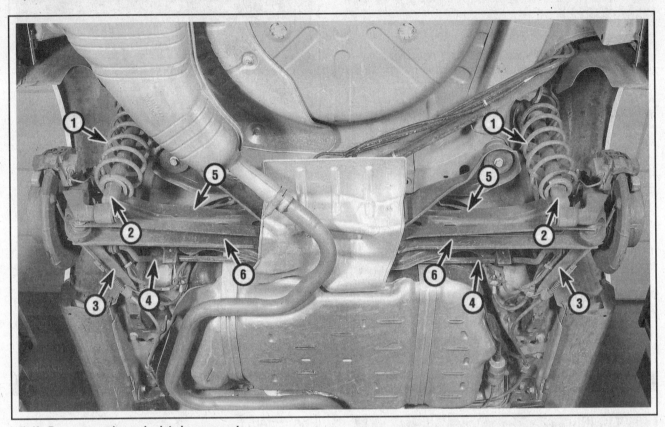

1.1b Rear suspension and related components

1. Coil spring
2. Shock absorber
3. Trailing arm
4. Stabilizer bar
5. Upper lateral control arm
6. Lower lateral control arm

2 Strut assembly (front) - removal, inspection and installation

> **☼☼ WARNING:**
> Always replace the struts and/or coil springs in pairs - never replace just one strut or one coil spring (this could cause dangerous handling peculiarities).

REMOVAL

▶ Refer to illustrations 2.3 and 2.5

1 Loosen the wheel lug nuts, raise the vehicle and support it securely on jackstands. Remove the wheel.

2 Disconnect the stabilizer bar link from the strut, then remove the brake hose from the struts brake hose bracket. On models equipped with ABS, remove the ABS harness from the same bracket.

3 Mark the relationship of the strut to the knuckle (these marks will be used during installation to ensure the camber is returned to its original setting). Remove the strut-to-knuckle nuts (see illustration) and knock the bolts out with a hammer and punch.

2.3 Remove the strut-to-steering knuckle nuts and bolts

10-4 SUSPENSION AND STEERING SYSTEMS

4 Separate the strut from the steering knuckle. Be careful not to overextend the inner CV joint. Also, don't let the steering knuckle fall outward, as the brake hose could be damaged.

5 Support the strut and spring assembly with one hand and remove the upper mounting nut (see illustration), then remove the assembly from the fenderwell.

INSPECTION

6 Check the strut body for leaking fluid, dents, cracks and other obvious damage that would warrant repair or replacement.

7 Check the coil spring for chips or cracks in the spring coating (this can cause premature spring failure due to corrosion). Inspect the spring seat for cuts, hardness and general deterioration.

8 If any undesirable conditions exist, proceed to the strut disassembly procedure (see Section 3).

INSTALLATION

9 Guide the strut assembly up into the fenderwell and insert the upper mounting stud through the hole in the shock tower. Once the stud protrudes from the shock tower, install the nut so the strut won't fall back through. This is most easily accomplished with the help of an assistant, as the strut is quite heavy and awkward.

10 Slide the steering knuckle into the strut flange and insert the two bolts. Install the nuts, align the marks you made in Step 3, then tighten the nuts to the torque listed in this Chapter's Specifications.

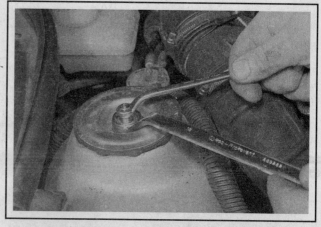

2.5 Unscrew the upper mounting nut while counterholding the damper rod with another wrench

11 Reattach the brake hose to the strut bracket and reconnect the stabilizer bar link. If equipped, install the speed sensor wiring harness to the strut bracket.

12 Install the wheel and lug nuts, then lower the vehicle and tighten the lug nuts to the torque listed in the Chapter 1 Specifications.

13 Tighten the upper mounting nut to the torque listed in this Chapter's Specifications.

14 Have the front wheel alignment checked and, if necessary, adjusted.

3 Strut/coil spring - replacement

✼✼✼ WARNING:

The manufacturer recommends replacing the damper shaft nut with a new one whenever it is removed.

➡Note: *You'll need a spring compressor for this procedure. Spring compressors are available on a daily rental basis at most auto parts stores or equipment yards.*

1 If the struts or coil springs exhibit the telltale signs of wear (leaking fluid, loss of damping capability, chipped, sagging or cracked coil springs) explore all options before beginning any work. The strut/coil spring assemblies are not serviceable and must be replaced if a problem develops. However, strut assemblies complete with springs may be available on an exchange basis, which eliminates much time and work. Whichever route you choose to take, check on the cost and availability of parts before disassembling your vehicle.

✼✼✼ WARNING:

Disassembling a strut is potentially dangerous and utmost attention must be directed to the job, or serious injury may result. Use only a high-quality spring compressor and carefully follow the manufacturer's instructions furnished with the tool. After removing the coil spring from the strut assembly, set it aside in a safe, isolated area.

DISASSEMBLY

▶ Refer to illustrations 3.3, 3.5a, 3.5b, 3.6, 3.7 and 3.8

2 Remove the strut and spring assembly (see Section 2). Mount the strut clevis bracket portion of the strut assembly in a vise.

✼✼✼ CAUTION:

Do not clamp any other portion of the strut assembly in the vise as it will be damaged. Line the vise jaws with wood or rags to prevent damage to the unit and don't tighten the vise excessively.

3 Following the tool manufacturer's instructions, install the spring compressor (which can be obtained at most auto parts stores or equipment yards on a daily rental basis) on the spring and compress it sufficiently to relieve all pressure from the upper spring seat (see illustration). This can be verified by wiggling the spring.

4 Hold the damper shaft from turning with a socket, and unscrew the damper shaft nut with a box-end wrench.

5 Remove the nut and upper mount (see illustrations). Lay the parts out in the exact order in which they are removed. Check the rubber portion of the upper mount for cracking and general deterioration. If there is any separation of the rubber, replace it.

SUSPENSION AND STEERING SYSTEMS

3.3 Install the spring compressor following the tool manufacturer's instructions; compress the spring until all pressure is relieved from the upper spring seat (you can verify this by wiggling the spring)

3.5a Remove the damper shaft nut . . .

6 Remove the upper spring seat from the damper shaft (see illustration). Check the rubber portion of the spring seat for cracking and hardness; replace it if necessary. Inspect the bearing in the spring seat for smooth operation. If it doesn't turn smoothly, replace it.

7 Slide the rubber bump stop off the damper shaft (see illustration). Check the bump stop for cracking and general deterioration. If there is any deterioration of the rubber, replace it.

8 Carefully lift the compressed spring from the assembly (see illustration) and set it in a safe place.

WARNING:
When removing the compressed spring, lift it off carefully and set it in a safe place. Keep the ends of the spring away from your body.

3.5b . . . and upper mount

REASSEMBLY

▸ Refer to illustration 3.10

9 Extend the damper rod to its full length and install the rubber bump stop.

➡ Note: If you are disassembling both struts, mark the springs LEFT and RIGHT so you don't mix them up (they're different).

3.6 Remove the upper spring seat . . .

3.7 . . . then slide the rubber bump stop off of the damper shaft

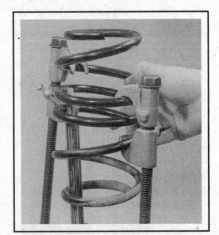

3.8 Carefully remove the compressed spring from the strut

10-6 SUSPENSION AND STEERING SYSTEMS

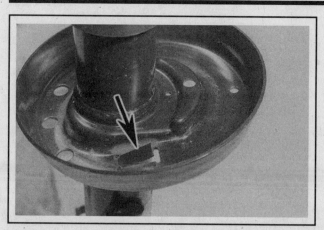

3.10 When installing the spring, make sure the end rests against the raised stop

10 Carefully place the compressed coil spring onto the lower seat of the damper, with the end of the spring resting against the raised stop (see illustration).

11 Install the upper insulator and spring seat.

12 Install the upper mount and mounting nut, then tighten it to the torque listed in this Chapter's Specifications. Remove the spring compressor tool.

13 Install the strut/spring assembly (see Section 2).

4 Stabilizer bar, bushings and links (front) - removal and installation

♦ Refer to illustrations 4.2a, 4.2b and 4.3

※※ WARNING:

The manufacturer recommends replacing all the stabilizer bar fasteners with new ones whenever they are removed.

1 Loosen the front wheel lug nuts, raise the front of the vehicle, support it securely on jackstands and remove the front wheels.

FOUR-CYLINDER MODELS

Bushings and links

➡Note: Stabilizer bar removal involves lowering the subframe, removing the stabilizer bar out the left wheelwell, and if one becomes damaged it is most likely the result of an accident that was severe enough to damage other major components (such as the subframe itself). Damage this severe will require the services of an auto body shop. For this reason, front stabilizer bar removal and installation is not covered in this manual.

2 Remove the nuts that attach the upper and lower ends of the stabilizer links to the strut/coil spring assembly and to the stabilizer bar (see illustrations). Detach the links.

3 Remove the bolts from the stabilizer bar bushing retainers (see illustration). Remove the retainers from the bushings, prying them off, if necessary.

4 Inspect the retainer bushings for cracks and tears. If either bushing is broken, damaged, distorted or worn, replace both of them. If the ballstuds on the links are loose or otherwise worn, replace the links.

5 Install the links, tightening the link nuts to the torque listed in this Chapter's Specifications.

6 Install the retainer bushings. Clean the areas on the stabilizer bar where the bushings are located. Lubricate the inside and outside of the new bushings with vegetable oil (used in cooking) to simplify reassembly.

※※ CAUTION:

Don't use petroleum or mineral-based lubricants or brake fluid - they will lead to deterioration of the bushings. These bushings are split so that you can install them without having to slide them onto the ends of the stabilizer bar. Install the bushings with the slit in each bushing facing towards the rear of the vehicle.

7 Install the retainers and bolts, tightening the bolts to the torque listed in this Chapter's Specifications.

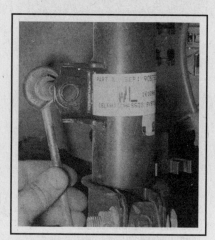

4.2a Remove the nut and detach the upper link from the strut . . .

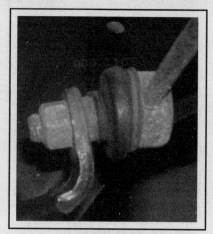

4.2b . . . then detach the link from the stabilizer bar

4.3 Remove the stabilizer bar bracket bolts

SUSPENSION AND STEERING SYSTEMS

V6 MODELS

Bushings, links and stabilizer bar

8 Remove the nuts that attach the upper and lower ends of the stabilizer links to the strut/coil spring assembly and to the stabilizer bar (see illustration 4.2). Detach the links.

9 Remove the bolts from the stabilizer bar bushing retainers (see illustration 4.3).

10 On vehicles equipped with a manual transaxle, remove the rear exhaust manifold pipe (see Chapter 6).

11 Maneuver the stabilizer out through the left front wheelwell and remove it from the vehicle.

12 Pull the bushing and retainers off the stabilizer bar and inspect the bushings for cracks, hardness and other signs of deterioration. If the bushings are damaged, replace them.

13 Installation is the reverse of removal, noting the following points:

   a) *Lubricate the inside and outside of the new bushings with vegetable oil (used in cooking) to simplify reassembly.*

✱✱ CAUTION:

Don't use petroleum or mineral-based lubricants or brake fluid - they will lead to deterioration of the bushings. These bushings are split so that you can install them without having to slide them onto the ends of the stabilizer bar. Install the bushings with the slit in each bushing facing towards the rear of the vehicle.

   b) *Install the link bolts, insulators and nuts, tightening the bolts to the torque listed in this Chapter's Specifications.*
   c) *On vehicles equipped with a manual transaxle, install the exhaust manifold pipe* (see Chapter 4).

14 Install the wheels and lug nuts, then lower the vehicle. Tighten the wheel lug nuts to the torque listed in the Chapter 1 Specifications.

5 Control arm - removal, inspection and installation

✱✱ WARNING:

The manufacturer recommends replacing all the control arm mounting fasteners with new ones whenever they are removed.

REMOVAL

▶ **Refer to illustrations 5.2a, 5.2b and 5.3**

1 Loosen the wheel lug nuts on the side to be disassembled. Apply the parking brake, raise the front of the vehicle, support it securely on jackstands and remove the wheel.

2 Remove the pinch bolt securing the balljoint to the control arm. Use a prybar to disconnect the control arm balljoint from the steering knuckle (see illustrations).

3 Remove the bolts that attach the control arm to the subframe (see illustration).

4 Remove the control arm.

INSPECTION

5 Check the control arm for distortion and the bushings for wear, replacing parts as necessary. Do not attempt to straighten a bent control arm. If the bushings are cracked or show signs of wear, take the control arm to an automotive machine shop and have the bushings replaced.

INSTALLATION

6 Installation is the reverse of removal, tighten all of the fasteners to the torque values listed in this Chapter's Specifications.

➡ **Note: Before tightening the control arm pivot bolts, raise the outer end of the control arm with a floor jack to simulate normal ride height.**

7 Install the wheel and lug nuts, lower the vehicle and tighten the lug nuts to the torque listed in the Chapter 1 Specifications.

8 It's a good idea to have the front wheel alignment checked and, if necessary, adjusted after this job has been performed.

5.2a To detach the control arm balljoint from the steering knuckle, remove the nut and bolt . . .

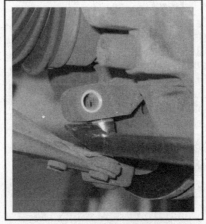

5.2b . . . then pry the balljoint out of the steering knuckle with a large prybar or screwdriver

5.3 Control arm-to-subframe mounting bolts

10-8 SUSPENSION AND STEERING SYSTEMS

6 Balljoints - check and replacement

CHECK

1 Raise the front of the vehicle and support it securely on jackstands. Apply the parking brake and block the rear wheels to keep the vehicle from rolling off the jackstands.

2 Place a large prybar under the balljoint and resting on the wheel, then try to pry the balljoint up while feeling for movement between the balljoint and steering knuckle. Now, pry between the control arm and the steering knuckle and try to lever the control arm down while feeling for movement between the balljoint and steering knuckle. If any movement is evident in either check, the balljoint is worn.

3 Have an assistant grasp the tire at the top and bottom and move the top of the tire in-and-out. Touch the balljoint stud nut. If any looseness is felt, suspect a worn balljoint stud or a widened hole in the steering knuckle boss. If the latter problem exists, the steering knuckle should be replaced as well as the balljoint.

4 Separate the control arm from the steering knuckle (see Section 5). Using your fingers (don't use pliers), try to twist the stud in the socket. If the stud turns, replace the balljoint.

REPLACEMENT

5 Raise the vehicle, support it securely on jackstands and remove the front wheel, if you haven't already done so. Remove the control arm (see Section 5).

6 Drill out the three rivets securing the balljoint, remove the balljoint and clean the control arm. Install the new balljoint against the mating surface of the control arm and secure it with the supplied fasteners (nuts and bolts are normally supplied with the new balljoint). Tighten the balljoint fasteners to the torque specified in the balljoint replacement kit instructions.

7 Install the control arm as described in Section 5.

8 Install the wheel and lug nuts, lower the vehicle and tighten the lug nuts to the torque listed in the Chapter 1 Specifications.

7 Steering knuckle and hub removal (front) - removal and installation

⁂ WARNING 1:

Dust created by the brake system is harmful to your health. Never blow it out with compressed air and don't inhale any of it. Do not, under any circumstances, use petroleum-based solvents to clean brake parts. Use brake system cleaner only.

⁂ WARNING 2:

The manufacturer recommends replacing the strut-to-steering knuckle fasteners with new ones whenever they are removed.

REMOVAL

1 Loosen the driveaxle/hub nut (see Chapter 8). Loosen the wheel lug nuts, raise the vehicle and support it securely on jackstands. Remove the wheel.

2 Remove the brake caliper and support it with a piece of wire as described in Chapter 9. Remove the caliper mounting bracket, then remove the brake disc from the hub. If the vehicle is equipped with ABS, remove the wheel speed sensor (see Chapter 9).

3 Mark the strut to the steering knuckle, then loosen, but do not remove the strut-to-steering knuckle bolts (see illustration 2.3).

4 Separate the tie-rod end from the steering knuckle arm (see Section 16).

5 Remove the balljoint-to-steering knuckle pinch bolt, then separate the balljoint from the steering knuckle (see Section 5).

6 Remove the driveaxle/hub nut and push the driveaxle from the hub as described in Chapter 8. Support the end of the driveaxle with a piece of wire.

7 The strut-to-knuckle bolts can now be removed.

8 Carefully separate the steering knuckle from the strut.

INSTALLATION

9 Guide the knuckle and hub assembly into position, inserting the driveaxle into the hub.

10 Push the knuckle into the strut flange and install the bolts and nuts, but don't tighten them yet.

11 Connect the balljoint to the knuckle and tighten the pinch bolt/nut to the torque listed in this Chapter's Specifications.

12 Attach the tie-rod to the steering knuckle arm (see Section 15). Tighten the strut bolt nuts and the tie-rod nut to the torque values listed in this Chapter's Specifications.

13 Place the brake disc on the hub and install the caliper mounting bracket and caliper as outlined in Chapter 9.

14 Install the driveaxle/hub nut and tighten it securely (final tightening will be carried out when the vehicle is lowered).

15 Install the wheel and lug nuts.

16 Lower the vehicle and tighten the lug nuts to the torque listed in the Chapter 1 Specifications. Tighten the driveaxle/hub nut to the torque listed in the Chapter 8 Specifications.

17 Have the front-end alignment checked and, if necessary, adjusted.

SUSPENSION AND STEERING SYSTEMS

8 Hub and bearing assembly (front)

Due to the special tools and expertise required to press the hub and bearing from the steering knuckle, this job should be left to a professional mechanic. However, the steering knuckle and hub may be removed and the assembly taken to an automotive machine shop or other qualified repair facility equipped with the necessary tools. See Section 7 for the steering knuckle and hub removal procedure.

9 Shock absorber/coil spring assembly (rear) - removal and installation

※※ WARNING 1:

Always replace the shock absorbers or coil springs in pairs - never replace just one of them.

※※ WARNING 2:

The manufacturer recommends replacing the shock absorber/coil spring assembly mounting bolts with new ones whenever they are removed.

REMOVAL

▶ Refer to illustrations 9.3, 9.4a and 9.4b

1 Loosen the rear wheel lug nuts. Chock the front wheels to keep the vehicle from rolling, then raise the rear of the vehicle and support it securely on jackstands. Remove the rear wheels.
2 Remove the rear inner fender splash shield.
3 Remove the shock absorber-to-trailing arm mounting bolt and washer (see illustration).
4 Support the shock, remove the shock carrier-to-body lower mounting bolts, then loosen the shock carrier-to-body upper mounting bolts and remove the shock (see illustrations).

9.3 Rear shock absorber-to-knuckle mounting bolt

INSTALLATION

5 Installation is the reverse of removal. Tighten the bolts to the torque listed in this Chapter's Specifications.

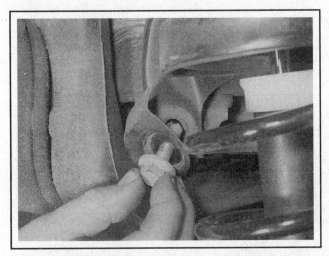

9.4a Remove the shock carrier's lower mounting bolts . . .

9.4b . . . then loosen the upper mounting bolts and remove the shock absorber/coil spring assembly

10-10 SUSPENSION AND STEERING SYSTEMS

10 Shock absorber/coil spring assembly (rear) - replacement

Refer to illustrations 10.3, 10.4a, 10.4b, 10.5a through 10.5f and 10.11

Note: You'll need a spring compressor for this procedure. Spring compressors are available on a daily rental basis at most auto parts stores or equipment yards.

1 If the shocks or coil springs exhibit the telltale signs of wear (leaking fluid, loss of damping capability, chipped, sagging or cracked coil springs) explore all options before beginning any work. The shock absorber/coil spring assemblies are not serviceable and must be replaced if a problem develops. However, shock assemblies complete with springs may be available on an exchange basis, which eliminates much time and work. Whichever route you choose to take, check on the cost and availability of parts before disassembling your vehicle.

✱✱ WARNING:

Disassembling a shock absorber/coil spring assembly is potentially dangerous and utmost attention must be directed to the job, or serious injury may result. Use only a high-quality spring compressor and carefully follow the manufacturer's instructions furnished with the tool. After removing the coil spring from the shock assembly, set it aside in a safe, isolated area.

2 Remove the shock absorber/coil spring assembly (see Section 9).

3 With the shock absorber/coil spring assembly resting on a bench, or clamped in a vice, follow the tool manufacturer's instructions, install the spring compressor on the spring and compress it sufficiently to relieve all pressure from the upper spring seat (see illustration). This can be verified by wiggling the spring.

4 Hold the shock damper shaft stationary with a socket, and unscrew the retaining nut with a box-end wrench (see illustrations).

5 Remove the shock absorber upper cup and insulator followed by the top carrier, lower insulator and sleeve, lower cup and bump stop (see illustrations).

6 If a new spring is to be installed, the original spring must be now be carefully released from the compressor. If it is to be reused, the spring can be left in compression.

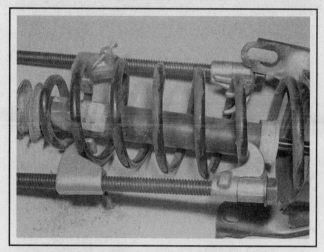

10.3 Install the spring compressor following the tool manufacturer's instructions; compress the spring until all pressure is relieved from the upper spring seat (you can verify this by wiggling the spring)

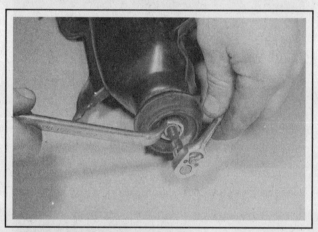

10.4a Loosen the retaining nut . . .

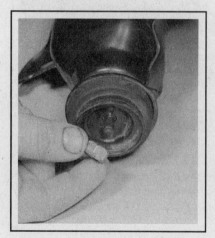

10.4b . . . then remove the nut from the top of the shock damper shaft

10.5a Remove the shock absorber upper cup . . .

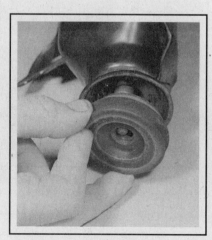

10.5b . . . and insulator . . .

SUSPENSION AND STEERING SYSTEMS   10-11

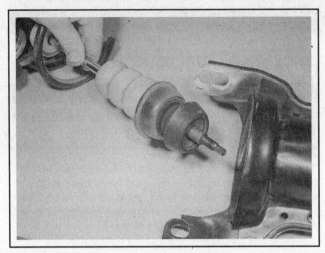

10.5c ... followed by the top shock carrier ...

10.5d ... and the lower insulator and sleeve ...

7 With the shock/coil spring assembly now completely disassembled, examine all the components for wear or damage. Replace the components as necessary. Carefully lift the compressed spring from the assembly and set it in a safe place.

WARNING:
When removing the compressed spring, lift it off carefully and set it in a safe place. Keep the ends of the spring away from your body.

→Note: If you are disassembling both shock/coil spring assembly's, mark the springs LEFT and RIGHT so you don't mix them up (they're different).

8 Examine the shock for signs of fluid leakage. Check the shock damper shaft for signs of pitting along its entire length, and check the shock body for signs of damage. Test the operation of the shock, while holding it in an upright position, by moving the damper shaft through a full stroke, and then through short strokes of 2 to 4 inches. In both cases, the resistance felt should be smooth and continuous. If the resistance is jerky, uneven, or if there is any visible sign of wear or damage to the shock, replacement is necessary.

9 Extend the shock damper shaft to its full length and install the rubber bump stop, followed by the lower cup, lower insulator and sleeve.

10 Carefully place the compressed coil spring onto the lower seat of the shock, with the end of the spring resting in the lowest part of the seat.

11 Position the upper spring seat onto the top carrier, then install the top carrier on the shock damper shaft and coil spring, making sure that the end of the coil spring is resting in the recessed portion of the upper seat (see illustration).

12 Install the upper insulator, cup and retaining nut and tighten it to the torque listed in this Chapter's Specifications. Remove the spring compressor tool.

13 Install the shock absorber/coil spring assembly (see Section 10).

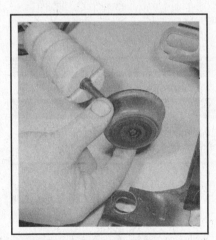

10.5e ... lower cup ...

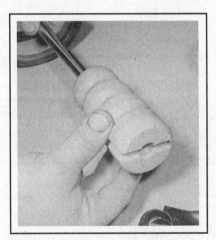

10.5f ... and bump stop

10.11 When installing the spring, make sure the end fits into the recessed portion of the upper seat

10-12 SUSPENSION AND STEERING SYSTEMS

11 Stabilizer bar, bushings and links (rear) - removal and installation

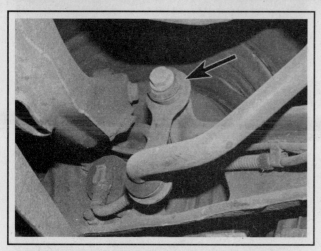

11.3 Remove the stabilizer bar link-to-trailing arm bolt

♦ Refer to illustrations 11.3 and 11.4

※ WARNING:
The manufacturer recommends replacing all the stabilizer bar fasteners with new ones whenever they are removed.

1 Loosen the rear wheel lug nuts. Chock the front wheels to keep the vehicle from rolling, then raise the rear of the vehicle and support it securely on jackstands. Remove the rear wheels.
2 Remove the hangers for the exhaust system (see Chapter 4), and the fuel tank heat shield (see illustration 12.3).
3 Detach the left and right side stabilizer bar link-to-trailing arm bolts (see illustration).
4 Unbolt the stabilizer bar bushing clamp fasteners from the rear suspension crossmember (see illustration), then remove the stabilizer bar.
5 Pull the brackets off the stabilizer bar and inspect the bushings for cracks, hardness and other signs of deterioration. If the bushings are damaged, replace them.
6 Installation is the reverse of removal, noting the following points:
   a) *Lubricate the inside and outside of the new bushings with vegetable oil (used in cooking) to simplify reassembly.*

※ CAUTION:
Don't use petroleum or mineral-based lubricants or brake fluid - they will lead to deterioration of the bushings. These bushings are split so that you can install them without having to slide them onto the ends of the stabilizer bar. Install the bushings with the slit in each bushing facing towards the rear of the vehicle.

   b) *Install the link bolts, insulators and nuts, tightening the bolts to the torque listed in this Chapter's Specifications.*

7 Install the wheel and lug nuts, lower the vehicle and tighten the lug nuts to the torque listed in the Chapter 1 Specifications.

11.4 To remove the rear stabilizer bar clamp, unscrew the nut on the other side of this stud, then swing the clamp down and unhook it

12 Suspension arms (rear) - removal and installation

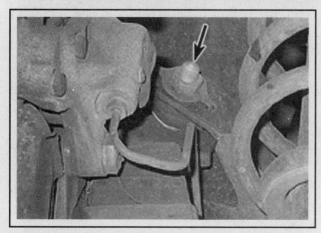

12.2 Remove the bolt securing the upper lateral control arm to the trailing arm

※ WARNING:
The manufacturer recommends replacing all suspension fasteners with new ones whenever they are removed.

1 Loosen the rear wheel lug nuts. Chock the front wheels to keep the vehicle from rolling, then raise the rear of the vehicle and support it securely on jackstands. Remove the rear wheels.

UPPER LATERAL CONTROL ARM

♦ Refer to illustrations 12.2, 12.3 and 12.4

2 Remove the upper control arm-to-trailing arm mounting bolt and nut (see illustration).

SUSPENSION AND STEERING SYSTEMS

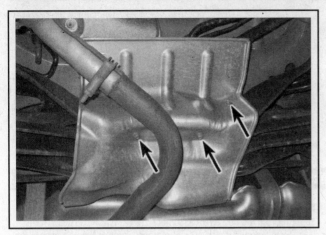

12.3 Remove the fasteners securing the fuel tank heat shield

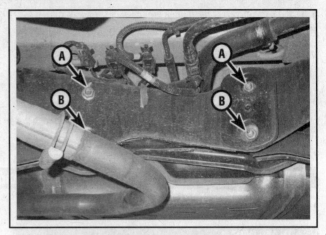

12.4 Upper (A) and lower (B) lateral control arm pivot bolts

3 Remove the fuel tank heat shield from the rear suspension crossmember (see illustration).

4 Remove the fuel tank (see Chapter 4), then remove the pivot bolt from the inner end of the arm (see illustration). Remove the arm from the vehicle.

5 Inspect the control arm pivot bushing for signs of deterioration. If it is in need of replacement, take the control arm to an automotive machine shop to have the bushing replaced.

6 Installation is the reverse of removal. Tighten the fasteners to the torque listed in this Chapter's Specifications.

LOWER LATERAL CONTROL ARM

♦ Refer to illustration 12.7

7 Remove the lower control arm-to-trailing arm mounting bolt and nut (see illustration).

8 Remove the fuel tank heat shield from the rear suspension crossmember (see illustration 12.3).

9 Remove the fuel tank (see Chapter 4), then remove the pivot bolt from the inner end of the arm (see illustration 12.4).

➥ Note: The rear stabilizer bar bushings may have to be removed to allow removal of the bolt.

Remove the arm from the vehicle.

10 Inspect the control arm pivot bushing for signs of deterioration. If it is in need of replacement, take the control arm to an automotive machine shop to have the bushing replaced.

11 Installation is the reverse of removal. Tighten the fasteners to the torque listed in this Chapter's Specifications.

TRAILING ARM

♦ Refer to illustration 12.19

12 Remove the brake shoe or disc brake assembly. If equipped, remove the ABS wheel speed sensor and unbolt the harness brackets from the trailing arm (see Chapter 9).

13 Unbolt the brake hose and brake line brackets from the trailing arm. Unbolt the parking brake cable brackets.

14 Remove the hub and bearing assembly (see Section 13).

15 Separate the brake backing plate from the trailing arm and suspend it out of the way with a piece of wire or string.

16 Remove the shock absorber/coil spring-to-trailing arm mounting bolt (see Section 9).

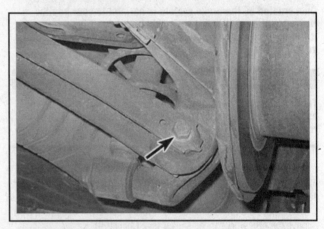

12.7 Remove the bolt securing the lower lateral control arm to the trailing arm

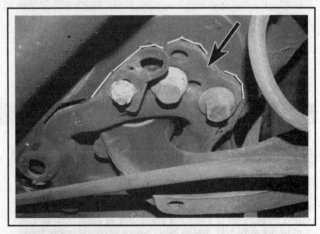

12.19 Mark the relationship of the trailing arm mounting bracket in relation to the vehicle chassis, then remove the mounting bolts

17 Remove the rear stabilizer bar link-to-trailing arm bolt (see Section 11).

18 Support the trailing arm with a jack, then remove the bolts securing the upper and lower lateral control arms to the trailing arm (see illustrations 12.2 and 12.7).

19 Mark the relationship of the trailing arm mounting bracket in relation to the vehicle chassis (see illustration), then remove the bracket

10-14 SUSPENSION AND STEERING SYSTEMS

bolts and carefully lower the trailing arm down and remove it from the vehicle.

20 Inspect the trailing arm pivot bushing for signs of deterioration. If it is in need of replacement, take the trailing arm to an automotive machine shop to have the bushing replaced.

21 Installation is the reverse of removal, noting the following points:

   a) Raise the trailing arm with the jack and align the mounting bracket with the matchmarks on the vehicle chassis and install the mounting bolts but don't tighten them yet.
   b) Install and tighten the upper and lower lateral arm-to-trailing arm mounting fasteners and the trailing arm bracket-to-body bolts to the torque listed in this Chapter's Specifications.
   c) Tighten all other fasteners to the proper torque specifications
   d) It won't be necessary to bleed the brakes unless a hydraulic fitting was loosened.
   e) Have the rear wheel alignment checked and, if necessary, adjusted.

13 Hub and bearing assembly (rear) - removal and installation

▸ Refer to illustration 13.3

※ WARNING 1:

Dust created by the brake system is harmful to your health. Never blow it out with compressed air and don't inhale any of it. Do not, under any circumstances, use petroleum-based solvents to clean brake parts. Use brake system cleaner only.

※ WARNING 2:

The manufacturer recommends replacing the hub and bearing assembly mounting nuts with new ones whenever they are removed.

1 Loosen the rear wheel lug nuts. Chock the front wheels to keep the vehicle from rolling, then raise the rear of the vehicle and support it securely on jackstands. Remove the rear wheels.

2 Remove the brake shoe or disc brake assembly. If equipped, also remove the ABS wheel speed sensor (see Chapter 9).

3 Remove the hub nuts (see illustration) and detach the hub from the knuckle.

4 Check the hub bearing for wear or damage. Spin it with your fingers and check for rough, loose or noisy rotation. The bearing can't be replaced separately, so if the bearing is bad or any other problems are found, replace the hub as an assembly.

5 Installation is the reverse of removal. Tighten the hub nuts to the torque listed in this Chapter's Specifications.

6 Install the wheel and lug nuts, lower the vehicle and tighten the lug nuts to the torque listed in the Chapter 1 Specifications.

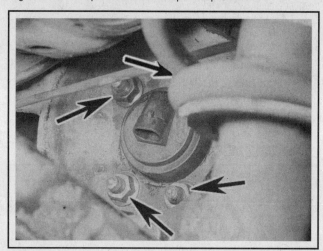

13.3 Remove the four nuts securing the hub to the trailing arm

14 Steering wheel - removal and installation

※ WARNING 1:

These models are equipped with a Supplemental Restraint System (SRS), more commonly known as airbags. Always disable the airbag system before working in the vicinity of any airbag system component to avoid the possibility of accidental deployment of the airbag(s), which could cause personal injury (see Chapter 12).

※ WARNING 2:

Do not use a memory saving device to preserve the PCM or radio memory when working on or near airbag system components.

※ WARNING 3:

The manufacturer recommends replacing the steering wheel mounting nut with a new one whenever it is removed.

REMOVAL

▸ Refer to illustrations 14.3, 14.4 and 14.9

1 Park the vehicle with the wheels pointing straight ahead. Disconnect the cable from the negative terminal of the battery (see Chapter 5, Section 1).

2 Disable the airbag system (see Chapter 12).

SUSPENSION AND STEERING SYSTEMS   10-15

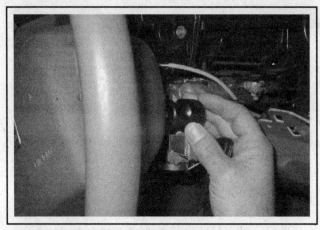

14.3 The airbag is secured by spring clips that engage four posts on the airbag (two posts per side); to release them, insert a flat-blade screwdriver into each slot and twist, then repeat on the other side

14.4 Lift the airbag module off and disconnect the electrical connectors (to detach the airbag connector, flip up the locking tab, then pull the connector straight off)

3   Remove the steering column covers (see Chapter 11). Starting with the steering wheel in the 3 o'clock position, detach the airbag module by inserting a flat-bladed screwdriver into the slots in the backside of the steering wheel, then twist the screwdriver to spread the retaining springs, which will release the airbag retaining posts (see illustration). Turn the steering wheel to the 9 o'clock position and repeat the procedure. Return the steering wheel to the original position with the wheels pointing straight ahead.

4   Lift the airbag module carefully away from the steering wheel and disconnect the electrical connectors (see illustration).

5   Remove the airbag module.

※※ WARNING:

When carrying the airbag module, keep the driver's side of it away from your body, and when you set it down (in an isolated area), have the driver's side facing up.

6   Remove the steering wheel nut, then mark the relationship of the steering wheel to the shaft.

7   Remove the steering wheel using a steering wheel puller (see illustration). The puller screw must be contacting the steering wheel bolt or shaft. Disconnect any remaining electrical connectors.

※※ CAUTION 1:

Don't thread the bolts of the puller into the steering wheel more than five turns, as they could contact the airbag clockspring and damage it.

※※ CAUTION 2:

While the steering wheel is removed, DO NOT turn the steering shaft. If you do so, the airbag clockspring could be damaged. Make a mark indicating the relationship of the steering wheel hub to the steering shaft

8   Remove the puller and lift the steering wheel off the shaft, feeding the wiring harness through the hole in the wheel.

9   If it is necessary to remove the clockspring, unplug the electrical connector for the clockspring, then detach the clockspring from the clips and remove it from the steering wheel (see illustration).

14.9 Disconnect the clockspring electrical connector

INSTALLATION

10   When installing the clockspring, make absolutely sure that the airbag clockspring is centered with the arrow on the clockspring pointing up. This shouldn't be a problem as long as you have not turned the steering shaft while the wheel was removed. If for some reason the shaft was turned, center the clockspring as follows:

  a) *Rotate the clockspring clockwise until it stops (don't apply too much force, though).*
  b) *Rotate the clockspring counterclockwise about 2-1/2 turns until the arrow on the clockspring points straight up.*

11   Installation is the reverse of removal, noting the following points:

  a) *Make sure the airbag clockspring is centered before installing the steering wheel.*
  b) *When installing the steering wheel, align the marks on the shaft and the steering wheel hub.*
  c) *Install then tighten a NEW steering wheel nut to the torque listed in this Chapter's Specifications.*
  d) *Install the airbag module on the steering wheel and push it into place until the retaining posts engage with the retaining springs.*
  e) *Enable the airbag system (see Chapter 12).*

10-16 SUSPENSION AND STEERING SYSTEMS

15 Steering column - removal and installation

※ WARNING 1:

These models are equipped with airbags. Always disable the airbag system before working in the vicinity of any airbag system component to avoid the possibility of accidental deployment of the airbag(s), which could cause personal injury (see Chapter 12).

※ WARNING 2:

Do not use a memory saving device to preserve the PCM's memory when working on or near airbag system components.

REMOVAL

▶ Refer to illustration 15.7

1 Park the vehicle with the wheels in the straight-ahead position. Disconnect the cable from the negative terminal of the battery (see Chapter 5, Section 1). Disable the airbag system (see Chapter 12).
2 Remove the steering column covers (see Chapter 11).
3 Remove the steering wheel and the airbag clockspring (see Section 14).
4 Remove the lower instrument panel trim (under the steering column), the knee bolster and the heater/air conditioning duct (see Chapter 11).
5 Disconnect the electrical connecter for the shift interlock solenoid from the ignition lock cylinder.
6 Remove the steering column switches (see Chapter 12).
7 Mark the relationship of the steering column shaft to the intermediate shaft, then remove the pinch bolt (see illustration).
8 Mark and disconnect any electrical connectors that would interfere with removal.
9 Remove the steering column mounting fasteners, then guide the column out from the instrument panel.

INSTALLATION

10 Guide the column into position, connecting the steering shaft with the intermediate shaft. Be sure to align the marks made in Step 7.
11 Install the mounting fasteners, tightening them to the torque listed in this Chapter's Specifications.
12 Install the pinch bolt and tighten it to the torque listed in this Chapter's Specifications.
13 The remainder of installation is the reverse of the removal procedure. Refer to Section 14 for the clockspring, steering wheel and airbag module installation details.

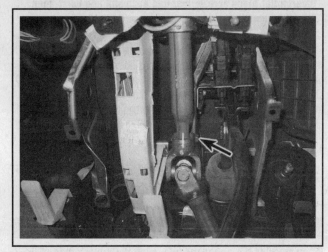

15.7 Mark the relationship of the two shafts, then remove the pinch bolt

16 Tie-rod ends - removal and installation

REMOVAL

▶ Refer to illustrations 16.2, 16.3 and 16.4

1 Loosen the wheel lug nuts, raise the front of the vehicle and support it securely on jackstands. Apply the parking brake and block the rear wheels to keep the vehicle from rolling off the jackstands. Remove the wheel.
2 Loosen the tie-rod end jam nut (see illustration).

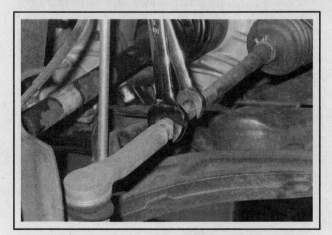

16.2 Using two wrenches, loosen the jam nut

SUSPENSION AND STEERING SYSTEMS

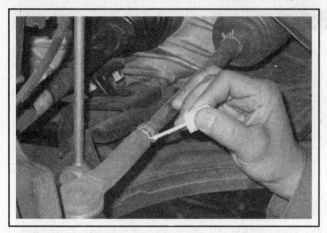

16.3 Mark the position of the tie-rod end in relation to the threads

16.4 Disconnect the tie-rod end from the steering knuckle arm with a puller

3 Mark the relationship of the tie-rod end to the threaded portion of the tie-rod. This will ensure the toe-in setting is restored when reassembled (see illustration).

4 Remove the cotter pin and loosen the nut from the tie-rod end ballstud a few turns. Disconnect the tie-rod end ballstud from the steering knuckle arm with a puller (see illustration).

5 Remove the nut from the ballstud, separate the tie-rod end from the steering knuckle, then unscrew the tie-rod end from the tie-rod.

INSTALLATION

6 Thread the tie-rod end onto the tie-rod to the marked position and connect the tie-rod end to the steering arm. Install the nut on the ballstud and tighten it to the torque listed in this Chapter's Specifications. Install a new cotter pin.

➡**Note: If necessary, tighten the nut a little more to allow insertion of the cotter pin. Never loosen the nut to align the cotter pin holes.**

7 Tighten the jam nut securely and install the wheel. Lower the vehicle and tighten the lug nuts to the torque listed in the Chapter 1 Specifications.

8 Have the front end alignment checked and, if necessary, adjusted.

17 Steering gear boots - removal and installation

▶ Refer to illustrations 17.3a and 17.3b

1 Loosen the lug nuts, raise the vehicle and support it securely on jackstands. Remove the wheel.

2 Remove the tie-rod end and jam nut (see Section 16).

3 Remove the outer steering gear boot clamp with a pair of pliers (see illustration). Cut off the inner boot clamp with a pair of diagonal cutters (see illustration). Slide off the boot.

4 Before installing the new boot, wrap the threads and serrations on the end of the steering rod with a layer of tape so the small end of the new boot isn't damaged.

5 Slide the new boot into position on the steering gear until it seats in the groove in the steering rod and install new clamps.

6 Remove the tape and install the tie-rod end (see Section 16).

7 Install the wheel and lug nuts. Lower the vehicle and tighten the lug nuts to the torque listed in the Chapter 1 Specifications.

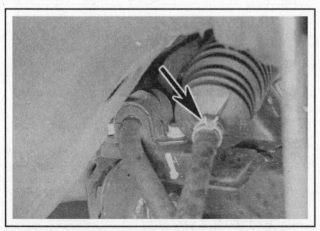

17.3a The outer ends of the steering gear boots are secured by band-type clamps; they're easily released with a pair of pliers

17.3b The inner ends of the of the steering gear boots are retained by boot clamps which may be cut off and discarded

10-18 SUSPENSION AND STEERING SYSTEMS

18 Steering gear - removal and installation

18.3 Remove the rear mount through-bolt and mount-to-frame bolt

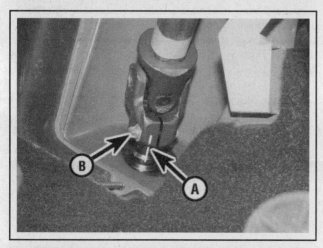

18.4 Mark the relationship of the universal joint to the steering gear input shaft (A) and remove the U-joint pinch bolt (B)

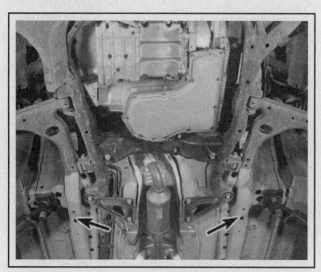

18.5 Support the vehicle with jackstands placed on the unibody structure, behind the subframe

✲✲ WARNING 1:

Make sure the steering shaft is not turned while the steering gear is removed or you could damage the airbag system clockspring. To prevent the shaft from turning, place the ignition key in the LOCK position or thread the seat belt through the steering wheel and clip it into place.

✲✲ WARNING 2:

The manufacturer recommends replacing the subframe bolts, cage nuts and steering gear mounting fasteners with new ones whenever they are removed.

REMOVAL

▶ Refer to illustrations 18.3, 18.4, 18.5, 18.14, 18.15 and 18.16

✲✲ WARNING:

Do not place any part of your body under the transaxle assembly, engine or subframe when it's supported only by a hoist or other lifting device.

1 Disconnect the cable from the negative battery terminal (see Chapter 5, Section 1).

2 Drain the power steering fluid from the remote power steering reservoir. This can be accomplished with a suction gun or large syringe, or by disconnecting the fluid hose and draining the fluid into a container.

3 Remove the transaxle rear mount through-bolt and the mount-to-frame bolt (see illustration).

4 From inside the vehicle under the dashboard, mark the relationship of the U-joint to the steering gear input shaft, then remove the pinch bolt securing the U-joint to the steering gear input shaft (see illustration).

5 Loosen the front wheel lug nuts, raise the front of the vehicle and support it securely on jackstands. Remove both front wheels.

➡ Note: The jackstands must be behind the front suspension subframe, not supporting the vehicle by the subframe (see illustration).

6 Detach the tie-rod ends from the steering knuckles (see Section 16).

7 On V6 models and four-cylinder models with a manual transaxle, remove the rear exhaust manifold pipe (see Chapter 6, Section 19).

8 Place a drain pan under the steering gear and detach the power steering pressure and return lines. Cap the ends to prevent excessive fluid loss and contamination.

9 Attach an engine hoist or an engine support fixture to the engine lifting hooks.

10 Using two floor jacks, support the subframe. Position one jack on each side of the subframe, midway between the front and rear mounting points.

11 Disconnect the stabilizer bar links from the struts (see Section 4).

12 On four-cylinder models with a manual transaxle, disconnect the frame-to-shifter link (see Chapter 7A).

13 Remove the bolts securing the upper transaxle mount to the body.

SUSPENSION AND STEERING SYSTEMS 10-19

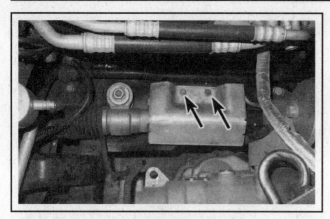

18.14 Remove the bolts securing the heat shield

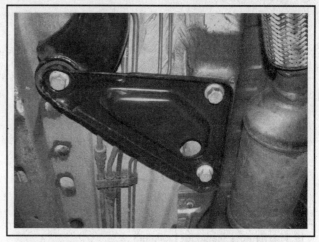

18.15 Remove the bolts securing the subframe supports

14 If equipped, remove the steering gear heat shield (see illustration), then remove the steering gear mounting bolts.

15 Remove the bolts securing the subframe supports (see illustration), then remove the supports.

16 Carefully mark the position of the subframe in relation to the vehicle chassis. With the jacks sufficiently supporting the subframe, loosen the four subframe-to-chassis mounting bolts (see illustration). Do not remove the bolts.

17 Lower the jacks until the subframe is sufficiently resting on the bolts that were loosened, enough to allow the steering gear to be removed through the left wheel opening.

INSTALLATION

18 Installation is the reverse of removal, noting the following points:
  a) Replace the steering gear mounting bolts/nuts with new ones. Tighten them to the torque listed in this Chapter's Specifications.
  b) Replace the subframe bolts and cage nuts with new ones. Tighten the subframe mounting bolts to the torque listed in this Chapter.
  c) Fill the power steering pump with the recommended fluid (see Chapter 1), bleed the system (see Section 18) and recheck the fluid level. Check for leaks.
  d) Run the engine and check for proper operation and leaks. Shut off the engine and recheck fluid levels.

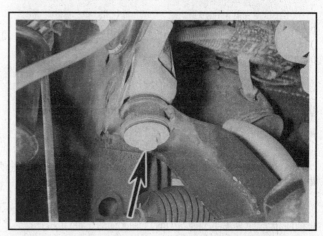

18.16 Remove the subframe-to-chassis mounting bolts

  e) Reconnect the negative battery cable (see Chapter 5, Section 1).
  f) Have the front end alignment checked and, if necessary, adjusted.

19 Power steering pump - removal and installation

REMOVAL

1 Disconnect the cable from the negative battery terminal (see Chapter 5, Section 1).

2 Using a large syringe or suction gun, suck as much fluid out of the power steering fluid reservoir as possible. Place a drain pan under the vehicle to catch any fluid that spills out when the hoses are disconnected.

Four-cylinder engine

▶ Refer to illustration 19.3

3 Disconnect the return hose from the power steering pump, then using a flare-nut wrench, unscrew the pressure line fitting from the pump (see illustration).

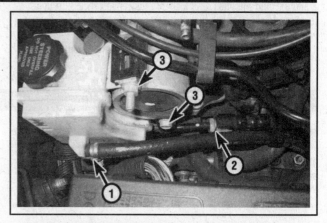

19.3 Power steering pump details

1　Return hose and clamp
2　Pressure hose
3　Mounting bolts

10-20 SUSPENSION AND STEERING SYSTEMS

19.11a Remove the pulley fasteners . . .

19.11b . . . then remove the pulley

4 Unscrew and remove the two mounting bolts, and remove the power steering pump (see illustration 19.3).

5 Remove the retaining clips and detach the reservoir from the pump.

6 Remove the reservoir-to-pump O-ring. Check the condition of the O-ring, replacing it if necessary.

V6 engine

♦ **Refer to illustrations 19.11a and 19.11b**

7 Remove the air filter housing (see Chapter 4).

8 Loosen the three bolts securing the power steering pump pulley to the pump.

9 Remove the drivebelt (see Chapter 1).

10 Disconnect the return hose from the power steering pump, then using a flare-nut wrench, unscrew the pressure line fitting from the pump.

11 Remove the power steering pump pulley (see illustrations), then remove three bolts securing the pump to the mounting bracket and remove the pump.

12 If you're installing a new pump, it may be necessary to transfer the pulley hub to the new pump. You'll need a special puller to remove the pulley hub, from the old pump, and another special tool to install it on the new pump. These tools are available at most auto parts stores.

➥**Note: Before removing the pulley hub from the old pump, measure the distance the pulley hub is pressed onto the pump shaft; the hub must be reinstalled to the same depth.**

INSTALLATION

13 Installation is the reverse of removal, noting the following points:
   a) *Tighten the bolts and fittings securely.*
   b) *The O-ring on the high-pressure outlet should be replaced.*
   c) *Install the drivebelt as outlined in Chapter 1.*
   d) *Fill the power steering reservoir with the recommended fluid* (see Chapter 1). *Bleed the power steering hydraulic system as described in Section 20.*

20 Power steering system - bleeding

1 The power steering system must be bled whenever a line is disconnected. Bubbles can be seen in power steering fluid that has air in it and the fluid will often have a tan or milky appearance. Low fluid level can cause air to mix with the fluid, resulting in a noisy pump as well as foaming of the fluid.

2 Open the hood and check the fluid level in the reservoir, adding the specified fluid necessary to bring it up to the proper level (see Chapter 1).

3 Start the engine and slowly turn the steering wheel several times from left-to-right and back again. Do not turn the wheel completely from lock-to-lock. Check the fluid level, topping it up as necessary until it remains steady and no more bubbles are visible.

21 Subframe - removal and installation

✱✱ WARNING:

The manufacturer recommends replacing the subframe bolts, cage nuts and steering gear mounting fasteners with new ones whenever they are removed.

1 Disconnect the cable from the negative battery terminal (see Chapter 5, Section 1).

2 Remove the transaxle rear mount (see illustration 18.3).

3 Disconnect the steering column U-joint from the steering gear input shaft (see Section 18).

4 Loosen the front wheel lug nuts, raise the front of the vehicle and support it securely on jackstands. Remove both front wheels.

➥**Note: The jackstands must be behind the front suspension subframe, not supporting the vehicle by the subframe (see illustration 18.5).**

5 To support the radiator during subframe removal, fasten the radiator to the upper radiator support.

SUSPENSION AND STEERING SYSTEMS

6 Detach the tie-rod ends from the steering knuckles (see Section 16).

7 Disconnect the stabilizer bar links from the struts (see Section 4).

8 On four-cylinder models with a manual transaxle, disconnect the frame-to-shifter link (see Chapter 7A).

9 On V6 models and four-cylinder models with a manual transaxle, remove the rear exhaust manifold pipe (see Chapter 6, Section 19).

10 Disconnect the control arms from the steering knuckles (see Section 5).

11 If equipped, remove the steering gear heat shield, then remove the steering gear mounting bolts from the subframe (see Section 18).

➡ **Note:** *The power steering gear will have to be supported using rope, later, when the subframe assembly is being lowered.*

12 Remove the bolts securing the front transaxle mount to the subframe support.

13 Carefully mark the position of the subframe in relation to the vehicle chassis.

14 Using two floor jacks, support the subframe. Position one jack on each side of the subframe, midway between the front and rear mounting points.

15 Roll an engine hoist into position and attach it to the engine with a couple pieces of heavy-duty chain. If the engine is equipped with lifting brackets, use them. If not, you'll have to fasten the chain to some substantial part of the engine - one that is strong enough to take the weight, but in a location that will provide good balance. If you're attaching the chain to a stud on the engine, or are using a bolt passing through the chain and into a threaded hole, place a washer between the nut or bolt head and the chain, and tighten the nut or bolt securely. Take up the slack in the chain, but don't lift the engine.

※ WARNING:

DO NOT place any part of your body under the engine when it's supported only by a hoist or other lifting device.

16 Remove the bolts securing the subframe supports (see illustration 18.15), then remove the supports.

17 With the jacks sufficiently supporting the subframe, remove the four subframe-to-chassis mounting bolts (see illustration 18.16).

18 Lower the jacks until the subframe is sufficiently resting on the ground.

INSTALLATION

19 Installation is the reverse of removal, noting the following points:
   a) *Replace the steering gear mounting bolts/nuts with new ones. Tighten them to the torque listed in Chapter's Specifications.*
   b) *Replace the subframe bolts and cage nuts with new ones. Align the reference marks on the subframe then tighten the subframe mounting bolts to the torque listed in this Chapter.*
   c) *Reconnect the negative battery cable (see Chapter 5, Section 1).*
   d) *Have the front end alignment checked and, if necessary, adjusted.*

22 Wheels and tires - general information

▶ **Refer to illustration 22.1**

1 All vehicles covered by this manual are equipped with metric-sized fiberglass or steel belted radial tires (see illustration). Use of other size or type of tires may affect the ride and handling of the vehicle. Don't mix different types of tires, such as radials and bias belted, on the same vehicle as handling may be seriously affected. It's recommended that tires be replaced in pairs on the same axle, but if only one tire is being replaced, be sure it's the same size, structure and tread design as the other.

2 Because tire pressure has a substantial effect on handling and wear, the pressure on all tires should be checked at least once a month or before any extended trips (see Chapter 1).

3 Wheels must be replaced if they are bent, dented, leak air, have elongated bolt holes, are heavily rusted, out of vertical symmetry or if the lug nuts won't stay tight. Wheel repairs that use welding or peening are not recommended.

4 Tire and wheel balance is important in the overall handling, braking and performance of the vehicle. Unbalanced wheels can adversely affect handling and ride characteristics as well as tire life. Whenever a tire is installed on a wheel, the tire and wheel should be balanced by a shop with the proper equipment.

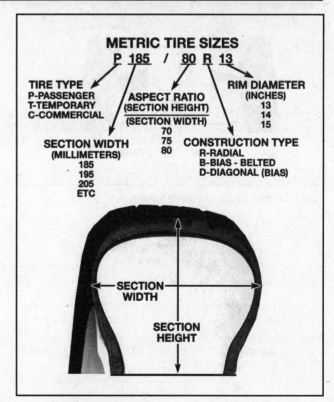

22.1 Metric tire size code

23 Wheel alignment - general information

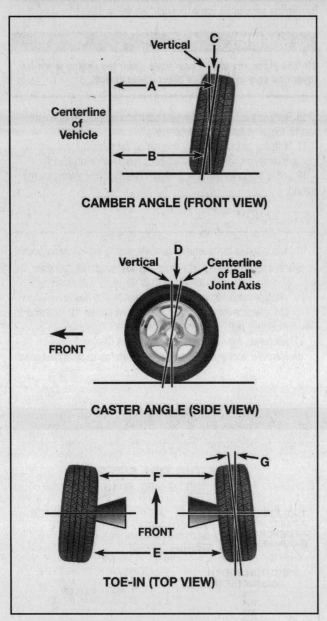

23.1 Camber, caster and toe-in angles

A minus B = C (degrees camber)
D = degrees caster
E minus F = toe-in (measured in inches)
G = toe-in (expressed in degrees)

▶ **Refer to illustration 23.1**

A wheel alignment refers to the adjustments made to the wheels so they are in proper angular relationship to the suspension and the ground. Wheels that are out of proper alignment not only affect vehicle control, but also increase tire wear. The front end angles normally measured are camber, caster and toe-in (see illustration). Toe-in and camber are adjustable; if the caster is not correct, check for bent components. Rear toe-in is also adjustable.

Getting the proper wheel alignment is a very exacting process, one in which complicated and expensive machines are necessary to perform the job properly. Because of this, you should have a technician with the proper equipment perform these tasks. We will, however, use this space to give you a basic idea of what is involved with a wheel alignment so you can better understand the process and deal intelligently with the shop that does the work.

Toe-in is the turning in of the wheels. The purpose of a toe specification is to ensure parallel rolling of the wheels. In a vehicle with zero toe-in, the distance between the front edges of the wheels will be the same as the distance between the rear edges of the wheels. The actual amount of toe-in is normally only a fraction of an inch. On the front end, toe-in is controlled by the tie-rod end position on the tie-rod. On the rear end, it's controlled by the position of the suspension trailing arm bracket on the chassis. Incorrect toe-in will cause the tires to wear improperly by making them scrub against the road surface.

Camber is the tilting of the wheels from vertical when viewed from one end of the vehicle. When the wheels tilt out at the top, the camber is said to be positive (+). When the wheels tilt in at the top the camber is negative (-). The amount of tilt is measured in-degrees from vertical and this measurement is called the camber angle. This angle affects the amount of tire tread which contacts the road and compensates for changes in the suspension geometry when the vehicle is cornering or traveling over an undulating surface. On the front end it is adjusted by altering the position of the strut upper mount in the strut tower.

Caster is the tilting of the front steering axis from the vertical. A tilt toward the rear is positive caster and a tilt toward the front is negative caster.

SUSPENSION AND STEERING SYSTEMS 10-23

Specifications

General
Power steering fluid type ..... See Chapter 1

Torque specifications — Ft-lbs (unless otherwise indicated) / Nm

Front suspension

Item	Ft-lbs	Nm
Strut		
Damper shaft nut*	40	55
Strut upper mounting nut*	40	55
Strut-to-steering knuckle bolts/nuts*		
2001 and earlier models		
Step 1	37	50
Step 2	66	90
Step 3	Tighten an additional 45 to 60-degrees	
2002 and later models		
Step 1	37	50
Step 2	73	100
Step 3	Tighten an additional 30 to 45-degrees	
Stabilizer bar		
Stabilizer bar link nuts*	48	65
Stabilizer bar bracket bolts*	15	20
Control arm		
Arm-to-subframe*		
Step 1	66	90
Step 2	Tighten an additional 45 to 60-degrees	
Balljoint-to-steering knuckle pinch bolt/nut*	75	100
Subframe		
Subframe-to-body bolts*		
Step 1	66	90
Step 2	Tighten an additional 45 to 60-degrees	
Subframe support bracket bolts*		
Step 1	66	90
Step 2	Tighten an additional 45 to 60-degrees	
Driveaxle/hub nut	See Chapter 8	

Rear suspension

Item	Ft-lbs	Nm
Upper/lower lateral control arms-to-suspension crossmember*		
Step 1	66	90
Step 2	Tighten an additional 60 to 75-degrees	
Rear hub nuts*		
Step 1	37	50
Step 2	Tighten an additional 30 to 45-degrees	
Shock carrier-to-body mounting bolts*	40	55
Shock absorber lower mounting bolt*		
Step 1	110	150
Step 2	Tighten an additional 30 to 45-degrees	
Shock absorber damper shaft nut*	15	20
Stabilizer bar		
Clamp bolts*	41	55
Bar link-to-trailing arm bolts*	41	55

10-24 SUSPENSION AND STEERING SYSTEMS

Torque specifications	Ft-lbs (unless otherwise indicated)	Nm
Rear suspension (continued)		
Trailing arm		
Bracket-to-body bolts*		
Step 1	66	90
Step 2	Tighten additional 60 to 75-degrees	
Steering system		
Power steering pump-to-engine mounting fasteners		
Four cylinder models	15	20
V6 models	18	25
Power steering pulley bolts	15	20
Steering gear mounting bolts/nuts*		
Step 1	35	45
Step 2	Tighten an additional 90-degrees	
Tie-rod end-to-steering knuckle nut*	44	60
Intermediate shaft-to-steering column pinch-bolt*	22	30
Steering column mounting fasteners*	22	30
Steering wheel mounting nut*	30	40
Rear hub nuts*		
Step 1	37	50
Step 2	Tighten an additional 30 to 45-degrees	

Fastener(s) must be replaced

11 BODY

1 General information
2 Body - maintenance
3 Vinyl trim - maintenance
4 Upholstery and carpets - maintenance
5 Body repair - minor damage
6 Body repair - major damage
7 Hinges and locks - maintenance
8 Windshield and fixed glass - replacement
9 Hood - removal, installation and adjustment
10 Hood latch handle and release cable - removal and installation
11 Bumpers - removal and installation
12 Front fender - removal and installation
13 Radiator grille - removal and installation
14 Cowl cover - removal and installation
15 Door trim panels - removal and installation
16 Door - removal, installation and adjustment
17 Door latch, lock cylinder and handles - removal and installation
18 Door window glass - removal and installation
19 Door window glass regulator - removal and installation
20 Door outer panel - removal and installation
21 Mirrors - removal and installation
22 Liftgate - removal, installation and adjustment
23 Liftgate latch, lock cylinder and handle - removal and installation
24 Center console - removal and installation
25 Dashboard trim panels
26 Steering column covers - removal and installation
27 Seats - removal and installation

11-2 BODY

1 General information

> **※ WARNING:**
> The models covered by this manual are equipped with Supplemental Restraint systems (SRS), more commonly known as airbags. Always disarm the airbag system before working in the vicinity of any airbag system component to avoid the possibility of accidental deployment of the airbag, which could cause personal injury (see Chapter 12). Do not use a memory saving device to preserve the PCM's memory when working on or near airbag system components.

These models feature a "unibody" layout, using a floor pan with integral side frame rails which support the body components, front and rear suspension systems and other mechanical components.

Certain components are particularly vulnerable to accident damage and can be unbolted and repaired or replaced. Among these parts are the body moldings, bumpers, front fenders, the hood and trunk lid, doors and all glass.

The doors, front fenders and bumper covers are made of a polymer and are designed to withstand minor impacts without damage. The hood, quarter panels, roof and rear deck (trunk lid) or liftgate are steel.

Only general body maintenance practices and body panel repair procedures within the scope of the do-it-yourselfer are included in this Chapter.

2 Body - maintenance

1 The condition of your vehicle's body is very important, because the resale value depends a great deal on it. It's much more difficult to repair a neglected or damaged body than it is to repair mechanical components. The hidden areas of the body, such as the wheel wells, the frame and the engine compartment, are equally important, although they don't require as frequent attention as the rest of the body.

2 Once a year, or every 12,000 miles, it's a good idea to have the underside of the body steam-cleaned. All traces of dirt and oil will be removed and the area can then be inspected carefully for rust, damaged brake lines, frayed electrical wires, damaged cables and other problems.

3 At the same time, clean the engine and the engine compartment with a steam cleaner or water-soluble degreaser.

4 The wheel wells should be given close attention, since undercoating can peel away and stones and dirt thrown up by the tires can cause the paint to chip and flake, allowing rust to set in. If rust is found, clean down to the bare metal and apply an anti-rust paint.

5 The body should be washed about once a week. Wet the vehicle thoroughly to soften the dirt, then wash it down with a soft sponge and plenty of clean soapy water. If the surplus dirt is not washed off very carefully, it can wear down the paint.

6 Spots of tar or asphalt thrown up from the road should be removed with a cloth soaked in kerosene. Scented lamp oil is available in most hardware stores and the smell is easier to work with than straight kerosene.

7 Once every six months, wax the body and chrome trim. If a chrome cleaner is used to remove rust from any of the vehicle's plated parts, remember that the cleaner also removes part of the chrome, so use it sparingly. On any plated parts where chrome cleaner is used, use a good paste wax over the plating for extra protection.

3 Vinyl trim - maintenance

Don't clean vinyl trim with detergents, caustic soap or petroleum-based cleaners. Plain soap and water works just fine, with a soft brush to clean dirt that may be ingrained. Wash the vinyl as frequently as the rest of the vehicle.

After cleaning, application of a high quality rubber and vinyl protectant will help prevent oxidation and cracks. The protectant can also be applied to weatherstripping, vacuum lines and rubber hoses, which often fail as a result of chemical degradation, and to the tires.

4 Upholstery and carpets - maintenance

1 Every three months remove the floormats and clean the interior of the vehicle (more frequently if necessary). Use a stiff whisk broom to brush the carpeting and loosen dirt and dust, then vacuum the upholstery and carpets thoroughly, especially along seams and crevices.

2 Dirt and stains can be removed from carpeting with basic household or automotive carpet shampoos available in spray cans. Follow the directions and vacuum again, then use a stiff brush to bring back the "nap" of the carpet.

3 Most interiors have cloth or vinyl upholstery, either of which can be cleaned and maintained with a number of material-specific cleaners or shampoos available in auto supply stores. Follow the directions on the product for usage, and always spot-test any upholstery cleaner on an inconspicuous area (bottom edge of a backseat cushion) to ensure that it doesn't cause a color shift in the material.

4 After cleaning, vinyl upholstery should be treated with a protectant.

➡ **Note:** Make sure the protectant container indicates the product can be used on seats - some products may make a seat too slippery.

BODY 11-3

> ※※ **CAUTION:**
> Do not use protectant on steering wheels.

5 Leather upholstery requires special care. It should be cleaned regularly with saddlesoap or leather cleaner. Never use alcohol, gasoline, nail polish remover or thinner to clean leather upholstery.

6 After cleaning, regularly treat leather upholstery with a leather conditioner, rubbed in with a soft cotton cloth. Never use car wax on leather upholstery.

7 In areas where the interior of the vehicle is subject to bright sunlight, cover leather seating areas of the seats with a sheet if the vehicle is to be left out for any length of time.

5 Body repair - minor damage

TPO FLEXIBLE PANELS (FRONT AND REAR BUMPER COVERS)

➡ **Note 1:** The following repair procedure applies to the bumper covers, fender splash shields and rocker panels, all of which are made of this material.

➡ **Note 2:** Below is a list of the equipment and materials necessary to perform the following repair procedures. Although a specific brand of material may be mentioned, it should be noted that equivalent products from other manufacturers may be used instead.

> Wax, grease and silicone removing solvent
> Cloth-backed body tape
> Sanding discs
> Drill motor with three-inch disc holder
> Hand sanding block
> Rubber squeegees
> Sandpaper
> Non-porous mixing palette
> Wood paddle or putty knife
> Curved tooth body file
> Flexible parts repair material

1 Remove the damaged panel, if necessary or desirable. In most cases, repairs can be carried out with the panel installed.

2 Clean the area(s) to be repaired with a wax, grease and silicone removing solvent applied with a water-dampened cloth.

3 If the damage is structural, that is, if it extends through the panel, clean the backside of the panel area to be repaired as well. Wipe dry.

4 Sand the rear surface about 1-1/2 inches beyond the break.

5 Cut two pieces of fiberglass cloth large enough to overlap the break by about 1-1/2 inches. Cut only to the required length.

6 Mix the adhesive from the 3M #5900 kit according to the instructions included with the kit, and apply a layer of the mixture approximately 1/8-inch thick on the backside of the panel. Overlap the break by at least 1-1/2 inches

7 Apply one piece of fiberglass cloth to the adhesive and cover the cloth with additional adhesive. Apply a second piece of fiberglass cloth to the adhesive and immediately cover the cloth with additional adhesive in sufficient quantity to fill the weave.

8 Allow the repair to cure for 20 to 30 minutes at 60-degrees to 80-degrees F.

9 If necessary, trim the excess repair material at the edge.

10 Remove all of the paint film over and around the area(s) to be repaired. The repair material should not overlap the painted surface.

11 With a drill motor and a sanding disc (or a rotary file), cut a "V" along the break line approximately 1/2-inch wide. Remove all dust and loose particles from the repair area.

12 Mix and apply the repair material. Apply a light coat first over the damaged area; then continue applying material until it reaches a level slightly higher than the surrounding finish.

13 Cure the mixture for 20 to 30 minutes at 60-degrees to 80-degrees F.

14 Roughly establish the contour of the area being repaired with a body file. If low areas or pits remain, mix and apply additional adhesive.

15 Block sand the damaged area with sandpaper to establish the actual contour of the surrounding surface.

16 If desired, the repaired area can be temporarily protected with several light coats of primer. Because of the special paints and techniques required for flexible body panels, it is recommended that the vehicle be taken to a paint shop for completion of the body repair.

RIGID PLASTIC PANELS (FRONT FENDERS AND DOORS)

➡ **Note 1:** Repairs are most effective when the temperature is approximately 60 to 80-degrees F.

➡ **Note 2:** The following repair procedure is for minor scratches and gouges. Repair of more serious damage should be left to a dealer service department or qualified auto body shop.

➡ **Note 3:** Below is a list of the equipment and materials necessary to perform the following repair procedures. Although a specific brand of material may be mentioned, it should be noted that equivalent products from other manufacturers may be used instead.

> Sanding discs
> Drill motor with three-inch disc holder
> Hand sanding block
> Rubber squeegees
> Sandpaper
> Non-porous mixing palette
> Wood paddle or putty knife
> Rigid plastic repair material

17 Clean the repair area. Start with soap and water, then finish with isopropyl alcohol.

18 Using 80-grit sandpaper, taper the scratch or gouge about 1-1/2 inches around the scratch or gouge. Be sure to penetrate well into the plastic material; a D-A-type power sander and/or coarser-grit sandpaper may help.

19 Using 220-grit sandpaper, feather-edge the paint around the repair area. Sand the repair area with 180-grit sandpaper.

20 Wipe the repair area clean with a clean tack rag or other clean, dry rag.

These photos illustrate a method of repairing simple dents. They are intended to supplement Body repair - minor damage in this Chapter and should not be used as the sole instructions for body repair on these vehicles.

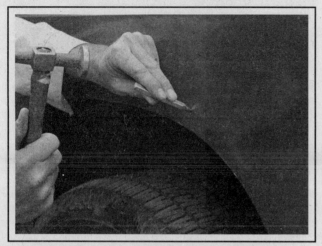

1 If you can't access the backside of the body panel to hammer out the dent, pull it out with a slide-hammer-type dent puller. In the deepest portion of the dent or along the crease line, drill or punch hole(s) at least one inch apart . . .

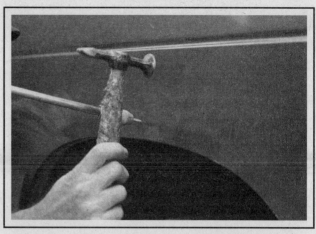

2 . . . then screw the slide-hammer into the hole and operate it. Tap with a hammer near the edge of the dent to help 'pop' the metal back to its original shape. When you're finished, the dent area should be close to its original contour and about 1/8-inch below the surface of the surrounding metal

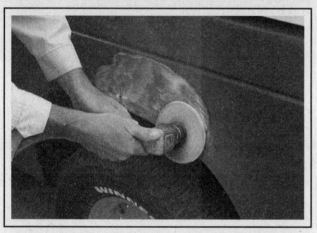

3 Using coarse-grit sandpaper, remove the paint down to the bare metal. Hand sanding works fine, but the disc sander shown here makes the job faster. Use finer (about 320-grit) sandpaper to feather-edge the paint at least one inch around the dent area

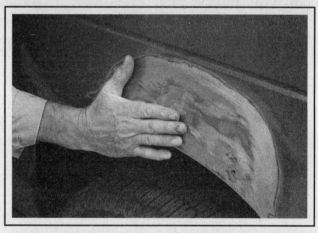

4 When the paint is removed, touch will probably be more helpful than sight for telling if the metal is straight. Hammer down the high spots or raise the low spots as necessary. Clean the repair area with wax/silicone remover

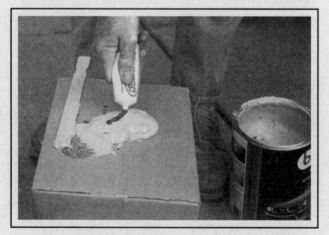

5 Following label instructions, mix up a batch of plastic filler and hardener. The ratio of filler to hardener is critical, and, if you mix it incorrectly, it will either not cure properly or cure too quickly (you won't have time to file and sand it into shape)

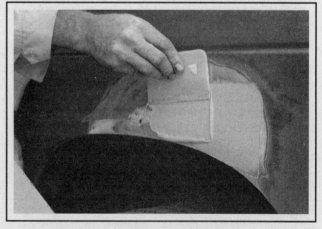

6 Working quickly so the filler doesn't harden, use a plastic applicator to press the body filler firmly into the metal, assuring it bonds completely. Work the filler until it matches the original contour and is slightly above the surrounding metal

7 Let the filler harden until you can just dent it with your fingernail. Use a body file or Surform tool (shown here) to rough-shape the filler

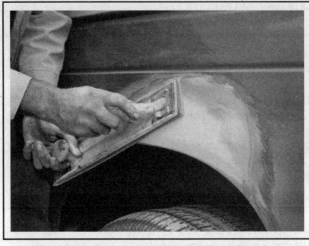

8 Use coarse-grit sandpaper and a sanding board or block to work the filler down until it's smooth and even. Work down to finer grits of sandpaper - always using a board or block - ending up with 360 or 400 grit

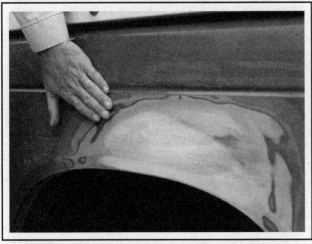

9 You shouldn't be able to feel any ridge at the transition from the filler to the bare metal or from the bare metal to the old paint. As soon as the repair is flat and uniform, remove the dust and mask off the adjacent panels or trim pieces

10 Apply several layers of primer to the area. Don't spray the primer on too heavy, so it sags or runs, and make sure each coat is dry before you spray on the next one. A professional-type spray gun is being used here, but aerosol spray primer is available inexpensively from auto parts stores

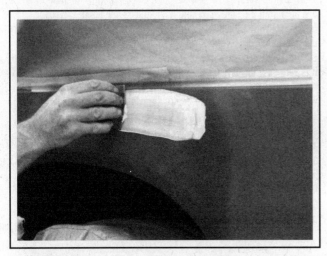

11 The primer will help reveal imperfections or scratches. Fill these with glazing compound. Follow the label instructions and sand it with 360 or 400-grit sandpaper until it's smooth. Repeat the glazing, sanding and respraying until the primer reveals a perfectly smooth surface

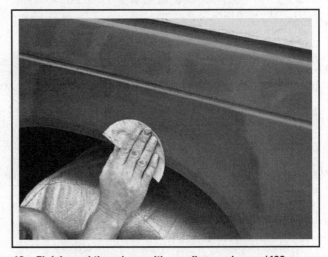

12 Finish sand the primer with very fine sandpaper (400 or 600-grit) to remove the primer overspray. Clean the area with water and allow it to dry. Use a tack rag to remove any dust, then apply the finish coat. Don't attempt to rub out or wax the repair area until the paint has dried completely (at least two weeks)

11-6 BODY

➡ **Note: The filler will adhere better to a smooth surface than a rough surface.**

21 According to label instructions, mix 3M Rigid Plastic Repair Material (RPRM), or equivalent, and apply it to the repair area with a squeegee or plastic spreader. Build the material slightly higher than the surrounding undamaged surface. Allow the material to cure approximately 20 to 30 minutes.

22 Using a sanding block, sand the repair area with 180-grit sandpaper, followed by 240-grit sandpaper.

23 Fill scratches and pinholes in the repair area with a light coat of 3M RPRM and sand lightly, as described in the previous Step.

24 Apply a double wet coat of Glasurit Glassohyd Sealer, or equivalent, over the repair area. Allow it to dry for 1 to 2 hours. Sand lightly with 320-grit sandpaper, then wipe with a dry cloth. Because of the special paints used on these panels, it's recommended that you allow a dealer service department or qualified auto body shop to apply the final paint.

STEEL PANELS (HOOD, TRUNK LID, QUARTER PANELS AND ROOF)

Repair of rust holes or gashes

25 Remove all paint from the affected area and from an inch or so of the surrounding metal using a sanding disk or wire brush mounted in a drill motor. If these are not available, a few sheets of sandpaper will do the job just as effectively.

26 With the paint removed, you will be able to determine the severity of the corrosion and decide whether to replace the whole panel, if possible, or repair the affected area. New body panels are not as expensive as most people think and it is often quicker to install a new panel than to repair large areas of rust.

27 Remove all trim pieces from the affected area except those which will act as a guide to the original shape of the damaged body, such as headlight shells, etc. Using metal snips or a hacksaw blade, remove all loose metal and any other metal that is badly affected by rust. Hammer the edges of the hole in to create a slight depression for the filler material.

28 Wire brush the affected area to remove the powdery rust from the surface of the metal. If the back of the rusted area is accessible, treat it with rust inhibiting paint.

29 Before filling is done, block the hole in some way. This can be done with sheet metal riveted or screwed into place, or by stuffing the hole with wire mesh.

30 Once the hole is blocked off, the affected area can be filled and painted. See the following subsection on *filling and painting*.

Filling and painting

31 Many types of body fillers are available, but generally speaking, body repair kits which contain filler paste and a tube of resin hardener are best for this type of repair work. A wide, flexible plastic or nylon applicator will be necessary for imparting a smooth and contoured finish to the surface of the filler material. Mix up a small amount of filler on a clean piece of wood or cardboard (use the hardener sparingly). Follow the manufacturer's instructions on the package, otherwise the filler will set incorrectly.

32 Using the applicator, apply the filler paste to the prepared area. Draw the applicator across the surface of the filler to achieve the desired contour and to level the filler surface. As soon as a contour that approximates the original one is achieved, stop working the paste. If you continue, the paste will begin to stick to the applicator. Continue to add thin layers of paste at 20-minute intervals until the level of the filler is just above the surrounding metal.

33 Once the filler has hardened, the excess can be removed with a body file. From then on, progressively finer grades of sandpaper should be used, starting with a 180-grit paper and finishing with 600-grit wet-or-dry paper. Always wrap the sandpaper around a flat rubber or wooden block, otherwise the surface of the filler will not be completely flat. During the sanding of the filler surface, the wet-or-dry paper should be periodically rinsed in water. This will ensure that a very smooth finish is produced in the final stage.

34 At this point, the repair area should be surrounded by a ring of bare metal, which in turn should be encircled by the finely feathered edge of good paint. Rinse the repair area with clean water until all of the dust produced by the sanding operation is gone.

35 Spray the entire area with a light coat of primer. This will reveal any imperfections in the surface of the filler. Repair the imperfections with fresh filler paste or glaze filler and once more smooth the surface with sandpaper. Repeat this spray-and-repair procedure until you are satisfied that the surface of the filler and the feathered edge of the paint are perfect. Rinse the area with clean water and allow it to dry completely.

36 The repair area is now ready for painting. Spray painting must be carried out in a warm, dry, windless and dust free atmosphere. These conditions can be created if you have access to a large indoor work area, but if you are forced to work in the open, you will have to pick the day very carefully. If you are working indoors, dousing the floor in the work area with water will help settle the dust which would otherwise be in the air. If the repair area is confined to one body panel, mask off the surrounding panels. This will help minimize the effects of a slight mismatch in paint color. Trim pieces such as chrome strips, door handles, etc., will also need to be masked off or removed. Use masking tape and several thicknesses of newspaper for the masking operations.

37 Before spraying, shake the paint can thoroughly, then spray a test area until the spray painting technique is mastered. Cover the repair area with a thick coat of primer. The thickness should be built up using several thin layers of primer rather than one thick one. Using 600-grit wet-or-dry sandpaper, rub down the surface of the primer until it is very smooth. While doing this, the work area should be thoroughly rinsed with water and the wet-or-dry sandpaper periodically rinsed as well. Allow the primer to dry before spraying additional coats.

38 Spray on the top coat, again building up the thickness by using several thin layers of paint. Begin spraying in the center of the repair area and then, using a circular motion, work out until the whole repair area and about two inches of the surrounding original paint is covered. Remove all masking material 10 to 15 minutes after spraying on the final coat of paint. Allow the new paint at least two weeks to harden, then use a very fine rubbing compound to blend the edges of the new paint into the existing paint. Finally, apply a coat of wax.

6 Body repair - major damage

1 Major damage must be repaired by an auto body shop specifically equipped to perform unibody repairs. These shops have the specialized equipment required to do the job properly.

2 If the damage is extensive, the body must be checked for proper alignment or the vehicle's handling characteristics may be adversely affected and other components may wear at an accelerated rate.

3 Due to the fact that some of the major body components (hood, fenders, doors, etc.) are separate and replaceable units, any seriously damaged components should be replaced rather than repaired. Sometimes the components can be found in a wrecking yard that specializes in used vehicle components, often at considerable savings over the cost of new parts.

7 Hinges and locks - maintenance

Once every 3,000 miles, or every three months, the hinges and latch assemblies on the doors, hood and trunk (or liftgate) should be given a few drops of light oil or lock lubricant. The door latch strikers should also be lubricated with a thin coat of grease to reduce wear and ensure free movement. Lubricate the door and trunk (or liftgate) locks with spray-on graphite lubricant.

8 Windshield and fixed glass - replacement

Replacement of the windshield and fixed glass requires the use of special fast-setting adhesive/caulk materials and some specialized tools and techniques. These operations should be left to a dealer service department or a shop specializing in glass work.

9 Hood - removal, installation and adjustment

➡ Note: The hood is somewhat awkward to remove and install, at least two people should perform this procedure.

REMOVAL AND INSTALLATION

♦ Refer to illustration 9.3

1 Open the hood, then place blankets or pads over the fenders and cowl area of the body. This will protect the body and paint as the hood is lifted off.

2 Disconnect any cables or wires that will interfere with removal. Disconnect the windshield washer tubing from the nozzles on the hood.

3 Make marks around the hood hinge to ensure proper alignment during installation (see illustration).

4 Have an assistant support the weight of the hood and remove the hinge-to-hood bolts and lift off the hood.

5 Installation is the reverse of removal. Align the hinge bolts with the marks made in Step 3.

ADJUSTMENT

♦ Refer to illustrations 9.9a and 9.9b

6 Fore-and-aft and side-to-side adjustment of the hood is done by moving the hood hinge after loosening the bolts (see Step 3). Fore-and-aft or side-to-side adjustments are made by loosening the hinge-to-hood bolts. Up-and-down adjustments at the rear of the hood are made by loosening the hinge-to-body bolts.

7 Mark around the entire hinge so you can determine the amount of movement.

8 Loosen the bolts and move the hood into correct alignment. Move it only a little at a time. Tighten the hinge bolts and carefully lower the hood to check the position.

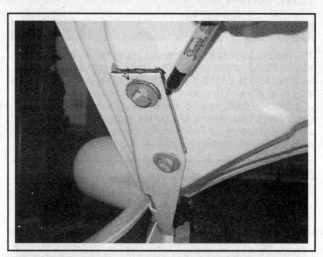

9.3 Draw alignment marks around the hood hinges to ensure proper alignment of the hood when it's reinstalled

11-8 BODY

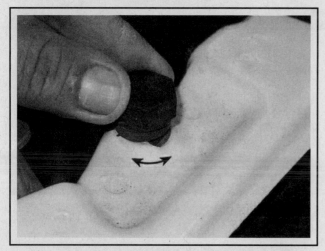

9.9a To adjust the vertical height of the leading edge of the hood so that it's flush with the fenders, turn each hood bumper clockwise (to lower the hood) or counterclockwise (to raise the hood)

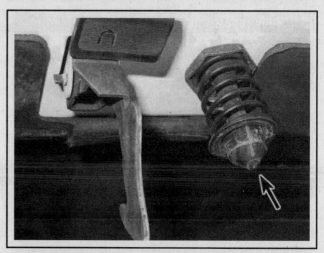

9.9b To adjust the tension of the hood latch, loosen the locknut and rotate the latch post counterclockwise to lessen the tension, or clockwise to increase it, then tighten the locknut

9 Adjust the hood bumpers on the radiator support so the hood, when closed, is flush with the fenders (see illustration). It may be necessary to adjust the length of the latch rod if a substantial height adjustment has been made (see illustration).

10 The hood latch assembly, as well as the hinges, should be periodically lubricated with white, lithium-base grease to prevent binding and wear.

10 Hood latch handle and release cable - removal and installation

▶ Refer to illustrations 10.3, 10.4, 10.5, 10.6, 10.8, 10.10a, 10.10b, 10.11 and 10.15

※ WARNING:

The models covered by this manual are equipped with Supplemental Restraint systems (SRS), more commonly known as airbags. Always disarm the airbag system before working in the vicinity of any airbag system component to avoid the possibility of accidental deployment of the airbag, which could cause personal injury (see Chapter 12). Do not use a memory saving device to preserve the PCM's memory when working on or near airbag system components.

1 Open the hood, then place blankets or pads over the fenders and cowl area of the body.
2 Remove the radiator grille (see Section 13).
3 Disconnect the hood latch cable by rotating the spring, providing slack and disconnect the cable from the hook in the spring (see illustration).
4 Remove the spring from the radiator support (see illustration).
5 Remove the hood latch cable bracket from the radiator support (see illustration).
6 Remove the rear hood seal (see illustration).
7 Remove the cowl cover (see Section 14).
8 Remove the insulator panel from below the driver's side of the instrument panel (see illustration).

10.3 First, rotate the spring to release cable tension and disconnect the cable from the hook in the spring using needle-nose pliers

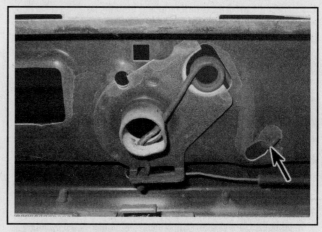

10.4 Use pliers to remove the spring end from the locating hole in the radiator support structure

BODY 11-9

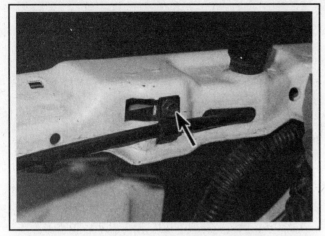

10.5 Remove the hood latch cable bracket mounting bolt and separate the cable from the radiator support structure

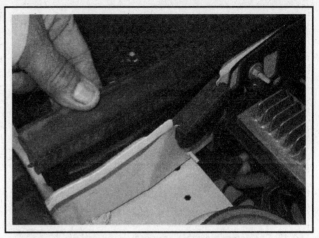

10.6 Remove the rear hood seal from the engine compartment cowl

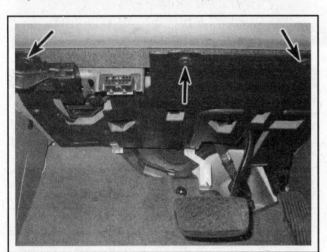

10.8 Pull out the pin retainers then remove the insulator panel

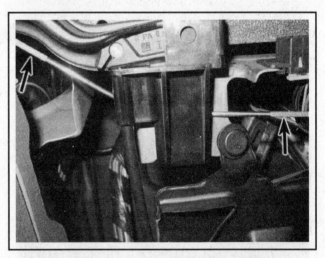

10.10a Use two thin screwdrivers inserted under the release clips . . .

9 Remove the lower heating duct from under the driver's side of the instrument panel.

10 Use two small screwdrivers to release the hood latch handle retainer clips (see illustrations).

11 Slide the hood latch handle forward and down (see illustration) to release the handle from the bracket.

12 Attach a piece of thin wire or string to the latch end of the cable.

13 Push the rubber grommet through the firewall and pull the hood release cable into the vehicle.

14 With the new cable attached to the wire or string, pull the wire or

10.10b . . . then release the clips by prying the clips down simultaneously

10.11 Slide the hood latch handle forward and down

11-10 BODY

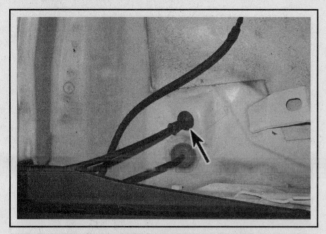

10.15 Check that the cable grommet in the engine compartment firewall is seated and will not leak

string back through the firewall until the new cable reaches the spring.

15 Check that the cable grommet is properly seated in the engine compartment firewall (see illustration).

16 Working in the passenger compartment, reinstall the hood latch handle into the bracket, sliding the handle onto its bracket until the retainer clips snap into the notches in the bracket.

17 The remainder of installation is the reverse of removal.

11 Bumpers - removal and installation

11.2a Remove the pin retainers from the inner fender splash shield at the bumper cover

✳ WARNING:

The models covered by this manual are equipped with Supplemental Restraint systems (SRS), more commonly known as airbags. Always disarm the airbag system before working in the vicinity of any airbag system component to avoid the possibility of accidental deployment of the airbag, which could cause personal injury (see Chapter 12). Do not use a memory saving device to preserve the PCM's memory when working on or near airbag system components.

1 Apply the parking brake, raise the vehicle and support it securely on jackstands.

FRONT BUMPER

▶ Refer to illustrations 11.2a, 11.2b, 11.2c, 11.4, 11.7 and 11.8

2 Remove the pin retainers from the inner fender splash shield (see illustrations).

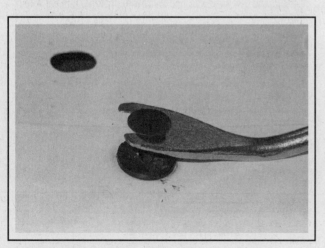

11.2b To release the pin retainer, lift up on the center section using a panel tool or screwdriver . . .

11.2c . . . then pry the retainer out

BODY 11-11

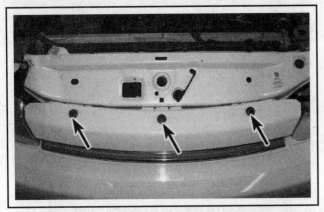

11.4 Remove the pin retainers from the upper section of the bumper cover

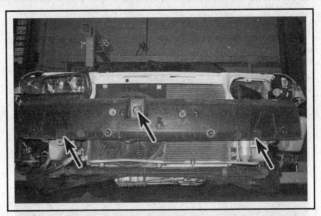

11.7 Location of the energy absorber pin retainers

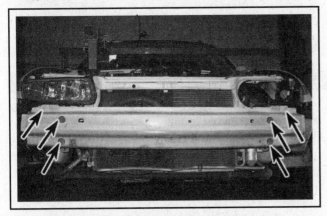

11.8 Location of the front bumper mounting nuts

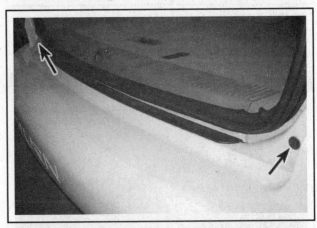

11.12 Remove the two pin retainers from the corners of the bumper assembly

3 Working below the bumper, remove the pin retainers from the lower section of the bumper cover and the radiator support structure.
4 Remove the pin retainers from the upper section of the bumper cover (see illustrations).
5 Disconnect the fog lamp electrical connector.
6 Remove the bumper cover from the fender.

➡ **Note: Some models are equipped with pin retainers while other models use a locking tab to retain the bumper cover to the fender.**

7 Remove the energy absorber from the front bumper (see illustration).
8 Remove the bumper mounting nuts (see illustration).
9 Remove the front bumper.
10 Installation is the reverse of removal.

REAR BUMPER

▸ **Refer to illustrations 11.12, 11.13, 11.17 and 11.18**

11 Open the liftgate.
12 On wagons, remove the pin retainers on the top of the bumper cover next to the tail light (see illustration).
13 Remove the pin retainers from the rear bumper extension (see illustration).
14 Remove the pin retainers from the inner fender splash shield.

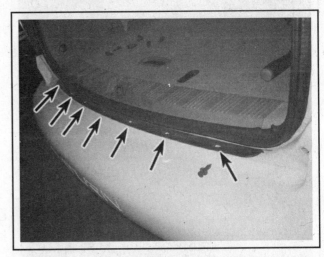

11.13 Remove the pin retainers from the rear bumper extension

Pull the splash shield away from the rear bumper area.
15 Remove the rear bumper cover mounting screws from the quarter panel.
16 Detach the rear bumper cover locking tabs from the quarter panel.

11-12 BODY

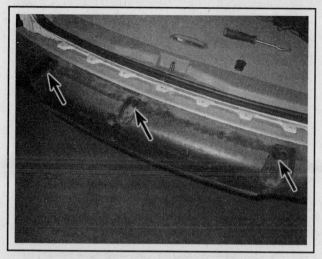

11.17 Location of the energy absorber pin retainers

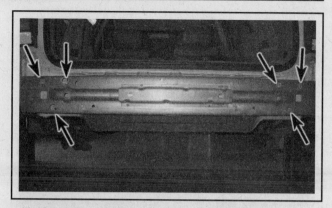

11.18 Location of the rear bumper mounting nuts

17 Remove the energy absorber from the rear bumper (see illustration).
18 Remove the rear bumper mounting nuts (see illustration).
19 Installation is the reverse of removal.

12 Front fender - removal and installation

▶ Refer to illustrations 12.3, 12.4, 12.7a, 12.7b, 12.12a, 12.12b, 12.12c and 12.12d

1 Raise the vehicle, support it securely on jackstands and remove the front wheel(s).
2 Open the front and rear doors and remove the rocker panel molding.
3 Remove the pin retainers from the top section of the rocker panel (see illustration).
4 Remove the pin retainers from the bottom of the rocker panel (see illustration).
5 Remove the pin retainers from the front and rear wheel fender splash shields at the rocker panel molding (see illustration 11.2a).
6 Remove the rocker panel molding by angling it down and away from the body.
7 Remove the inner fender splash shield (see illustrations).

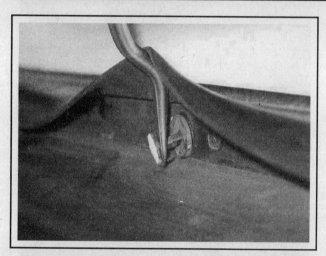

12.3 Lift the weatherstrip and remove the pin retainers from the upper edge of the rocker panel

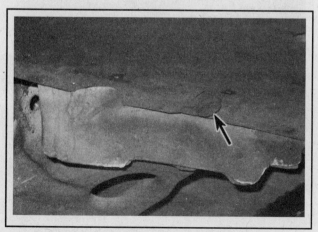

12.4 Remove the pin retainers from the bottom of the rocker panel

12.7a Remove the pin retainers from the inner fender splash shield from each side of the vehicle - left side shown, right side similar

BODY   11-13

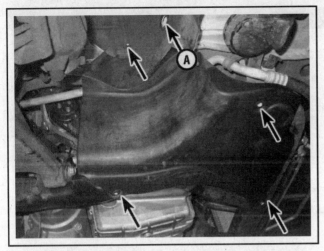

12.7b Working on the right side of the vehicle, first remove the lower splash shield pin retainers before removing the inner fender splash shield - one bolt can be removed through the access hole (A) in the inner fender splash shield

8 Remove the radiator grille (see Section 13).
9 Remove the headlight housings (see Chapter 12).
10 On 2000 through 2002 models, remove the side marker lights (see Chapter 12).
11 Remove the pin retainers from the fender to the bumper cover and separate the cover from the fender.
12 Remove the bolts from the fender (see illustrations).

➥**Note: Be sure to remove the lower fender bolts that are exposed after the rocker panel molding has been removed (see illustration).**

13 Lift off the fender. It's a good idea to have an assistant support the fender while it's being moved away from the vehicle to prevent damage to the surrounding body panels.
14 Installation is the reverse of removal. Check the alignment of the fender to the hood and front edge of the door before final tightening of the fender fasteners.

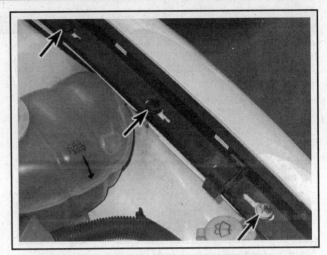

12.12a Location of the upper fender mounting bolts

12.12b Location of the fender mounting bolt located behind the front door

12.12c On 2000 through 2002 models, remove the side marker light to access this fender bolt

12.12d Location of the lower fender mounting bolts

11-14 BODY

13 Radiator grille - removal and installation

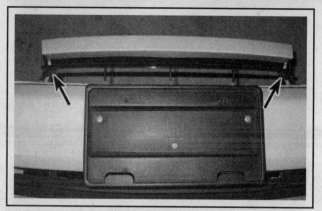

13.2 On 2002 and earlier models there is a screw securing each side of the grille to the bumper cover

▶ **Refer to illustration 13.2**

1 Remove the grille retainer pins from the top of the bumper cover (see illustrations 11.2a and 11.2c).
2 On 2002 and earlier models, remove the grille retaining screws (see illustration).
3 Remove the grille from the front bumper cover.
4 Installation is the reverse of removal.

14 Cowl cover - removal and installation

1 Remove the wiper arms (see Chapter 12) and the nuts from the wiper arm shafts.
2 Remove the hood seal from the rear of the engine compartment.
3 Lift the cowl cover from the vehicle.
4 If the vent tray needs to be removed, first remove the wiper motor linkage assembly, then remove the vent tray (see Chapter 12).
5 Installation is the reverse of removal.

15 Door trim panels - removal and installation

※ **WARNING:**

The models covered by this manual are equipped with Supplemental Restraint systems (SRS), more commonly known as airbags. Always disarm the airbag system before working in the vicinity of any airbag system component to avoid the possibility of accidental deployment of the airbag, which could cause personal injury (see Chapter 12). Do not use a memory saving device to preserve the PCM's memory when working on or near airbag system components.

REMOVAL

Front doors

▶ **Refer to illustrations 15.2, 15.3, 15.4, 15.6, 15.9 and 15.10**

1 Remove the upper mirror trim panel, if equipped (see Section 21).
2 Remove the door handle grip cover (see illustration).
3 Remove the door trim panel mounting screw located behind handle (see illustration).

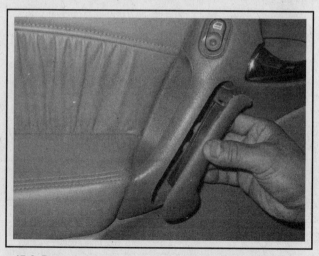

15.2 Remove the grip cover from the door handle by pulling evenly to unsnap the plastic clips . . .

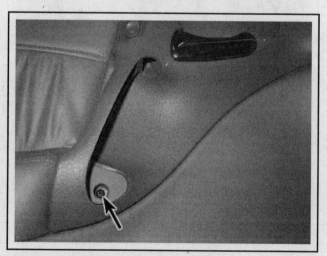

15.3 . . . and remove the screw underneath

BODY 11-15

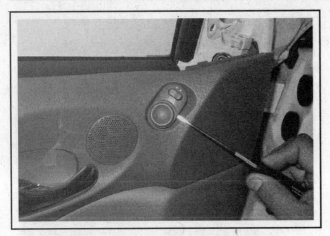

15.4 Carefully pry the switches from the door panel and disconnect the electrical connectors

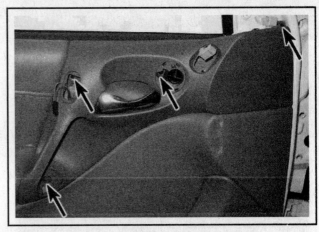

15.6 Location of the door handle mounting screws

4 Remove the door lock switch and power mirror switch, if equipped (see illustration).
5 Remove the tweeter speaker cover (see illustration).
6 Remove the door handle assembly mounting screws (see illustration).
7 Lift the door handle off the door partially and disconnect the electrical connectors from behind.
8 On manual crank windows, remove the regulator handle from the door (see illustrations 15.18a and 15.18b).
9 Remove the lower fasteners from the door trim panel (see illustration).
10 Remove the door trim panel using a door panel removal tool (see illustration). Start from the bottom of the trim panel and work around the perimeter until all fasteners have been released from the door.
11 Lift the trim panel up to disengage the panel from the upper door ridge, unplug any electrical connectors, and remove the panel.
12 For access to the door outside handle or the door window regulator inside the door, raise the window fully, then carefully peel back the plastic watershield.

Rear doors

▶ Refer to illustrations 15.13, 15.14, 15.16, 15.17, 15.18 and 15.19

13 Remove the door handle grip plastic plug (see illustration).

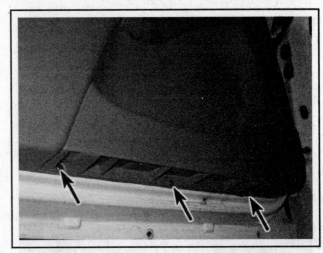

15.9 Remove the door trim panel screws located on the bottom of the panel

14 Remove the tweeter speaker or cover (see illustration).
15 Remove the door lock switch, if equipped, or the plastic plug from the door handle (see illustration 15.4).
16 Remove the door handle and trim panel mounting screws located

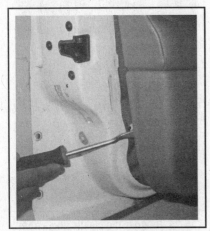

15.10 Using a door panel tool, start from the bottom and release any clips attached to the door

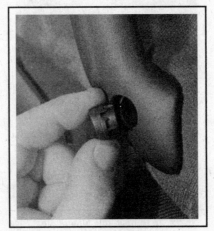

15.13 Pry out the plastic plug to access the trim panel screw

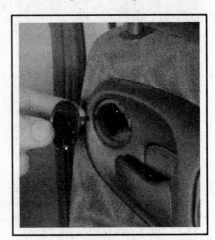

15.14 Remove the tweeter cover to access another trim panel screw

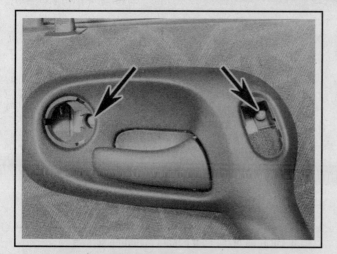

15.16 Location of the door handle mounting screws

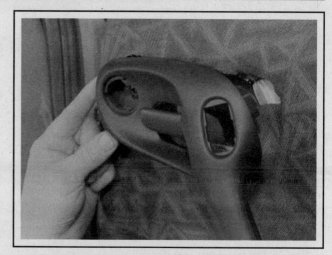

15.17 Remove the door handle from the door trim panel

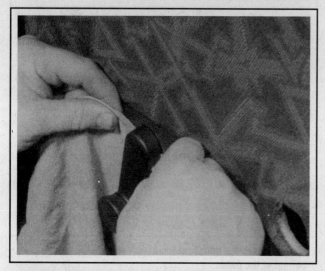

15.18 Position a shop towel between the handle and the trim panel, then work the shop towel back-and-forth to release the spring clip and the handle from the splined shaft

behind the tweeter, the handle plugs and/or switch (see illustration).

17 Remove the door handle (see illustration).

18 On manual-crank windows, remove the regulator handle from the door (see illustration).

19 Remove the upper extension panel (see illustration).

20 Remove the door trim panel using a door panel removal tool (see illustration 15.10). Start from the bottom of the trim panel and work around the perimeter until the fasteners have been released from the door.

21 Unplug any electrical connectors, lift the trim panel up to disengage the panel from the upper door ridge and remove the panel.

22 For access to the door outside handle or the door window regulator inside the door, raise the window fully, then carefully peel back the plastic watershield.

Liftgate

▶ Refer to illustrations 15.23, 15.24 and 15.25

23 Remove the liftgate panel pin retainers (see illustration).

24 Remove the plastic plugs from the handle trim and remove the panel mounting screws (see illustration).

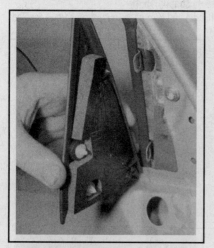

15.19 Use a panel tool to pry the extension panel from the upper door

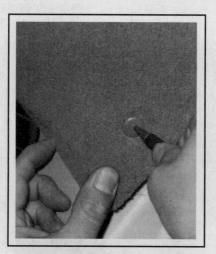

15.23 To remove this type of pin retainer, push the center in, then pry out the retainer

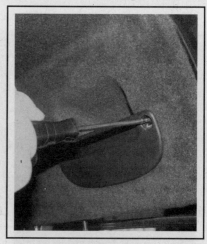

15.24 Remove the plastic plugs from the liftgate handle trim cover, then remove the screws

BODY 11-17

25 Using a screwdriver or trim removal tool, pry out the clips and remove the trim panel from the liftgate (see illustration).

INSTALLATION

26 Prior to installation of the door trim panels and/or the liftgate trim panel, be sure to reinstall any clips in the panel which may have come out when you removed the panel.

27 Position the wire harness connectors for the power door lock switch and the power window switch (if equipped) on the back of the panel, then place the panel in position in the door. Press the door panel into place until the clips are seated.

28 The remainder of the installation is the reverse of removal.

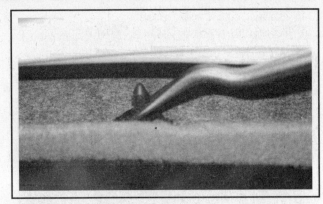

15.25 Carefully pry the clips using a panel tool to release the liftgate trim panel

16 Door - removal, installation and adjustment

※※ WARNING:

The models covered by this manual are equipped with Supplemental Restraint systems (SRS), more commonly known as airbags. Always disarm the airbag system before working in the vicinity of any airbag system component to avoid the possibility of accidental deployment of the airbag, which could cause personal injury (see Chapter 12). Do not use a memory saving device to preserve the PCM's memory when working on or near airbag system components.

→Note 1: These models are equipped with bolt-on door hinges or weld-on door hinges. The following procedure details door removal for bolt-on doors. Weld-on doors require special door hinge pin removal tools. Have the weld-on doors replaced and adjusted by a dealer or other qualified automotive repair facility.

→Note 2: The door is heavy and somewhat awkward to remove and install - at least two people should perform this procedure.

REMOVAL AND INSTALLATION

♦ Refer to illustrations 16.9a and 16.9b

1 Raise the window completely in the door. Open the door all the way and support it on jacks or blocks covered with rags to prevent damaging the paint.

2 Disconnect the cable from the negative battery terminal (see Chapter 5, Section 1).

3 Remove the door trim panel and watershield as described in Section 15.

4 Remove the door outer panel (see Section 20).

5 Disconnect all electrical connections, ground wires and harness retaining clips from the door.

→Note: It is a good idea to label all connections to aid the reassembly process.

6 Detach the rubber conduit, grommet and wiring harness from the door.

7 Remove the door stop strut bolt.

8 Mark around the door hinge reinforcement plates with a pen or a scribe to facilitate realignment during reassembly.

9 With an assistant holding the door, remove the hinge-to-door nuts (see illustrations) and lift the door off.

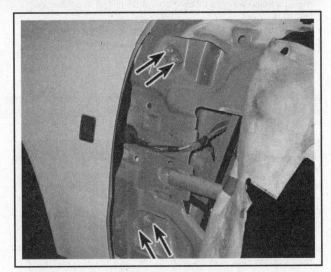

16.9a Location of the front door hinge nuts

16.9b Location of the rear door hinge nuts

11-18 BODY

10 Installation is the reverse of removal.
11 Reconnect the battery (see Chapter 5, Section 1).

ADJUSTMENT

♦ **Refer to illustration 16.15**

12 Having proper door-to-body alignment is a critical part of a well-functioning door assembly. First check the door hinge pins for excessive play. Fully open the door and lift up and down on the door without lifting the body. If a door has 1/16-inch or more excessive play, the hinges should be replaced.

13 Door-to-body alignment adjustments are made by loosening the hinge-to-body bolts or hinge-to-door nuts and moving the door. Proper body alignment is achieved when the top of the doors are parallel with the roof section, the front door is flush with the fender, the rear door is flush with the rear quarter panel and the bottom of the doors are aligned with the lower rocker panel. If these goals can't be reached by adjusting the hinge-to-body or hinge-to-door bolts, body alignment shims may have to be purchased and inserted behind the hinges to achieve correct alignment.

14 To adjust the door-closed position, mark around the striker plate to provide a reference point, then check that the door latch is contacting

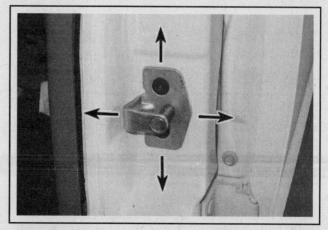

16.15 Adjust the door lock striker by loosening the mounting screws and gently tapping the striker in the desired direction

the center of the latch striker. If not, adjust the up and down position first.

15 Finally adjust the latch striker sideways position, so that the door outer panel is flush with the center pillar or rear quarter panel and provides positive engagement with the latch mechanism (see illustration).

17 Door latch, lock cylinder and handles - removal and installation

※ WARNING:

The models covered by this manual are equipped with Supplemental Restraint systems (SRS), more commonly known as airbags. Always disarm the airbag system before working in the vicinity of any airbag system component to avoid the possibility of accidental deployment of the airbag, which could cause personal injury (see Chapter 12). Do not use a memory saving device to preserve the PCM's memory when working on or near airbag system components.

※ CAUTION:

Wear gloves when working inside the door openings to protect against cuts from sharp metal edges.

➡ Note: The following procedure requires removal of the door outer panel to access the handle mechanism, the door lock cylinder and linkage. However, if only the door latch removal is required, the door outer panel can remain in place.

FRONT DOOR

♦ **Refer to illustrations 17.2a, 17.2b, 17.4, 17.5a, 17.5b, 17.7, 17.8, 17.10, 17.11 and 17.12**

1 Raise the window, then remove the door trim panel and watershield (see Section 15).
2 Working on the outside of the door, remove the door handle cover (see illustrations).

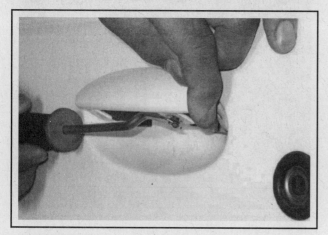

17.2a First, remove the pin retainer from the door handle cover . . .

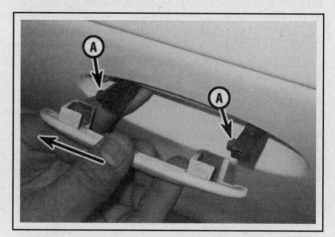

17.2b . . . then slide the door handle cover off the locator pins

BODY   11-19

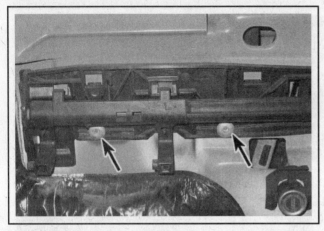

17.4 Location of the front door handle mounting rivets

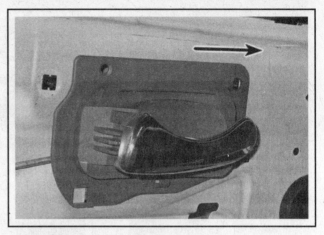

17.5a Slide the door handle forward...

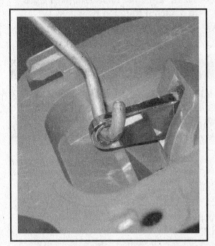

17.5b ...then remove the actuating rod by rotating the handle

17.7 Pry the lock rod retainers off the rods, then detach the rods

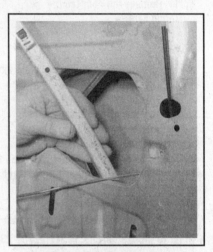

17.8 Remove the window run channel from the door access hole

> ❋❋ **CAUTION:**
> Take care not to scratch the paint on the outside of the door. Wide masking tape applied around the handle opening before beginning the procedure can help avoid scratches.

3 Remove the door outer panel (see Section 20).
4 Remove the rivets (see illustration), disconnect the handle linkage and remove the handle mechanism from the door.
5 Working on the inside of the door, slide the door handle forward, pivot the assembly away from the door and unlatch the actuating rod from the door handle lever (see illustrations).
6 Pull the front lock rod out and disconnect it from the latch mechanism. Remove the lock rod through the access hole in the door.
7 Unhook the door handle actuating rods from the latch mechanism (see illustration).
8 Remove the bolt from the lower section of the door jamb, reach inside the door and lower the window run channel. Remove the window run channel from the door (see illustration).
9 Pull the upper weatherstrip toward the rear to make room to raise the window glass.

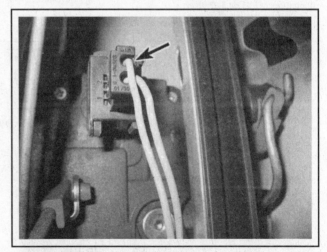

17.10 Disconnect the electrical connector from the latch mechanism

10 Disconnect the electrical connectors from the latch mechanism (see illustration).

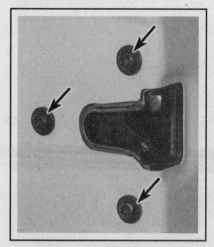

17.11 Remove the door latch mounting screws

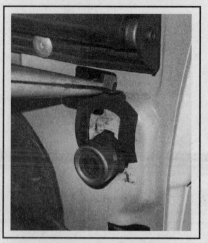

17.12 Remove the retaining clip from the lock cylinder, then push the lock cylinder through the door

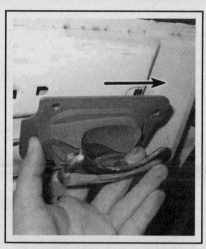

17.19 Slide the door handle forward to detach it from the door

11 Remove the latch mounting bolts (see illustration) and angle the latch mechanism out of the door access hole.

12 Remove the lock cylinder rod from the latch mechanism (if not already removed), remove the clip and remove the lock cylinder from the door (see illustration).

13 Installation is the reverse of removal.

➡14 Note: It's a good idea to replace the lock rod retainers with new ones whenever they are removed.

REAR DOOR

◆ Refer to illustrations 17.19, 17.22, 17.23 and 17.24

15 Raise the window, then remove the door trim panel and watershield (see Section 15).

16 Remove the door handle cover from the exterior of the door (see illustrations 17.2a and 17.2b).

※ CAUTION:

Take care not to scratch the paint on the outside of the door. Wide masking tape applied around the handle opening before beginning the procedure can help avoid scratches.

17 Remove the door outer panel (see Section 20).

18 Remove the rivets (see illustration 17.4), disconnect the handle linkage and remove the handle mechanism from the door.

19 Working inside the door, slide the door handle forward, pivot the assembly away from the door and unlatch the actuating rod from the door handle lever (see illustration).

20 Pull the front lock rod out and disconnect it from the latch mechanism. Remove the lock rod through the access hole in the door.

21 Unhook the door handle actuating rod from the latch mechanism.

22 Remove the bolts from the window run channel (see illustration), reach inside the door and remove the foam insulator and remove the

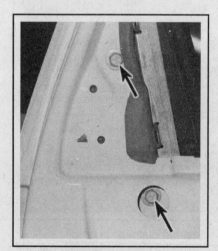

17.22 Remove the window run channel bolts

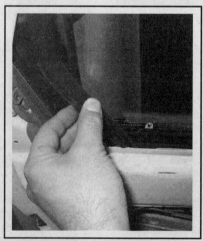

17.23 Pull the rear of the weatherstrip out of its channel

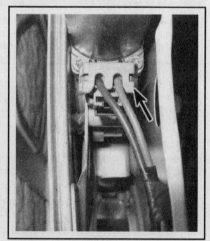

17.24 Disconnect the electrical connector from the latch mechanism

BODY 11-21

weatherstrip mounting screws.

23 Pull the rear of the weatherstrip out of its channel to make room to raise the window glass (see illustration).

→Note: **Early models are equipped with a one-piece weatherstrip while later models are equipped with a two-piece weatherstrip.**

24 Disconnect the electrical connectors from the latch mechanism (see illustration).

25 Mark each actuating rod for correct reinstallation and disconnect them from the latch mechanism.

26 Remove the latch mounting bolts and angle the latch mechanism out of the door access hole.

27 Installation is the reverse of removal.

→Note: **It's a good idea to replace the lock rod retainers with new ones whenever they are removed.**

18 Door window glass - removal and installation

WARNING:

The models covered by this manual are equipped with Supplemental Restraint systems (SRS), more commonly known as airbags. Always disarm the airbag system before working in the vicinity of any airbag system component to avoid the possibility of accidental deployment of the airbag, which could cause personal injury (see Chapter 12). Do not use a memory saving device to preserve the PCM's memory when working on or near airbag system components.

CAUTION:

Wear gloves when working inside the door openings to protect against cuts from sharp metal edges.

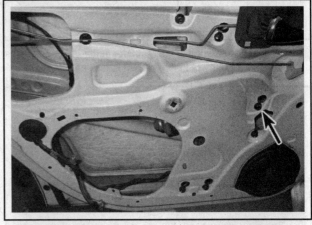

18.6 Remove the upper right mounting bolt from the door window regulator

FRONT DOOR GLASS

▶ Refer to illustration 18.6

1 Remove the door trim panel and the plastic watershield (see Section 15).
2 Lower the window glass all the way down into the door.
3 Remove the bolt from the lower section of the door jamb, reach inside the door and lower the window run channel. Remove the window run channel from the door.
4 Remove the weatherstrip from the front door interior glass channel.
5 Raise the window just enough to access the window retaining bolts through the holes in the door frame.
6 Remove the upper right window regulator mounting bolt (see illustration).
7 Pull the rear section of the window regulator arm away from the door toward the access hole and lift the glass by pulling it up and out.
8 Installation is the reverse of removal.

REAR DOOR GLASS

9 Remove the door trim panel and the plastic watershield (see Section 15).
10 Lower the window glass all the way down into the door.
11 Remove the bolts from the window run channel (see illustration 17.22), reach inside the door and lower the window run channel. Remove the window run channel from the door.
12 Remove the screws from the weatherstrip and pull the upper weatherstrip from the interior glass channel (see illustration 17.23).

→Note: **Early models are equipped with a one-piece weatherstrip design while later models are equipped with an updated two-piece weatherstrip design.**

13 Raise the window just enough to access the window and lift the glass by pulling it up and out.
14 Installation is the reverse of removal.

19 Door window glass regulator - removal and installation

WARNING:

The models covered by this manual are equipped with Supplemental Restraint systems (SRS), more commonly known as airbags. Always disarm the airbag system before working in the vicinity of any airbag system component to avoid the possibility of accidental deployment of the airbag, which could cause personal injury (see Chapter 12). Do not use a memory saving device to preserve the PCM's memory when working on or near airbag system components.

CAUTION:

Wear gloves when working inside the door openings to protect against cuts from sharp metal edges.

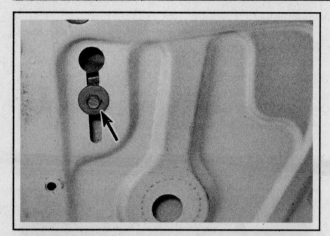

19.6a Mark the position of the regulator guide rail mounting bolt before loosening to insure correct reassembly

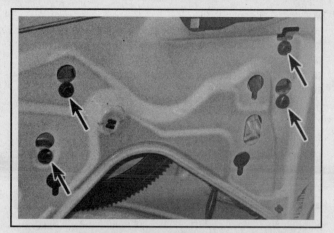

19.6b Location of the front window regulator mounting bolts

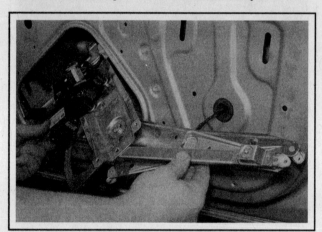

19.7 Removing the front door window regulator

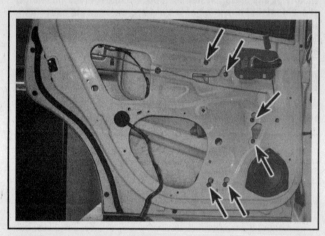

19.15 Location of the rear window regulator mounting bolts

FRONT

♦ **Refer to illustrations 19.6a, 19.6b and 19.7**

1 Remove the door trim panel and the plastic watershield (see Section 15).
2 Lower the window glass all the way down.
3 Remove the door window glass (see Section 18).
4 On power window models, be sure to remove the key from the ignition switch.
5 On power window models, disconnect the electrical connector from the window regulator motor.
6 Remove the regulator/motor assembly mounting bolts (see illustrations).
7 Remove the window regulator/motor assembly from the door (see illustration).
8 Lubricate the rollers and wear points on the regulator with white grease before installation.
9 Installation is the reverse of removal.

REAR

♦ **Refer to illustrations 19.15 and 19.16**

10 Remove the door trim panel and the plastic watershield (see Section 15).
11 Lower the window glass all the way down.
12 Remove the door window glass (see Section 18).
13 On power window models, be sure to remove the key from the ignition switch.
14 On power window models, disconnect the electrical connector

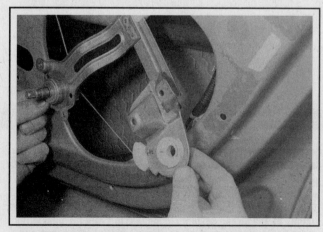

19.16 Removing the rear door window regulator

BODY   11-23

from the window regulator motor.

15 Remove the regulator/motor assembly mounting bolts (see illustration).

16 Remove the window regulator/motor assembly from the door (see illustration).

17 Lubricate the rollers and wear points on the regulator with white grease before installation.

18 Installation is the reverse of removal.

20 Door outer panel - removal and installation

WARNING:

The models covered by this manual are equipped with Supplemental Restraint systems (SRS), more commonly known as airbags. Always disarm the airbag system before working in the vicinity of any airbag system component to avoid the possibility of accidental deployment of the airbag, which could cause personal injury (see Chapter 12). Do not use a memory saving device to preserve the PCM's memory when working on or near airbag system components.

CAUTION:

Wear gloves when working inside the door openings to protect against cuts from sharp metal edges.

FRONT DOOR

▶ Refer to illustrations 20.3a, 20.3b, 20.6 and 20.9

1 Remove the outside mirror (see Section 21).

2 Working on the outside of the door, remove the door handle (see Section 17).

3 Remove the front outer sealing strip (see illustrations).

4 Remove the door outer panel mounting bolts. Follow the reverse of the tightening sequence (see illustration 20.10).

5 Lift the outer door panel up and off the door bracket mounts.

6 Remove the outer watershield to access the door components (see illustration).

7 Installation is the reverse of removal.

8 Make sure the outer door alignment pins are correctly centered in the holes in the door structure before installing the door outer panel.

9 Apply thread locking compound to the door outer panel mounting bolts. Tighten the bolts in the correct sequence (see illustration) and to the torque listed in this Chapter's Specifications.

20.3a First, remove the screw that retains the outer sealing strip . . .

20.3b . . . the pry the sealing strip from the upper door using a panel tool

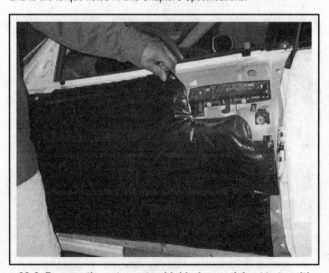

20.6 Remove the outer watershield - be careful not to tear it!

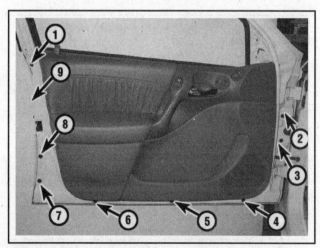

20.9 Tightening sequence for the front door outer panel mounting bolts - number 9 bolt is for 2004 and later models

11-24 BODY

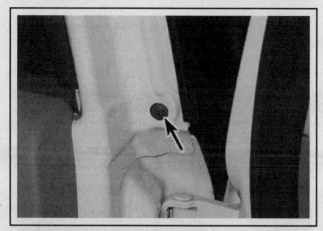

20.11 Remove the sealing strip mounting screw and separate the sealing strip from the outer door using a panel tool

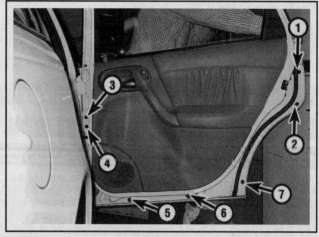

20.16 Tightening sequence for the rear door outer panel mounting bolts

REAR DOOR

♦ Refer to illustrations 20.11 and 20.16

10 Working on the outside of the door, remove the door handle (see Section 17).
11 Remove the sealing strip from the outer door along the window glass (see illustration).
12 Remove the door outer panel mounting bolts. Follow the reverse of the tightening sequence (see illustration 20.16).
13 Lift the outer door panel up and off the door bracket mounts.
14 Installation is the reverse of removal.
15 Make sure the outer door alignment pins are correctly centered in the net holes in the door structure before installing the door outer panel.
16 Apply thread locking compound to the door outer panel mounting bolts. Tighten the bolts in the correct sequence (see illustration) and to the torque listed in this Chapter's Specifications.

21 Mirrors - removal and installation

WARNING:

The models covered by this manual are equipped with Supplemental Restraint systems (SRS), more commonly known as airbags. Always disarm the airbag system before working in the vicinity of any airbag system component to avoid the possibility of accidental deployment of the airbag, which could cause personal injury (see Chapter 12). Do not use a memory saving device to preserve the PCM's memory when working on or near airbag system components.

OUTSIDE MIRRORS

♦ Refer to illustrations 21.1 and 21.3

1 Remove the front door interior sail panel (see illustration).
2 Remove the foam insulator, if equipped.
3 If you're removing a power mirror, disconnect the electrical connector (see illustration).
4 Remove the three mirror retaining nuts and detach the mirror from the vehicle.
5 Installation is the reverse of removal.

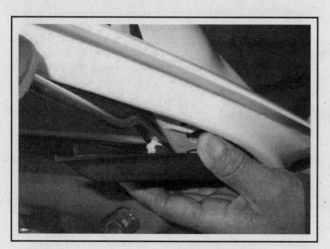

21.1 Use a panel tool to pry the sail panel off the interior door

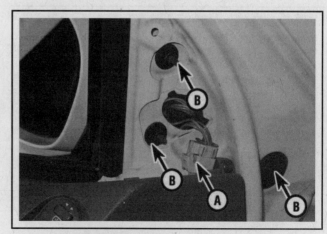

21.3 Disconnect the power mirror electrical connector (A) and remove the mounting nuts (B) - one mounting nut behind panel cover

BODY   11-25

INSIDE MIRROR

6 Remove the mirror retaining screw and slide the mirror up and off its base.

7 Installation is the reverse of removal. Tighten the screw securely.
8 If the mount plate itself has come off the windshield, adhesive kits are available at auto parts stores to resecure it. Follow the instructions included with the kit.

22 Liftgate - removal, installation and adjustment

➡ **Note: The liftgate is heavy and somewhat awkward to remove and install - at least two people should perform this procedure.**

REMOVAL AND INSTALLATION

▶ **Refer to illustrations 22.2a, 22.2b, 22.4 and 22.5**

1 Have an assistant hold the liftgate in the open position.
2 Disconnect all electrical connections, ground wires and harness retaining clips from the liftgate (see illustrations).

➡ **Note: It is a good idea to label all connections to aid the reassembly process.**

3 From the liftgate side, detach the rubber conduit between the body and the liftgate. Then pull the wiring harness through the conduit hole and remove it from the liftgate.
4 Detach the support struts from the liftgate (see illustration).
5 With an assistant holding the liftgate, remove the hinge pins and lift the door off (see illustration).

➡ **Note: Draw a reference line around the hinges before removing the bolts.**

6 Installation is the reverse of removal.

ADJUSTMENT

▶ **Refer to illustration 22.9**

7 Having proper door-to-body alignment is a critical part of a well-functioning liftgate assembly. First, check the liftgate hinge pins for excessive play. Fully open the liftgate and lift up and down on the liftgate door without lifting the body. If a door has 1/16-inch or more excessive play, the hinges should be replaced.
8 Liftgate-to-body alignment adjustments are made by loosening the hinge-to-body bolts and moving the liftgate. Proper body alignment is achieved when the top of the liftgate is parallel with the roof section and the sides of the liftgate are flush with the rear quarter panels and the bottom of the liftgate is aligned with the lower door sill. If these goals can't be reached by adjusting the hinge-to-body bolts, body alignment shims may have to be purchased and inserted behind the hinges to achieve correct alignment.

22.2a Remove the high mount brake light cover

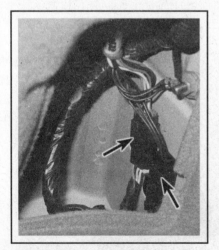

22.2b Disconnect the electrical connectors and ground wires from the liftgate

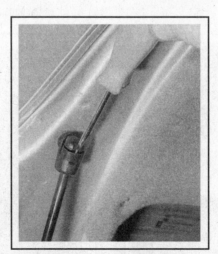

22.4 Pry the clip out and slide the support strut from the ballstud

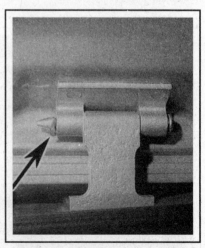

22.5 Remove the hinge pins and separate the liftgate from the body

11-26 BODY

22.9 Adjust the liftgate door striker by loosening the mounting screws and gently tapping the striker in the desired direction

9 To adjust the door-closed position, scribe a line or mark around the striker plate to provide a reference point, then check that the door latch is contacting the center of the latch striker. If not, adjust the latch striker sideways position, so that the door panel is flush with the rear quarter panel and provides positive engagement with the latch mechanism (see illustration).

23 Liftgate latch, lock cylinder and handle - removal and installation

LIFTGATE LATCH

▶ **Refer to illustrations 23.2a, 23.2b and 23.4**

1 Open the liftgate door and remove the trim panel as described in Section 15.
2 Disconnect the control cable from the liftgate latch assembly (see illustrations).
3 Remove the liftgate latch remote control electrical connectors.
4 Remove the screws securing the latch to the door (see illustration). Remove the latch assembly through the door opening.
5 Installation is the reverse of removal.

LIFTGATE LOCK CYLINDER AND HANDLE

▶ **Refer to illustrations 23.8 and 23.9**

6 Open the liftgate door and remove the door trim panel and water-

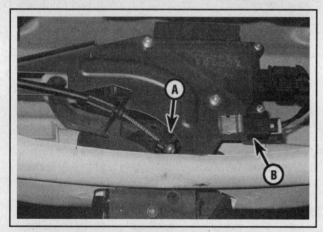

23.2a Location of the liftgate latch control cable (A) and the latch electrical connector (B)

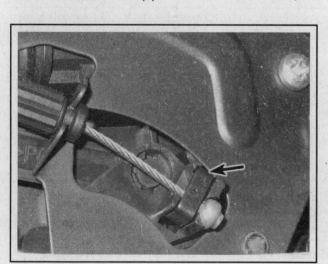

23.2b Use a screwdriver to release the spring clip from the cable end

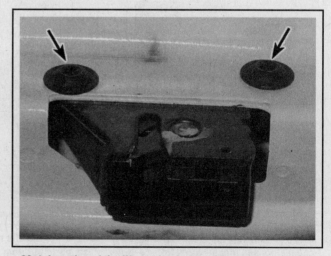

23.4 Location of the liftgate latch mounting screws

BODY   11-27

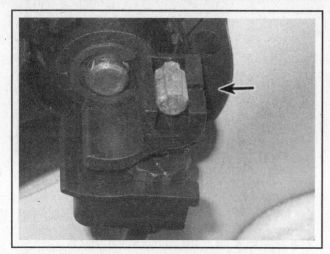

23.8 Rotate the cable lock counterclockwise to release tension and slide the cable end through the slotted portion in the housing

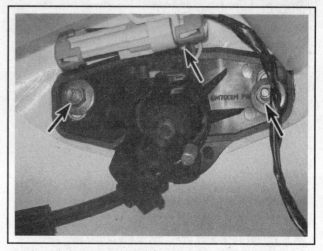

23.9 Location of the liftgate lock cylinder mounting nuts and license plate lamp electrical connector

shield as described in Section 15.
  7 Disconnect the electrical connector from the lock cylinder.
  8 Disconnect the lock cylinder actuating cable (see illustration).
  9 Remove the liftgate lock cylinder mounting nuts (see illustration).

Remove the assembly from the liftgate.
  10 Disconnect the license plate lamp electrical connector.
  11 Remove the nuts from the license plate handle and remove the handle from the liftgate.
  12 The remainder of the installation is the reverse of removal.

24 Center console - removal and installation

▶ Refer to illustrations 24.2, 24.4, 24.5, 24.6, 24.8 and 24.9

⁂ WARNING:

The models covered by this manual are equipped with Supplemental Restraint systems (SRS), more commonly known as airbags. Always disarm the airbag system before working in the vicinity of any airbag system component to avoid the possibility of accidental deployment of the airbag, which could cause personal injury (see Chapter 12). Do not use a memory saving device to preserve the PCM's memory when working on or near airbag system components.

  1 Disconnect the cable from the negative battery terminal (see Chapter 5).
  2 Remove the trim panels from each side of the console (see illustration).
  3 Remove the power window switches (see Chapter 12).
  4 Detach the parking brake boot from the console (see illustration).

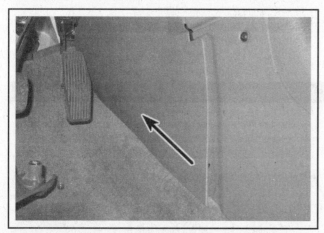

24.2 Slide the trim panel back and out from the center console area

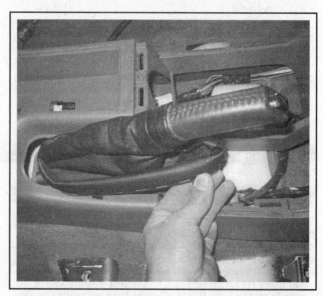

24.4 Detach the parking brake boot from the center console

11-28 BODY

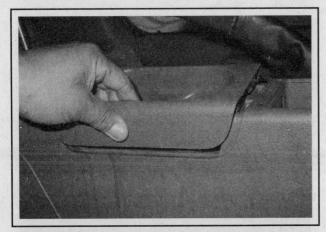

24.5 Pry the cupholder panel out of the center console

24.6 Disconnect the center console wiring harness

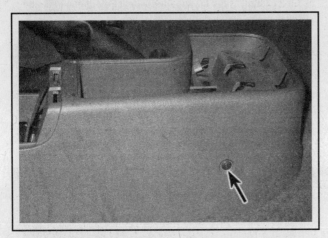

24.8 Location of the console rear mounting screw - one on each side of the console

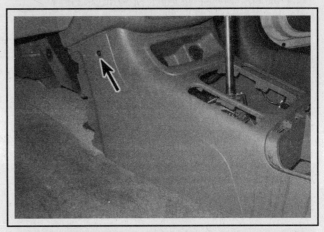

24.9 Location of the console forward mounting screw - one on each side

5   Remove the cupholder assembly from the console (see illustration). Use a flat-bladed screwdriver and remove the cupholder assembly from the housing.

6   Disconnect the console wiring harness from below the cupholder housing (see illustration).

7   Remove the shift lever boot (see Chapter 7A or 7B).

➥ Note: On automatic transaxle models, pull up on the PRNDL cover and disconnect the lamp connector. On manual transaxle models, pull up on the boot and lift it over the shift handle.

8   Slide the front seats forward and remove the console rear mounting screws (see illustration).

9   Remove the console front mounting screws (see illustration).

10 Lift the center console from the vehicle.

11 Installation is the reverse of removal.

12 Reconnect the battery (see Chapter 5, Section 1).

25 Dashboard trim panels

⚠ WARNING:

The models covered by this manual are equipped with Supplemental Restraint systems (SRS), more commonly known as airbags. Always disarm the airbag system before working in the vicinity of any airbag system component to avoid the possibility of accidental deployment of the airbag, which could cause personal injury (see Chapter 12). Do not use a memory saving device to preserve the PCM's memory when working on or near airbag system components.

INSTRUMENT CLUSTER BEZEL

♦ Refer to illustrations 25.4a, 25.4b, 25.4c and 25.5

1   Disconnect the cable from the negative battery terminal (see Chapter 5).

2   Remove the knee bolster (see Steps 13 through 16).

3   Remove the steering column covers (see Section 26).

BODY 11-29

25.4a Remove the upper mounting screws from the instrument cluster bezel

25.4b Location of the lower right mounting screw

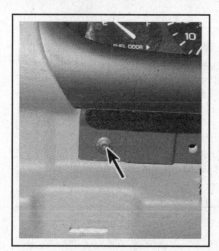

25.4c Location of the lower left mounting screw

25.5 Carefully pry the bezel from the instrument panel

25.9 Pull the center trim panel off of the instrument panel . . .

4 Remove the screws at the top of the instrument cluster bezel (see illustration) and each side of the steering column (see illustration).

5 Grasp the bezel with both hands and pull straight out to disengage the bezel from the instrument panel (see illustration).

6 Installation is the reverse of the removal procedure. Make sure the clips are engaged properly before pushing the bezel firmly into place.

7 Reconnect the battery (see Chapter 5, Section 1).

CENTER TRIM PANEL

♦ Refer to illustrations 25.9 and 25.10

8 Disconnect the cable from the negative battery terminal (see Chapter 5).

9 Pull the center trim panel out of the instrument panel (see illustration).

10 Disconnect the electrical connectors from the accessory switches mounted in the center trim panel (see illustration).

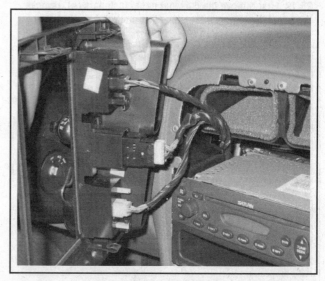

25.10 . . . then disconnect the electrical connectors from the back

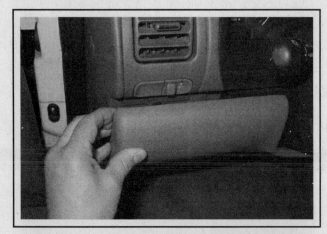

25.13 Carefully pull the knee bolster off the instrument panel

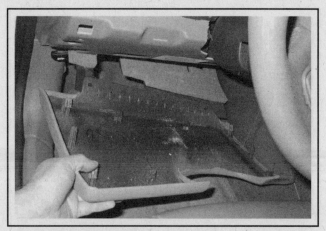

25.14 Remove the pin retainers from the rear of the sound insulator and remove the knee bolster and sound insulator as one complete assembly

➥Note: It will be necessary to pull the center trim panel assembly out slightly to gain access to the remaining electrical connectors.

11 Installation is the reverse of removal. Make sure the clips are engaged properly before pushing the panel firmly into place.

12 Reconnect the battery (see Chapter 5, Section 1).

KNEE BOLSTER

◆ **Refer to illustrations 25.13 and 25.14**

13 Pull the knee bolster toward the front seat to release the clips from the instrument panel (see illustration).

14 Separate the sound insulator from the lower dash. Do not disconnect the insulator from the knee bolster but remove it as a complete assembly (see illustration).

15 Remove the knee bolster from the dash area.

16 Installation is the reverse of removal. Make sure the clips are engaged properly before pushing the knee bolster firmly into place.

GLOVE BOX

◆ **Refer to illustrations 25.19, 25.20, 25.21 and 25.22**

17 Disconnect the cable from the negative battery terminal (see Chapter 5).

18 Remove the glove box lamp assembly. Disconnect the electrical connector from the lamp assembly.

19 Remove the closeout panel fasteners (see illustration) and slide the closeout panel toward the front seat to disengage it from the interior clips.

20 Remove the glove box mounting screws. Unhook the two plastic hinges from the instrument panel (see illustration) and remove the door.

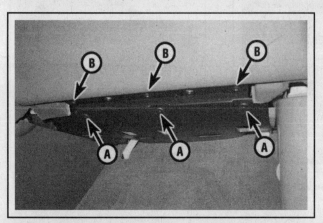

25.19 First remove the closeout panel fasteners (A), remove the panel then remove the glove box mounting screws (B)

25.20 Lower the glove box and slide the hinges out of the locating holes in the glove box housing

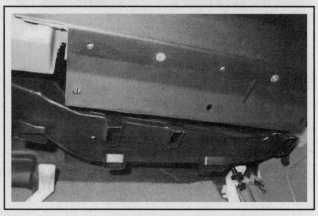

25.21 Remove the pin retainer from the lower heater duct

BODY 11-31

21 Remove the pin retainer from the lower heater duct (see illustration). Separate the lower heater duct from the glove box.

22 Remove the glove box mounting fasteners (see illustration).

23 Slide the glove box from the instrument panel, disconnect the electrical connectors from the Body Control Module (BCM) (see Chapter 6).

24 Installation is the reverse of removal.

25 Reconnect the battery (see Chapter 5, Section 1).

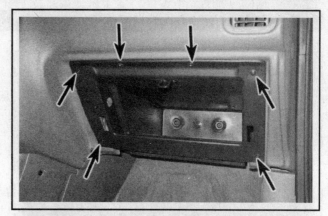

25.22 Remove the glove box mounting fasteners

26 Steering column covers - removal and installation

♦ Refer to illustrations 26.2a and 26.2b

WARNING:

The models covered by this manual are equipped with Supplemental Restraint systems (SRS), more commonly known as airbags. Always disarm the airbag system before working in the vicinity of any airbag system component to avoid the possibility of accidental deployment of the airbag, which could cause personal injury (see Chapter 12). Do not use a memory saving device to preserve the PCM's memory when working on or near airbag system components.

1 On tilt steering columns, move the column to the lowest position.

2 Remove the screws, then separate the halves and remove the upper and lower steering column covers (see illustrations).

3 Installation is the reverse of the removal procedure.

26.2a Remove the steering column cover mounting screws . . .

26.2b . . . and separate the upper and lower halves from each other

11-32 BODY

27 Seats - removal and installation

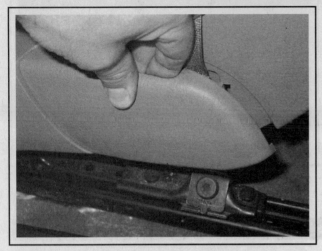

27.1 Remove the front recliner trim panel from the outer seat pad

27.3 Remove the seat belt mounting bolt

27.4 Slide the seat to the rear of the vehicle and unhook the seat tracks

⚠ WARNING:

The models covered by this manual are equipped with Supplemental Restraint systems (SRS), more commonly known as airbags. Always disarm the airbag system before working in the vicinity of any airbag system component to avoid the possibility of accidental deployment of the airbag, which could cause personal injury (see Chapter 12). Do not use a memory saving device to preserve the PCM's memory when working on or near airbag system components.

FRONT SEAT

Refer to illustrations 27.1, 27.3, 27.4 and 27.5

1 Remove the front recliner trim panel (see illustration).
2 On power seat models, remove the seat belt webbing trim panel from the side of the seat.
3 Remove the bolt from the seat belt (see illustration).
4 Slide the seat to the rear of the vehicle to unhook the seat tracks (see illustration).
5 Tilt the seat to the rear, disconnect any electrical connectors (see illustration) and lift the seat from the vehicle.
6 Installation is the reverse of removal.

REAR SEAT

▶ **Refer to illustrations 27.7, 27.10 and 27.11**

⚠ CAUTION:

The seat back assembly is equipped with a "snap fit" bracket that allows the rear seat to lock into position. Do not use excessive force to pry on the bracket or it may become damaged. Move the bracket only enough to lift the seat back from the vehicle. If the bracket has moved away from the seat, carefully bend it into position before installing the rear seat back.

27.5 Disconnect the seat electrical connector

BODY   11-33

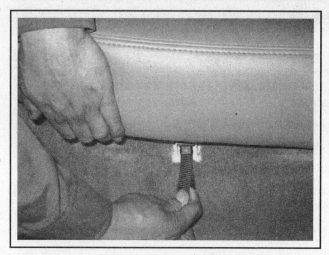

27.7 Pull on the rear seat strap while lifting the cushion to disengage the rear seat from the forward latch

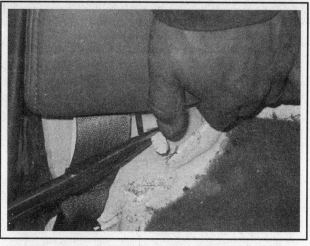

27.10 Locate the metal tab, bend it out of the way and slide the seat latch over the metal tab

7 Pull on the rear seat strap while pulling up on the cushion to disengage the rear seat from the forward latch (see illustration).

8 Push the seat cushion to the rear of the vehicle to unhook the rear latch.

9 Lift the rear seat cushion while sliding the seat belt through the cushion opening. Move the rear seat out of the vehicle.

10 Push on the seat back cushion, locate the metal tab and bend it out of the way (see illustration). Move the latch over the metal tab.

11 Use a flat-bladed screwdriver to separate the outer bracket from the pivot pin (see illustration). Lift up on the rear seat back and lift it from the vehicle.

12 Installation is the reverse of removal. Guide the seat back assembly onto the center pivot pin. Align the outer pivot pin with the ramp center on the bracket. Push down on the seat back to engage the bracket. Raise the seat slightly to lock the seat back assembly into position.

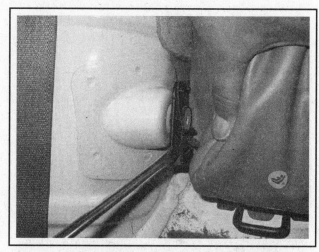

27.11 Pry the seat bracket retainer in, toward the seat to release the "snap fit" bracket from the pivot pin

Specifications

Torque specifications

	Ft-lbs (unless otherwise indicated)	Nm
Door outer panel mounting bolts	53 in-lbs	6

Notes

12 CHASSIS ELECTRICAL SYSTEM

Section

1 General information
2 Electrical troubleshooting - general information
3 Fuses and fusible links - general information
4 Circuit breakers - general information
5 Relays - general information and testing
6 Steering column switches - replacement
7 Ignition switch and key lock cylinder - replacement
8 Dashboard and center console switches - replacement
9 Instrument cluster - removal and installation
10 Wiper motors - check and replacement
11 Radio and speakers - removal and installation
12 Antenna - removal and installation
13 Rear window defogger - check and repair
14 Headlight bulb - replacement
15 Headlights - adjustment
16 Headlight housing - replacement
17 Horn - replacement
18 Bulb replacement
19 Electric side view mirrors - general information
20 Cruise control system - general information
21 Power window system - general information
22 Power door lock system - general information
23 Daytime Running Lights (DRL) - general information
24 Airbag system - general information and precautions
25 Wiring diagrams - general information

12-2 CHASSIS ELECTRICAL SYSTEM

1 General information

The electrical system is a 12-volt, negative ground type. Power for the lights and all electrical accessories is supplied by a lead/acid-type battery, which is charged by the alternator.

This Chapter covers repair and service procedures for the various electrical components not associated with the engine. Information on the battery, alternator and starter motor can be found in Chapter 5.

It should be noted that when portions of the electrical system are serviced, the negative battery cable should be disconnected from the battery to prevent electrical shorts and/or fires.

2 Electrical troubleshooting - general information

▶ Refer to illustrations 2.5a and 2.5b

1 A typical electrical circuit consists of an electrical component, any switches, relays, motors, fuses, fusible links or circuit breakers related to that component and the wiring and connectors that link the component to both the battery and the chassis. To help you pinpoint an electrical circuit problem, wiring diagrams are included at the end of this Chapter.

2 Before tackling any troublesome electrical circuit, first study the appropriate wiring diagrams to get a complete understanding of what makes up that individual circuit. Noting whether other components related to the circuit are operating correctly, for instance, can often narrow down the location of potential trouble spots. If several components or circuits fail at one time, chances are the problem is in a fuse or ground connection, because several circuits are often routed through the same fuse and ground connections.

3 Electrical problems usually stem from simple causes, such as loose or corroded connections, a blown fuse, a melted fusible link or a failed relay. Visually inspect the condition of all fuses, wires and connections in a problem circuit before troubleshooting the circuit.

4 If test equipment and instruments are going to be utilized, use the diagrams to plan ahead of time where you will make the necessary connections in order to accurately pinpoint the trouble spot.

5 For electrical troubleshooting, you'll need a circuit tester, voltmeter or a 12-volt bulb with a set of test leads, a continuity tester and a jumper wire, preferably with a circuit breaker incorporated, which can be used to bypass electrical components (see illustrations). Before attempting to locate a problem with test instruments, use the wiring diagram(s) to decide where to make the connections.

VOLTAGE CHECKS

▶ Refer to illustration 2.6

6 Voltage checks should be performed if a circuit is not functioning properly. Connect one lead of a circuit tester to either the negative

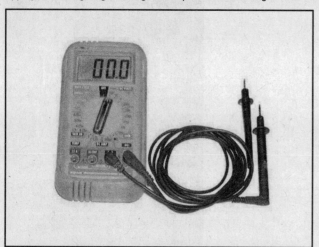

2.5a The most useful tool for electrical troubleshooting is a digital multimeter that can check volts, amps, and test continuity

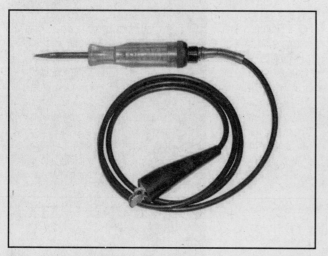

2.5b A simple test light is a very handy tool for testing voltage

2.6 In use, a basic test light's lead is clipped to a known good ground, then the pointed probe can test connectors, wires or electrical sockets - if the bulb lights, the circuit being tested has battery voltage

battery terminal or a known good ground. Connect the other lead to a connector in the circuit being tested, preferably nearest to the battery or fuse (see illustration). If the bulb of the tester lights, voltage is present, which means that the part of the circuit between the connector and the battery is problem free. Continue checking the rest of the circuit in the same fashion. When you reach a point at which no voltage is present, the problem lies between that point and the last test point with voltage. Most of the time the problem can be traced to a loose connection.

➡ Note: Keep in mind that some circuits receive voltage only when the ignition key is in the ACC or RUN position.

FINDING A SHORT

7 One method of finding shorts in a live circuit is to remove the fuse and connect a test light in place of the fuse terminals (fabricate two jumper wires with small spade terminals, plug the jumper wires into the fuse box and connect the test light). There should be no voltage present in the circuit. Move the suspected wiring harness from side-to-side while watching the test light. If the bulb goes on, there is a short to ground somewhere in that area, probably where the insulation has rubbed through.

GROUND CHECK

8 Perform a ground test to check whether a component is properly grounded. Disconnect the battery and connect one lead of a continuity tester or multimeter (set to the ohm scale), to a known good ground. Connect the other lead to the wire or ground connection being tested. If the resistance is low (less than 5 ohms), the ground is good. If the bulb on a self-powered test light does not go on, the ground is not good.

CONTINUITY CHECK

▶ Refer to illustration 2.9

9 A continuity check determines whether there are any breaks in a circuit, i.e. whether it can no longer carry current from the voltage source to ground. With the circuit off (no power in the circuit), a self-powered continuity tester or multimeter can be used to check the circuit. Connect the test leads to both ends of the circuit (or to the "power" end and a good ground), and if the test light comes on the circuit is passing current properly (see illustration). If the resistance is low (less than 5 ohms), there is continuity; if the reading is 10,000 ohms or higher, there is a break somewhere in the circuit. The same procedure can be used to test a switch, by connecting the continuity tester to the switch terminals. With the switch turned to ON, the test light should come on (or low resistance should be indicated on a meter).

FINDING AN OPEN CIRCUIT

10 When diagnosing for possible open circuits, it is often difficult to locate them by sight because the connectors hide oxidation or terminal

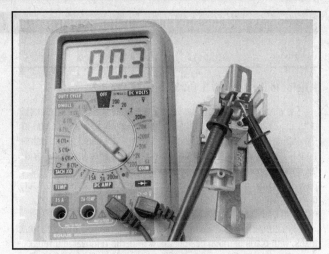

2.9 With a multimeter set to the ohm scale, resistance can be checked across two terminals - when checking for continuity, a low reading indicates continuity, a high reading or infinity indicates high resistance or lack of continuity

misalignment. Merely wiggling a connector on a sensor or in the wiring harness may correct the open circuit condition. Remember this when an open circuit is indicated when troubleshooting a circuit. Intermittent problems may also be caused by oxidized or loose connections.

11 Electrical troubleshooting is simple if you keep in mind that all electrical circuits are basically electricity running from the battery, through the wires, switches, relays, fuses and fusible links to each electrical component (light bulb, motor, etc.) and to ground, from which it is passed back to the battery. Any electrical problem is an interruption in the flow of electricity to and from the battery.

CONNECTORS

12 Most electrical connections on these vehicles are made with multiwire plastic connectors. The mating halves of many connectors are secured with locking clips molded into the plastic connector shells. The mating halves of large connectors, such as some of those under the instrument panel, are held together by a bolt through the center of the connector.

13 To separate a connector with locking clips, use a small screwdriver to pry the clips apart carefully, then separate the connector halves. Pull only on the shell, never pull on the wiring harness as you may damage the individual wires and terminals inside the connectors. Look at the connector closely before trying to separate the halves. Often the locking clips are engaged in a way that is not immediately clear. Additionally, many connectors have more than one set of clips.

14 Each pair of connector terminals has a male half and a female half. When you look at the end view of a connector in a diagram, be sure to understand whether the view shows the harness side or the component side of the connector. Connector halves are mirror images of each other, and a terminal that is shown on the right side end-view of one half will be on the left side end view of the other half.

12-4 CHASSIS ELECTRICAL SYSTEM

3 Fuses and fusible links - general information

FUSES

▶ **Refer to illustrations 3.1a, 3.1b, 3.1c and 3.2**

The electrical circuits of the vehicle are protected by a combination of fuses, circuit breakers and fusible links. Fuse blocks are located in the engine compartment and under the instrument panel (see illustrations). Each of the fuses is designed to protect a specific circuit, and the various circuits are identified on the fuse panel cover. If the fuse panel cover is difficult to read, or missing, you can also refer to your owner's manual, which includes a complete guide to all fuses and relays in both fuse/relay boxes.

Miniaturized fuses are employed in the fuse blocks. These compact fuses, with blade terminal design, allow fingertip removal and replacement. If an electrical component fails, always check the fuse first. The best way to check a fuse is with a test light. Check for power at the exposed terminal tips of each fuse. If power is present on one side of the fuses but not the other, the fuse is blown. A blown fuse can also be confirmed by visually inspecting it (see illustration).

Be sure to replace blown fuses with the correct type. Fuses of different ratings are physically interchangeable, but only fuses of the proper rating should be used. Replacing a fuse with one of a higher or lower value than specified is not recommended. Each electrical circuit needs a specific amount of protection. The amperage value of each fuse is molded into the fuse body.

If the replacement fuse immediately fails, don't replace it again until the cause of the problem is isolated and corrected. In most cases, this will be a short circuit in the wiring caused by a broken or deteriorated wire.

FUSIBLE LINKS

Some circuits are protected by fusible links. The links are used in circuits which are not ordinarily fused, or which carry high current, such as the circuit between the alternator and the starter motor. Fusible links, which are usually four wire gauges smaller in size than the circuit that they protect, are designed to melt if the circuit is subjected to more current than it was designed to carry. If you have to replace a blown fusible link, make sure that you replace it with one of the same specification. If the replacement fusible link blows in the same circuit, make sure that you troubleshoot the circuit in which the fusible link melted BEFORE installing another fusible link.

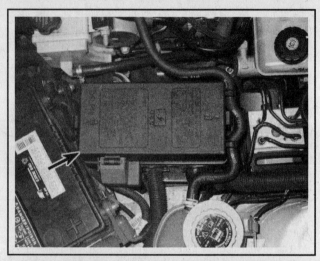

3.1a The engine compartment fuse/relay box is mounted on the left side of the engine compartment; all fuses and relays are listed by location and function on the underside of the fuse/relay box cover

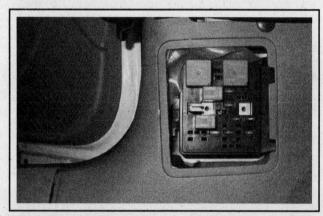

3.1b There are two fuse/relay boxes inside the vehicle; one is in the left kick panel . . .

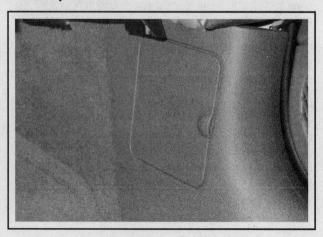

3.1c . . . and the other is in the right kick panel

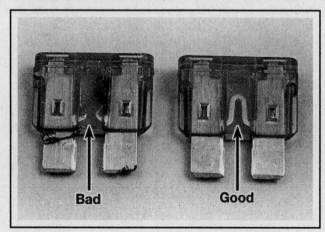

3.2 When a fuse blows, the element between the terminals melts - the fuse on the left is blown, the fuse on the right is good

CHASSIS ELECTRICAL SYSTEM   12-5

4 Circuit breakers - general information

Circuit breakers protect certain circuits, such as the power windows or heated seats. Depending on the vehicle's accessories, there may be one or two circuit breakers, located in the fuse/relay box in the engine compartment.

Because the circuit breakers reset automatically, an electrical overload in a circuit-breaker-protected system will cause the circuit to fail momentarily, then come back on. If the circuit does not come back on, check it immediately.

For a basic check, pull the circuit breaker up out of its socket on the fuse panel, but just far enough to probe with a voltmeter. The breaker should still contact the sockets.

With the voltmeter negative lead on a good chassis ground, touch each end prong of the circuit breaker with the positive meter probe. There should be battery voltage at each end. If there is battery voltage only at one end, the circuit breaker must be replaced.

Some circuit breakers must be reset manually.

5 Relays - general information and testing

GENERAL INFORMATION

1 Several electrical accessories in the vehicle, such as the fuel injection system, horns, starter, and fog lamps use relays to transmit the electrical signal to the component. Relays use a low-current circuit (the control circuit) to open and close a high-current circuit (the power circuit). If the relay is defective, that component will not operate properly. Most relays are mounted in the engine compartment fuse/relay box, with some specialized relays located above the interior fuse box in the dash (see illustrations 3.1a, 3.1b and 3.1c). If a faulty relay is suspected, it can be removed and tested using the procedure below or by a dealer service department or a repair shop. Defective relays must be replaced as a unit.

TESTING

♦ Refer to illustrations 5.2a and 5.2b

2 Most of the relays used in these vehicles are of a type often called "ISO" relays, which refers to the International Standards Organization. The terminals of ISO relays are numbered to indicate their usual circuit connections and functions. There are two basic layouts of termi-

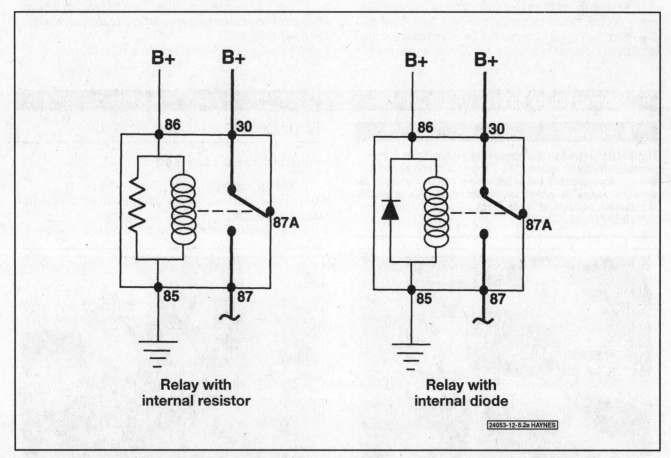

5.2a Typical ISO relay designs, terminal numbering and circuit connections

12-6 CHASSIS ELECTRICAL SYSTEM

5.2b Most relays are marked on the outside to easily identify the control circuits and the power circuits - four terminal type shown

nals on the relays used in these vehicles (see illustrations).

3 Refer to the wiring diagram for the circuit to determine the proper connections for the relay you're testing. If you can't determine the correct connection from the wiring diagrams, however, you may be able to determine the test connections from the information that follows.

4 Two of the terminals are the relay control circuit and connect to the relay coil. The other relay terminals are the power circuit. When the relay is energized, the coil creates a magnetic field that closes the larger contacts of the power circuit to provide power to the circuit loads.

5 Terminals 85 and 86 are normally the control circuit. If the relay contains a diode, terminal 86 must be connected to battery positive (B+) voltage and terminal 85 to ground. If the relay contains a resistor, terminals 85 and 86 can be connected in either direction with respect to B+ and ground.

6 Terminal 30 is normally connected to the battery voltage (B+) source for the circuit loads. Terminal 87 is connected to the circuit leading to the component being powered. If the relay has several alternate terminals for load or ground connections, they usually are numbered 87A, 87B, 87C, and so on.

7 Use an ohmmeter to check continuity through the relay control coil.

  a) *Connect the meter according to the polarity shown in illustration 5.2a for one check; then reverse the ohmmeter leads and check continuity in the other direction.*
  b) *If the relay contains a resistor, resistance will be indicated on the meter, and should be the same value with the ohmmeter in either direction.*
  c) *If the relay contains a diode, resistance should be higher with the ohmmeter in the forward polarity direction than with the meter leads reversed.*
  d) *If the ohmmeter shows infinite resistance in both directions, replace the relay.*

8 Remove the relay from the vehicle and use the ohmmeter to check for continuity between the relay power circuit terminals. There should be no continuity between terminal 30 and 87 with the relay de-energized.

9 Connect a fused jumper wire to terminal 86 and the positive battery terminal. Connect another jumper wire between terminal 85 and ground. When the connections are made, the relay should click.

10 With the jumper wires connected, check for continuity between the power circuit terminals. Now, there should be continuity between terminals 30 and 87.

11 If the relay fails any of the above tests, replace it.

6 Steering column switches - replacement

※※ WARNING:

The models covered by this manual are equipped with a Supplemental Restraint System (SRS), more commonly known as airbags. Always disarm the airbag system before working in the vicinity of any airbag system component to avoid the possibility of accidental deployment of the airbag, which could cause personal injury (see Section 24).

TURN SIGNAL/MULTI-FUNCTION SWITCH

▶ Refer to illustrations 6.2 and 6.3

1 Remove the steering column covers (see Chapter 11).
2 Remove the turn signal/multi-function switch from the steering column (see illustration).
3 Disconnect the electrical connector from the multi-function

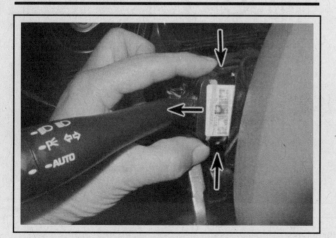

6.2 To release the turn signal/multi-function switch from the steering column, depress these two locking tabs and pull it out

6.3 After removing the turn signal/multi-function switch from the steering column, depress this locking tab and unplug the electrical connector

CHASSIS ELECTRICAL SYSTEM   12-7

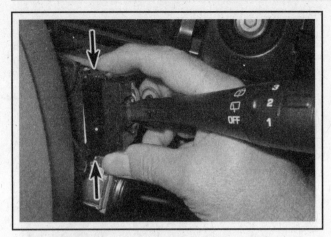

6.6 To release the windshield wiper/washer switch from the steering column, depress these two locking tabs and pull it out

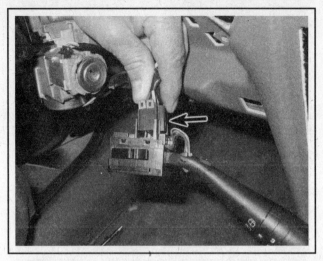

6.7 After removing the windshield wiper/washer switch from the steering column, depress this locking tab and unplug the electrical connector

switch (see illustration).
4 Installation is the reverse of removal.

WINDSHIELD WIPER/WASHER SWITCH

▶ **Refer to illustrations 6.6 and 6.7**

5 Remove the steering column covers (see Chapter 11).
6 Remove the windshield wiper/washer switch from the steering column (see illustration).
7 Disconnect the electrical connector from the windshield wiper/washer switch (see illustration).
8 Installation is the reverse of removal.

7 Ignition switch and key lock cylinder - replacement

※ WARNING:

The models covered by this manual are equipped with a Supplemental Restraint System (SRS), more commonly known as airbags. Always disarm the airbag system before working in the vicinity of any airbag system component to avoid the possibility of accidental deployment of the airbag, which could cause personal injury (see Section 24).

1 Disconnect the cable from the negative terminal of the battery (see Chapter 5, Section 1).

IGNITION SWITCH

▶ **Refer to illustration 7.3**

2 Remove the steering column covers (see Chapter 11).
3 Disconnect the two electrical connectors from the ignition switch (see illustration).
4 Remove the two switch mounting screws and remove the ignition switch.
5 Installation is the reverse of removal. Reconnect the battery (see Chapter 5, Section 1).

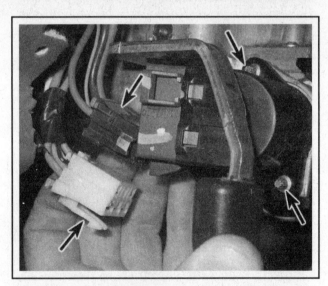

7.3 To remove the ignition switch from the steering column, disconnect the two electrical connectors and remove these two reverse-Torx head screws

12-8 CHASSIS ELECTRICAL SYSTEM

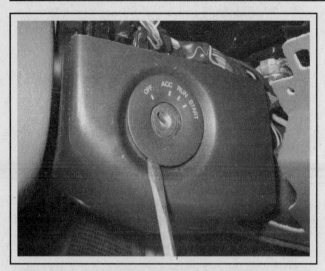

7.6 Carefully pry off the trim bezel around the key lock cylinder with a small screwdriver

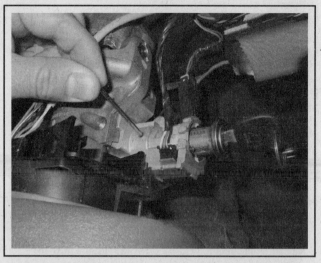

7.9a To remove the key lock cylinder, rotate the cylinder to the START position, depress the cylinder locking button by inserting a thin awl or punch through the hole in the top of the lock cylinder housing and pushing it in until it stops . . .

KEY LOCK CYLINDER

▶ Refer to illustrations 7.6 and 7.9a, 7.9b and 7.10

6 Remove the lock cylinder trim bezel from the lower steering column cover (see illustration).
7 Remove the steering column covers (see Chapter 11).
8 Remove the windshield wiper/washer switch (see Section 6).
9 Insert the ignition key into the lock cylinder. Rotate the key lock cylinder to the START position, insert a 1/8-inch awl or punch through the hole in the top of the casting that houses the key lock cylinder (see illustration) and pull out the cylinder (see illustration).

10 Before installing the lock cylinder, make sure that the actuator blade inside the key lock cylinder bore is still in the RUN position (see illustration). If it isn't, rotate it to this position with a pair of needle-nose pliers. The key lock cylinder will not fit all the way into the bore if the actuator blade is in any other position.
11 Installation is otherwise the reverse of removal. Reconnect the battery (see Chapter 5, Section 1).
12 When you're done, verify that the ignition switch operates correctly in the OFF, ACC, RUN and START positions.

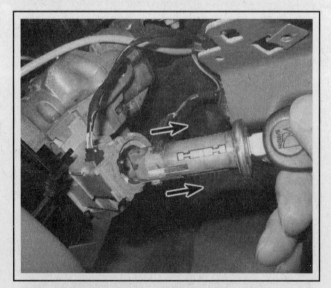

7.9b . . . then pull out the key lock cylinder

7.10 Before installing the key lock cylinder into its bore, make sure that the actuator blade is in the RUN position; if it's been moved, rotate it to the RUN position (shown) with a pair of needle-nose pliers (the key lock cylinder won't seat correctly unless the actuator blade is in this position)

CHASSIS ELECTRICAL SYSTEM   12-9

8 Dashboard and center console switches - replacement

> **WARNING:**
> The models covered by this manual are equipped with a Supplemental Restraint System (SRS), more commonly known as airbags. Always disarm the airbag system before working in the vicinity of any airbag system component to avoid the possibility of accidental deployment of the airbag, which could cause personal injury (see Section 24).

SWITCHES IN THE CENTER TRIM PANEL

1 The center trim panel houses up to five switches: driver's seat heater switch, fog lamp switch, hazard flasher switch, traction control switch and passenger's seat heater switch. All of these switches (except for the hazard flasher switch) are controls for optional accessories. To remove any of these switches, you must first remove the center trim panel (see Chapter 11).

Left heated front seat switch and fog lamp switch

▸ Refer to illustrations 8.2a and 8.2b

➡Note: *The accompanying illustrations depict the removal of the left heated front seat switch from the center trim panel, but the procedure applies to the fog lamp switch, if equipped, as well.*

2 Release the tabs (see illustration) and push the heated front seat/fog lamp switches out of the center trim panel from the backside (see illustration).

3 Installation is the reverse of removal.

Hazard/turn signal flasher and hazard flasher switch

▸ Refer to illustrations 8.8a and 8.8b

4 The turn signal lights and all exterior lights that are part of the hazard flasher system are controlled from a single electronic flasher and switch unit, which is mounted in the center of the center trim panel.

5 When the flasher unit is functioning properly, an audible click can be heard during its operation. If the turn signal indicator on the instrument cluster flashes more rapidly than normal when you activate the turn signal stalk to indicate a left or right turn, then the filament of a turn signal bulb is probably blown.

6 If BOTH turn signals on one side fail to blink correctly, or at all, the problem might be a blown fuse, a faulty flasher unit, a broken turn signal or hazard flasher switch or a loose or open connection.

7 Check the fuse for the turn signal and side marker lights. This fuse is located in the engine compartment fuse/relay box. If the turn signal fuse has blown, check the wiring for a short before installing a new fuse.

8 To replace the hazard flasher and switch assembly, release the tabs on both sides of the unit (see illustration) and push the switch/flasher assembly out of the center trim panel from the backside (see illustration).

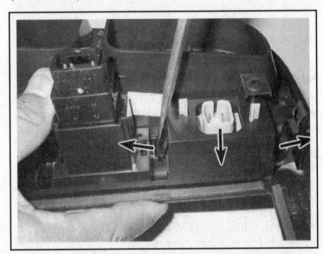

8.2a To remove the switches for the left heated front seat and the fog lamps, use a small screwdriver or a pick to release the tabs on both sides of the switch housing . . .

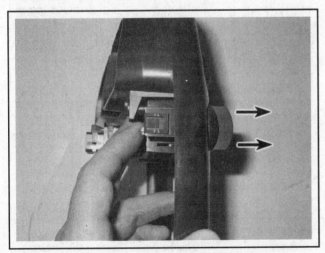

8.2b . . . then push the switch assembly out of the center trim panel from the backside (this procedure applies to the traction control switch and the right heated front seat switch as well)

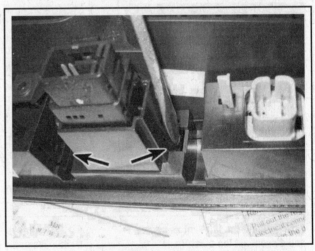

8.8a To remove the hazard flasher switch/flasher unit from the center trim panel, release the tabs on both sides of the housing with a small screwdriver or pick . . .

12-10 CHASSIS ELECTRICAL SYSTEM

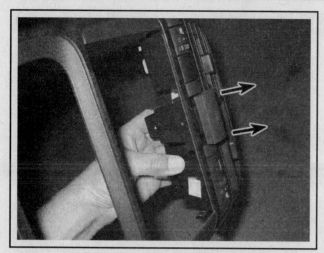

8.8b ... then push the switch out of the center trim panel from the backside

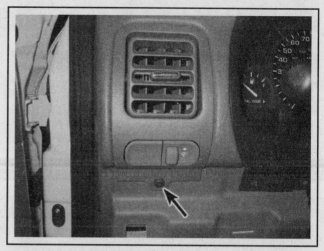

8.13 To detach the left vent assembly from the dashboard, remove this retaining screw

9 Make SURE that the replacement unit is identical to the original. Compare the old one to the new one before installing it.

10 Installation is the reverse of removal.

Traction control switch and right heated front seat switch

➡ **Note: On vehicles equipped with BOTH traction control and heated seats, the traction control switch and the right heated front seat switch are an integral part of the same assembly and cannot be replaced separately. If either switch is defective, you must replace both of them.**

11 These switches are removed in the same manner as the left heated front seat switch and the fog lamp switch (see Steps 2 and 3).

INSTRUMENT PANEL DIMMER SWITCH (RHEOSTAT)

▸ **Refer to illustrations 8.13 and 8.14**

12 Remove the knee bolster (see Chapter 11).

13 Remove the left vent/dimmer switch assembly retaining screw (see illustration).

14 Pull out the vent/dimmer switch assembly and disconnect the electrical connector (see illustration).

15 Depress the locking tabs and remove the dimmer switch from the vent trim panel.

16 Installation is the reverse of removal.

SWITCHES IN THE CENTER CONSOLE

Power window switches

▸ **Refer to illustrations 8.17 and 8.18**

17 Carefully pry the power window switch out of the center console with a panel removal tool or a screwdriver (see illustration).

18 Disconnect the electrical connector from the power window switch (see illustration).

19 Installation is the reverse of removal.

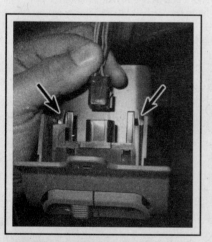

8.14 Pull out the left vent assembly and disconnect the electrical connector, then depress the two locking tabs and slide the switch out of the vent panel

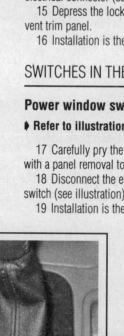

8.17 To remove a power window switch assembly from the center console, carefully pry it loose with a panel removal tool or a screwdriver

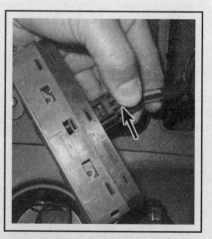

8.18 After removing the power window switch, depress the locking tab on the side of the electrical connector and unplug the connector

CHASSIS ELECTRICAL SYSTEM   12-11

9 Instrument cluster - removal and installation

▶ Refer to illustrations 9.4a, 9.4b and 9.4c

⚠ WARNING:
The models covered by this manual are equipped with a Supplemental Restraint System (SRS), more commonly known as airbags. Always disarm the airbag system before working in the vicinity of any airbag system component to avoid the possibility of accidental deployment of the airbag, which could cause personal injury (see Section 24).

1 Remove the knee bolster and the steering column covers (see Chapter 11).
2 Tilt the steering wheel to its lowest position.
3 Remove the instrument cluster bezel (see Chapter 11).
4 Remove the instrument cluster retaining screws (see illustrations), then pull out the cluster (see illustration) and remove it.
5 When installing the instrument cluster, make sure that the cluster side of the electrical connector is aligned with, and plugged into, the dash side of the connector before installing the cluster retaining screws. Installation is otherwise the reverse of removal.

9.4a To remove the instrument cluster, remove these two screws from the left side . . .

9.4b . . . remove these two from the right . . .

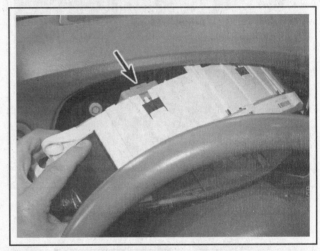
9.4c . . . and pull out the cluster; when installing the cluster, make sure that the electrical connector on the backside of the cluster is aligned with, and plugged into, the dash side of the connector before installing the retaining screws

10 Wiper motors - check and replacement

WIPER MOTOR CIRCUIT CHECK

➡ **Note:** When checking for voltage, probe a grounded 12-volt test light to each terminal at a connector until it lights; this verifies voltage (power) at the terminal. If the following checks fail to locate the problem, have the system diagnosed by a dealer service department or other properly equipped repair facility.

1 If the wipers work slowly, make sure the battery is in good condition and has a strong charge (see Chapter 5). If the battery is in good condition, remove the wiper motor (see below) and operate the wiper arms by hand. Check for binding linkage and pivots. Lubricate or repair the linkage or pivots as necessary. Reinstall the wiper motor. If the wipers still operate slowly, check for loose or corroded connections, especially the ground connection. If all connections look OK, replace the motor.

2 If the wipers fail to operate when activated, check the fuse (see Section 3). If the fuse is OK, connect a jumper wire between the wiper motor's ground terminal and ground, then retest. If the motor works now, repair the ground connection. If the motor still doesn't work, turn the wiper switch to the HI position and check for voltage at the motor.

➡ **Note:** The hood seal and cowl cover must be removed to access the electrical connector (see Chapter 11).

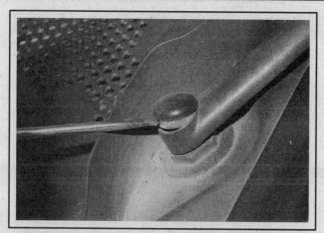

10.6a To remove the windshield wiper arms, carefully pry off each trim cap with a small screwdriver and remove the wiper arm retaining nut . . .

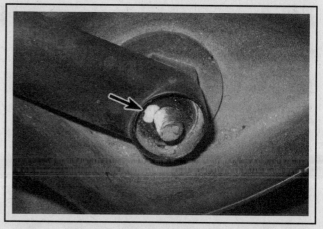

10.6b . . . then mark the relationship of each arm to its shaft BEFORE removing the arm

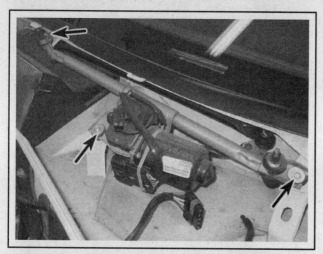

10.8 Disconnect the electrical connector, then remove these three bolts to detach the windshield wiper motor and linkage from the cowl

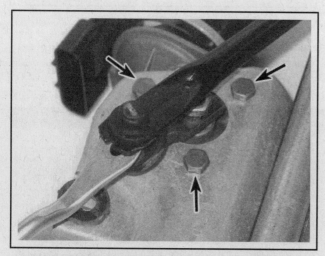

10.11 To separate the windshield wiper motor from the linkage assembly, pry the link arm loose from the spherical bearing on the end of the wiper motor's actuator arm, then remove the three motor mounting bolts

3 If there's voltage at the connector, remove the motor and check it off the vehicle with fused jumper wires from the battery. If the motor now works, check for binding linkage (see Step 1). If the motor still doesn't work, replace it. If there's no voltage to the motor, check for voltage at the wiper control relays. If there's voltage at the wiper control relays and no voltage at the wiper motor, have the switch tested. If the switch is OK, the wiper control relay is probably bad. See Section 5 for relay testing.

4 If the interval (delay) function is inoperative, check the continuity of all the wiring between the switch and wiper control module.

5 If the wipers stop at the position they're in when the switch is turned off (fail to park), check for voltage at the park feed wire of the wiper motor connector when the wiper switch is OFF but the ignition is ON. If no voltage is present, check for an open circuit between the wiper motor and the fuse panel.

WIPER MOTOR REPLACEMENT

Windshield wiper motor

▶ Refer to illustrations 10.6a, 10.6b, 10.8 and 10.11

6 Pry off the windshield wiper trim caps (see illustration) and remove the windshield wiper retaining nuts and washers. Mark the position of each wiper arm in relation to its shaft (see illustration), then remove the wiper arms. Store the trim caps, retaining nuts and washers in a plastic bag.

7 Remove the hood seal and the cowl cover (see Chapter 11).

8 Disconnect the windshield wiper motor electrical connector (see illustration).

9 Remove the windshield wiper motor and linkage assembly mounting bolts (see illustration 10.8).

10 Remove the windshield wiper motor and linkage assembly from the cowl area.

11 Pry the link arm loose from the spherical bearing on the end of the motor's actuator arm (see illustration), then remove the three motor mounting bolts and remove the motor.

12 Installation is the reverse of removal. Be sure to align the marks you made on the windshield wiper arms and the wiper arm shafts.

CHASSIS ELECTRICAL SYSTEM   12-13

10.13 To remove the rear window wiper arm, flip open the trim cap, remove the nut and washer, mark the relationship of the wiper arm to the wiper motor shaft and pull the arm off the shaft

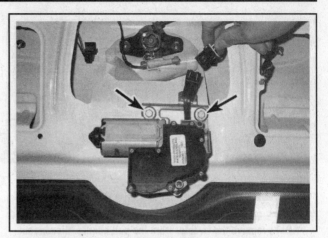

10.15 To remove the rear wiper motor from the liftgate on station wagon models, disconnect the electrical connector and remove the motor mounting bolts

Rear window wiper motor (station wagon models)

♦ Refer to illustrations 10.13 and 10.15

13 Flip open the protective cover for the nut that secures the wiper arm to the wiper motor shaft, then remove the nut (see illustration). Remove the washer, mark the relationship of the wiper arm to the wiper motor shaft (see illustration 10.6b) and remove the wiper arm.

14 Open the liftgate and remove the liftgate trim panel (see Chapter 11).
15 Disconnect the electrical connector from the rear wiper motor (see illustration).
16 Remove the motor mounting bolts (see illustration 10.15).
17 Installation is the reverse of removal. Be sure to align the marks you made on the windshield wiper arm and the wiper arm shaft.

11 Radio and speakers - removal and installation

WARNING:
The models covered by this manual are equipped with a Supplemental Restraint System (SRS), more commonly known as airbags. Always disarm the airbag system before working in the vicinity of any airbag system component to avoid the possibility of accidental deployment of the airbag, which could cause personal injury (see Section 24).

RADIO

♦ Refer to illustrations 11.2 and 11.3

1 Remove the center trim panel (see Chapter 11).
2 Remove the radio mounting screws (see illustration) and pull the radio out from the dash.
3 Disconnect the antenna cable, the electrical connectors and the ground strap from the back of the radio (see illustration) and remove the radio from the dash.
4 Installation is the reverse of removal.

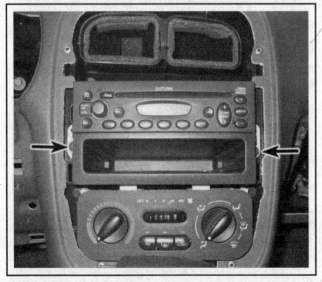

11.2 To detach the radio from the dash, remove these two screws

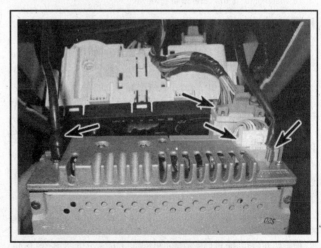

11.3 Pull out the radio out from the dash and disconnect the antenna cable, the electrical connectors and the ground strap

12-14 CHASSIS ELECTRICAL SYSTEM

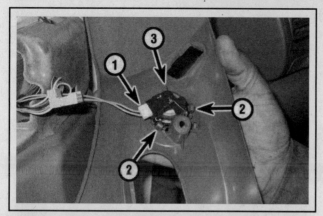

11.6 To remove the tweeter from the pull handle, disconnect the electrical connector (1), spread the two locking tangs (2) apart slightly and pull the tweeter out of the pull handle. When installing the tweeter, align it with the mounting hole and push it into place until the two locking tangs snap into place. The locator rib (3) is there simply to make sure that the tweeter is centered correctly

SPEAKERS

Front door speakers

Upper front door speakers (tweeters)

▶ Refer to illustration 11.6

5 Remove the front door pull handle (see Chapter 11).
6 Disconnect the electrical connector from the tweeter (see illustration).
7 Spread the two locking tangs slightly (see illustration 11.6) and pull the tweeter out of the pull handle.
8 Installation is the reverse of removal.

Lower front door speakers

▶ Refer to illustration 11.10

9 Remove the front door trim panel (see Chapter 11).
10 Remove the speaker mounting screws (see illustration).
11 Pull out the speaker and disconnect the electrical connector.
12 Installation is the reverse of removal.

Rear door speakers

Upper rear door speakers (tweeters)

▶ Refer to illustration 11.14

13 Remove the rear door pull handle (see Chapter 11).
14 Disconnect the electrical connector from the tweeter (see illustration).
15 Spread the two locking tangs slightly (see illustration 11.14) and pull the tweeter out of the pull handle.
16 Installation is the reverse of removal.

Lower rear door speakers

17 Remove the rear door trim panel (see Chapter 11).
18 Remove the speaker mounting screws (see illustration 11.10) and pull the speaker out of its receptacle in the door.
19 Disconnect the electrical connector and remove the speaker from the vehicle.
20 Installation is the reverse of removal.

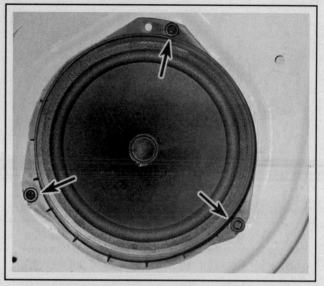

11.10 To remove a door speaker, remove these mountings screws, then pull out the speaker and disconnect the electrical connector

11.14 To remove the tweeter from the rear door pull handle, disconnect the electrical connector, spread the two locking tangs slightly and pull the tweeter out

Rear speakers (subwoofers)

Note: Not all models are equipped with rear speakers, which are the subwoofers for an optional sound system.

Sedans

21 Remove the rear window shelf (see Chapter 11).
22 Remove the subwoofer mounting screws and pull out the subwoofer.
23 Disconnect the electrical connector from the subwoofer and remove the subwoofer.
24 Installation is the reverse of removal.

Station wagons

25 Remove the subwoofer grille from the subwoofer.
26 Disconnect the electrical connector from the subwoofer.
27 Remove the subwoofer mounting bolts and remove the subwoofer.
28 Installation is the reverse of removal.

CHASSIS ELECTRICAL SYSTEM 12-15

12 Antenna - removal and installation

SEDANS

1 Unscrew the antenna mast from its mounting base and remove it.
2 Open the access panel for the vehicle jack, which is located in the left side of the trunk, and remove the jack.
3 Remove the antenna mounting bracket nuts and remove the antenna and mounting bracket assembly through the hole in the rear quarter panel.
4 Disconnect the antenna lead from the antenna base lead.
5 Installation is the reverse of removal.

STATION WAGONS

▶ Refer to illustrations 12.6 and 12.7

6 Unscrew the antenna mast from its base and remove it (see illustration).
7 Remove the rear headliner molding and pull down the rear edge of the headliner far enough to access the antenna base (see illustration).
8 Disconnect the coaxial cable from the antenna base lead.
9 Remove the large nut that secures the antenna mounting base to the roof and remove the base.
10 Installation is the reverse of removal.

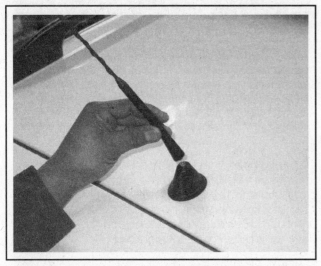

12.6 To remove the antenna from its mounting base, simply unscrew it

12.7 To access the antenna mounting base on a station wagon, remove the rear headliner molding and pull down the rear edge of the headliner; to remove the antenna base, simply unplug the coaxial cable from the base and unscrew the large retaining nut

13 Rear window defogger - check and repair

1 The rear window defogger consists of a number of horizontal elements baked onto the glass surface.
2 Small breaks in the element can be repaired without removing the rear window.

CHECK

▶ Refer to illustrations 13.4, 13.5 and 13.7

3 Turn the ignition switch and defogger system switches to the ON position. Using a voltmeter, place the positive probe against the defogger grid positive terminal and the negative probe against the ground terminal. If battery voltage is not indicated, check the fuse, defogger switch and related wiring. If voltage is indicated, but all or part of the defogger doesn't heat, proceed with the following tests.
4 When measuring voltage during the next two tests, wrap a piece of aluminum foil around the tip of the voltmeter positive probe and press the foil against the heating element with your finger (see illustration). Place the negative probe on the defogger grid ground terminal.

13.4 When measuring the voltage at the rear window defogger grid, wrap a piece of aluminum foil around the positive probe of the voltmeter and press the foil against the wire with your finger

12-16 CHASSIS ELECTRICAL SYSTEM

13.5 To determine if a heating element has broken, check the voltage at the center of each element; if the voltage is 5 or 6-volts, the element is unbroken, but if the voltage is 10 or 12-volts, the element is broken between the center and the ground side. If there is no voltage, the element is broken between the center and the positive side

5   Check the voltage at the center of each heating element (see illustration). If the voltage is 5 or 6-volts, the element is okay (there is no break). If the voltage is zero, the element is broken between the center of the element and the positive end. If the voltage is 10 to 12-volts the element is broken between the center of the element and ground. Check each heating element.

6   Connect the negative lead to a good body ground. The reading should stay the same. If it doesn't, the ground connection is bad.

13.7 To find the break, place the voltmeter negative lead against the defogger ground terminal, place the voltmeter positive lead with the foil strip against the heating element at the positive terminal end and slide it toward the negative terminal end. The point at which the voltmeter reading changes abruptly is the point at which the element is broken

7   To find the break, place the voltmeter negative probe against the defogger ground terminal. Place the voltmeter positive probe with the foil strip against the heating element at the positive terminal end and slide it toward the negative terminal end. The point at which the voltmeter deflects from several volts to zero is the point at which the heating element is broken (see illustration).

REPAIR

▶ Refer to illustration 13.13

8   Repair the break in the element using a repair kit specifically recommended for this purpose, available at most auto parts stores. Included in this kit is plastic conductive epoxy.

9   Prior to repairing a break, turn off the system and allow it to cool off for a few minutes.

10   Lightly buff the element area with fine steel wool, then clean it thoroughly with rubbing alcohol.

11   Use masking tape to mask off the area being repaired.

12   Thoroughly mix the epoxy, following the instructions provided with the repair kit.

13   Apply the epoxy material to the slit in the masking tape, overlapping the undamaged area about 3/4-inch on either end (see illustration).

14   Allow the repair to cure for 24 hours before removing the tape and using the system.

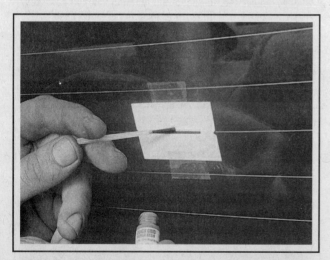

13.13 To use a defogger repair kit, apply masking tape to the inside of the window at the damaged area, then brush on the special conductive coating

CHASSIS ELECTRICAL SYSTEM 12-17

14 Headlight bulb - replacement

▶ Refer to illustrations 14.2 and 14.3

✳ WARNING:

Halogen gas-filled bulbs are under pressure and can shatter if the surface is scratched or the bulb is dropped. Wear eye protection and handle the bulbs carefully, grasping only the base whenever possible. Do not touch the surface of the bulb with your fingers because the oil from your skin could cause it to overheat and fail prematurely. If you do touch the bulb surface, clean it with rubbing alcohol.

1 If you're replacing the left headlight bulb, remove the fan control module (see illustration 3.3 in Chapter 5).
2 Disconnect the electrical connector from the headlight assembly (see illustration).
3 To remove the bulb from the headlight assembly, rotate it 1/4-turn counterclockwise (see illustration) and pull it out.
4 To install a new bulb, align the three tabs on the headlight mounting flange with the grooves in the mounting base (see illustration 14.3).
5 Installation is otherwise the reverse of removal.

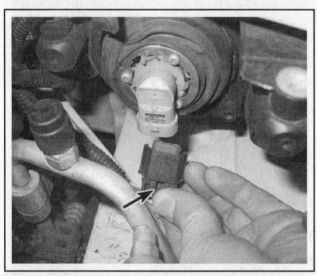

14.2 To release the headlight bulb electrical connector, depress this locking tab and pull down on the connector

14.3 To remove a headlight bulb from the headlight housing, rotate it counterclockwise and pull it out of the housing. To install the headlight bulb, make sure the three tabs on the mounting flange are aligned with the three grooves in the mounting base, then give the bulb a clockwise turn

15 Headlights - adjustment

▶ Refer to illustrations 15.1 and 15.3

➡ **Note:** *The headlights must be aimed correctly. If adjusted incorrectly they could blind the driver of an oncoming vehicle and cause a serious accident or seriously reduce your ability to see the road. The headlights should be checked for proper aim every 12 months and any time a new headlight is installed or front end body work is performed. It should be emphasized that the following procedure is only an interim step that will provide temporary adjustment until a properly equipped shop can adjust the headlights.*

1 The vertical adjustment screws are located behind each headlight housing (see illustration). (There are no horizontal adjustment screws.)
2 There are several methods for adjusting the headlights. The simplest method requires masking tape, a blank wall and a level floor.

15.1 Vertical adjustment screws for the right headlight assembly (the vertical adjustment screws for the left headlight assembly are identical)

12-18 CHASSIS ELECTRICAL SYSTEM

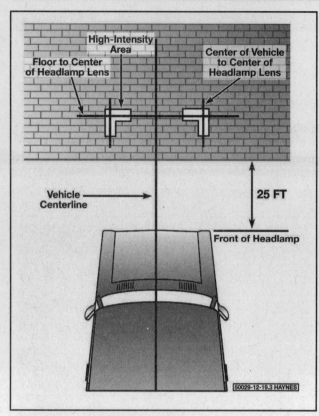

15.3 Headlight adjustment details

3   Position masking tape vertically on the wall in reference to the vehicle centerline and the centerlines of both headlights (see illustration).

4   Position a horizontal tape line in reference to the centerline of all the headlights.

➥**Note: It might be easier to position the tape on the wall with the vehicle parked only a few inches away.**

5   Adjustment should be made with the vehicle parked 25 feet from the wall, sitting level, the gas tank half-full and no heavy load in the vehicle.

6   Starting with the low beam adjustment, position the high intensity zone so it is two inches below the horizontal line. Adjustment is made by turning the adjusting screw clockwise to raise the beam and counterclockwise to lower the beam.

7   With the high beams on, the high intensity zone should be vertically centered with the exact center just below the horizontal line.

➥**Note: It might not be possible to position the headlight aim exactly for both high and low beams. If a compromise must be made, keep in mind that the low beams are the most used and have the greatest effect on safety.**

8   Have the headlights adjusted by a dealer service department or service station at the earliest opportunity.

16 Headlight housing - replacement

2000 THROUGH 2002 MODELS

◆ Refer to illustrations 16.2 and 16.3

1   Disconnect the electrical connector from the headlight bulb (see illustration 14.2) and remove the turn signal bulb socket from the headlight housing (see Section 18).

2   Remove the bolt and screw that secure the headlight housing to the upper radiator crossmember (see illustration).

3   Pull out the lower front corner of the headlight housing and slide out the headlight housing at an angle (see illustration).

4   Installation is the reverse of removal.

5   Adjust the headlights when you're done (see Section 15).

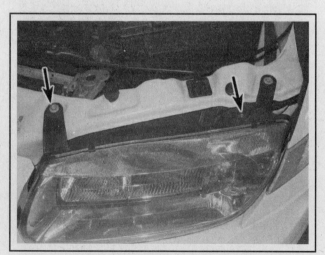

16.2 To detach the headlight housing from the upper radiator crossmember, remove this bolt and the Phillips screw

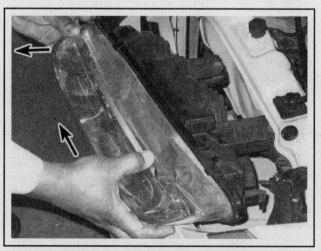

16.3 To remove the headlight housing, pull out the lower front corner of the headlight housing and slide out the headlight housing at an angle

CHASSIS ELECTRICAL SYSTEM   12-19

2003 AND LATER MODELS

6 Remove the upper grille (see Chapter 11).
7 Remove the two headlight housing bolts.
8 Rotate the locking lever to release the ball-and-socket retainers.
9 Remove the headlight housing from the headlight housing pocket and disconnect the electrical connectors.
10 Installation is the reverse of removal.
11 Adjust the headlights when you're done.

17 Horn - replacement

▶ **Refer to illustration 17.2**

1 Raise the front of the vehicle and place it securely on jackstands.
2 Locate the horn at the right front corner of the vehicle (see illustration).
3 Disconnect the electrical connector from the horn.
4 Remove the horn mounting nut and separate the horn from its mounting bracket.
5 Installation is the reverse of removal.

17.2 The horn is located in the void ahead of the right front wheel, right behind the front bumper cover; to remove the horn, disconnect the electrical connector and remove the mounting nut

18 Bulb replacement

FRONT TURN SIGNAL BULBS

▶ **Refer to illustration 18.2**

1 Open the hood and locate the front turn signal bulb socket. There are three light sockets on each headlight housing. The turn signal bulb sockets are the outer units (the ones with the amber lenses).
2 Remove the turn signal bulb socket from the headlight housing (see illustration).
3 To remove the turn signal bulb from the bulb socket, simply pull it straight out of the socket.
4 Installation is the reverse of removal.

FRONT SIDEMARKER BULBS

▶ **Refer to illustration 18.6**

5 Open the hood and locate the front sidemarker bulb socket. The front sidemarker bulbs are located inside the housings with the amber lenses that are located at the front corners of the vehicle, next to the headlight housings. It isn't necessary to remove the housing for access to the bulb unless your hands are too big to fit in the opening. If you

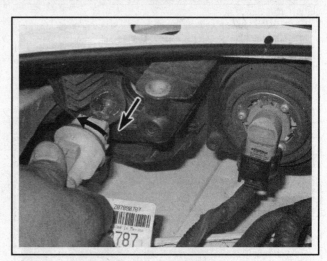

18.2 To remove a turn signal bulb socket, turn it counterclockwise and pull it out (it's not necessary to disconnect the electrical connector unless you're planning to replace the bulb socket itself)

12-20 CHASSIS ELECTRICAL SYSTEM

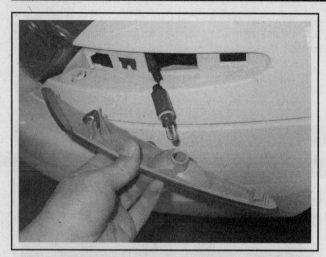

18.6 To remove a front sidemarker bulb socket, turn it 1/4-turn counterclockwise and pull it out (lens assembly removed for clarity only; for most people it's unnecessary to remove the lens to access the bulb socket)

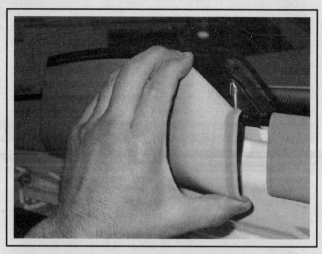

18.16 Carefully pull on the outer edges of the trim cover for the center-mount brake light assembly and remove the cover (station wagon shown, sedans similar)

have to detach the housing, pull outward on the front edge until the clip releases, then pull it forward to detach the hook at the rear of the housing.

6 Remove the bulb socket from the front sidemarker light housing (see illustration).

7 To remove the sidemarker bulb from the bulb socket, simply pull it straight out of the socket.

8 Installation is the reverse of removal.

FOG LIGHT BULBS

9 The fog lights, if equipped, are located in the lower corners of the front bumper cover. They're easily accessed from underneath the bumper cover.

10 Raise the front of the vehicle and place it securely on jackstands.

11 Rotate the bulb socket counterclockwise while pressing the socket in firmly, then pull the socket and bulb out of the fog light housing.

12 Disconnect the electrical connector from the socket.

13 Remove the bulb from the socket.

14 Installation is the reverse of removal.

CENTER-MOUNT BRAKE LIGHT BULBS

▸ Refer to illustrations 18.16 and 18.17

15 Locate the center-mount brake light assembly inside the vehicle. On sedans, it's located on the rear edge of the headliner, just ahead of the rear window. Pull the high-mount brake light assembly trim cover down by its edges and remove it. On station wagons, it's located on the lower edge of the rear window, which is in the liftgate.

16 Remove the high-mount brake light assembly trim cover (see illustration).

17 To remove a bulb socket, rotate it 1/4-turn counterclockwise and pull it out (see illustration).

18 Each high-mount brake light bulb and socket is a one-piece assembly, so don't try to separate the bulb from the socket. If you're replacing a bulb, the new bulb will include a new socket.

19 Installation is the reverse of removal.

BRAKE/TAIL/TURN/BACK-UP LIGHT BULBS

▸ Refer to illustrations 18.20, 18.21 and 18.22

20 On sedans, open the trunk and pull back the interior trim in the

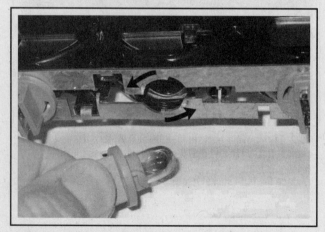

18.17 To remove a center-mount brake light bulb socket, rotate it 1/4-turn counterclockwise and pull it out

18.20 On station wagon models, open this little access panel to get to the brake/tail/turn/back-up light bulb sockets

CHASSIS ELECTRICAL SYSTEM 12-21

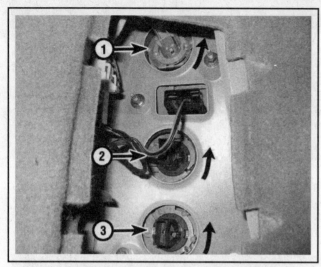

18.21 To remove the socket for the brake/taillight bulb (1), turn signal bulb (2) or back-up light bulb (3), rotate it counterclockwise and pull it out of the taillight assembly

18.22 To remove a taillight, turn signal or back-up light bulb from its socket, push it into the socket, rotate it 1/8-turn counterclockwise, then pull it out. When installing the bulb, align the locator pins on the side of the bulb with the slots in the socket, insert the bulb into the socket, push it into the socket and rotate it 1/8-turn clockwise

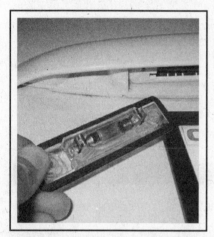

18.24 To remove the license plate light bulb and lens assembly, remove the two retaining screws and pull it down; to replace the bulb, simply pull out the old bulb and install a new one

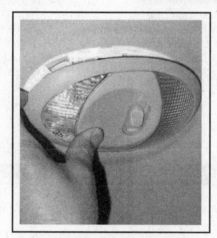

18.27 Carefully pry off the dome light lens by pulling firmly on the edges with your fingers

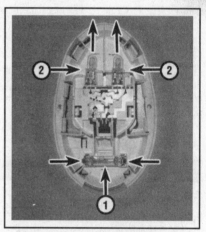

18.28 To disengage the rear dome light bulb (1) from its metal contacts, squeeze the two contacts together and pull the bulb straight down; to remove either of the front dome light bulbs, pull it straight out of its socket

left or right rear corner of the trunk. On station wagons, open the liftgate and remove the trim panel (see illustration).

21 Remove the socket for the brake/taillight, turn signal light or back-up light bulb that you want to replace (see illustration).

22 Remove the bulb from its socket (see illustration).

23 Installation is the reverse of removal.

LICENSE PLATE LIGHT BULBS

▸ Refer to illustration 18.24

24 Remove the license plate lens screws and pull down the license plate lens and light bulb as a single assembly (see illustration).

25 To replace a bulb, simply pull it out.
26 Installation is the reverse of removal.

DOME LIGHT ASSEMBLIES

▸ Refer to illustrations 18.27 and 18.28

➥Note: The following procedure depicts bulb replacement for the rear dome light assembly, but the procedure for removing one of the dome light bulbs from the front dome light assembly is identical.

27 Pry off the dome light lens (see illustration).
28 Remove the light bulb (see illustration).
29 Installation is the reverse of removal.

12-22 CHASSIS ELECTRICAL SYSTEM

19 Electric side view mirrors - general information

1 Most electric rear view mirrors use two motors to move the glass; one for up and down adjustments and one for left-right adjustments.

2 The control switch has a selector portion that sends voltage to the left or right side mirror. With the ignition ON but the engine OFF, roll down the windows and operate the mirror control switch through all functions (left-right and up-down) for both the left and right side mirrors.

3 Listen carefully for the sound of the electric motors running in the mirrors.

4 If the motors can be heard but the mirror glass doesn't move, there's a problem with the drive mechanism inside the mirror.

5 If the mirrors do not operate and no sound comes from the mirrors, check the fuse (see Section 3).

6 If the fuse is OK, remove the mirror control switch. Have the switch continuity checked by a dealership service department or other qualified automobile repair facility.

7 Make sure the mirror is properly grounded.

8 If the mirror still doesn't work, remove the mirror and check the wires at the mirror for voltage.

9 If there's not voltage in each switch position, check the circuit between the mirror and control switch for opens and shorts.

10 If there's voltage, remove the mirror and test it off the vehicle with jumper wires. Replace the mirror if it fails this test.

20 Cruise control system - general information

1 The cruise control system maintains vehicle speed with an electrically-operated motor located in the engine compartment, which is connected to the throttle lever by a cable. The system consists of the cruise brake switch, the cruise clutch switch (manual transaxle), the cruise control module, the cruise control switches, the cruise control cable, the Powertrain Control Module (PCM) and the brake light switch. The cruise control system requires special testers and diagnostic procedures that are beyond the scope of this manual. Listed below are some general procedures that may be used to locate common problems.

2 Check the fuses (see Section 3).

3 Have an assistant operate the brake pedal while you check the operation of the brake lights (voltage from the brake light switch deactivates the cruise control).

4 If the brake lights don't come on, or if they stay on all the time, correct the problem and retest the cruise control system.

5 Visually inspect the control cable between the cruise control motor and the throttle body for free movement. Replace it if necessary.

6 Test drive the vehicle to determine if the cruise control is now working. If it isn't, take it to a dealer service department or an automotive electrical specialist for further diagnosis.

21 Power window system - general information

1 The power window system operates electric motors, mounted in the doors, which lower and raise the windows. The system consists of the control switches, the motors, regulators, glass mechanisms and associated wiring.

2 The power windows can be lowered and raised from the master control switch by the driver or by remote switches located at the individual windows. Each window has a separate motor, which is reversible. The position of the control switch determines the polarity and therefore the direction of operation.

3 The circuit is protected by a fuse and a circuit breaker. Each motor is also equipped with an internal circuit breaker; this prevents one stuck window from disabling the whole system.

4 The power window system will only operate when the ignition switch is ON. In addition, many models have a window lockout switch at the master control switch which, when activated, disables the switches at the rear windows and, sometimes, the switch at the passenger's window also. Always check these items before troubleshooting a window problem.

5 These procedures are general in nature, so if you can't find the problem using them, take the vehicle to a dealer service department or other properly equipped repair facility.

6 If the power windows won't operate, always check the fuse and circuit breaker first.

7 If only the rear windows are inoperative, or if the windows only operate from the master control switch, check the rear window lockout switch for continuity in the unlocked position. Replace it if it doesn't have continuity.

8 Check the wiring between the switches and fuse panel for continuity. Repair the wiring, if necessary.

9 If only one window is inoperative from the master control switch, try the other control switch at the window.

➡ **Note: This doesn't apply to the driver's door window.**

10 If the same window works from one switch, but not the other, check the switch for continuity.

11 If the switch tests OK, check for a short or open in the circuit between the affected switch and the window motor.

12 If one window is inoperative from both switches, remove the trim panel from the affected door and check for voltage at the switch and at the motor while the switch is operated.

13 If voltage is reaching the motor, disconnect the glass from the regulator (see Chapter 11). Move the window up and down by hand while checking for binding and damage. Also check for binding and damage to the regulator. If the regulator is not damaged and the window moves up and down smoothly, replace the motor. If there's binding or damage, lubricate, repair or replace parts, as necessary.

14 If voltage isn't reaching the motor, check the wiring in the circuit for continuity between the switches and motors. You'll need to consult the wiring diagram for the vehicle. If the circuit is equipped with a relay, check that the relay is grounded properly and receiving voltage.

CHASSIS ELECTRICAL SYSTEM   12-23

22 Power door lock system - general information

1 A power door lock system operates the door lock actuators mounted in each door. The system consists of the switches, actuators, a control unit and associated wiring. Diagnosis can usually be limited to simple checks of the wiring connections and actuators for minor faults that can be easily repaired.

2 Power door lock systems are operated by bi-directional solenoids located in the doors. The lock switches have two operating positions: Lock and Unlock. When activated, the switch sends a ground signal to the door lock control unit to lock or unlock the doors. Depending on which way the switch is activated, the control unit reverses polarity to the solenoids, allowing the two sides of the circuit to be used alternately as the feed (positive) and ground side.

3 Some vehicles may have an anti-theft system incorporated into the power locks. If you are unable to locate the trouble using the following general Steps, consult a dealer service department or other qualified repair shop.

4 Always check the circuit protection first. Some vehicles use a combination of circuit breakers and fuses.

5 Operate the door lock switches in both directions (Lock and Unlock) with the engine off. Listen for the click of the solenoids operating.

6 Test the switches for continuity. Remove the switches and have them checked by a dealer service department or other qualified automobile repair facility.

7 Check the wiring between the switches, control unit and solenoids for continuity. Repair the wiring if there's no continuity.

8 Check for a bad ground at the switches or the control unit.

9 If all but one lock solenoids operate, remove the trim panel from the door with the problem (see Chapter 11) and check for voltage at the solenoid while the lock switch is operated. One of the wires should have voltage in the Lock position; the other should have voltage in the Unlock position.

10 If the inoperative solenoid is receiving voltage, replace the solenoid.

11 If the inoperative solenoid isn't receiving voltage, check the relay for an open or short in the wire between the lock solenoid and the control unit.

23 Daytime Running Lights (DRL) - general information

The Daytime Running Lights (DRL) system illuminates the headlights whenever the engine is running. The only exception is with the engine running and the parking brake engaged. Once the parking brake is released, the lights will remain on as long as the ignition switch is on, even if the parking brake is later applied.

The DRL system supplies reduced power to the headlights so they won't be too bright for daytime use, while prolonging headlight life.

24 Airbag system - general information and precautions

GENERAL INFORMATION

1 All models are equipped with two front airbags, formally known as the Supplemental Inflatable Restraint (SIR) system. This system is designed to protect the driver and the front seat passenger (and rear passengers on models equipped with side-impact airbags) from serious injury in the event of a frontal collision and, if equipped with side-impact airbags, hits from the side. It consists of an array of external and internal (inside the SDM) information sensors (decelerometers), the Inflatable Restraint Sensing and Diagnostic Module (SDM), the inflator modules (a driver's airbag in the steering wheel and a passenger airbag in the dash) and the wiring and connectors tying all these components together. An optional pair of side-impact airbags, also known as "roof rail" or "side curtain" airbags, is available for protection against side impacts. The side-impact airbags, if equipped, are located along the left and right edges of the headliner, above the doors.

AIRBAG/INFLATOR MODULES

Driver's airbag/inflator module

2 The airbag inflator module in the steering wheel contains a housing incorporating the cushion (airbag), an initiating device, and a canister of gas-generating material. The initiator is part of the inflator module deployment loop. When a collision occurs, the SDM sends current through the deployment loop to the initiator. Current passing through the initiator ignites the material in the canister, producing a rapidly expanding gas, which inflates the airbag almost instantaneously. Seconds after the airbag inflates, it deflates almost as quickly through airbag vent holes and/or the airbag fabric. Each inflator module is equipped with a shorting bar, which is located on the module's electrical connector. The shorting bar shorts the circuit for the inflator module deployment loop to prevent accidental deployment of the airbag when it's disconnected.

3 When the SDM sends current to the initiator, it travels through the airbag circuit to the steering column. From there, a "clockspring" on the steering wheel delivers the current to the module initiator. This clockspring assembly, which is the final segment of the airbag ignition circuit, functions as the bridge between the end of the airbag circuit on the (fixed) steering column and the beginning of the circuit on the (rotating) steering wheel. It's designed to maintain a closed circuit between the steering column and the steering wheel regardless of the position of the steering wheel. For this reason, removing and installing the clockspring is critical to the performance of the driver's side airbag. For information on how to remove and install the driver's side airbag, refer to "Steering wheel - removal and installation" in Chapter 10.

Passenger's airbag/inflator module

4 The passenger's airbag/inflator module is mounted above the glove compartment. It's similar in design to the driver's airbag except that it doesn't use a clockspring. When deployed by the SDM, the passenger's airbag bursts through the dashboard above the glovebox.

12-24 CHASSIS ELECTRICAL SYSTEM

Although this area looks like it's simply part of the dashboard, it's actually a trim cover with a perforated seam that allows the cover to separate from the dash when the passenger's airbag inflates.

Side impact airbag/inflator ("roof rail") modules

5   The (optional) side-impact airbag/inflator ("roof rail") modules are mounted along the outer edges of the headliner, right above the door openings. They extend from the "A-pillar" (front windshield pillar) to the "C-pillar" (rear window pillar). Each module consists of a housing, an inflatable airbag, an initiator and a canister of gas-generating material. Each roof rail module employs its own side impact sensor (SIS), which contains a sensing device that monitors changes in vehicle acceleration and velocity. This data is sent to the SDM, which compares it with its program. When the data exceeds a certain threshold, the SDM determines that the vehicle has been hit hard enough on one side or the other to warrant deployment of the roof rail on that side. The SDM doesn't deploy the roof rail airbags on both sides, just on the side being hit. Then the SDM sends current to the roof rail initiator to inflate the airbag, ripping open the headliner trim as it deploys to protect the occupant(s) on the left or right side of the vehicle. Side impact airbag/inflator modules are long enough to protect the driver and a left-side rear-seat passenger, or a front seat passenger and right-side rear-seat passenger.

INFLATABLE RESTRAINT SENSING AND DIAGNOSTIC MODULE (SDM)

6   The SDM is the computer module that controls the airbag system. Besides a microprocessor, the SDM also includes an array of sensors. Some of them are inside the SDM itself. Other external sensors are located throughout the vehicle. All of the sensors, internal and external, send a continuous voltage signal to the SDM, which compares this data to values stored in its memory. When these signals exceed a threshold value, i.e. when the SDM determines that the vehicle is decelerating more quickly than the threshold value, the SDM allows current to flow through the circuit to the appropriate airbag module(s), which initiates deployment of the airbag(s).

7   For more information about the airbag system in your vehicle, refer to your owner's manual.

DISARMING THE SYSTEM AND OTHER PRECAUTIONS

✱✱✱ WARNING:
Failure to follow these precautions could result in accidental deployment of the airbag and personal injury.

8   Whenever working in the vicinity of the steering wheel, instrument panel or any of the other SIS system components, the system must be disarmed. To disarm the system:
   a) *Point the wheels straight ahead and turn the key to the Lock position.*
   b) *Disconnect the cable from the negative battery terminal. Refer to Chapter 5, Section 1 for the disconnecting procedure.*
   c) *Wait at least two minutes for the back-up power supply to be depleted.*

9   Whenever handling an airbag module, always keep the airbag opening (the trim side) pointed away from your body. Never place the airbag module on a bench or other surface with the airbag opening facing the surface. Always place the airbag module in a safe location with the airbag opening (the upholstered side) facing up.

10   Never measure the resistance of any SIS component or use any electrical test equipment on any of the wiring or components. An ohmmeter has a built-in battery supply that could accidentally deploy the airbag.

11   Never dispose of a live airbag/inflator module. Return it to a dealer service department or other qualified repair shop for safe deployment and disposal.

12   Never use electrical welding equipment in the vicinity of any airbag components. The connectors for the system are easy to spot because they're bright YELLOW. Do NOT disconnect or tamper with these connectors, or you run the risk of setting a Diagnostic Trouble Code (DTC) in the SDM. Like the PCM, the SDM has a malfunction indicator light, known as the AIR BAG indicator light, on the instrument cluster. When you turn the ignition key to ON, the SDM checks out all of the SIS components and circuits. If everything is okay, the AIR BAG indicator light goes off, just like the PCM's Malfunction Indicator Light (MIL). But if there's a problem somewhere, the light stays on, and will remain on until the problem is repaired and the DTC(s) cleared from the SDM's memory.

25 Wiring diagrams - general information

Since it isn't possible to include all wiring diagrams for every year covered by this manual, the following diagrams are those that are typical and most commonly needed.

Prior to troubleshooting any circuits, check the fuse and circuit breakers (if equipped) to make sure they're in good condition. Make sure the battery is properly charged and check the cable connections (see Chapter 1).

When checking a circuit, make sure that all connectors are clean, with no broken or loose terminals. When unplugging a connector, do not pull on the wires. Pull only on the connector housings themselves.

CHASSIS ELECTRICAL SYSTEM 12-25

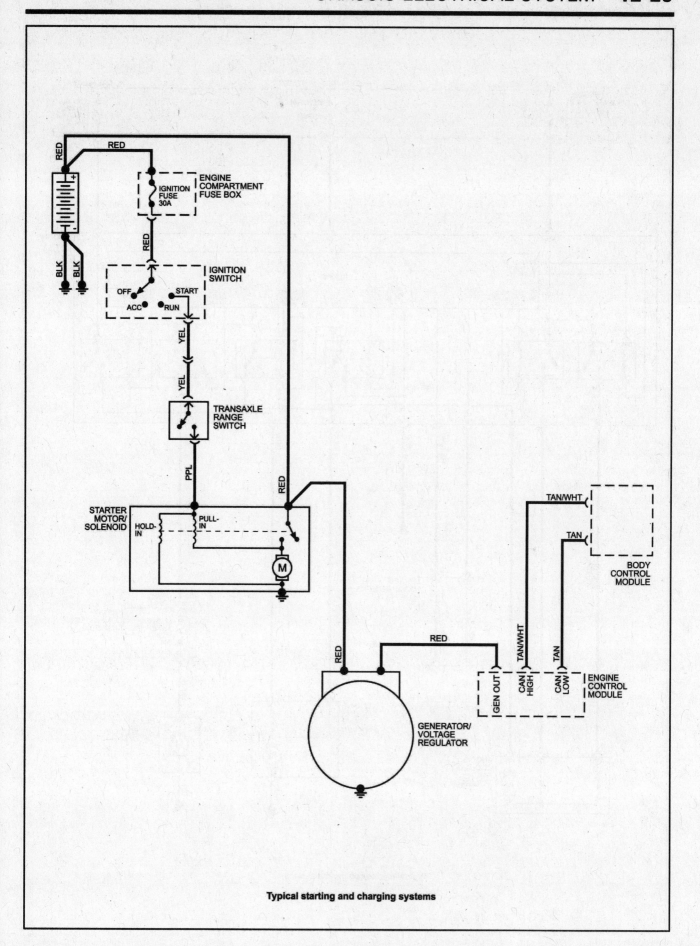

Typical starting and charging systems

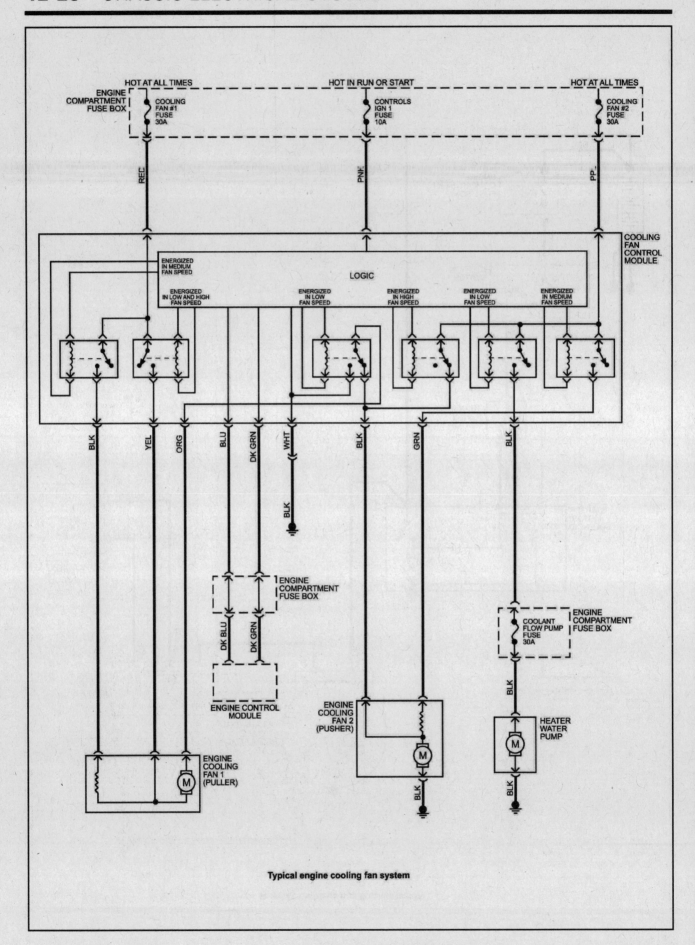

Typical engine cooling fan system

CHASSIS ELECTRICAL SYSTEM   12-27

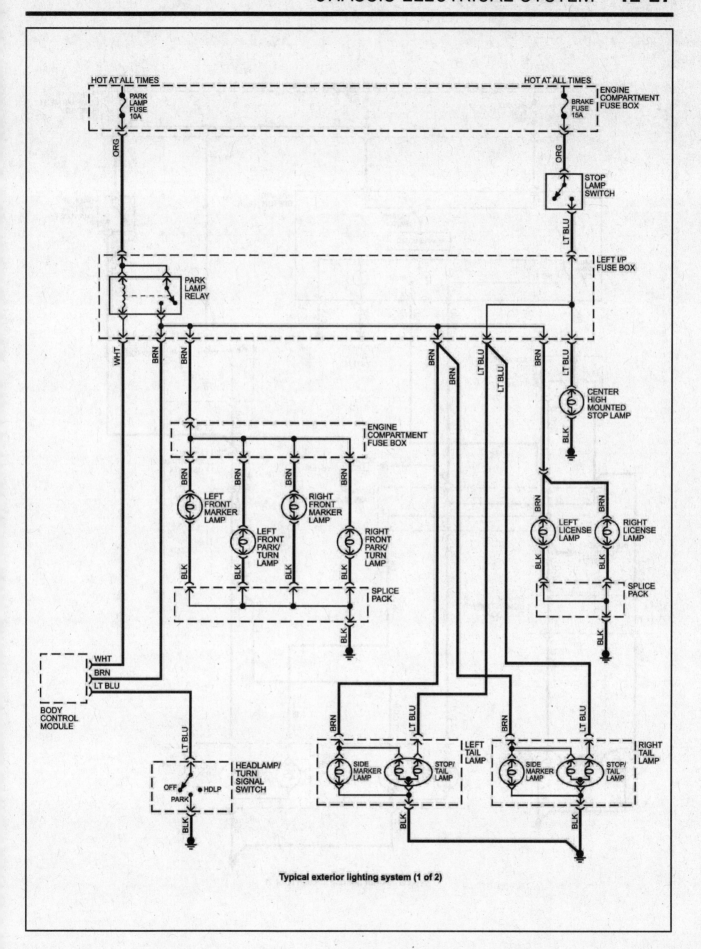

Typical exterior lighting system (1 of 2)

12-28 CHASSIS ELECTRICAL SYSTEM

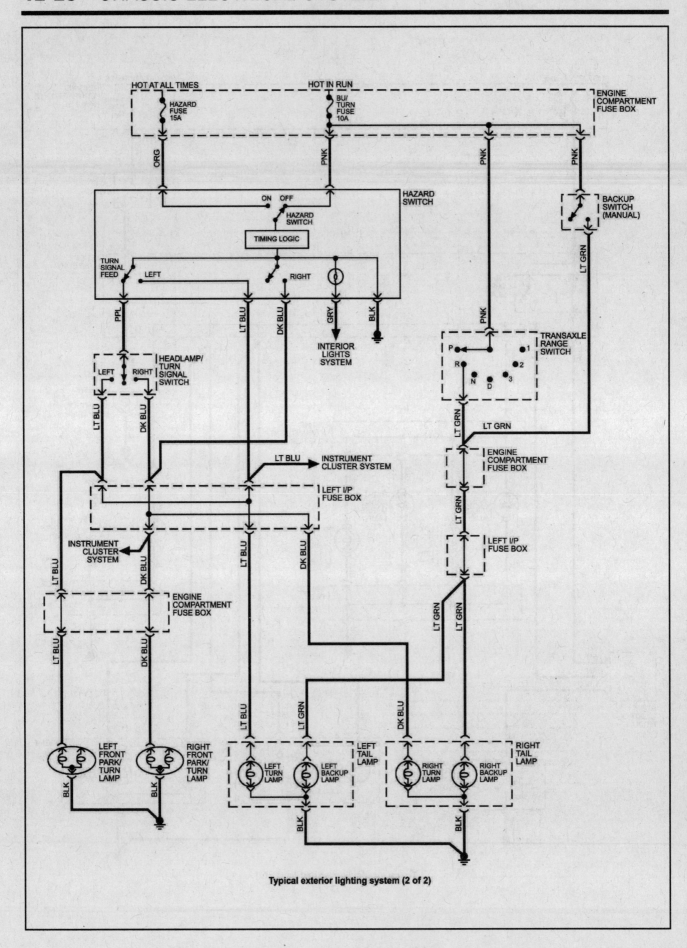

Typical exterior lighting system (2 of 2)

CHASSIS ELECTRICAL SYSTEM 12-29

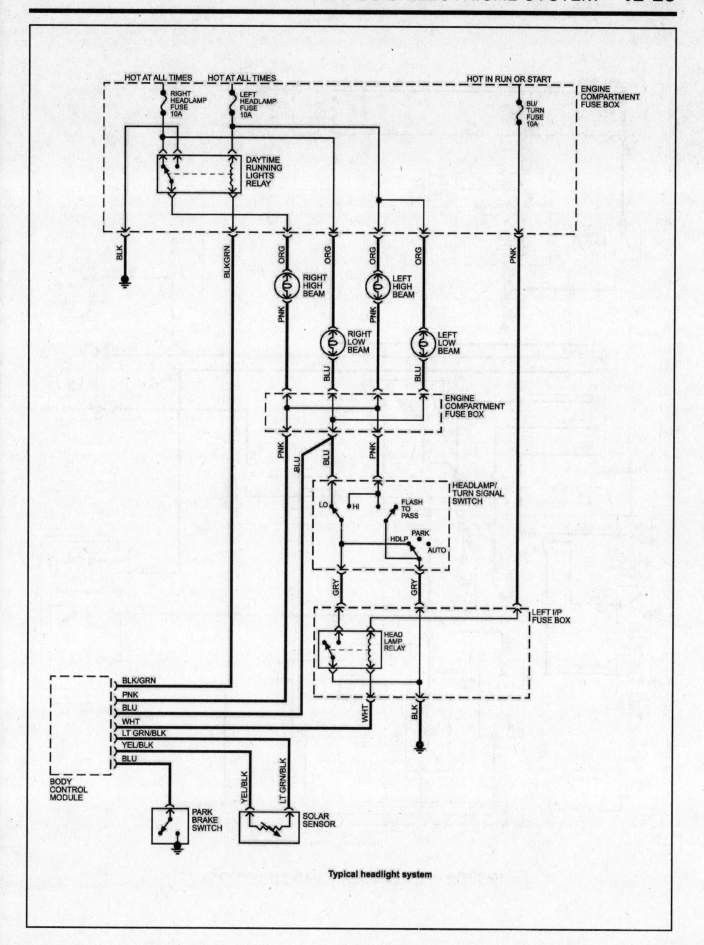

Typical headlight system

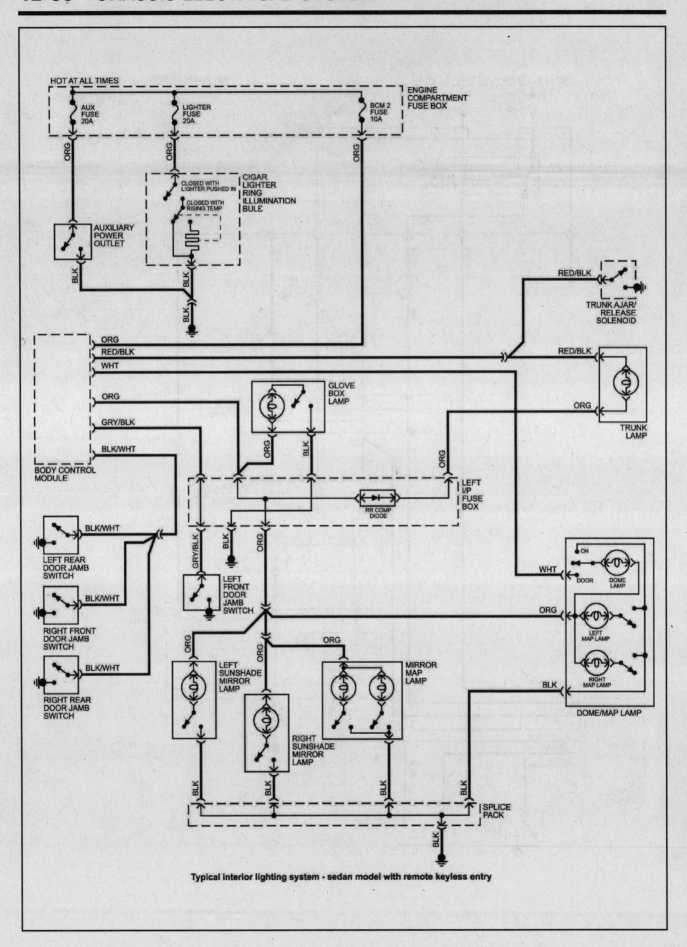

Typical interior lighting system - sedan model with remote keyless entry

CHASSIS ELECTRICAL SYSTEM 12-31

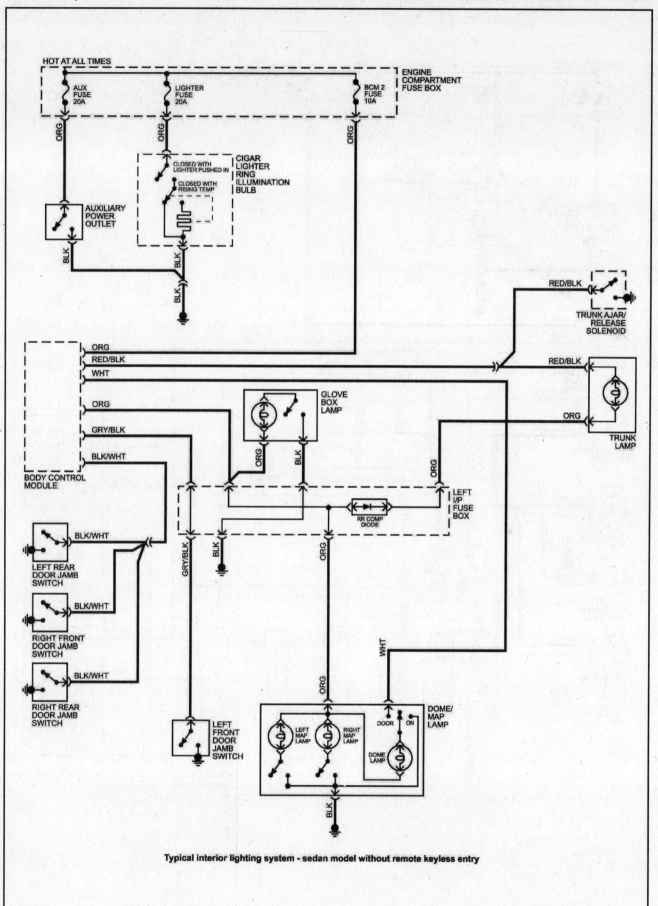

Typical interior lighting system - sedan model without remote keyless entry

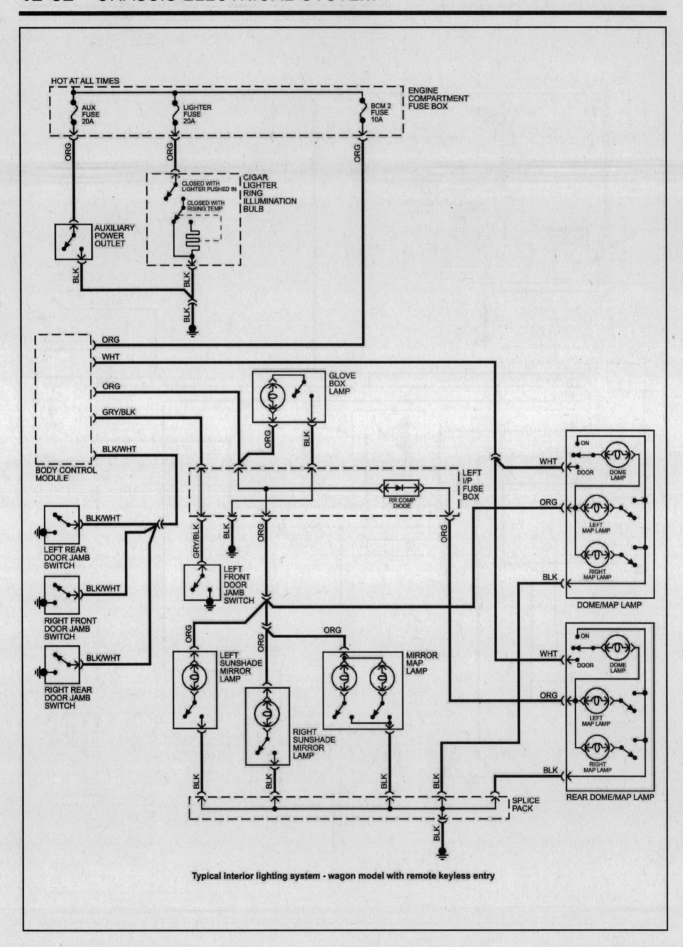

Typical interior lighting system - wagon model with remote keyless entry

CHASSIS ELECTRICAL SYSTEM 12-33

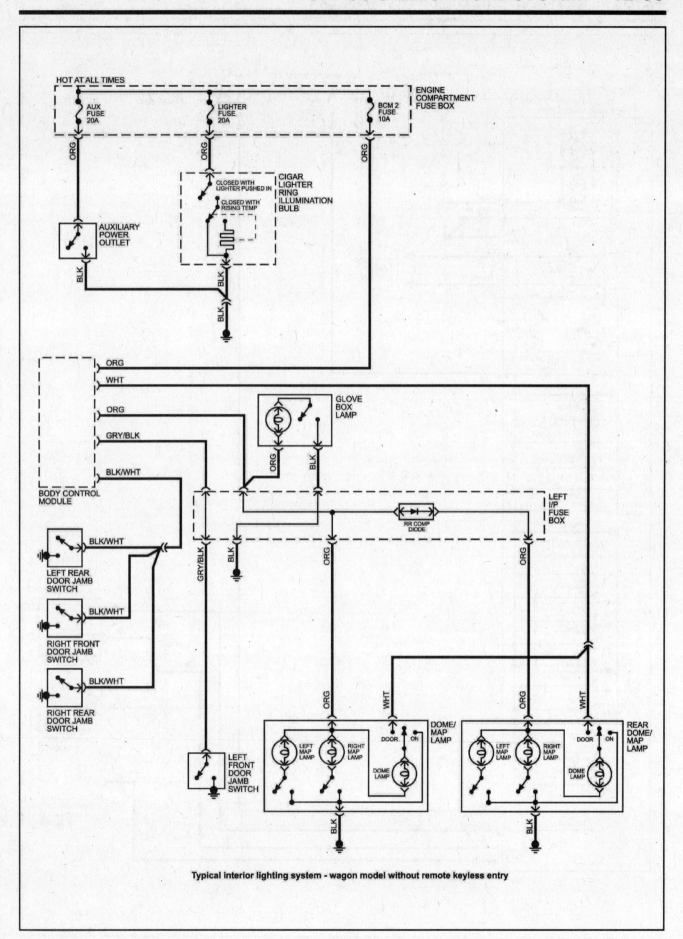

Typical interior lighting system - wagon model without remote keyless entry

12-34 CHASSIS ELECTRICAL SYSTEM

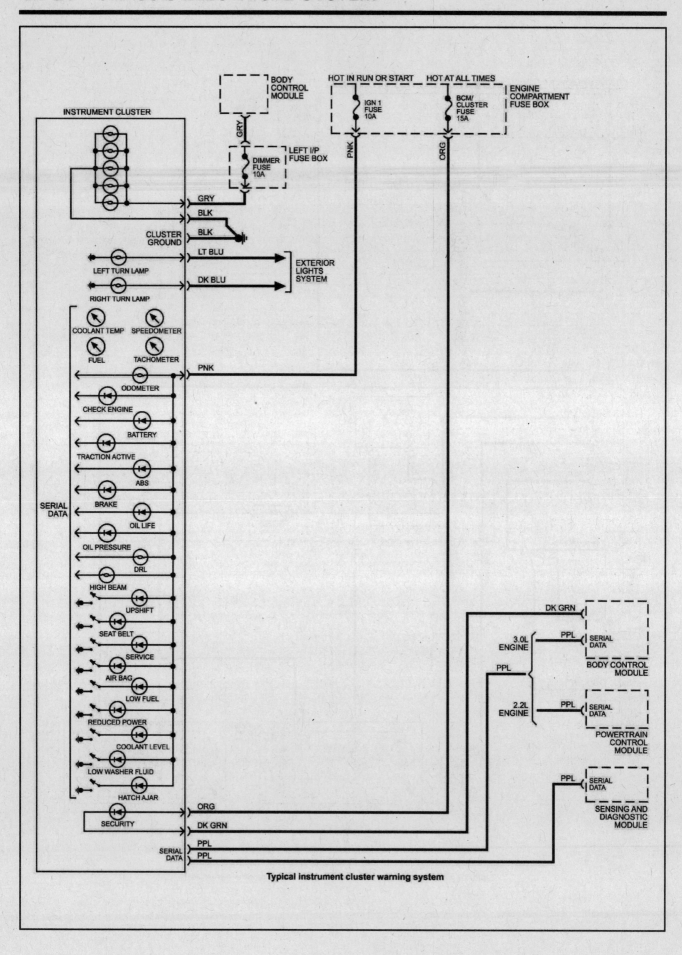

Typical instrument cluster warning system

CHASSIS ELECTRICAL SYSTEM 12-35

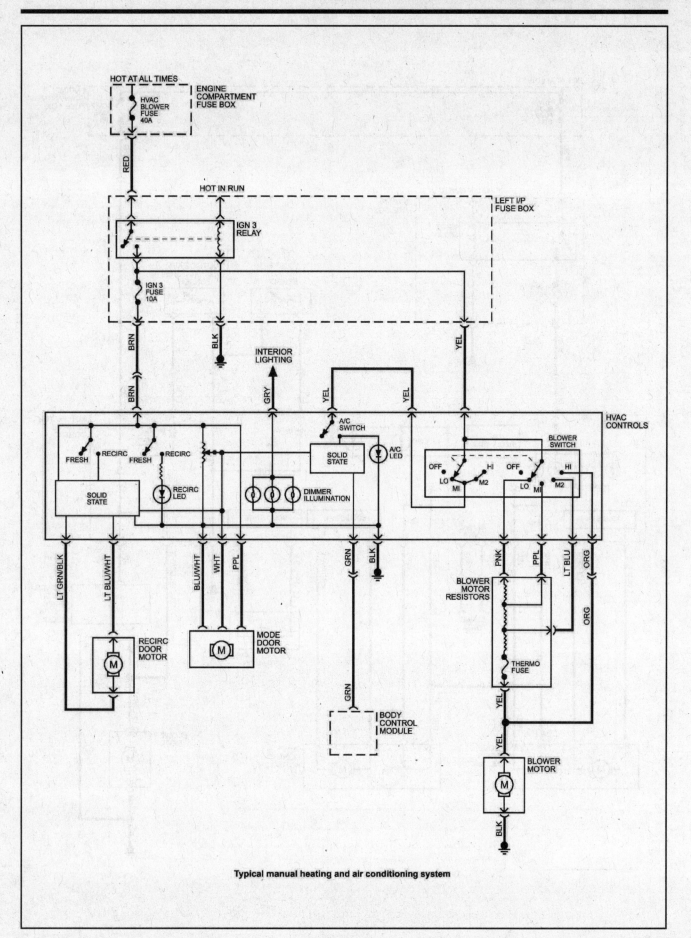

Typical manual heating and air conditioning system

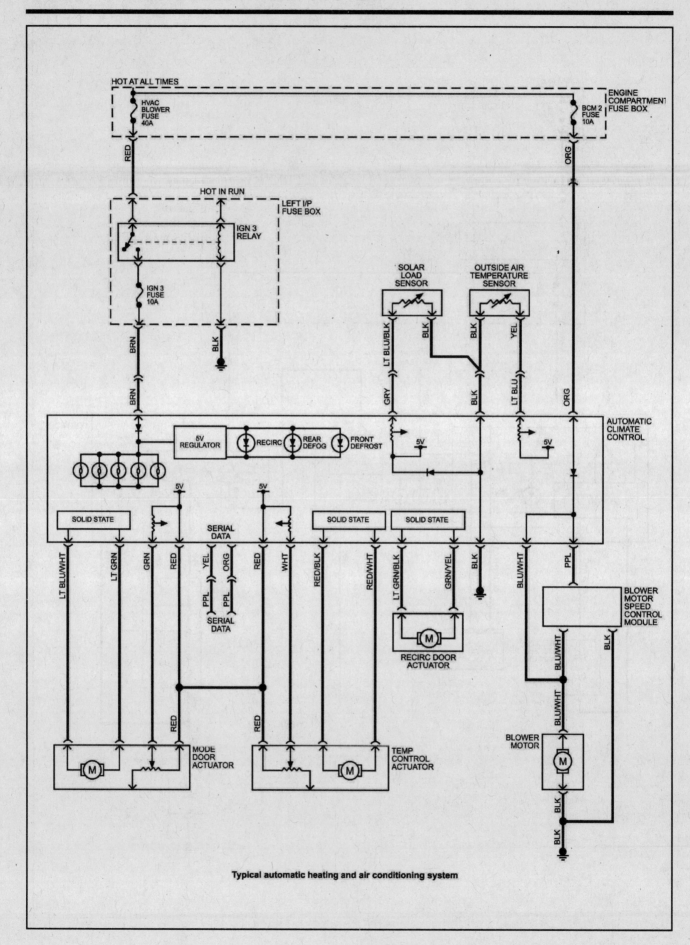

Typical automatic heating and air conditioning system

CHASSIS ELECTRICAL SYSTEM 12-37

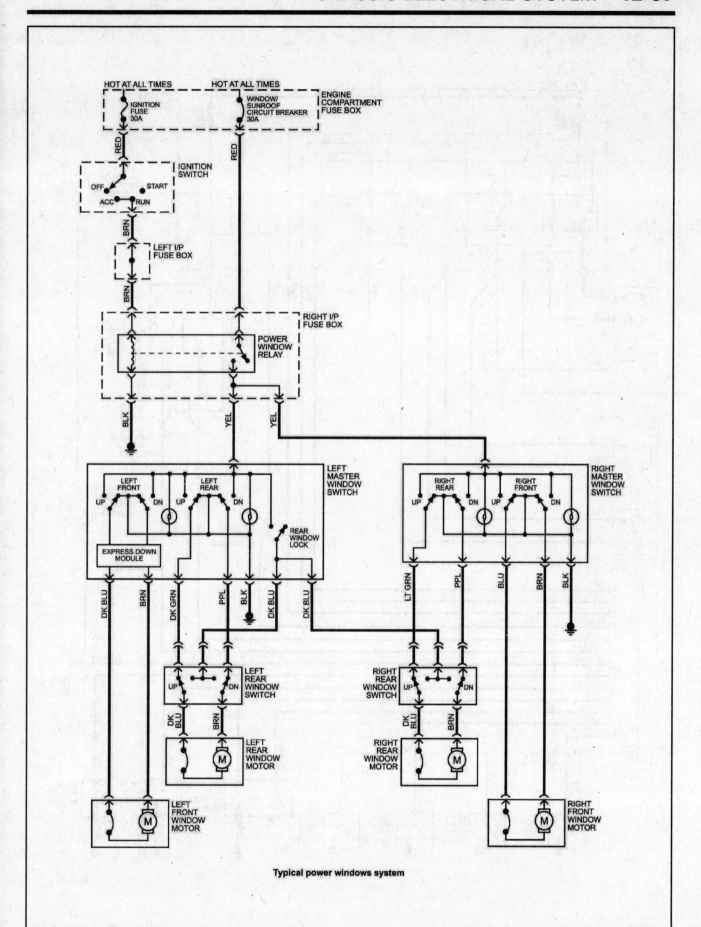

Typical power windows system

12-38 CHASSIS ELECTRICAL SYSTEM

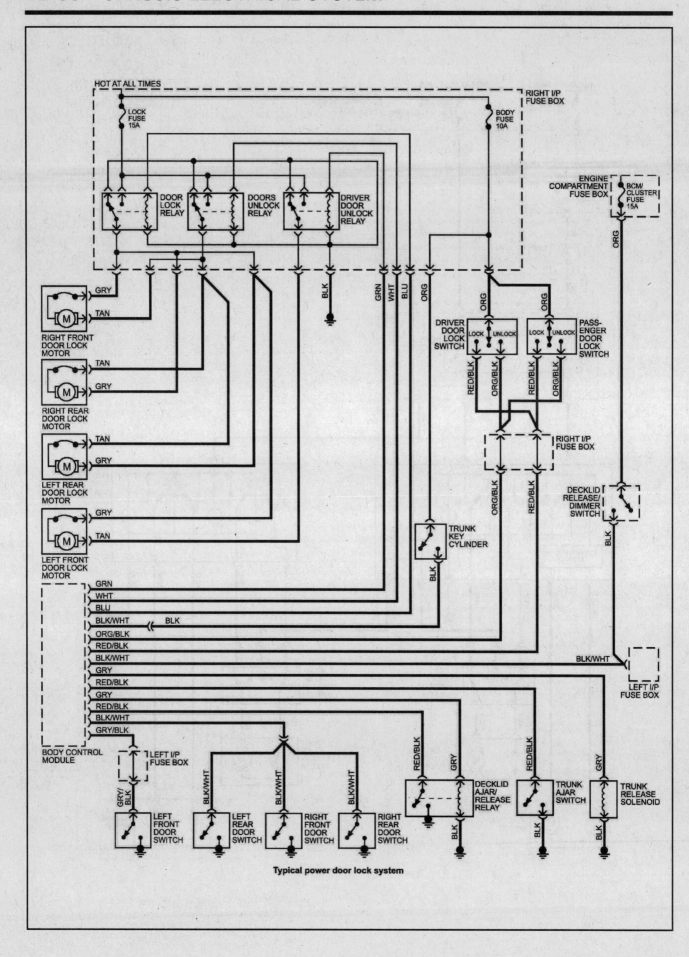

Typical power door lock system

CHASSIS ELECTRICAL SYSTEM 12-39

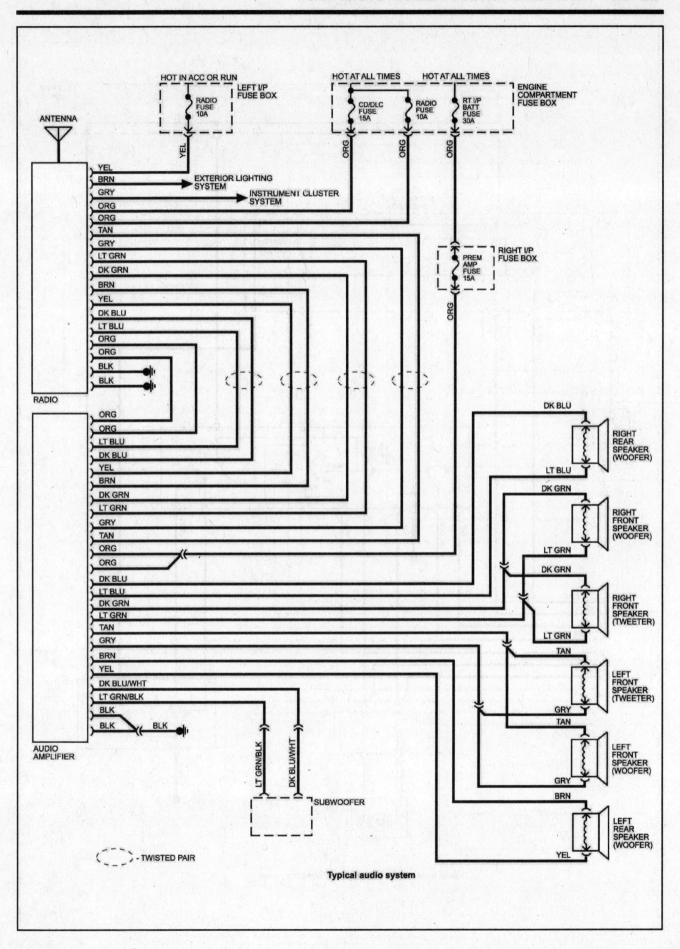

Typical audio system

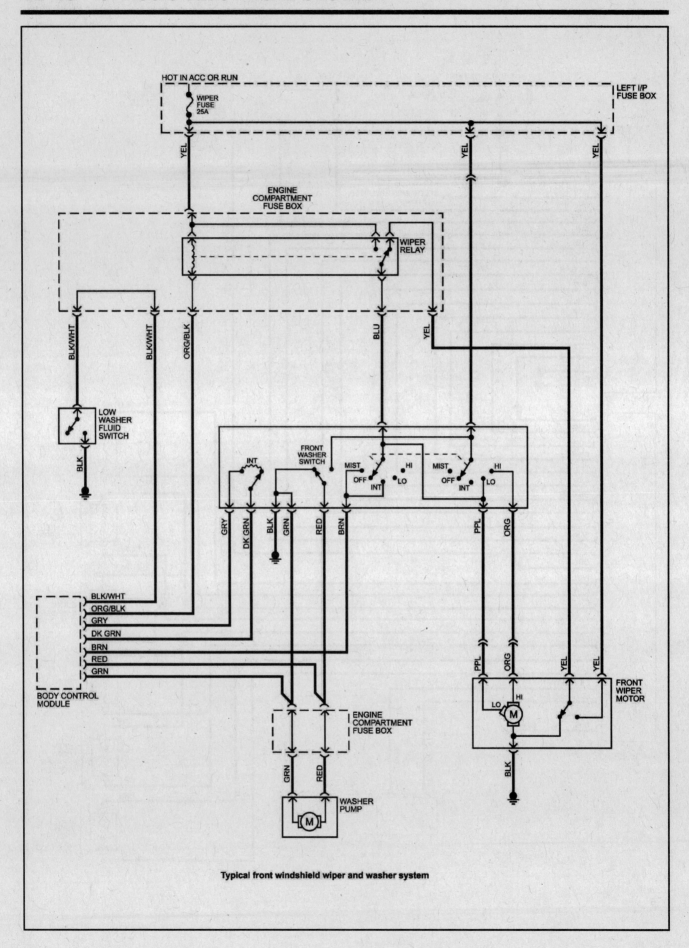

Typical front windshield wiper and washer system

CHASSIS ELECTRICAL SYSTEM 12-41

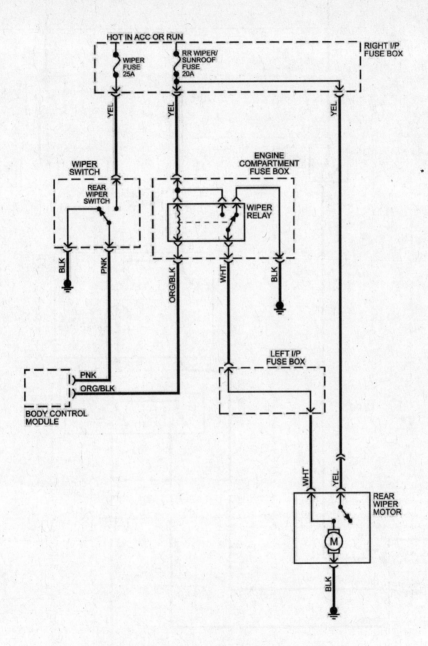

Typical rear windshield wiper and washer system (wagon only)

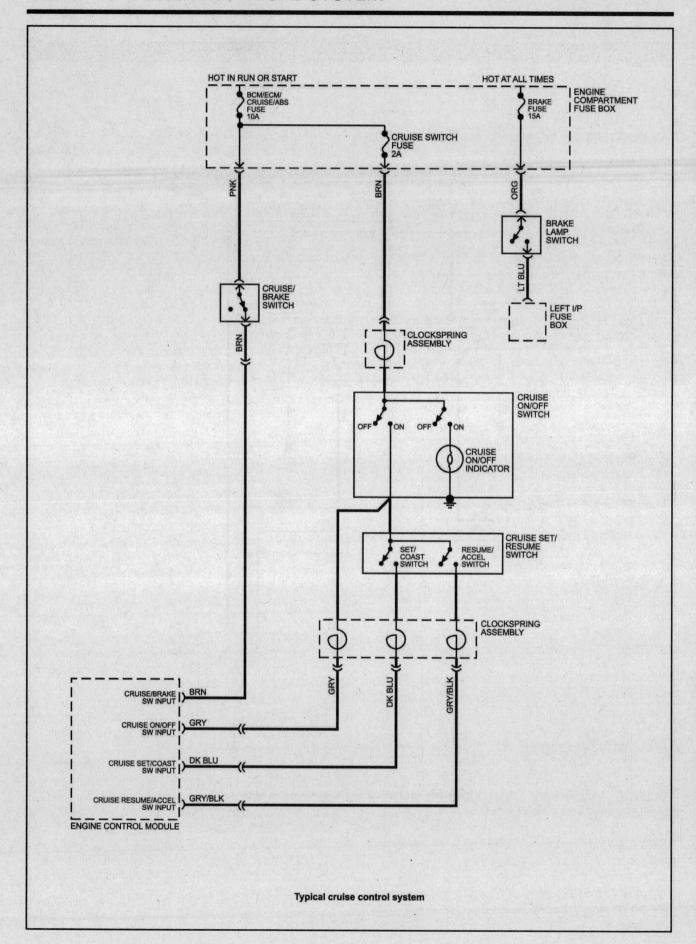

Typical cruise control system

GLOSSARY GL-1

GLOSSARY

AIR/FUEL RATIO: The ratio of air-to-gasoline by weight in the fuel mixture drawn into the engine.

AIR INJECTION: One method of reducing harmful exhaust emissions by injecting air into each of the exhaust ports of an engine. The fresh air entering the hot exhaust manifold causes any remaining fuel to be burned before it can exit the tailpipe.

ALTERNATOR: A device used for converting mechanical energy into electrical energy.

AMMETER: An instrument, calibrated in amperes, used to measure the flow of an electrical current in a circuit. Ammeters are always connected in series with the circuit being tested.

AMPERE: The rate of flow of electrical current present when one volt of electrical pressure is applied against one ohm of electrical resistance.

ANALOG COMPUTER: Any microprocessor that uses similar (analogous) electrical signals to make its calculations.

ARMATURE: A laminated, soft iron core wrapped by a wire that converts electrical energy to mechanical energy as in a motor or relay. When rotated in a magnetic field, it changes mechanical energy into electrical energy as in a generator.

ATMOSPHERIC PRESSURE: The pressure on the Earth's surface caused by the weight of the air in the atmosphere. At sea level, this pressure is 14.7 psi at 32°F (101 kPa at 0°C).

ATOMIZATION: The breaking down of a liquid into a fine mist that can be suspended in air.

AXIAL PLAY: Movement parallel to a shaft or bearing bore.

BACKFIRE: The sudden combustion of gases in the intake or exhaust system that results in a loud explosion.

BACKLASH: The clearance or play between two parts, such as meshed gears.

BACKPRESSURE: Restrictions in the exhaust system that slow the exit of exhaust gases from the combustion chamber.

BAKELITE: A heat resistant, plastic insulator material commonly used in printed circuit boards and transistorized components.

BALL BEARING: A bearing made up of hardened inner and outer races between which hardened steel balls roll.

BALLAST RESISTOR: A resistor in the primary ignition circuit that lowers voltage after the engine is started to reduce wear on ignition components.

BEARING: A friction reducing, supportive device usually located between a stationary part and a moving part.

BIMETAL TEMPERATURE SENSOR: Any sensor or switch made of two dissimilar types of metal that bend when heated or cooled due to the different expansion rates of the alloys. These types of sensors usually function as an on/off switch.

BLOWBY: Combustion gases, composed of water vapor and unburned fuel, that leak past the piston rings into the crankcase during normal engine operation. These gases are removed by the PCV system to prevent the buildup of harmful acids in the crankcase.

BRAKE PAD: A brake shoe and lining assembly used with disc brakes.

BRAKE SHOE: The backing for the brake lining. The term is, however, usually applied to the assembly of the brake backing and lining.

BUSHING: A liner, usually removable, for a bearing; an anti-friction liner used in place of a bearing.

CALIPER: A hydraulically activated device in a disc brake system, which is mounted straddling the brake rotor (disc). The caliper contains at least one piston and two brake pads. Hydraulic pressure on the piston(s) forces the pads against the rotor.

CAMSHAFT: A shaft in the engine on which are the lobes (cams) which operate the valves. The camshaft is driven by the crankshaft, via a belt, chain or gears, at one half the crankshaft speed.

CAPACITOR: A device which stores an electrical charge.

CARBON MONOXIDE (CO): A colorless, odorless gas given off as a normal byproduct of combustion. It is poisonous and extremely dangerous in confined areas, building up slowly to toxic levels without warning if adequate ventilation is not available.

CARBURETOR: A device, usually mounted on the intake manifold of an engine, which mixes the air and fuel in the proper proportion to allow even combustion.

CATALYTIC CONVERTER: A device installed in the exhaust system, like a muffler, that converts harmful byproducts of combustion into carbon dioxide and water vapor by means of a heat-producing chemical reaction.

CENTRIFUGAL ADVANCE: A mechanical method of advancing the spark timing by using flyweights in the distributor that react to centrifugal force generated by the distributor shaft rotation.

GLOSSARY

CHECK VALVE: Any one-way valve installed to permit the flow of air, fuel or vacuum in one direction only.

CHOKE: A device, usually a moveable valve, placed in the intake path of a carburetor to restrict the flow of air.

CIRCUIT: Any unbroken path through which an electrical current can flow. Also used to describe fuel flow in some instances.

CIRCUIT BREAKER: A switch which protects an electrical circuit from overload by opening the circuit when the current flow exceeds a predetermined level. Some circuit breakers must be reset manually, while most reset automatically.

COIL (IGNITION): A transformer in the ignition circuit which steps up the voltage provided to the spark plugs.

COMBINATION MANIFOLD: An assembly which includes both the intake and exhaust manifolds in one casting.

COMBINATION VALVE: A device used in some fuel systems that routes fuel vapors to a charcoal storage canister instead of venting them into the atmosphere. The valve relieves fuel tank pressure and allows fresh air into the tank as the fuel level drops to prevent a vapor lock situation.

COMPRESSION RATIO: The comparison of the total volume of the cylinder and combustion chamber with the piston at BDC and the piston at TDC.

CONDENSER: 1. An electrical device which acts to store an electrical charge, preventing voltage surges. 2. A radiator-like device in the air conditioning system in which refrigerant gas condenses into a liquid, giving off heat.

CONDUCTOR: Any material through which an electrical current can be transmitted easily.

CONTINUITY: Continuous or complete circuit. Can be checked with an ohmmeter.

COUNTERSHAFT: An intermediate shaft which is rotated by a mainshaft and transmits, in turn, that rotation to a working part.

CRANKCASE: The lower part of an engine in which the crankshaft and related parts operate.

CRANKSHAFT: The main driving shaft of an engine which receives reciprocating motion from the pistons and converts it to rotary motion.

CYLINDER: In an engine, the round hole in the engine block in which the piston(s) ride.

CYLINDER BLOCK: The main structural member of an engine in which is found the cylinders, crankshaft and other principal parts.

CYLINDER HEAD: The detachable portion of the engine, usually fastened to the top of the cylinder block and containing all or most of the combustion chambers. On overhead valve engines, it contains the valves and their operating parts. On overhead cam engines, it contains the camshaft as well.

DEAD CENTER: The extreme top or bottom of the piston stroke.

DETONATION: An unwanted explosion of the air/fuel mixture in the combustion chamber caused by excess heat and compression, advanced timing, or an overly lean mixture. Also referred to as "ping".

DIAPHRAGM: A thin, flexible wall separating two cavities, such as in a vacuum advance unit.

DIESELING: A condition in which hot spots in the combustion chamber cause the engine to run on after the key is turned off.

DIFFERENTIAL: A geared assembly which allows the transmission of motion between drive axles, giving one axle the ability to turn faster than the other.

DIODE: An electrical device that will allow current to flow in one direction only.

DISC BRAKE: A hydraulic braking assembly consisting of a brake disc, or rotor, mounted on an axle, and a caliper assembly containing, usually two brake pads which are activated by hydraulic pressure. The pads are forced against the sides of the disc, creating friction which slows the vehicle.

DISTRIBUTOR: A mechanically driven device on an engine which is responsible for electrically firing the spark plug at a predetermined point of the piston stroke.

DOWEL PIN: A pin, inserted in mating holes in two different parts allowing those parts to maintain a fixed relationship.

DRUM BRAKE: A braking system which consists of two brake shoes and one or two wheel cylinders, mounted on a fixed backing plate, and a brake drum, mounted on an axle, which revolves around the assembly.

DWELL: The rate, measured in degrees of shaft rotation, at which an electrical circuit cycles on and off.

ELECTRONIC CONTROL UNIT (ECU): Ignition module, module, amplifier or igniter. See Module for definition.

ELECTRONIC IGNITION: A system in which the timing and firing of the spark plugs is controlled by an electronic control unit, usually called a module. These systems have no points or condenser.

END-PLAY: The measured amount of axial movement in a shaft.

GLOSSARY GL-3

ENGINE: A device that converts heat into mechanical energy.

EXHAUST MANIFOLD: A set of cast passages or pipes which conduct exhaust gases from the engine.

FEELER GAUGE: A blade, usually metal, or precisely predetermined thickness, used to measure the clearance between two parts.

FIRING ORDER: The order in which combustion occurs in the cylinders of an engine. Also the order in which spark is distributed to the plugs by the distributor.

FLOODING: The presence of too much fuel in the intake manifold and combustion chamber which prevents the air/fuel mixture from firing, thereby causing a no-start situation.

FLYWHEEL: A disc shaped part bolted to the rear end of the crankshaft. Around the outer perimeter is affixed the ring gear. The starter drive engages the ring gear, turning the flywheel, which rotates the crankshaft, imparting the initial starting motion to the engine.

FOOT POUND (ft. lbs. or sometimes, ft.lb.): The amount of energy or work needed to raise an item weighing one pound, a distance of one foot.

FUSE: A protective device in a circuit which prevents circuit overload by breaking the circuit when a specific amperage is present. The device is constructed around a strip or wire of a lower amperage rating than the circuit it is designed to protect. When an amperage higher than that stamped on the fuse is present in the circuit, the strip or wire melts, opening the circuit.

GEAR RATIO: The ratio between the number of teeth on meshing gears.

GENERATOR: A device which converts mechanical energy into electrical energy.

HEAT RANGE: The measure of a spark plug's ability to dissipate heat from its firing end. The higher the heat range, the hotter the plug fires.

HUB: The center part of a wheel or gear.

HYDROCARBON (HC): Any chemical compound made up of hydrogen and carbon. A major pollutant formed by the engine as a byproduct of combustion.

HYDROMETER: An instrument used to measure the specific gravity of a solution.

INCH POUND (inch lbs.; sometimes in.lb. or in. lbs.): One twelfth of a foot pound.

INDUCTION: A means of transferring electrical energy in the form of a magnetic field. Principle used in the ignition coil to increase voltage.

INJECTOR: A device which receives metered fuel under relatively low pressure and is activated to inject the fuel into the engine under relatively high pressure at a predetermined time.

INPUT SHAFT: The shaft to which torque is applied, usually carrying the driving gear or gears.

INTAKE MANIFOLD: A casting of passages or pipes used to conduct air or a fuel/air mixture to the cylinders.

JOURNAL: The bearing surface within which a shaft operates.

KEY: A small block usually fitted in a notch between a shaft and a hub to prevent slippage of the two parts.

MANIFOLD: A casting of passages or set of pipes which connect the cylinders to an inlet or outlet source.

MANIFOLD VACUUM: Low pressure in an engine intake manifold formed just below the throttle plates. Manifold vacuum is highest at idle and drops under acceleration.

MASTER CYLINDER: The primary fluid pressurizing device in a hydraulic system. In automotive use, it is found in brake and hydraulic clutch systems and is pedal activated, either directly or, in a power brake system, through the power booster.

MODULE: Electronic control unit, amplifier or igniter of solid state or integrated design which controls the current flow in the ignition primary circuit based on input from the pick-up coil. When the module opens the primary circuit, high secondary voltage is induced in the coil.

NEEDLE BEARING: A bearing which consists of a number (usually a large number) of long, thin rollers.

OHM: (Ω) The unit used to measure the resistance of conductor-to-electrical flow. One ohm is the amount of resistance that limits current flow to one ampere in a circuit with one volt of pressure.

OHMMETER: An instrument used for measuring the resistance, in ohms, in an electrical circuit.

OUTPUT SHAFT: The shaft which transmits torque from a device, such as a transmission.

OVERDRIVE: A gear assembly which produces more shaft revolutions than that transmitted to it.

OVERHEAD CAMSHAFT (OHC): An engine configuration in which the camshaft is mounted on top of the cylinder head and operates the valve either directly or by means of rocker arms.

GL-4 GLOSSARY

OVERHEAD VALVE (OHV): An engine configuration in which all of the valves are located in the cylinder head and the camshaft is located in the cylinder block. The camshaft operates the valves via lifters and pushrods.

OXIDES OF NITROGEN (NOx): Chemical compounds of nitrogen produced as a byproduct of combustion. They combine with hydrocarbons to produce smog.

OXYGEN SENSOR: Use with the feedback system to sense the presence of oxygen in the exhaust gas and signal the computer which can reference the voltage signal to an air/fuel ratio.

PINION: The smaller of two meshing gears.

PISTON RING: An open-ended ring with fits into a groove on the outer diameter of the piston. Its chief function is to form a seal between the piston and cylinder wall. Most automotive pistons have three rings: two for compression sealing; one for oil sealing.

PRELOAD: A predetermined load placed on a bearing during assembly or by adjustment.

PRIMARY CIRCUIT: the low voltage side of the ignition system which consists of the ignition switch, ballast resistor or resistance wire, bypass, coil, electronic control unit and pick-up coil as well as the connecting wires and harnesses.

PRESS FIT: The mating of two parts under pressure, due to the inner diameter of one being smaller than the outer diameter of the other, or vice versa; an interference fit.

RACE: The surface on the inner or outer ring of a bearing on which the balls, needles or rollers move.

REGULATOR: A device which maintains the amperage and/or voltage levels of a circuit at predetermined values.

RELAY: A switch which automatically opens and/or closes a circuit.

RESISTANCE: The opposition to the flow of current through a circuit or electrical device, and is measured in ohms. Resistance is equal to the voltage divided by the amperage.

RESISTOR: A device, usually made of wire, which offers a preset amount of resistance in an electrical circuit.

RING GEAR: The name given to a ring-shaped gear attached to a differential case, or affixed to a flywheel or as part of a planetary gear set.

ROLLER BEARING: A bearing made up of hardened inner and outer races between which hardened steel rollers move.

ROTOR: 1. The disc-shaped part of a disc brake assembly, upon which the brake pads bear; also called, brake disc. 2. The device mounted atop the distributor shaft, which passes current to the distributor cap tower contacts.

SECONDARY CIRCUIT: The high voltage side of the ignition system, usually above 20,000 volts. The secondary includes the ignition coil, coil wire, distributor cap and rotor, spark plug wires and spark plugs.

SENDING UNIT: A mechanical, electrical, hydraulic or electromagnetic device which transmits information to a gauge.

SENSOR: Any device designed to measure engine operating conditions or ambient pressures and temperatures. Usually electronic in nature and designed to send a voltage signal to an on-board computer, some sensors may operate as a simple on/off switch or they may provide a variable voltage signal (like a potentiometer) as conditions or measured parameters change.

SHIM: Spacers of precise, predetermined thickness used between parts to establish a proper working relationship.

SLAVE CYLINDER: In automotive use, a device in the hydraulic clutch system which is activated by hydraulic force, disengaging the clutch.

SOLENOID: A coil used to produce a magnetic field, the effect of which is to produce work.

SPARK PLUG: A device screwed into the combustion chamber of a spark ignition engine. The basic construction is a conductive core inside of a ceramic insulator, mounted in an outer conductive base. An electrical charge from the spark plug wire travels along the conductive core and jumps a preset air gap to a grounding point or points at the end of the conductive base. The resultant spark ignites the fuel/air mixture in the combustion chamber.

SPLINES: Ridges machined or cast onto the outer diameter of a shaft or inner diameter of a bore to enable parts to mate without rotation.

TACHOMETER: A device used to measure the rotary speed of an engine, shaft, gear, etc., usually in rotations per minute.

THERMOSTAT: A valve, located in the cooling system of an engine, which is closed when cold and opens gradually in response to engine heating, controlling the temperature of the coolant and rate of coolant flow.

TOP DEAD CENTER (TDC): The point at which the piston reaches the top of its travel on the compression stroke.

TORQUE: The twisting force applied to an object.

TORQUE CONVERTER: A turbine used to transmit power from a

GLOSSARY GL-5

driving member to a driven member via hydraulic action, providing changes in drive ratio and torque. In automotive use, it links the driveplate at the rear of the engine to the automatic transmission.

TRANSDUCER: A device used to change a force into an electrical signal.

TRANSISTOR: A semi-conductor component which can be actuated by a small voltage to perform an electrical switching function.

TUNE-UP: A regular maintenance function, usually associated with the replacement and adjustment of parts and components in the electrical and fuel systems of a vehicle for the purpose of attaining optimum performance.

TURBOCHARGER: An exhaust driven pump which compresses intake air and forces it into the combustion chambers at higher than atmospheric pressures. The increased air pressure allows more fuel to be burned and results in increased horsepower being produced.

VACUUM ADVANCE: A device which advances the ignition timing in response to increased engine vacuum.

VACUUM GAUGE: An instrument used to measure the presence of vacuum in a chamber.

VALVE: A device which control the pressure, direction of flow or rate of flow of a liquid or gas.

VALVE CLEARANCE: The measured gap between the end of the valve stem and the rocker arm, cam lobe or follower that activates the valve.

VISCOSITY: The rating of a liquid's internal resistance to flow.

VOLTMETER: An instrument used for measuring electrical force in units called volts. Voltmeters are always connected parallel with the circuit being tested.

WHEEL CYLINDER: Found in the automotive drum brake assembly, it is a device, actuated by hydraulic pressure, which, through internal pistons, pushes the brake shoes outward against the drums.

GL-6 GLOSSARY

NOTES

MASTER INDEX

A

ABOUT THIS MANUAL, 0-5
ACCELERATOR CABLE, REMOVAL AND INSTALLATION, 4-13
ACCELERATOR PEDAL POSITION (APP) SENSOR (V6 MODELS), REPLACEMENT, 6-10
ACKNOWLEDGEMENTS, 0-2
AIR CONDITIONING
and heating system, check and maintenance, 3-18
compressor, removal and installation, 3-20
condenser, removal and installation, 3-22
pressure sensor, replacement, 3-23
receiver-drier, removal and installation, 3-21
AIR FILTER CHECK AND REPLACEMENT, 1-22
AIR FILTER HOUSING, REMOVAL AND INSTALLATION, 4-12
AIR INJECTION REACTION (AIR) SYSTEM, GENERAL INFORMATION AND COMPONENT REPLACEMENT, 6-24
AIRBAG SYSTEM, GENERAL INFORMATION AND PRECAUTIONS, 12-23
ALTERNATOR, REMOVAL AND INSTALLATION, 5-11
ANTENNA, REMOVAL AND INSTALLATION, 12-15
ANTIFREEZE, GENERAL INFORMATION, 3-4
ANTI-LOCK BRAKE SYSTEM (ABS), GENERAL INFORMATION AND SPEED SENSOR REMOVAL AND INSTALLATION, 9-2
AUTOMATIC TRANSAXLE, 7B-1 THROUGH 7B-8
diagnosis, general, 7B-2
driveaxle oil seals, replacement, 7B-3
fluid change, 1-24
overhaul, general information, 7B-8
removal and installation, 7B-6
shift
    cable, removal, installation and adjustment, 7B-3
    lever and shift interlock solenoid, replacement, 7B-5
Transaxle Control Module (TCM), removal and installation, 7B-6
Transmission Range (TR) switch, replacement, 6-20
AUTOMOTIVE CHEMICALS AND LUBRICANTS, 0-19

B

BACK-UP LIGHT SWITCH, REPLACEMENT, 7A-2
BALANCE SHAFT CHAIN AND BALANCE SHAFTS (FOUR-CYLINDER ENGINE), REMOVAL, INSPECTION AND INSTALLATION, 2A-10
BALLJOINTS, CHECK AND REPLACEMENT, 10-8
BATTERY
cables, replacement, 5-5
check
    maintenance and charging, 1-12
    removal and installation, 5-3
precautions and disconnection, 5-1

IND-2 MASTER INDEX

BLOWER MOTOR RESISTOR AND BLOWER MOTOR, REPLACEMENT, 3-15
BODY REPAIR
major damage, 11-7
minor damage, 11-3
BODY, 11-1 THROUGH 11-34
BODY, MAINTENANCE, 11-2
BOOSTER BATTERY (JUMP) STARTING, 0-16
BRAKES, 9-1 THROUGH 9-20
Anti-lock Brake System (ABS), general information and speed sensor removal and installation, 9-2
caliper, removal and installation, 9-6
check, 1-15
disc, inspection, removal and installation, 9-8
fluid
    change, 1-23
    level check, 1-7
hoses and lines, inspection and replacement, 9-17
hydraulic system, bleeding, 9-17
light switch, replacement, 9-19
master cylinder, removal and installation, 9-15
pads, replacement, 9-3
parking brake
    adjustment, 9-19
    shoes, replacement, 9-9
power brake booster, removal and installation, 9-18
shoes, replacement, 9-9
wheel cylinder, removal and installation, 9-15
BULB REPLACEMENT, 12-19
BUMPERS, REMOVAL AND INSTALLATION, 11-10
BUYING PARTS, 0-7

C

CABLE REPLACEMENT
accelerator, 4-13
automatic transaxle, 7B-3
battery, 5-5
CALIPER, DISC BRAKE, REMOVAL AND INSTALLATION, 9-6
CAMSHAFT POSITION (CMP) SENSOR (V6 MODELS), REPLACEMENT, 6-11
CAMSHAFTS AND CAM FOLLOWERS, V6 ENGINE, REMOVAL, INSPECTION AND INSTALLATION, 2B-12
CAMSHAFTS AND HYDRAULIC LASH ADJUSTERS, FOUR-CYLINDER ENGINE, REMOVAL, INSPECTION AND INSTALLATION, 2A-14
CAPACITIES, LUBRICANTS AND FLUIDS, 1-2
CATALYTIC CONVERTERS, GENERAL INFORMATION, CHECK AND REPLACEMENT, 6-26

CENTER CONSOLE
removal and installation, 11-27
switches, replacement, 12-9
CHARGING SYSTEM
alternator, removal and installation, 5-11
check, 5-10
general information and precautions, 5-10
CHASSIS ELECTRICAL SYSTEM, 12-1 THROUGH 12-42
CHEMICALS AND LUBRICANTS, 0-19
CIRCUIT BREAKERS, GENERAL INFORMATION, 12-5
CLUTCH
components, removal, inspection and installation, 8-4
description and check, 8-2
fluid level check, 1-7
hydraulic system, bleeding, 8-3
master cylinder, removal and installation, 8-2
release cylinder, replacement, 8-3
start switch, replacement, 8-6
CLUTCH AND DRIVEAXLES, 8-1 THROUGH 8-12
COIL PACK, IGNITION, REMOVAL AND INSTALLATION, 5-8
COMPRESSOR, AIR CONDITIONING, REMOVAL AND INSTALLATION, 3-20
CONDENSER, AIR CONDITIONING, REMOVAL AND INSTALLATION, 3-22
CONTROL ARM, FRONT, REMOVAL, INSPECTION AND INSTALLATION, 10-7
CONVERSION FACTORS, 0-17
COOLANT LEVEL CHECK, 1-6
COOLANT TEMPERATURE (ECT) SENSOR, REPLACEMENT, 6-10
COOLING SYSTEM
antifreeze, general information, 3-4
blower motor resistor and blower motor replacement, 3-15
check, 1-14
coolant expansion tank, removal and installation, 3-8
coolant temperature sending unit, check and replacement, 3-15
engine oil cooler (V6 models), removal and installation, 3-13
fans, engine cooling, check and replacement, 3-6
general information, 3-2
radiator, removal and installation, 3-9
servicing (draining, flushing and refilling), 1-22
thermostat, check and replacement, 3-4
water pump check, 3-10
water pump, heater core pump (V6 models) and auxillary pump (V6 models), 3-10
COOLING, HEATING AND AIR CONDITIONING SYSTEMS, 3-1 THROUGH 3-24
COWL COVER, REMOVAL AND INSTALLATION, 11-14
CRANKSHAFT, REMOVAL AND INSTALLATION, 2C-18

MASTER INDEX

CRANKSHAFT POSITION (CKP) SENSOR, REPLACEMENT, 6-11

CRANKSHAFT PULLEY AND FRONT OIL SEAL, REMOVAL AND INSTALLATION
four-cylinder engine, 2A-13
V6 engine, 2B-11

CRUISE CONTROL SYSTEM, GENERAL INFORMATION, 12-22

CYLINDER COMPRESSION CHECK, 2C-4

CYLINDER HEAD, REMOVAL AND INSTALLATION
four-cylinder engine, 2A-17
V6 engine, 2B-15

D

DASHBOARD AND CENTER CONSOLE SWITCHES, REPLACEMENT, 12-9

DASHBOARD TRIM PANELS, 11-28

DAYTIME RUNNING LIGHTS (DRL), GENERAL INFORMATION, 12-23

DEFOGGER, REAR WINDOW, CHECK AND REPAIR, 12-15

DIAGNOSIS, 0-21

DIAGNOSTIC TROUBLE CODES, 6-5

DISC BRAKE
caliper, removal and installation, 9-6
disc, inspection, removal and installation, 9-8
pads, replacement, 9-3

DOOR
latch, lock cylinder and handles, removal and installation, 11-18
outer panel, removal and installation, 11-23
removal, installation and adjustment, 11-17
trim panels, removal and installation, 11-14
window glass regulator, removal and installation, 11-21
window glass, removal and installation, 11-21

DRIVEAXLE
boot
  check, 1-16
  replacement, 8-8
oil seals, replacement, 7B-3
removal and installation, 8-6

DRIVEBELT CHECK AND REPLACEMENT, 1-21

DRIVEPLATE, REMOVAL AND INSTALLATION
four-cylinder engine, 2A-17
V6 engine, 2B-19

DRUM BRAKE
shoes/parking brake shoes, replacement, 9-9
wheel cylinder, removal and installation, 9-15

E

ELECTRIC SIDE VIEW MIRRORS, GENERAL INFORMATION, 12-22

ELECTRICAL TROUBLESHOOTING, GENERAL INFORMATION, 12-2

EMISSIONS AND ENGINE CONTROL SYSTEMS, 6-1 THROUGH 6-34

ENGINE COOLANT TEMPERATURE (ECT) SENSOR, REPLACEMENT, 6-12

ENGINE COOLING FANS, CHECK AND REPLACEMENT, 3-6

ENGINE ELECTRICAL SYSTEMS, 5-1 THROUGH 5-14

ENGINE FRONT COVER, FOUR-CYLINDER ENGINE, REMOVAL AND INSTALLATION, 2A-6

ENGINE MOUNTS, CHECK AND REPLACEMENT
four-cylinder engine, 2A-24
V6 engine, 2B-19

ENGINE OIL AND FILTER CHANGE, 1-10

ENGINE OIL COOLER (V6 MODELS), REMOVAL AND INSTALLATION, 3-13

ENGINE OIL LEVEL CHECK, 1-6

ENGINE OVERHAUL
disassembly sequence, 2C-14
general information, 2C-2
reassembly sequence, 2C-23
rebuilding alternatives, 2C-6

ENGINE REMOVAL, METHODS AND PRECAUTIONS, 2C-7

ENGINE, GENERAL OVERHAUL PROCEDURES, 2C-1 THROUGH 2C-26
crankshaft, removal and installation, 2C-18
cylinder compression check, 2C-4
engine overhaul
  disassembly sequence, 2C-14
  general information, 2C-2
  reassembly sequence, 2C-23
engine rebuilding alternatives, 2C-6
engine removal, methods and precautions, 2C-7
engine, removal and installation, 2C-8
initial start-up and break-in after overhaul, 2C-23
oil pressure check, 2C-4
pistons and connecting rods, removal and installation, 2C-14
vacuum gauge diagnostic checks, 2C-5

ENGINE, IN-VEHICLE REPAIR PROCEDURES
Four-cylinder engine, 2A-1 through 2A-24
  balance shaft chain and balance shafts, removal, inspection and installation, 2A-10
  camshafts and hydraulic lash adjusters, removal, inspection and installation, 2A-14

IND-4 MASTER INDEX

crankshaft pulley and front oil seal, removal and installation, 2A-13
cylinder head, removal and installation, 2A-17
engine front cover, removal and installation, 2A-6
exhaust manifold, removal and installation, 2A-5
flywheel/driveplate, removal, inspection and installation, 2A-20
intake manifold, removal and installation, 2A-4
oil pan, removal and installation, 2A-18
oil pump, removal, inspection and installation, 2A-19
powertrain mounts, check and replacement, 2A-24
rear main oil seal, replacement, 2A-20
repair operations possible with the engine in the vehicle, 2A-2
timing chain and sprockets, removal, inspection and installation, 2A-7
Top Dead Center (TDC) for number one piston, locating, 2A-2
valve cover, removal and installation, 2A-3

V6 engine, 2B-1 through 2B-22
camshafts and cam followers, removal, inspection and installation, 2B-12
crankshaft pulley and front oil seal, removal and installation, 2B-11
cylinder heads, removal and installation, 2B-15
driveplate, removal and installation, 2B-19
engine mounts, check and replacement, 2B-19
exhaust manifold, removal and installation, 2B-5
intake manifold, removal and installation, 2B-4
oil pan, removal and installation, 2B-16
oil pump, removal, inspection and installation, 2B-17
rear main oil seal, replacement, 2B-19
repair operations possible with the engine in the vehicle, 2B-2
timing belt and sprockets, removal, inspection and installation, 2B-6
timing belt cover, removal and installation, 2B-5
Top Dead Center (TDC) for number one piston, locating, 2B-2
valve cover, removal and installation, 2B-3

ENGINE, REMOVAL AND INSTALLATION, 2C-8
EVAPORATIVE EMISSIONS CONTROL (EVAP) SYSTEM, GENERAL INFORMATION AND COMPONENT REPLACEMENT, 6-28
EXHAUST GAS RECIRCULATION (EGR) SYSTEM (V6 MODELS), GENERAL INFORMATION AND COMPONENT REPLACEMENT, 6-31
EXHAUST MANIFOLD, REMOVAL AND INSTALLATION
four-cylinder engine, 2A-5
V6 engine, 2B-5

EXHAUST SYSTEM
check, 1-20
servicing, general information, 4-21
EXPANSION TANK, REMOVAL AND INSTALLATION, 3-8

F

FANS, ENGINE COOLING, CHECK AND REPLACEMENT, 3-6
FAULT FINDING, 0-21
FENDER, FRONT, REMOVAL AND INSTALLATION, 11-12
FILTER REPLACEMENT
engine air, 1-22
engine oil, 1-10
interior ventilation, 1-17
FIRING ORDER, 1-29
FLUID LEVEL CHECKS, 1-5
brake and clutch fluid, 1-7
engine coolant, 1-6
engine oil, 1-6
power steering fluid, 1-8
windshield washer fluid, 1-8
FLUIDS AND LUBRICANTS
capacities, 1-28
recommended, 28
FLYWHEEL/DRIVEPLATE, REMOVAL, INSPECTION AND INSTALLATION
four-cylinder engine, 2A-20
V6 engine, 2B-16
FOUR-CYLINDER ENGINE, 2A-1 THROUGH 2A-24
balance shaft chain and balance shafts, removal, inspection and installation, 2A-10
camshafts and hydraulic lash adjusters, removal, inspection and installation, 2A-14
crankshaft pulley and front oil seal, removal and installation, 2A-13
cylinder head, removal and installation, 2A-17
engine front cover, removal and installation, 2A-6
exhaust manifold, removal and installation, 2A-5
flywheel/driveplate, removal, inspection and installation, 2A-20
intake manifold, removal and installation, 2A-4
oil pan, removal and installation, 2A-18
oil pump, removal, inspection and installation, 2A-19
powertrain mounts, check and replacement, 2A-24
rear main oil seal, replacement, 2A-20
repair operations possible with the engine in the vehicle
four-cylinder engine, 2A-2
timing chain and sprockets, removal, inspection and installation, 2A-7
Top Dead Center (TDC) for number one piston, locating
four-cylinder engine, 2A-2

MASTER INDEX IND-5

valve cover, removal and installation, 2A-2
FRACTION/DECIMAL/MILLIMETER EQUIVALENTS, 0-18
FRONT OIL SEAL, ENGINE, REMOVAL AND INSTALLATION
four-cylinder engine, 2A-13
V6 engine, 2B-11
FUEL
filter replacement, 1-25
general information and precautions, 4-2
lines and fittings, general information, 4-4
pressure regulator, removal and installation, 4-17
pressure relief procedure, 4-2
pump/fuel level sensor module, removal and installation, 4-9
pump/fuel level sensor, component replacement, 4-11
pump/fuel pressure, check, 4-3
rail and injectors, removal and installation, 4-18
Sequential Fuel Injection (SFI) system
    general check, 4-15
    general information, 4-14
system check, 1-20
tank
    cleaning and repair, general information, 4-9
    removal and installation, 4-7
FUEL AND EXHAUST SYSTEMS, 4-1 THROUGH 4-22
FUSES AND FUSIBLE LINKS, GENERAL INFORMATION, 12-4

G

GENERAL ENGINE OVERHAUL PROCEDURES, 2C-1 THROUGH 2C-26
crankshaft, removal and installation, 2C-18
cylinder compression check, 2C-4
engine overhaul
    disassembly sequence, 2C-14
    general information, 2C-2
    reassembly sequence, 2C-23
engine rebuilding alternatives, 2C-6
engine removal, methods and precautions, 2C-7
engine, removal and installation, 2C-8
initial start-up and break-in after overhaul, 2C-23
oil pressure check, 2C-4
pistons and connecting rods, removal and installation, 2C-14
vacuum gauge diagnostic checks, 2C-5

H

HEADLIGHTS
adjustment, 12-17
bulb, replacement, 12-17
housing, replacement, 12-18
HEATER CORE PUMP AND AUXILIARY PUMP, V6 MODELS, REPLACEMENT, 3-10
HEATER CORE, REPLACEMENT, 3-17
HEATER/AIR CONDITIONER CONTROL ASSEMBLY, REMOVAL AND INSTALLATION, 3-16
HINGES AND LOCKS, MAINTENANCE, 11-7
HOOD LATCH HANDLE AND RELEASE CABLE, REMOVAL AND INSTALLATION, 11-8
HOOD, REMOVAL, INSTALLATION AND ADJUSTMENT, 11-7
HORN, REPLACEMENT, 12-19
HUB AND BEARING ASSEMBLY, REMOVAL AND INSTALLATION
front, 10-9
rear, 10-14
HYDRAULIC SYSTEM, BLEEDING
brake, 9-17
clutch, 8-3

I

IDLE AIR CONTROL (IAC) VALVE (FOUR-CYLINDER MODELS), REPLACEMENT, 6-23
IGNITION SYSTEM
check, 5-7
coil pack, removal and installation, 5-8
control module (four-cylinder models), replacement, 5-8
general information, 5-6
switch and key lock cylinder, replacement, 12-7
INFORMATION SENSORS, GENERAL INFORMATION, 6-3
INITIAL START-UP AND BREAK-IN AFTER OVERHAUL, 2C-23
INJECTORS, FUEL, REMOVAL AND INSTALLATION, 4-18
INLET AIR TEMPERATURE (IAT) SENSOR, REPLACEMENT, 6-13
INSTRUMENT CLUSTER, REMOVAL AND INSTALLATION, 12-11
INTAKE MANIFOLD RUNNER CONTROL (IMRC) SYSTEM (V6 MODELS), GENERAL INFORMATION AND COMPONENT REPLACEMENT, 6-23
INTAKE MANIFOLD, REMOVAL AND INSTALLATION
four-cylinder engine, 2A-4
V6 engine, 2B-4
INTERIOR VENTILATION FILTER REPLACEMENT, 1-17
INTRODUCTION TO THE SATURN L-SERIES, 0-5

J

JACKING AND TOWING, 0-15
JUMP STARTING, 0-16

K

KEY LOCK CYLINDER, IGNITION, REPLACEMENT, 12-7
KNOCK SENSOR, REPLACEMENT, 6-13

L

LIFTGATE
latch, lock cylinder and handle, removal and installation, 11-26
removal, installation and adjustment, 11-25
LOCK CYLINDER, IGNITION KEY, REPLACEMENT, 12-7
LUBRICANTS AND CHEMICALS, 0-19
LUBRICANTS AND FLUIDS
capacities, 1-28
recommended, 1-28

M

MAINTENANCE, ROUTINE, 1-1 THROUGH 1-30
schedule, 1-2
techniques, tools and working facilities, 0-7
MANIFOLD ABSOLUTE PRESSURE (MAP) SENSOR, REPLACEMENT, 6-14
MANUAL TRANSAXLE, 7A-1 THROUGH 7A-6
back-up light switch, replacement, 7A-2
lubricant level check and change, 1-25
overhaul, general information, 7A-5
removal and installation, 7A-3
shift linkage assembly, removal and installation, 7A-2
MASS AIR FLOW (MAF) SENSOR (V6 MODELS), REPLACEMENT, 6-16
MASTER CYLINDER, REMOVAL AND INSTALLATION
brake, 9-15
clutch, 8-2
MEMORY SAVERS, GENERAL INFORMATION, 5-2
MIRRORS, ELECTRIC SIDE VIEW, GENERAL INFORMATION, 12-22
MIRRORS, REMOVAL AND INSTALLATION, 11-24
MODULE, IGNITION (FOUR-CYLINDER MODELS), REPLACEMENT, 5-8

O

OIL AND FILTER CHANGE, ENGINE, 1-10

OIL COOLER (V6 MODELS), REMOVAL AND INSTALLATION, 3-13
OIL PAN, REMOVAL AND INSTALLATION
four-cylinder engine, 2A-18
V6 engine, 2B-16
OIL PRESSURE CHECK, 2C-4
OIL PUMP, REMOVAL, INSPECTION AND INSTALLATION
four-cylinder engine, 2A-19
V6 engine, 2B-17
ON BOARD DIAGNOSTIC (OBD) SYSTEM AND DIAGNOSTIC TROUBLE CODES (DTCS), 6-2
OUTPUT ACTUATORS, GENERAL INFORMATION, 6-4
OUTPUT SHAFT SPEED (OSS) SENSOR, REPLACEMENT, 6-16
OWNER'S MANUAL AND VECI LABEL INFORMATION, 1-5
OXYGEN SENSORS, GENERAL INFORMATION AND REPLACEMENT, 6-17

P

PADS, DISC BRAKE, REPLACEMENT, 9-3
PARKING BRAKE
adjustment, 9-19
shoes, replacement, 9-9
PARTS, REPLACEMENT, BUYING, 0-7
PISTONS AND CONNECTING RODS, REMOVAL AND INSTALLATION, 2C-14
POSITIVE CRANKCASE VENTILATION (PCV) SYSTEM, GENERAL INFORMATION, INSPECTION AND COMPONENT REPLACEMENT, 6-32
POWER BRAKE BOOSTER, REMOVAL AND INSTALLATION, 9-18
POWER DOOR LOCK SYSTEM, GENERAL INFORMATION, 12-23
POWER STEERING
fluid level check, 1-8
pump, removal and installation, 10-19
system, bleeding, 10-20
POWER WINDOW SYSTEM, GENERAL INFORMATION, 12-22
POWERTRAIN CONTROL MODULE (PCM), REMOVAL AND INSTALLATION, 6-21
POWERTRAIN MOUNTS, CHECK AND REPLACEMENT
four-cylinder engine, 2A-24
V6 engine, 2B-19
PRESSURE REGULATOR, FUEL, REMOVAL AND INSTALLATION, 4-17
PRESSURE SENSOR, AIR CONDITIONING, REPLACEMENT, 3-23

MASTER INDEX IND-7

R

RADIATOR GRILLE, REMOVAL AND INSTALLATION, 11-14
RADIATOR, REMOVAL AND INSTALLATION, 3-9
RADIO AND SPEAKERS, REMOVAL AND INSTALLATION, 12-13
REAR MAIN OIL SEAL, REPLACEMENT
four-cylinder engine, 2A-20
V6 engine, 2B-19
REAR WINDOW DEFOGGER, CHECK AND REPAIR, 12-15
RECEIVER-DRIER, AIR CONDITIONING, REMOVAL AND INSTALLATION, 3-21
RECOMMENDED LUBRICANTS AND FLUIDS, 1-28
REGULATOR, WINDOW GLASS, REMOVAL AND INSTALLATION, 11-18
RELAYS, GENERAL INFORMATION AND TESTING, 12-5
RELEASE CYLINDER, CLUTCH, REPLACEMENT, 8-3
REPAIR OPERATIONS POSSIBLE WITH THE ENGINE IN THE VEHICLE
four-cylinder engine, 2A-2
V6 engine, 2B-2
REPLACEMENT PARTS, BUYING, 0-7
ROTOR, BRAKE, INSPECTION, REMOVAL AND INSTALLATION, 9-8
ROUTINE MAINTENANCE SCHEDULE, 1-2
ROUTINE MAINTENANCE, 1-1 THROUGH 1-30

S

SAFETY FIRST!, 0-20
SCHEDULED MAINTENANCE, 1-2
SEAT BELT CHECK, 1-15
SEATS, REMOVAL AND INSTALLATION, 11-32
SEQUENTIAL FUEL INJECTION (SFI) SYSTEM
general check, 4-15
general information, 4-14
SHIFT CABLE, AUTOMATIC TRANSAXLE, REMOVAL, INSTALLATION AND ADJUSTMENT, 7B-3
SHIFT LEVER AND SHIFT INTERLOCK SOLENOID, AUTOMATIC TRANSAXLE, REPLACEMENT, 7B-5
SHOCK ABSORBER/COIL SPRING ASSEMBLY (REAR)
removal and installation, 10-9
replacement, 10-10
SHOES, DRUM BRAKE, REPLACEMENT, 9-9
SIDE VIEW MIRRORS, ELECTRIC, GENERAL INFORMATION, 12-22
SLAVE CYLINDER, CLUTCH, REPLACEMENT, 8-2
SPARK PLUG
check and replacement, 1-26
type and gap, 1-28
SPEAKERS, REMOVAL AND INSTALLATION, 12-13
STABILIZER BAR, BUSHINGS AND LINKS, REMOVAL AND INSTALLATION
front, 10-6
rear, 10-12
STARTER MOTOR
and circuit, check, 5-11
removal and installation, 5-12
STARTING SYSTEM, GENERAL INFORMATION AND PRECAUTIONS, 5-11
STEERING
column covers, removal and installation, 11-31
column switches, replacement, 12-6
column, removal and installation, 10-16
gear boots, removal and installation, 10-17
gear, removal and installation, 10-18
knuckle and hub removal (front), removal and installation, 10-8
wheel, removal and installation, 10-14
STEERING, SUSPENSION AND DRIVEAXLE BOOT CHECK, 1-18
STRUT ASSEMBLY (FRONT), REMOVAL, INSPECTION AND INSTALLATION, 10-3
STRUT/COIL SPRING, REPLACEMENT, 10-4
SUBFRAME, REMOVAL AND INSTALLATION, 10-20
SUSPENSION AND STEERING SYSTEMS, 10-1 THROUGH 10-24
SUSPENSION ARMS (REAR), REMOVAL AND INSTALLATION, 10-10

T

TEMPERATURE SENDING UNIT, CHECK AND REPLACEMENT, 3-15
TENSIONER REPLACEMENT, DRIVEBELT, 1-21
THERMOSTAT, CHECK AND REPLACEMENT, 3-4
THROTTLE BODY, INSPECTION, REMOVAL AND INSTALLATION, 4-16
THROTTLE POSITION (TP) SENSOR, REPLACEMENT, 6-19
TIE-ROD ENDS, REMOVAL AND INSTALLATION, 10-16
TIMING BELT AND SPROCKETS, V6 ENGINE, REMOVAL, INSPECTION AND INSTALLATION, 2B-6
TIMING BELT COVER, V6 ENGINE, REMOVAL AND INSTALLATION, 2B-5
TIMING CHAIN AND SPROCKETS (FOUR-CYLINDER ENGINE), REMOVAL, INSPECTION AND INSTALLATION, 2A-7
TIRE AND TIRE PRESSURE CHECKS, 1-9

TIRE ROTATION, 1-15
TOOLS AND WORKING FACILITIES, 0-7
TOP DEAD CENTER (TDC) FOR NUMBER ONE PISTON, LOCATING
four-cylinder engine, 2A-2
V6 engine, 2B-2
TORQUE SPECIFICATIONS
brake caliper bolts, 9-20
cylinder head bolts
    four-cylinder engine, 2A-22
    V6 engine, 2B-20
spark plugs, 1-29
thermostat housing bolts, 3-23
water pump bolts, 3-23
wheel lug nuts, 1-29
Other torque specifications can be found in the Chapter that deals with the particular component being serviced
TOWING, 0-15
TRANSAXLE, AUTOMATIC, 7B-1 THROUGH 7B-8
diagnosis, general, 7B-2
driveaxle oil seals, replacement, 7B-3
fluid change, 1-24
overhaul, general information, 7B-8
removal and installation, 7B-6
shift
    cable, removal, installation and adjustment, 7B-3
    lever and shift interlock solenoid, replacement, 7B-5
Transaxle Control Module (TCM), removal and installation, 7B-6
Transmission Range (TR) switch, replacement, 6-20
TRANSAXLE, MANUAL, 7A-1 THROUGH 7A-6
back-up light switch, replacement, 7A-12
lubricant level check and change, 1-25
overhaul, general information, 7A-5
removal and installation, 7A-3
shift linkage assembly, removal and installation, 7A-2
TRANSMISSION RANGE (TR) SWITCH, REPLACEMENT, 6-20
TROUBLE CODE CHART, 6-5
TROUBLESHOOTING, 0-21
TUNE-UP AND ROUTINE MAINTENANCE, 1-1 THROUGH 1-30
TUNE-UP, GENERAL INFORMATION, 1-5

U

UNDERHOOD HOSE CHECK AND REPLACEMENT, 1-17
UPHOLSTERY AND CARPETS, MAINTENANCE, 11-2

V

V6 ENGINE, 2B-1 THROUGH 2B-22
camshafts and cam followers, removal, inspection and installation, 2B-12
crankshaft pulley and front oil seal, removal and installation, 2B-11
cylinder heads, removal and installation, 2B-15
driveplate, removal and installation, 2B-19
engine mounts, check and replacement, 2B-19
exhaust manifold, removal and installation, 2B-5
intake manifold, removal and installation, 2B-4
oil pan, removal and installation, 2B-16
oil pump, removal, inspection and installation, 2B-17
rear main oil seal, replacement, 2B-19
repair operations possible with the engine in the vehicle, 2B-2
timing belt and sprockets, removal, inspection and installation, 2B-6
timing belt cover, removal and installation, 2B-5
Top Dead Center (TDC) for number one piston, locating, 2B-2
valve cover, removal and installation, 2B-3
VACUUM GAUGE DIAGNOSTIC CHECKS, 2C-5
VALVE COVER, REMOVAL AND INSTALLATION
four-cylinder engine, 2A-2
V6 engine, 2B-3
VEHICLE IDENTIFICATION NUMBERS, 0-6
VINYL TRIM, MAINTENANCE, 11-2

W

WATER PUMP
check, 3-10
replacement, 3-10
WHEEL ALIGNMENT, GENERAL INFORMATION, 10-22
WHEEL CYLINDER, REMOVAL AND INSTALLATION, 9-15
WHEELS AND TIRES, GENERAL INFORMATION, 10-21
WINDOW GLASS REGULATOR, REMOVAL AND INSTALLATION, 11-18
WINDSHIELD AND FIXED GLASS, REPLACEMENT, 11-7
WINDSHIELD WIPER BLADE INSPECTION AND REPLACEMENT, 1-12
WIPER MOTORS, CHECK AND REPLACEMENT, 12-11
WIRING DIAGRAMS, GENERAL INFORMATION, 12-24
WORKING FACILITIES, 0-7